THEORY OF COMPUTATION

THEORY OF COMPUTATION

D. P. ACHARJYA

Associate Professor
School of Computing Sciences
VIT University
Vellore, TN

MJP PUBLISHERS

Cataloguing-in-Publication   Data

Acharjya   (1969-   ).
Theory of Computation / by D. P. Acharjya. –
Chennai : MJP Publishers, 2010
   xx, 396 p.; 24 cm.
   Includes References and Index.
   ISBN 978-81-8094-076-7 (pbk.)
   1. Computation, Theory of 2. Mathematics-Principles
I. Title.
004.0151 dc22 ACH MJP 072

ISBN 978-81-8094-076-7
© Publishers, 2010
All rights reserved
Printed and bound in India

MJP PUBLISHERS
47, Nallathambi Street
Triplicane
Chennai 600 005

Publisher : J.C. Pillai
Managing Editor : C. Sajeesh Kumar
Marketing Manager : S.Y. Sekar
Project Editor : P. Parvath Radha
Acquisitions Editor : C. Janarthanan
Editorial Team : B. Ramalakshmi, M. Gnanasoundari,
Lissy John, N. Yamuna Devi, R. Magesh
CIP Data : Prof. K. Hariharan, Librarian
RKM Vivekananda College, Chennai.

This book has been published in good faith that the work of the author is original. All efforts have been taken to make the material error-free. However, the author and publisher disclaim responsibility for any inadvertent errors.

To
My beloved wife Asima

PREFACE

The major foundations of computer sciences are the limitations of mechanical computation, the mathematical description of computational networks, and the formal specification of languages, which are highly interrelated disciplines and which require a great deal of mathematical maturity. This book makes it possible for the average student to deal with all these concepts by developing the standard mathematical models of computational devices, as well as by investigating the cognitive and generative capabilities of such machines.

The mathematical theory of a simple class of algorithms that are important in computer sciences and mathematics is called the theory of finite automata. It has primarily two applications in computer science, one is modeling of finite state systems and the other is description of regular sets of finite words. It has many emerging applications such as computer algorithms and programs, verification and specification of protocols, designing of compilers, uses of regular expressions to professional programmers, searching for patterns in texts, natural language processing, etc. Algorithms are recipes that tell us how to solve problems. Initially mathematicians were interested in the dividing line between what was and what was not algorithmic. Certainly, this implied you could write a program to solve the problem, but it was observed that many of the programs would not run quickly or efficiently. This leads to two approaches to classifying algorithms. One is according to their running times and is known as complexity theory whereas the other is according to the types of memory used in implementing them and is known as language theory. In language theory, the simplest algorithms are those that can be implemented by finite automata.

In this book, I have tried to emphasize on mathematical reasoning and problem-solving techniques in order to provide the students learning the theory of computation with the essentials of computation. Each chapter begins with a clear statement of definition, concepts, principles and theorems with illustrative and other descriptive materials. This is followed by sets of solved examples, chapter review questions and exercises. The solved examples serve to illustrate and amplify the material. This has been done to make the book more flexible and to stimulate further interest in the various topics. Once the fundamental concepts have been developed, more complex material and applications are presented.

The material in this book includes fundamental concepts, figures, tables, chapter review exercises, exercises and solved examples to help the reader master introductory theory of computation. This book is intended for one semester introductory course in theory of computation. The book is especially appropriate for the students of BE (Computer Science/IT), B. Tech. (Computer Science/IT), MCA and M. Sc. (Computer Science).

The topics discussed in this book include mathematical preliminaries, introduction to the notion of deterministic finite automata (DFA), non-deterministic finite automata (NFA) and their use to recognize strings in a language, Moore and Mealy machine and minimization of finite automata. Grammar and Chomsky hierarchy includes Chomsky classification, formal grammar and operations on languages. Regular languages and expressions include Arden's theorem, pumping lemma of regular languages and decision algorithm for regular sets. Also, chapters on context-free languages, push down automata, Turing machines and linear bounded automata, NP-completeness, computability and un-decidability, and LR(k) and LL(k) grammars are included.

Although there are many books on this topic, this book has been brought out keeping in mind that the reader will use them in practical applications related to computer science and information technology. It is hoped that the theoretical concepts present in this book will permit a student to understand most of the fundamental concepts. The text is so designed that the students who do not have a strong background in theory of computation will find it very useful to begin with, and the students with an exposure to theory of computation will also find the book very useful as some of exercises given are thought-provoking and help them for application-building.

I take this as a unique opportunity to express any deepest sense of gratitude to Prof. S. Nanda, NIT, Rourkela; Prof. S. Padhy, Utkal University; Prof. N. Parhi, NISER, Bhubaneswaar; Prof. B. K. Tripathy, VIT University and Dr. Md. N. Khan, IGIT, Sarang, who helped shape my scientific viewpoint and education style.

It could not have possible for me to write this book without the help of my friends, Sreekumar, B. D. Sahoo, A. Kumar, Anirban Mitra, A. Panda, S. Dehuri, and R. L. Hotta. This book grew out of notes that I presented for teaching at Sambalpur University and VIT for the past 7 years, and I wish to thank all my students. I trust and hope they will forgive me for not listing all their names.

I wish to thank Dr. G. Viswanathan, Chancellor, VIT University, India, whose constant encouragement and motivation helped me in completing this work successfully.

Last, but not the least, I would like to express my love to my wife, Asima, and my little daughter, Aditi. Their patience and support while preparing the manuscript cannot be thrown into oblivion.

No book—certainly no technical book—is the product of its authors alone. I am pleased to acknowledge here the contributions of several authors, which I have referred while preparing the manuscript. I shall be grateful to the readers for pointing out errors and omissions that, in spite of all care, might have crept in. I shall be delighted if this book is accepted and appreciated by the scholars of today. You can email your comments and suggestions to dpacharjya@gmail.com or dpacharjya@vit.ac.in .

Finally, I express my heartfelt thanks to the Director and the editorial team of MJP Publishers, Chennai, for their kind cooperation and publication with high accuracy.

D. P. ACHARJYA

CONTENTS

INTRODUCTION

The impact of advanced technology on the society does not facilitate the appreciation of the intellectual contents of the theory underlying it. People are so busy with the latest technology that they fail to wonder about the theory and concept behind it. Specially, people need not think of computing in general terms. As a result the intellectual contents of computation are rarely understood, though it is sure to become one of the most important aspects of the sciences. This is because it is the science of how machines can be made to be more intelligent and carry out intellectual processes. It is also known that a general-purpose digital computer can do any intellectual process that can be carried out mechanically by a human being. Moreover, the restrictions on what we have been able to make computers do so far clearly come far more from our weakness as users than from the intrinsic restrictions of the machines. We trust and hope that these restrictions can be drastically reduced by developing a mathematical science of computation. Moreover, a precise and well-defined terminology is required to get the subject clearly and more accurately.

In fact there are three directions in the mathematical science of computation, which include numerical analysis, theory of computability and the theory of finite automata. But the theory of finite automata is of greater use, since the theory of computation is a scientific discipline concerned with the study of general properties of computation. Most specifically, it aims to understand the nature of efficient computation. This book explores some of the more important concepts and questions concerning problems, and computation. The exploration reduces in many cases to a study of mathematical theories, such as formal language and automata theory. These theories provide abstract computation models that are easier to discover as their formalisms avoid immaterial details.

The field of computer science is concerned with the development of methodology for designing programs, and, with the development of computers, for executing programs. It is therefore essential to understand the concepts of mathematical computation, computers, and problems. Learning the theory of computation would help students learn the essentials of computation for the purpose of which this book has been organized into ten chapters arranged in increasing order of complexity. Also in many cases, new topics are discussed as refinements of old ones, and their study is motivated through their association to programs.

The purpose of this book is to emphasize on mathematical reasoning and problem-solving techniques that penetrate computer science. Computer Science is a cluster of related scientific and engineering disciplines concerned with the study and application of computations. These disciplines range from the theory of computer science to engineering disciplines concerned with specific applications. The theory of computer science can be divided into two subgroups. One deals with the theory of computation whereas the other is concerned with the theory of programming. The theory of computation is the study of inherent possibilities and limitations of efficient computation. By its nature, the subject is very close to mathematics, with progress made by conjectures, theorems, and proofs.

The topics to be discussed are mathematical preliminaries, finite state automata, grammar and Chomsky classification, regular languages, context free languages, pushdown automata, Turing machine, LR and LL grammar, computability and undecidability, and NP completeness.

Chapter 1 is an attempt to create a fundamental basis for a mathematical theory of computation. Before mentioning what is in the chapter, we discuss briefly the basic discrete mathematics followed by the notion of strings, and the role that strings have in presenting information. Then it relates the concept of languages to the notion of strings. Graphical representations are included wherever possible to improve readability.

Chapter 2 discusses the first fundamental model of computation, finite state machine. Results which use the finiteness of the number of states tend not to be very useful in dealing with present digital computers which have so many states that it is unfeasible for them to go through a substantial fraction of them in a reasonable time. It also discusses the finite state automaton with output. The chapter concludes by considering the notion of problems, the relationship between problems and fundamental concepts.

Grammar and Chomsky hierarchy are discussed in Chapter 3. The goal of this chapter is to provide a notion of a formal grammar that arises from the need to formalize the informal notions of grammar and language. Many formal grammars were invented: regular grammars, context-free grammars, context-sensitive grammars and unrestricted grammars. Surprisingly, these grammars can be placed in a natural hierarchy as investigated by Noam Chomsky in the late fifties. He investigated the deep relation between these grammars, the strings they generate and the models of computation. This chapter briefly addresses the impact of formal language for many computer science applications.

In Chapter 4, one type of language known as regular language which is the simplest of the Chomsky hierarchy, and regular expression which is one of the ways to describe regular languages have been discussed. Regular expressions are used to denote regular languages. They can represent regular languages and operations on them concisely. Finally, we discuss pumping lemma for regular languages, Myhill–Nerode theorem and decision algorithm for regular sets.

The next two chapters, Chapters 5 and 6, give an account of a model with an additional stack memory called as pushdown automata and the languages accepted by pushdown automata. Each pushdown automaton is essentially a finite state automaton that can access an auxiliary memory that behaves like pushdown storage of unlimited size. Most programming languages can be approximated by context-free grammar, and compilers for them have been developed based on properties of context-free languages. Finally, closure properties, decidable properties of context-free languages, and membership algorithm are discussed.

Chapter 7 deals with abstract computing machines, called Turing machines. It can be thought of as a desktop personal computer with a potentially infinite memory capacity, though it can only access this memory in small discrete chunks. It is a generalization of pushdown automaton that places no restriction on the auxiliary memory. The Turing machines are proposed for characterizing the programs in general and computability in particular. It is shown that a function is Turing-computable if and only if it is computable by a deterministic Turing machine. In addition to this, the universal Turing machine, variations of Turing machine, Post machine, and linear bounded automaton are discussed.

The earlier chapters mainly discuss the generative aspects of strings using productions and performing the membership test. In the design of programming languages and compilers, it is very much essential to study the parsing techniques. A typical programming language ALGOL has $LR(1)$ parser. These subclasses of context-free grammars, called $LR(k)$ grammars, are discussed in Chapter 8. Also, we study $LL(k)$ grammar that is a proper subset of the $LR(k)$ grammar. Keeping view of these, we restrict our study to parsing only.

In Chapter 9 we study computability and undecidability, as a branch of recursive function theory. It is a way of describing the functions that are computable in terms of given base functions, using recursive function definitions and conditional expressions. The results of the basic work in this theory, including the existence of unlimited register machine, Gödel numbering and the existence of unsolvable problems, have established a framework in which any theory of computation must fit. Unfortunately, much attention has been given to establish more and better unsolvability problems whereas only little attention paid to positive results and none to study the properties of the kinds of algorithms that are actually used.

Chapter 10 discusses the most notorious open problem P versus NP. This is a problem that has withstood attack since the dawn of the computer age and that has clear and overwhelming practical implications. Also, it is recognized as one of seven great mathematical challenges of the millennium. It is also clear that even after fifty years of research, no one has been able to rule out this problem. Over the last few decades, it is the theory of computation that has at least tried to accept that this problem is a difficult one, and has solved easier versions of it.

We provide the Theory of Computation as a fundamental science of computation that seeks to understand computational phenomena, be it natural, man-made or imaginative. This has been extremely successful and productive due to extensive research in the past few decades. This textbook not only provides material for study in the foundations of computing for graduate students in computer science and information technology, but also provides an insight into some more advanced topics in the subject and contains an invaluable collection of lectures for graduates on the subject. Topics and features include fundamental concepts, solved examples for graduate students, and a dozen homework sets and exercises. I trust and hope that the theoretical concepts present in this textbook will motivate its readers to seek out more in the subject.

1

MATHEMATICAL PRELIMINARIES

1.0 INTRODUCTION

Computer Science is a practical discipline. The mathematical study of models of computation is termed as theoretical computer science. This was originated in the 1930s, before the existence of modern computers. The famous logicians Church, Post, Turing and Kleene have made great contributions towards this subject. As the subject deals with mathematical study it is necessary to know the fundamental concepts of mathematics. Keeping this in view, the first chapter contains the mathematical preliminaries. It includes concepts of set theory, relations, functions and graph theory. Besides, it also contains mathematical induction, alphabets, strings, properties and operations on strings.

1.1 SET

A set is a collection of well-defined objects. Well-defined means the objects are distinct and distinguishable. By the form distinct, we mean that no object is repeated. Generally the sets are denoted by capital letters A, B, C, ... The elements are denoted by small letters a, b, c, ... If x is an element of the set A, we write $x \in A$ else we write $x \notin A$. A set can be described in various ways, which are given below.

Method of extension   Expressing the elements of a set within braces where elements are separated by commas is known as method of extension or tabular method or roaster method. For example,

$$A = \{2, 4, 6, 8, 10\}$$

Method of intension   Expressing the elements of a set by a rule is known as method of intension or set-builder method. Mathematically,

$$A = \{x \mid P(x)\}; P(x) \text{ is the property of } x.$$

For example $A = \{x \mid x = 2k; k \in N; 0 < k < 6\}$

Method of recursion   Expressing the elements of a set by a computational rule is known as method of recursion. For example

$$A = \{a_n : a_0 = 1, a_{n+1} = a_n + 5, n \in N\}$$

1.1.1 Types of Sets

A set which contains no element is called as an empty set. This is otherwise known as null set or void set. Generally denoted by for ϕ or $\{\}$. A set which contains only one element is known as a singleton set whereas a set which is having two elements is called a pair set.

A set having countable number of elements is known as finite set whereas a set which contains infinite number of elements is called an infinite set. The different types of sets with examples are given below.

Empty set, e.g. $S = \{x \mid x \neq x\}$

Singleton set, e.g. $S = \{a\}$

Pair set, e.g. $S = \{a, b\}$

Finite set, e.g. $S = \{a, b, c, d, x, y\}$

Infinite set, e.g. $I = \{1, 2, 3, 4 \ldots\}$

1.1.2 Subset and Superset

Set A is said to be a subset of a set B if A is fully contained in B; that is every element of set A is also an element of set B. We represent it as $A \subseteq B$. Mathematically,

$$A \subseteq B \equiv \{x \in A \Rightarrow x \in B\}$$

Set A is said to be a proper subset of a set B if every element of set A is an element of set B and set B has atleast one element which is not an element of set A. We represent it as $A \subset B$. Mathematically,

$$A \subset B \equiv \{x \in A \Rightarrow x \in B \text{ and } y \in B \Rightarrow y \notin A\}$$

Set A is said to be a superset of a set B if B is fully contained in A. We represent it as $A \supseteq B$. It is to be noted that empty set ϕ is a subset of every set and every set is a subset of itself.

1.1.3 Power Set

Let A be any set. The set of all subsets of A is said to be the power set of A and is denoted by 2^A or $P(A)$. Consider the example, $A = \{1, 2, 3\}$.

Therefore, $2^{|A|} = \{A, \phi, \{1\}, \{2\}, \{3\}, \{1, 2\}, \{1, 3\}, \{2, 3\}\}$

Hence, it is clear that if A contains 3 elements then the power set 2^A contains 8 elements. Thus, in general if A contains n elements, then the power set 2^A contains 2^n elements. Mathematically,

$$|A| = n \Rightarrow \left|2^A\right| = 2^{|A|} = 2^n$$

1.1.4 Set Operations

The basic set operations union ($\cup$), intersection ($\cap$), difference (–) and complementation ($'$) are defined as below, where U is the universal set. Any set which is superset of all the sets under consideration is known as universal set. It is to be noted that the universal set can be chosen arbitrarily, but once chosen it is fixed for the discussion.

$$(A \cup B) = \{x \mid x \in A \text{ or } x \in B\}$$

$$(A \cap B) = \{x \mid x \in A \text{ and } x \in B\}$$

$$(A - B) = \{x \mid x \in A \text{ and } x \notin B\}$$

$$A' = \{x \mid x \in U \text{ and } x \notin A\}$$

Properties of Sets   The important properties of sets for further manipulation are given below.

Commutative laws $(A \cup B) = (B \cup A)$ and $(A \cap B) = (B \cap A)$

Associative laws $A \cap (B \cap C) = (A \cap B) \cap C$ and

$A \cup (B \cup C) = (A \cup B) \cup C$

Distributive laws $A \cup (B \cap C) = (A \cup B) \cap (A \cup C)$ and

$A \cap (B \cup C) = (A \cap B) \cup (A \cap C)$

Idempotent laws $(A \cup A) = A$ and $(A \cap A) = A$

Identity laws $(A \cup \phi) = A$ and $(A \cap \phi) = \phi$

$(A \cup U) = U$ and $(A \cap U) = A$

Absorption laws $A \cup (A \cap B) = A$ and $A \cap (A \cup B) = A$

Involution laws $(A')' = A$

De Morgan's laws $(A \cup B)' = A' \cap B'$ and $(A \cap B)' = A' \cup B'$

1.1.5   Partition and Covering

A partition on the set A is defined as a set of non-empty subsets A_i each of which is pairwise disjoint and whose union yields the set A. We represent it as $\Pi(A)$. Each member A_i of the partition $\Pi(A)$ is known as a block. This is as shown in Figure 1.1a. Mathematically,

   i. $A_i \cap A_j = \phi$ for each pair (i, j) and $i \neq j$

   ii. $\bigcup\limits_i A_i = A$

A **covering** on the set A is defined as a set of non-empty subsets A_i, whose union yields the set A. The non-empty subsets need not be disjoint. This is depicted in Figure 1.1b. Mathematically,

i. $\bigcup_i A_i = A$

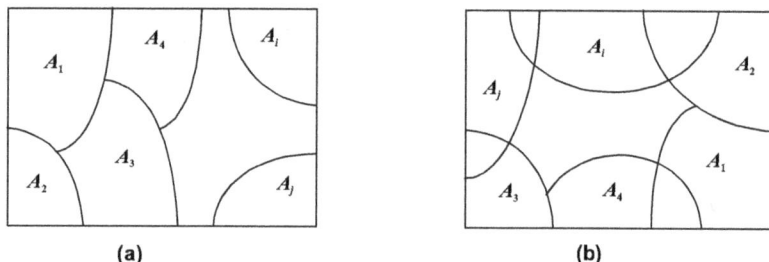

Figure 1.1   (a) Partition (b) Covering

1.1.6 Product of Sets

The Cartesian product of two sets A and B is the set of all order pairs whose first coordinate is an element of A and the second coordinate is an element of B. The product of sets is denoted as $(A \times B)$. Mathematically,

$$(A \times B) = \{(x, y) \mid x \in A \text{ and } y \in B\}$$

Similarly, the product of sets can be extended to n sets $A_1, A_2, \ldots, A_n$. We can define the Cartesian product as

$$(A_1 \times A_2 \times \ldots \times A_n) = \{(x_1, x_2, \ldots, x_n) \mid x_1 \in A_1, x_2 \in A_2, \ldots, x_n \in A_n\}$$

1.2   RELATION

The concept of relation is a basic concept in real life as well as in computer science. It has got tremendous applications in almost every fields such as social, economy, engineering and technology. The concept of relation in computer science plays a major tool to learn and understand it more clearly.

A relation R between two sets A and B is a subset of the Cartesian product $(A \times B)$. In other words any subset of the Cartesian product $(A \times B)$ is a relation R from the set A to the set B. Mathematically,

$$R \subseteq (A \times B) = \{(x, y) \mid x \in A \text{ and } x \in B\}$$

We write $x\, R\, y$, if and only if $(x, y) \in R$. If $(x, y) \notin R$ then $x \not\!R y$. Consider an example in which $A = \{1, 2, 6, 9\}$; $B = \{1, 3, 5, 7\}$ and $R \subseteq (A \times B) = \{(x, y) \mid x \leq y\}$.

Therefore, we get

$$R = \{(1, 1), (1, 3), (1, 5), (1, 7), (2, 3), (2, 5), (2, 7), (6, 7)\}$$

1.2.1 Properties of Relation

A relation R in a set A has the following important properties. Besides these, it has several other properties such as anti-reflexive and asymmetric.

Reflexive relation   A relation R in a set A is said to be a reflexive relation if $(x, x) \in R$ for all $x \in A$, i.e., $x R x \forall x \in A$.

Symmetric relation   A relation R in a set A is said to be a symmetric relation if $(x, y) \in R$ implies $(y, x) \in R$, i.e., $x R y \Rightarrow y R x$.

Transitive relation   A relation R in a set A is said to be a transitive relation if $(x, y) \in R$ and $(y, z) \in R$, then $(x, z) \in R$, i.e., $x R y$ and $y R z \Rightarrow x R z$.

Anti-symmetric relation   A relation R in a set A is said to be anti-symmetric relation if $(x, y) \in R$ and $(y, x) \in R$, then $x = y$, i.e., $x R y$ and $y R x \Rightarrow x = y$.

1.2.2 Equivalence Relation

A relation R in a set A is said to be an equivalence relation in A if and only if R is reflexive, symmetric and transitive. Consider the example "congruence modulo n" over the set of integers. Mathematically,

$$x R y : x \equiv y (\bmod n);\ x, y, n \in Z^{+}.$$

This implies that $x R y : (x - y)$ is divisible by n. Our claim is that the relation R is reflexive, symmetric and transitive.

Reflexive   For every $x \in Z^{+}$ we have $(x - x) = 0$. Therefore $(x - x)$ is divisible by the integer n, i.e., $x R x$ for all $x \in Z^{+}$.

Symmetric   Suppose that $x R y$. This implies that $(x - y)$ is divisible by n.

Therefore, $(x - y) = kn;\ k \in Z^{+} \Rightarrow (y - x) = (-k)n;\ k \in Z^{+}$

i.e., $(y - x)$ is also divisible by n and hence $y R x$.

Transitive   Suppose that $x R y$ and $y R z$. This implies that both $(x - y)$ and $(y - z)$ are divisible by n. Therefore we get

$$(x - y) = k_1\, n \text{ and } (y - z) = k_2\, n \text{ for } k_1,\ k_2 \in Z^{+}$$

On adding, we get $(x - z) = (k_1 + k_2)\, n$ with $(k_1 + k_2) \in Z^{+}$, i.e., $x R z$.

Therefore, the relation $x \equiv y \,(\bmod n)$ over the set of positive integers is an equivalence relation.

1.2.3 Partial Order Relation

A relation R in a set A is a partial order relation in A if and only if R is reflexive, transitive and anti-symmetric. Consider the example $x R y$ such that $x \leq y$ over the set of positive integers, i.e., $x R y : x \leq y;\ x, y \in Z^{+}$.

Reflexive   For every $x \in Z^+$ we have $x \le x$. This implies that $x R x$ for all $x \in Z^+$.

Transitive   Suppose that $x R y$ and $y R z$. Therefore, $x \le y$ and $y \le z$. This implies that $x \le z$, i.e., $x R z$.

Anti-symmetric   Suppose that $x R y$ and $y R x$. Therefore, $x \le y$ and $y \le x$. This implies that $x = y$. Therefore, the relation defined above is a partial order relation.

1.2.4 Closures of Relation

Sometimes a given relation R defined in a set A may not be reflexive, symmetric or transitive. By adding more order pairs to R we can make it reflexive, symmetric or transitive. But the smallest relation containing R having such properties is interesting. The different closures of relation are defined as below.

Reflexive closure   Let R be a relation defined on the set A. The reflexive closure $r(R)$ is defined as below.

    i.   If $(x, y) \in R$, then $(x, y) \in r(R)$ and for all $x \in A, (x, x) \in r(R)$

   ii.   Nothing is in $r(R)$ unless it is so follows from (i).

Symmetric closure   Let R be a relation defined on the set A. The symmetric closure $s(R)$ is defined as below.

    i.   If $(x, y) \in R$, then $(x, y) \in s(R)$ and $(y, x) \in s(R)$

   ii.   Nothing is in $s(R)$ unless it is so follows from (i).

Transitive closure   Let R be a relation defined on the set A. The transitive closure $t(R)$ or R^+ is defined as below.

    i.   If $(x, y) \in R$, then $(x, y) \in R^+$.

   ii.   If $(x, y) \in R, (y, z) \in R$ then $(x, z) \in R^+$

  iii.   Nothing is in R^+ unless it is so follows from (i) and (ii).

In other words the transitive closure R^+ is defined as $R^+ = R \cup R^2 \cup R^3 \cup ...$, where $R^n = (R^{n-1} \circ R)$. The reflexive and transitive closure of the relation R defined in a set A is denoted by R^* and is defined as $R^* = r(R) \cup R^+$. Consider the relation R on $\{1, 2, 3, 5\}$ as $R = \{(1, 1), (1, 3), (3, 5), (3, 3)\}$. The different closures are given as

$$r(R) = \{(1, 1), (1, 3), (3, 5), (3, 3), (2, 2), (5, 5)\}$$
$$s(R) = \{(1, 1), (1, 3), (3, 5), (3, 3), (5, 3), (3, 1)\}$$
$$R^+ = \{(1, 1), (1, 3), (3, 5), (3, 3), (1, 5)\}$$

1.2.5 Equivalence Classes

Let A be a non-empty set and let R be an equivalence relation in A. Further let x be an element of A, i.e., $x \in A$. The equivalence class of x with respect to R is denoted by $[x]$ and is defined as the set of elements $y \in A$ such that $(y, x) \in R$. As R is an equivalence relation in A, $[x]$ is a non-empty subset of A since $x \in [x]$. Mathematically we write,

$$[x] = \{y \mid y \in A \ \text{and} \ (y, x) \in R\}$$

Theorem *Let A be a non-empty set and R be an equivalence relation defined on A. Further let x and y be two arbitrary elements in A. Then $x \in [x]$ and if $y \in [x]$; then $[x] = [y]$.*

Proof Given that A is a non-empty set and R is an equivalence relation defined on A. Let x, $y \in A$. Since R is an equivalence relation in A, we must have $(x, x) \in R$ for all x belonging to A. This indicates that $x \in [x]$.

In order to show the second part let us assume that $y \in [x]$. Therefore, we get $(y, x) \in R$. This implies that $(x, y) \in R$ since R is symmetric.

Let $z \in [x]$; this implies that $(z, x) \in R$ and hence $(x, z) \in R$ as R is symmetric. Now $(y, x) \in R$ and $(x, z) \in R$ implies $(y, z) \in R$, since R is a transitive relation. Again $(y, z) \in R$ implies $(z, y) \in R$ as R is symmetric. Hence we get $z \in [y]$.

Therefore,

$$z \in [x] \Rightarrow z \in [y], \text{i.e.,} [x] \subseteq [y] \text{ (i)}$$

Similarly we will get

$$[y] \subseteq [x] \text{ (ii)}$$

Hence from equations (i) and (ii) we conclude that $[x] = [y]$.

Theorem *Let A be a non-empty set and R be an equivalence relation defined in A. Further let x, y be two arbitrary elements in A. Then [x] = [y] if and only if $(x, y) \in R$.*

Proof Given that A is a non-empty set and R is an equivalence relation defined on A. Let x, $y \in A$. Assume that $[x] = [y]$. Our claim is $(x, y) \in R$.

Since R is reflexive, we have $(x, x) \in R$. This indicates that $x \in [x]$. It further indicates that $x \in [y]$, since $[x] = [y]$. Therefore, we get $(x, y) \in R$.

Conversely, suppose that $(x, y) \in R$. Our claim is $[x] = [y]$.

Let $z \in [x]$. This implies $(z, x) \in R$. Now $(z, x) \in R; (x, y) \in R$ and hence $(z, y) \in R$, since R is transitive. Therefore, we get $z \in [y]$ and hence $[x] \subseteq [y]$.

Similarly it can be shown that $[y] \subseteq [x]$. Hence we conclude that $[x] = [y]$.

Theorem *Let R be an equivalence relation in A and x, $y \in A$. Then the equivalence classes [x] and [y] are either equal or disjoint.*

Proof Assume that R be an equivalence relation defined in A and x, $y \in A$.

Further assume that the equivalence classes $[x]$ and $[y]$ are not disjoint, i.e., $[x] \cap [y] \neq \phi$. Thus there exists at least one element $z \in [x] \cap [y]$. This indicates that $z \in [x]$ and $z \in [y]$. Hence, we get $(z, x) \in R$ and $(z, y) \in R$. Again as R is symmetric we get $(x, z) \in R$ and $(z, y) \in R$ and hence by transitive property $(x, y) \in R$.

Therefore, by previous theorem $[x] = [y]$. Hence, we conclude that two equivalence classes $[x]$ and $[y]$ are either disjoint or equal.

1.3   FUNCTION

The concept of function arises when we want to associate a unique value (or output) with a given argument (or input). Let A and B be two sets. A relation f from a set A to a set B is said to be a function from A to B if it satisfies the following conditions.

  i.   Dom $f = A$ and
  ii. If $(x, y) \in f$ and $(x, z) \in f$, then $y = z$.

In other words a function f from a set A to a set B is a rule which associates every element $x \in A$, to a unique element in B, which is denoted by $f(x)$. Generally the function is denoted by $f : A \rightarrow B$. $f(x)$ is called the image of $x \in A$ under f.

The function is otherwise known as mapping or total function. If the condition (i) is not satisfied then we call it as a partial function. A function can be represented either by giving the image of all elements of A or by a computational rule that computes $f(x)$ once x is given. Consider the examples

  i. $f : \{a, b, c\} \rightarrow \{1, 2, 3, 4\}$ defined by $f(a) = 2, f(b) = 3, f(c) = 4$
  ii. $f : R \rightarrow R$ defined by $f(x) = x^2 + 1, x \in R$ (Set of real numbers)

In the first case the function is defined by giving the images of all elements whereas the second case is defined by a computational rule.

1.3.1 Types of Function

The different types of function are injection, surjection and bijection. A brief discussion is given below.

Injection   A mapping $f : A \rightarrow B$ is said to be an one-one (injective) function if $f(x_1) = f(x_2)$ implies $x_1 = x_2$, where x_1, x_2 are belonging to A. In other words $x_1 \neq x_2$ in A implies $f(x_1) \neq f(x_2)$.

Surjection   A mapping $f : A \rightarrow B$ is said to be an onto (surjective) function, if for every element $y \in B$ there exists an element $x \in A$ such that $f(x) = y$. In other words a mapping $f : A \rightarrow B$ is said to be an onto function if range $(f) = B$, i.e., $f(A) = B$.

Bijection   A mapping $f : A \rightarrow B$ which is injective and surjective is said to be an one-one onto (bijective) function.

Consider the function $f : Z \rightarrow Z$ defined by $f(x) = x + 2; x \in Z$.

Assume that $f(x_1) = f(x_2)$. This implies that $x_1 + 2 = x_2 + 2$, i.e., $x_1 = x_2$. Therefore, $f(x) = x + 2$; $x \in Z$ is a one-one function. Similarly, $f(x) = x + 2$ implies that $y - 2 = x$, since $y = f(x)$. This indicates that for every element y in the co-domain set Z has a pre-image $(y - 2)$ in the domain set Z. Hence $f(x) = x + 2$ is onto function. This implies that $f: Z \rightarrow Z$ defined by $f(x) = x + 2$; $x \in Z$ is a bijective function.

1.3.2 Inverse of a Function

Let $f: A \rightarrow B$ is a bijective mapping. Then the mapping $f^{-1}: B \rightarrow A$ which associates to each element $y \in B$ the element $x \in A$, such that $f(x) = y$, is called the inverse mapping of the mapping $f: A \rightarrow B$. As $f^{-1}: B \rightarrow A$, so for $y \in B$ there exists $x \in A$ such that $f^{-1}(y) = x$.

Consider the mapping $f: Z \rightarrow Z$ defined by $f(x) = x + 5$; $x \in Z$

Assume that $f(x_1) = f(x_2)$. This implies that $x_1 + 5 = x_2 + 5$, i.e., $x_1 = x_2$. Therefore, $f(x) = x + 5$; $x \in Z$ is an one-one function. Similarly, $f(x) = x + 5$ implies that $y - 5 = x$, since $y = f(x)$. This indicates that for every element y in the co-domain set Z has a pre-image $(y - 5)$ in the domain set Z. Hence $f(x) = x + 5$ is an onto function. This implies that $f: Z \rightarrow Z$ defined by $f(x) = x + 5$; $x \in Z$ is a bijective function. To compute the inverse we have $f^{-1}(y) = x = y - 5$. In general $f^{-1}(x) = (x - 5)$ is the inverse mapping of the mapping $f(x) = x + 5$, $x \in Z$.

Note It is to be noted that inverse of a mapping is possible only when the mapping is bijective, i.e., one-one and onto.

1.4 GRAPH

Graph theory has applications in many areas like mathematics, engineering, computer science and communication science. The major developments of graph theory occurred by the ever-growing importance of computer science and its connection with graph theory. A graph G consists of a finite set of vertices V and a finite set of edges E. The number of vertices in the graph $G(V, E)$ is called its order whereas the number of edges in it is called its size. Mathematically,

$$G = (V, E), \text{ where } E = \{(v_i, v_j) \mid v_i, v_j \in V\}$$

Consider the graph G given below. The order of the graph is 5 whereas the size is 6.

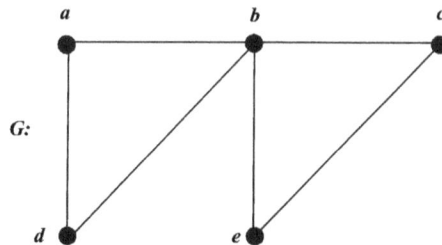

1.4.1 Types of Graph

The different types of graph are simple graph, multi graph, pseudo graph, digraph, weighted graph, complete graph, regular graph and connected graph.

Simple graph   A graph $G(V, E)$ that has no self loop or parallel edges is called a simple graph. Edges are said to be parallel if they have same pair of vertices. An edge is said to be a loop if the starting and terminating vertex are identical.

Multi graph   A graph $G(V, E)$ is known as a multi graph if it contains parallel edges, i.e., two or more edges between a pair of vertices. It is to be noted that every simple graph is a multi graph but the converse is not true.

Pseudo graph   A graph $G(V, E)$ is known as a pseudo graph if we allow both parallel edges and loops. It is to be noted that every simple and multi graph are pseudo graph but the converse is not true.

Digraph   A graph $G(V, E)$, where V is the set of vertices or nodes and E is the set of edges having direction. If (v_i, v_j) is an edge, then there is an edge from the vertex (v_i) to the vertex (v_j). A digraph is also called as directed graph.

Weighted graph   A graph (digraph) $G(V, E)$ is known as a weighted graph (digraph) if each edge of the graph has some weights.

Complete graph   A graph (digraph) $G(V, E)$ is said to be complete if each vertex u is adjacent to every other vertex v in G. That is there exists edges from any vertex to all other vertices.

Regular graph   A graph $G(V, E)$ is said to be regular if the degree of every vertex are equal. Mathematically, G is said to be regular if $deg\ (u) = deg\ (v)$ for all u, v belonging to the set of vertices V.

The degree of a vertex is defined as the number of edges connected to that vertex. If u be a vertex then the degree of u is denoted as $deg\ (u)$. In case of digraph, there are two degrees. These are in-degree and out-degree. The number of edges coming to a vertex u is known as in-degree of u, whereas the number of edges emanating from the vertex u is known as out-degree of u. In-degree and out-degree are generally denoted as $indeg\ (u)$ and $outdeg\ (u)$ respectively. In case of a loop it contributes 2 to the degree of a vertex.

Connected graph   A graph $G(V, E)$ is said to be connected if for every pair of distinct vertices u and v in G, there exists a path. A directed graph is said to be strongly connected if for every pair of distinct vertices u and v in G, there is a directed path from u to v and also from v to u. A directed graph in which every pair of distinct vertices has a path without taking the directions is termed as weakly connected.

A path in a graph is a sequence $v_1\ v_2\ v_3\ \dots\ v_k$ of vertices each adjacent to next, and a choice of an edge between each v_i to v_{i+1} so that no edge is chosen more than once. Consider the examples of different graphs shown in Figure 1.2.

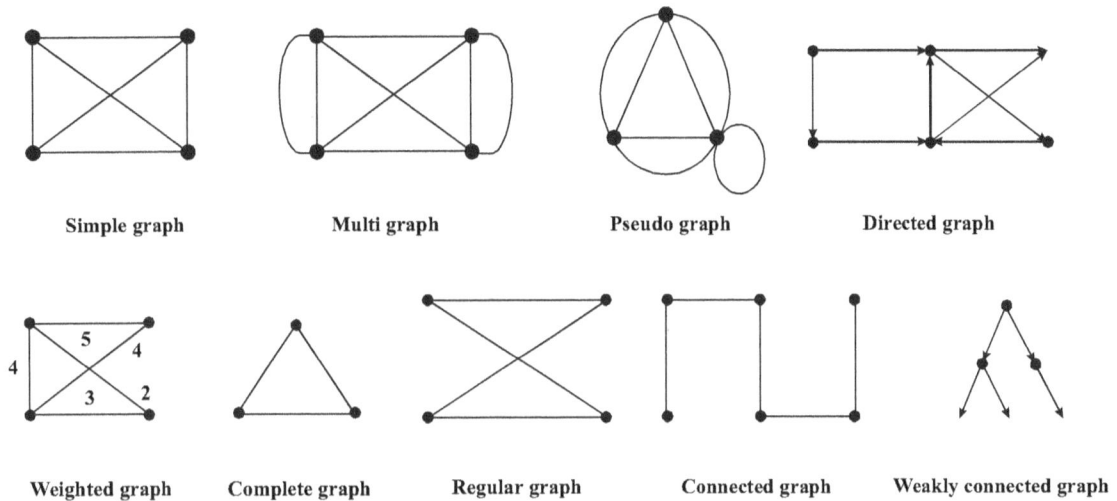

Figure 1.2 Types of graphs

1.4.2 Graph as a Relation

Mathematics deals with statements about objects and the relations between them. It is natural to say, for example, "equal to" is a relation between objects of a certain kind. In this section we study, digraphs that exhibit the properties of binary relations. As discussed earlier, a relation R on A is a subset of $(A \times A)$, where A is a set. This relation R on A can be represented by a digraph.

Each element of A is represented by a node of the directed graph and there exists a directed edge from u to v if and only if $(u,v) \in R$. Therefore, from the graph given below, it is clear that, there is no formal distinction between binary relations on a set A and digraphs with nodes from A. Consider the relation R as

$$R = \{(a, b), (e, b), (e, e), (b, d), (a, c), (c, d), (d, a), (d, e)\}$$

on $A = \{a, b, c, d, e\}$. The above relation R can be represented by a digraph as below.

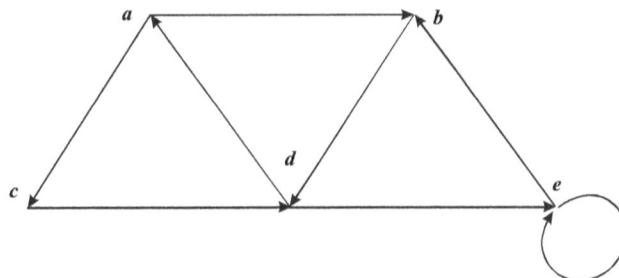

The directed graph representing a reflexive relation has a loop from each node to itself. This is due to $(x,x) \in R$ for all $x \in A$. For example, the directed graph given in Figure 1.3a represents a reflexive relation whereas Figure 1.3b does not. In the directed graph, whenever there is an edge between two nodes, there are edges between those nodes in both directions. This is due to $(x, y) \in R$ implies $(y, x) \in R$. Such type of directed graph represents a symmetric relation. For example, the directed graph given in Figure 1.3c represents a symmetric relation whereas Figure 1.3d does not. Whenever there is a sequence of edges leading from an element x to an element z, there is an edge directly from x to z. This is due to $(x, y) \in R, (y, z) \in R$ implies $(x, z) \in R$. Such type of graph represents transitive relation. For example the directed graph given in Figure 1.3e represents a transitive relation whereas the Figure 1.3f does not.

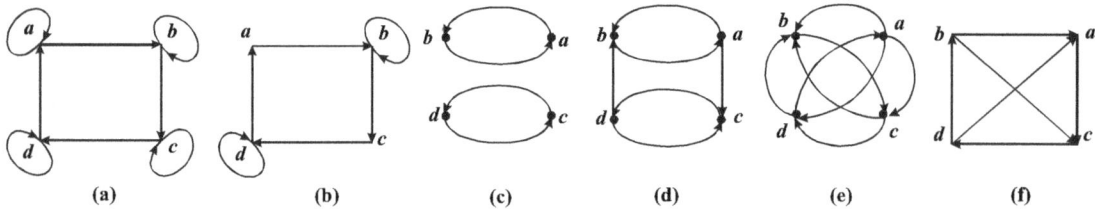

Figure 1.3 Graph as relation

1.5   TREE

In the previous section we have discussed on graphs. An important class of graphs is called trees. A connected acyclic graph is called a tree. A tree T is a finite set of one or more nodes such that

 i. There is a specially designated node called the root.

 ii. Remaining nodes are partitioned into k-disjoint sets $T_1, T_2, \ldots T_k$; $k > 0$,

where, each T_i, $i = 1, 2, \ldots k$ is a tree. $T_1, T_2, \ldots T_k$ are called subtrees of the root.

Consider the very common example of ancestor tree as given in Figure 1.4. This tree shows the ancestors of Aditi. Her parents are Debi and Ashima; Debi's parents are Gouri and Pramod who are also the grandparents of Aditi (on father's side); Ashima's parents are Prafulla and Jayanti who are also the grandparents of Aditi (on mother's side) and so on. The different properties of trees are given below that are used in developing transition systems and studying grammar rules.

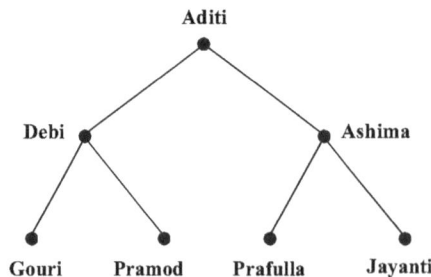

Figure 1.4 An ancestor tree

i. A tree is a connected graph with no cycles or loops.

ii. In a tree there exists one and only one path between every pair of vertices.

iii. If a graph contains unique path between every pair of vertices, then the graph is a tree.

iv. A tree with n nodes has $(n - 1)$ edges.

v. If a graph contains no circuits with n vertices and $(n - 1)$ edges, then the graph is a tree.

vi. A connected graph with n vertices and $(n - 1)$ edges is a tree.

1.5.1 Types of Trees

This section deals with different types of trees such as rooted tree, binary tree, full binary tree and complete binary tree (Figure 1.5)

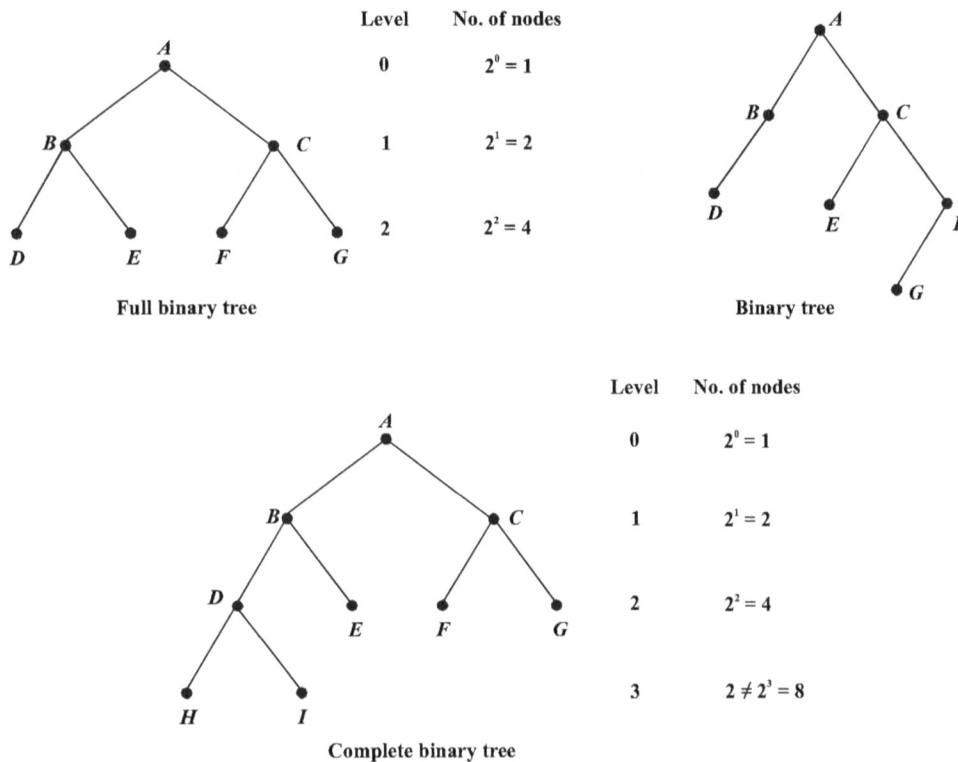

Figure 1.5 Types of trees

Rooted tree A tree is said to be rooted if it has one node, called the root that is distinguished from the other nodes. In Figure 1.4, the root is Aditi and hence called a rooted tree.

Binary tree A binary tree T is a finite set of nodes such that either T is empty or T contains a specially designated node called the root of T and the remaining nodes of T form two disjoint binary trees T_1 and T_2. The two disjoint binary trees are called the left subtree and right subtree.

Generally denoted by $L(T)$ and $R(T)$ respectively. This implies that, in a binary tree a node may have at most two children.

Full binary tree   A binary tree T is said to be a full binary tree if it contains maximum possible number of nodes in all level. This indicates that, for the level n of the tree it must contain 2^n number of nodes.

Complete binary tree   A binary tree T is said to be complete binary tree if it contains maximum possible nodes in all levels except the last level. Examples of different types of trees are shown in Figure 1.5.

1.6   DIFFERENT METHODS OF PROOF

Every proof is different because every proof is designed to establish a different result. In this section we will discuss two fundamental and powerful methods such as mathematical induction and pigeonhole principle.

1.6.1 Mathematical Induction

Suppose that n be a natural number. Our aim is to show that some statement P(n) involving n is true for any n. The following steps are used in mathematical induction.

1.   Suppose that $P(n)$ be a statement.
2.   Show that $P(1)$ and $P(2)$ are true. That is $P(n)$ is true for $n = 1$ and 2.
3.   Assume that $P(k)$ is true. That is $P(n)$ is true for $n = k$.
4.   Show that $P(k + 1)$ follows from $P(k)$.

The mathematical induction consists of three steps, i.e., basic step, induction hypothesis and induction step. The steps 1 and 2 are known as basic steps. Step 3 is known as induction hypothesis whereas step 4 is termed as induction step.

Consider the example $P(n) \equiv 1^2 + 2^2 + 3^2 + \ldots + n^2 = \dfrac{n(n+1)(2n+1)}{6}$

$P(1) \equiv 1^2 = \dfrac{1(1+1)(2+1)}{6}; P(2) \equiv 1^2 + 2^2 = 5 = \dfrac{2(2+1)(4+1)}{6}$. So, $P(1)$ and $P(2)$ are true. Assume that $P(k)$ is true. Therefore we have,

$$P(k) \equiv 1^2 + 2^2 + \ldots + k^2 = \frac{k(k+1)(2k+1)}{6}.$$

Now,

$$P(k+1) \equiv 1^2 + 2^2 + \cdots + k^2 + (k+1)^2 = \frac{k(k+1)(2k+1)}{6} + (k+1)^2$$

$$= \frac{(k+1)}{6}[k(2k+1) + 6(k+1)]$$

$$= \frac{(k+1)(k+2)(2k+3)}{6}$$

Hence, $P(n)$ is true for all n.

1.6.2 Pigeonhole Principle

If A and B are finite sets with $|A| > |B|$, then there is no one-to-one function from A to B. In other words, if we attempt to pair off the elements of A (the "Pigeons") with elements of B (the "Pigeonholes"), sooner or later we have to put more than one pigeon in a pigeonhole.

1.7   ALPHABETS AND STRINGS

In this section we will discuss the basic concepts related to alphabets and strings. This is the basic data structure to finite automaton and grammar. Strings are defined over an alphabet and are finite in length whereas alphabet may vary based upon the applications.

1.7.1 Alphabet

An alphabet is a finite non-empty set Σ of objects called symbols. For example

    i. $\Sigma = \{0,1\}$

    ii. $\Sigma = \{a,b,c,\ldots z\}$

    iii. $\Sigma = \{0,1,2,\ldots 9,a,b,c \ldots z\}$

1.7.2 String

An n-tuple of symbols of Σ is called a string or word over Σ. Instead of writing a string as $(a_1, a_2, a_3, \ldots a_n)$ we write simply $(a_1 a_2 a_3, \ldots a_n)$. Generally the strings are denoted by lower case letters like x, y, w. The string having no symbol is called the empty string or null string. It is denoted by $\in$ or Λ or λ. For example

    i.   010011; 11001; 1110 are strings over an alphabet $\Sigma = \{0,1\}$

    ii.   *ab, aabb, abab* are strings over an alphabet $\Sigma = \{a,b\}$

Length of a string   The number of symbols present in a string is known as length of the string. If w be a string, then the length of the string w is denoted by $|w|$. In case of a null string, the length of the string is zero. For example if

    i. $w = 1011001$,   then $|w| = 7$

    ii. $w = aabba$,   then $|w| = 5$

Reverse of a string   The reverse of a string is obtained by writing symbols of string in reverse order. If w be a string, then w^R or reverse (w) is the reverse string of w. Consider the following examples.

   i. $w = 000111$ then $w^R = 111000$

   ii. $w = abcd$ then $w^R = dcba$

   iii. $w = abx110y$ then $w^R = y011xba$

Palindrome   A string w is said to be palindrome if $w = w^R$. In other words palindrome is a string which is same whether written backward or forward. Consider the following examples.

   i $w = katak$ then $w^R = katak = w$

   ii. $w = Malayalam$ then $w^R = Malayalam$

   iii. $w = ababa$ then $w^R = ababa$

Substring   Let w be a string on alphabet Σ. A string u is said to be a substring of w, if it appears within the string w. For example $u = baba$ is a substring of $w = ababab$. Similarly, $u = 100$ is a substring of $w = 1100111$.

Concatenation   Let u and w be two strings of alphabet Σ. The concatenation of u and w is denoted by uw and is defined as a new string obtained by u followed by w. Concatenation with an empty string does not change the string. For example if $u = x_1 x_2 x_3 \cdots x_n$ and $w = y_1 y_2 y_3 \cdots y_m$ then the concatenation uw is given as below.

$$uw = x_1 x_2 x_3 \cdots x_n y_1 y_2 y_3 \cdots y_m$$

   Therefore it is clear that if $|u| = n$ and $|w| = m$, then $|uw| = (n + m)$.

1.7.3. Properties of Strings

String over an alphabet set Σ is a finite sequence of symbols of Σ. It has the following properties.

Associative law   Let u, v, w be three strings over an alphabet set Σ, i.e., $u, v, w \in \Sigma^*$.

   Then we have $u(vw) = (uv)w$

Existence of identity   For all string $u \in \Sigma^*$, there exist an identity element $\Lambda \in \Sigma^*$ such that $u\Lambda = u = \Lambda u$.

Cancellation laws   Let $u, v, w \in \Sigma^*$. Then if

   i. $uv = uw$ then $v = w$ [Left cancellation law]

   ii. $vu = wu$ then $v = w$ [Right cancellation law]

1.7.4 Kleene Closure

The set of all strings including the empty string on the alphabet is written as Σ^*. This is otherwise known as Kleene closure or Kleene star after the famous logician who was one of the founders of this concept. For example if

i. $\Sigma = \{1\}$, then $\Sigma^* = \{\Lambda, 1, 11, 111, \ldots\}$

ii. $\Sigma = \{a, b\}$ then $\Sigma^* = \{\Lambda, a, b, ab, ba, aa, bb, aab, \ldots\}$

Language   Let Σ be an alphabet and Σ^* be the set of all string on an alphabet Σ. Any subset of Σ^* is called a language over Σ or language with alphabet Σ. Generally language is denoted by L. Thus, $L \subseteq \Sigma^*$. If a language L consists of only empty string Λ, i.e., $L = \{\Lambda\}$, then it is a language on any alphabet. It is clear from the definition that, language L is a set. So, all set operation hold true for languages also. Let L_1 and L_2 be two languages over the alphabet Σ. Therefore the different set operations are defined as below.

i.   The union $(L_1 \cup L_2) = \{w \mid w \in L_1 \text{ or } w \in L_2\}$

ii.   The intersection $(L_1 \cap L_2) = \{w \mid w \in L_1 \text{ and } w \in L_2\}$

iii. The difference $(L_1 - L_2) = \{w \mid w \in L_1 \text{ and } w \notin L_2\}$

iv.   The complement $\overline{L_1} = \{w \mid w \in \Sigma^* \text{ and } w \notin L_1\}$

SOLVED EXAMPLES

Example 1

If $R = \{(2, 2), (2, 4), (4, 8), (4, 6)\}$ be a relation in the set $\{2, 4, 6, 8\}$, then find the transitive closure R^+.

Solution   Given $R = \{(2, 2), (2, 4), (4, 8), (4, 6)\}$ be a relation in $A = \{2, 4, 6, 8\}$. Therefore, $R^2 = R \circ R = \{(2, 2), (2, 4), (2, 6), (2, 8)\}$ and $R^3 = R^2 \circ R = \{(2, 2), (2, 4), (2, 6), (2, 8)\}$

Similarly, $R^4 = R^5 = \ldots = R^2 = \{(2, 2), (2, 4), (2, 6), (2, 8)\}$

$$R^+ = R \cup R^2$$

Thus the transitive closure R^+ is defined as $R^3 = R^2 \circ R$

i.e., $R^+ = \{(2, 2), (2, 4), (2, 6), (2, 8), (4, 8), (4, 6)\}$

Example 2

Let $R = \{(a, b), (b, c), (c, d), (c, c)\}$ be a relation in $\{a, b, c, d\}$. Find the reflexive and transitive closure R^*.

Solution   Given $R = \{(a, b), (b, c), (c, d), (c, c)\}$ be a relation in $A = \{a, b, c, d\}$. Therefore, $R^* = r(R) \cup R^+$, where $r(R)$ is the reflexive closure. Now

$r(R) = \{(a, b), (b, c), (c, d), (c, c), (a, a), (b, b), (d, d)\}$

$R^2 = R \circ R = \{(a, c), (b, c), (b, d), (c, c), (c, d)\}$

$$R^3 = R^2 \circ R = \{(a, c), (a, d), (b, c), (b, d), (c, c), (c, d)\}$$

Similarly, $R^4 = R^5 = \dots = R^3 = \{(a, c), (a, d), (b, c), (b, d), (c, c), (c, d)\}$. Thus we get $R^+ = R \cup R^2 \cup R^3 = \{(a, b), (b, c), (c, d), (c, c), (a, c), (b, d), (a, d)\}$. Hence the reflexive and transitive closure R^* is given as

$$R^* = r(R) \cup R^+$$
$$= \{(a,b),(b,c),(c,d),(c,c),(a,c),(b,d),(a,d),(a,a),(b,b),(d,d)\}$$

Example 3

Show that the relation $x \equiv y(\mathrm{mod}\,3)$ on the set of integers I is an equivalence relation.

Solution   Given that the relation is $x \equiv y(\mathrm{mod}\,3)$. It indicates that $(x - y)$ is divisible by 3. Therefore, we have $xRy:(x-y)=3k; \ k \in I$. Now in order to show it is an equivalence relation, we have to show that the relation is reflexive, symmetric and transitive.

Reflexive   For all $x \in I$ we have $(x - x) = 0 = 3k; \ k = 0 \in I$. Therefore it is clear that $x\, R\, x$ for all x belonging to I.

Symmetric   Suppose that xRy. This implies that $(x - y) = 3k, k \in I$,

i.e., $(y - x) = -3k = 3(-k); \ -k \in I$.

Therefore it is clear that yRx, i.e., $xRy \Rightarrow yRx$.

It indicates that the relation is symmetric.

Transitive   Suppose that xRy and yRz.

This implies that $(x - y) = 3k_1$; and $(y - z) = 3k_2; \ k_1, k_2 \in I$

$$\Rightarrow (x - y) + (y - z) = 3k_1 + 3k_2 = 3(k_1 + k_2);(k_1 + k_2) \in I$$

Therefore it is clear that $x\, R\, z$, i.e., xRy and $yRz \Rightarrow xRz$. It indicates that the relation is transitive and hence the relation is an equivalence relation.

Example 4

For the relation R on the set $\{5, 6, 7, 8, 9\}$ defined by the rule (x, y) if $x + 2y \le 20$, find R and R^{-1}.

Solution   Given R be a relation on the set $\{5, 6, 7, 8, 9\}$ defined by the rule (x, y) if $x + 2y \le 20$, i.e., $xRy:x+2y \le 20$.

Thus, $R = \{(5, 5), (5, 6), (5, 7), (6, 5), (6, 6), (6, 7), (7, 5), (7, 6), (8, 5), (8, 6), (9, 5)\}$

Therefore, $R^{-1} = \{(5, 5), (6, 5), (7, 5), (5, 6), (6, 6), (7, 6), (5, 7), (6, 7), (5, 8), (6, 8), (5, 9)\}$.

Example 5

Let A be the set of non-zero integers and R be a relation in A defined by $(a, b) R(c, d)$ if and only if $(a + d) = (b + c)$. Show that R is an equivalence relation.

Solution Let A be the set of non-zero integers and R be a relation in A defined as $(a, b) R(c, d)$ if and only if $(a + d) = (b + c)$.

Reflexive For all $a, b \in A$; we have $(a + b) = (b + a)$; i.e., $(a, b) R(a, b)$ and hence the relation is reflexive.

Symmetric Suppose that $(a, b) R(c, d)$. Therefore, $(a + d) = (b + c)$.

This implies that $(b + c) = (a + d)$, i.e., $(c + b) = (d + a)$ and hence $(c, d) R(a, b)$. Thus the given relation is symmetric.

Transitive Suppose that $(a, b) R(c, d)$ and $(c, d) R(e, f)$

This implies that $(a + d) = (b + c)$ and $(c + f) = (d + e)$.

On adding we get, $(a + d) + (c + f) = (b + c) + (d + e)$

i.e., $(a + f) + (c + d) = (b + e) + (c + d)$

i.e., $(a + f) = (b + e)$

Therefore, $(a, b) R(e, f)$ and hence the relation is transitive. So, the relation $(a, b) R(c, d)$ if and only if $(a + d) = (b + c)$ is an equivalence relation.

Example 6

Prove by mathematical induction that

$$3 + 7 + 11 + \ldots + (4n - 1) = n(2n + 1)$$

Solution Let $P(n) \equiv 3 + 7 + 11 + \ldots + (4n - 1) = n(2n + 1)$. For $n = 1, 2$ we have $P(1) \equiv 3 = 1(2 + 1)$ and $P(2) \equiv 3 + 7 = 10 = 2(2(2) + 1)$. Thus, $P(1)$ and $P(2)$ are true. Assume that the statement is true for $n = k$. Therefore,

$$P(k) \equiv 3 + 7 + 11 + \ldots + (4k - 1) = k(2k + 1)$$
$$P(k + 1) \equiv 3 + 7 + 11 + \ldots + (4k - 1) + (4k + 3) = k(2k + 1) + (4k + 3)$$
$$= 2k^2 + 5k + 3 = (k + 1)(2k + 3)$$

Hence, $P(k + 1)$ is true. Therefore, the statement $P(n)$ is true for all n.

Example 7

Prove by mathematical induction that

$$3 \times 8 + 6 \times 11 + \ldots + 3n(3n + 5) = 3n(n + 1)(n + 3)$$

Solution   Let $P(n) \equiv 3 \times 8 + 6 \times 11 + \ldots + 3n(3n + 5) = 3n(n + 1)(n + 3)$. For $n = 1$ and 2 we have $P(1) \equiv 3 \times 8 = 24 = 3\,(1 + 1)\,(1 + 3)$ and $P(2) \equiv 3 \times 8 + 6 \times 11 = 90 = 6\,(2 + 1)(2 + 3)$

Thus, $P(1)$ and $P(2)$ are true. Assume that $P(k)$ is true. Therefore,

$$P(k) \equiv 3 \times 8 + 6 \times 11 + \ldots + 3k\,(3k + 5) = 3k\,(k + 1)\,(k + 3)$$

$$P(k + 1) \equiv 3 \times 8 + 6 \times 11 + \ldots + 3k\,(3k + 5) + (3k + 3)(3k + 8)$$

$$= 3k\,(k + 1)\,(k + 3) + (3k + 3)(3k + 8)$$

$$= 3\,(k + 1)[\,k\,(k + 3) + (3k + 8)]$$

$$= 3\,(k + 1)(k + 2)(k + 4)$$

Hence, $P(k + 1)$ is true. Therefore the statement $P(n)$ is true for all n.

Example 8

Let Q be the set of rational numbers. Show that the function $f : Q \to Q$ defined by $f(x) = 2x + 3, x \in Q$ is a bijective function. Find $f^{-1}(0), f^{-1}(1)$ and $f^{-1}(2)$.

Solution   Given Q be the set of rational numbers and $f : Q \to Q$ defined by $f(x) = 2x + 3, x \in Q$. In order to show the function bijective we have to show that it is one-one and onto.

One-One   Suppose that $f(x_1) = f(x_2)$ for $x_1, x_2 \in Q$. It implies that $2x_1 + 3 = 2x_2 + 3$, i.e., $x_1 = x_2$

Therefore, $f(x) = 2x + 3, \ x \in Q$ is an one-one function.

Onto   Given that $f(x) = 2x + 3, \ x \in Q$. This implies that $x = \dfrac{y - 3}{2}$. Therefore, for every element y in the co-domain set Q has a pre-image x in the domain set Q. Hence $f(x) = 2x + 3$ is an onto function. Thus we conclude that $f : Q \to Q$ defined by $f(x) = 2x + 3, x \in Q$ is bijective. Now to compute the inverse we have

$$f^{-1}(y) = x = \frac{y - 3}{2}, \text{ i.e., } f^{-1}(y) = \frac{y - 3}{2}$$

$$\text{Therefore, } f^{-1}(0) = -\frac{3}{2}; \ f^{-1}(1) = -1; \ f^{-1}(2) = -\frac{1}{2}$$

Example 9

Let $A = R - \left\{\dfrac{2}{3}\right\}$ and $B = R - \left\{\dfrac{2}{3}\right\}$, where R is the set of real numbers. Let $f : A \to B$ defined by

$f(x) = \dfrac{2x - 1}{3x - 2}; x \in A$. Prove that f is bijective.

Solution   Given $A = R - \left\{\dfrac{2}{3}\right\}$ and $B = R - \left\{\dfrac{2}{3}\right\}$. Let $f : A \to B$ defined by $f(x) = \dfrac{2x-1}{3x-2}$; $x \in A$. Now we have to show that $f(x)$ is injective and surjective.

Injective   Suppose that $f(x_1) = f(x_2)$ for x_1, $x_2 \in A$. It implies that

$$\frac{2x_1 - 1}{3x_1 - 2} = \frac{2x_2 - 1}{3x_2 - 2}, \text{ i.e., } 6x_1x_2 - 3x_2 - 4x_1 + 2 = 6x_1x_2 - 3x_1 - 4x_2 + 2$$

i.e., $x_1 = x_2$

Therefore, $f(x)$ as defined above is an injective function.

Surjective   Given mapping is $f(x) = \dfrac{2x-1}{3x-2}$; $x \in A$. This implies that $x = \dfrac{2y-1}{3y-2}$

Therefore, for every element $y \in B$ has a pre-image $x \in A$ such that $y = f(x)$. Thus $f(x)$ as defined above is surjective. Hence it is clear that $f(x)$ defined as above is bijective.

Example 10

Let $A = \{a, b, c\}$. Determine the power set $P(A)$ of A.

Solution   Given that $A = \{a, b, c\}$.

Therefore, $P(A) = \{\phi, A, \{a\}, \{b\}, \{c\}, \{a, b\}, \{b, c\}, \{a, c\}\}$.

Example 11

Let A and B be two sets. Prove the De-Morgan's laws

   i. $(A \cup B)^c = A^c \cap B^c$

   ii. $(A \cap B)^c = A^c \cup B^c$

Solution   Let $x \in (A \cup B)^c$

   $\Leftrightarrow$ $x \notin (A \cup B)$

   $\Leftrightarrow$ $x \notin A$ *and* $x \notin B$

   $\Leftrightarrow$ $x \in A^c$ *and* $x \in B^c$

   $\Leftrightarrow$ $x \in A^c \cap B^c$

Therefore, $(A \cup B)^c = A^c \cap B^c$. Similarly the second relation can also be proved.

Example 12

Find the degree of every vertex for the following graphs.

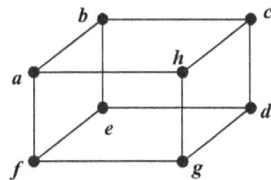 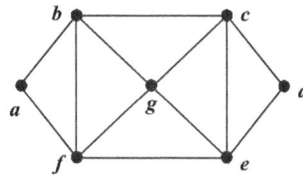

(i)   (ii)

Solution

1. According to the graph (i) given above, degree of every vertex is 3. That is $deg(a) = deg(b) = deg(c) = deg(d) = deg(e) = deg(f) = deg(g) = deg(h) = 3$.

2. According to the graph (ii) given above the degree of different vertices are given below.
    Here, $deg(a) = 2 = deg(d)$;

    $deg(b) = deg(c) = deg(e) = deg(f) = deg(g) = 4$

Example 13

Construct the binary tree for the arithmetic expressions.

   1. $(4x + xy)(3x + 2y)$ 2. $(xy + yz) + (zx + xq)$

Solution

1. Given expression is $(4x + xy) (3x + 2y)$. The binary tree representation is given in (i).

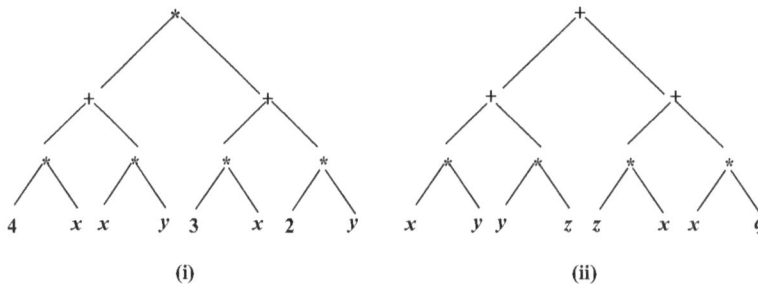

(i)   (ii)

2. Given expression is $(xy + yz) + (zx + xq)$. The binary tree representation is given (ii).

Example 14

Construct the following graphs.

   1. 3 regular but not complete   2. 3 regular and complete
   3. 2 regular but not complete   4. 2 regular and complete

Solution   The different graphs are given below.

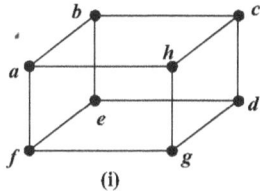

(i)	(ii)	(iii)	(iv)
3 regular but not complete	3 regular and complete	2 regular but not complete	2 regular and complete

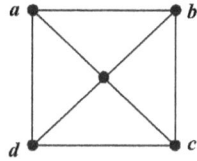

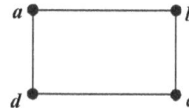

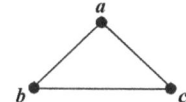

Example 15

Show that a relation is reflexive and circular if and only if it is reflexive, symmetric and transitive.

Solution   Let the relation R be reflexive and circular. A relation R is said to be circular if $(a,b) \in R, (b,c) \in R$ implies $(c,a) \in R$.

Our claim R is reflexive, symmetric and transitive. Therefore, we need to show R is symmetric and transitive.

Symmetric   Let $(a,b) \in R$.

Now $(a,a) \in R$ and $(a,b) \in R$ [∵ R is reflexive]

$\Rightarrow (b,a) \in R$ [∵ R is circular]

Transitive   Let $(a,b) \in R$ and $(b,c) \in R$

$\Rightarrow (c,a) \in R$ [∵ R is circular]

$\Rightarrow (a,c) \in R$ [∵ R is symmetric]

Thus, $(a,b) \in R$ and $(b,c) \in R$ implies $(a,c) \in R$. Hence R is transitive relation. Therefore, the relation R is reflexive, symmetric and transitive.

Conversely, suppose that R is reflexive, symmetric and transitive.

Our claim R is reflexive and circular. So we need to show R is circular only.

Circular   Let $(a,b) \in R$ and $(b,c) \in R$

$\Rightarrow (a,c) \in R$ [∵ R is transitive]

$\Rightarrow (c,a) \in R$ [∵ R is symmetric]

Therefore, $(a,b) \in R$ and $(b,c) \in R$ implies $(c,a) \in R$. Thus, the relation R is circular.

Example 16

Given $\Sigma = \{0,1\}$. Find the Kleene closure on the alphabet Σ.

Solution   Given alphabet $\Sigma = \{0,1\}$. The set of all strings including the null string on the alphabet Σ is defined as Kleene star. Therefore,

$$\Sigma^* = \{\wedge, 0, 1, 00, 11, 01, 10, \ldots\}$$

Example 17

Let $u = a^3 b a^2 b^3$ and $v = b^2 a^2 b a^4$. Find the following strings and their length.

1. u^2 2. uv 3. vu 4. v^2

5. uv^2 6. uvu 7. u^2v 8. vuv

Solution   Given that $u = a^3 b a^2 b^3$ and $v = b^2 a^2 b a^4$

1. $u^2 = uu = (a^3 b a^2 b^3)(a^3 b a^2 b^3) = a^3 b a^2 b^3 a^3 b a^2 b^3; \left|u^2\right| = 18.$

2. $uv = (a^3 b a^2 b^3)(b^2 a^2 b a^4) = a^3 b a^2 b^5 a^2 b a^4; \left|uv\right| = 18$

3. $vu = (b^2 a^2 b a^4)(a^3 b a^2 b^3) = b^2 a^2 b a^7 b a^2 b^3; \left|vu\right| = 18$

4. $v^2 = vv = (b^2 a^2 b a^4)(b^2 a^2 b a^4) = b^2 a^2 b a^4 b^2 a^2 b a^4; \left|v^2\right| = 18$

5. $uv^2 = (a^3 b a^2 b^3)(b^2 a^2 b a^4 b^2 a^2 b a^4) = a^3 b a^2 b^5 a^2 b a^4 b^2 a^2 b a^4; \left|uv^2\right| = 27$

6. $uvu = (a^3 b a^2 b^5 a^2 b a^4)(a^3 b a^2 b^3) = a^3 b a^2 b^5 a^2 b a^7 b a^2 b^3; \left|uvu\right| = 27$

7. $(u^2v) = (a^3 b a^2 b^3 a^3 b a^2 b^3)(b^2 a^2 b a^4) = a^3 b a^2 b^3 a^3 b a^2 b^5 a^2 b a^4; \ \left|u^2v\right| = 27$

8. $vuv = (b^2 a^2 b a^4)(a^3 b a^2 b^5 a^2 b a^4) = b^2 a^2 b a^7 b a^2 b^5 a^2 b a^4; \left|vuv\right| = 27$

Example 18

Given $\Sigma = \{a,b\}$. Obtain two languages L_1 and L_2. Find $(L_1 \cup L_2); (L_1 \cap L_2)$ and $(L_1 - L_2)$.

Solution   Given alphabet $\Sigma = \{a,b\}$. Any subset of Kleene star (Σ^*) is called a language. Therefore, we have $\Sigma^* = \{\wedge, a, b, aa, bb, ab, ba, \ldots\}$

Let us consider two languages L_1 and L_2 as below.

$L_1 = \{a, b, aa, ab, aabb, abab\}$ and $L_2 = \{b, bb, ab, abab, bbaa\}$

Therefore,

$(L_1 \cup L_2) = \{a, b, aa, ab, aabb, abab, bb, bbaa\}$

$(L_1 \cap L_2) = \{b, ab, abab\}$ and $(L_1 - L_2) = \{a, aa, aabb\}$.

Example 19

Let w be a string. Use mathematical induction to show $\left|w^n\right| = n\left|w\right|$ for all natural number n.

Solution   Let w be a string. Let $P(n) \equiv \left|w^n\right| = n\left|w\right|$. For $n = 1$ and 2 we have $P(1) \equiv |w| = 1|w|$ and $P(2) \equiv |w^2| = |ww| = |w| + |w| = 2|w|$.

Thus the statement is true for $n = 1$ and 2. Assume that $P(k)$ is true. This implies that $|w^k| = k|w|$. Therefore,

$$|w^{k+1}| = |w^k w| = |w^k| + |w| = k|w| + |w| = (k+1)|w|.$$

It indicates that $P(k + 1)$ is true and hence the statement is true for all natural number $n \in N$.

Example 20

Check which of the following strings are palindrome.

1. malayalam   2. *aabbaa*   3. *aaabb*

Solution

1. Let w = malayalam

   Therefore, w^R = malayalam = w. Hence the string "malayalam" is palindrome.

2. Let w = *aabbaa*

   Therefore, w^R = *aabbaa* = w. Hence the string "*aabbaa*" is palindrome.

3. Let w = *aaabb*

   Therefore, w^R = *bbaaa* ≠ w. Hence the string '*aaabb*' is not palindrome.

Example 21

Given a language $L = \{a^n b^n : n \geq 0\}$ defined over some alphabet Σ. Obtain L^3 and L^R.

Solution
Given language is $L = \{a^n b^n : n \geq 0\}$.

Therefore, $L^3 = \{a^n b^n a^m b^m a^p b^p : n \geq 0, m \geq 0, p \geq 0\}$.

It is to be noted that n, m, p are unrelated. For example the string "*aabbaaabbbab*" is in L^3. Similarly, the reverse of the language L defined over the same alphabet Σ is given as $L^R = \{b^n a^n : n \geq 0\}$.

Example 22

Given alphabet $\Sigma = \{a, b\}$. Obtain the star closure Σ^* and the positive closure Σ^+.

Solution
Given alphabet $\Sigma = \{a, b\}$. The set of all strings including the null string on the alphabet Σ is known as star closure Σ^*. Therefore,

$$\Sigma^* = \{\Lambda, a, b, ab, ba, aa, bb, aab, \ldots\}$$
$$\Sigma^+ = \Sigma^* - \{\Lambda\} = \{a, b, ab, ba, aa, bb, aab, \ldots\}$$

It is clear that both Σ^* and Σ^+ are infinite languages.

Example 23

Let $L = \{ba, bb, abb\}$. Identify the strings from the following strings that belong to L^*.

1. *abbbaabbbbbbba*   2. *bbbbabbbb*

3. *abbbbbabb*   4. *abababb*

Solution   Given that $L = \{ba, bb, abb\}$.

1.   Let $w = abbbaabbbbbbba$.

    This can be decomposed as $w = (abb)(ba)(abb)(bb)(bb)(ba)$. It is clear that each sub-string is in L and hence the string w is in L^*.

2.   Let $w = bbbbabbbb$.

    This can be decomposed as $w = (bb)(bb)(abb)(bb)$. It is clear that each sub-string is in L and hence the string w is in L^*.

3.   Let $w = abbbbbabb$.

    This can be decomposed as $w = (abb)(bb)(ba)(bb)$. It is clear that each sub-string is in L and hence the string w is in L^*.

4.   Let $w = abababb$.

    This can be decomposed as $w = a(ba)(ba)(bb)$. It is clear that each sub-string except '*a*' is in L, i.e., $a \notin L$. Therefore the given string w does not belong to L^*.

Example 24

Given $L = \{a^n b^{n+1} : n \geq 0\}$. Prove or disprove that $L = L^*$ for the given language L.

Solution   Given language $L = \{a^n b^{n+1} : n \geq 0\}$, i.e., $L = \{b, ab^2, a^2b^3, a^3b^4, \dots\}$. The star closure of a language L, i.e., L^* is defined as below.

$$L^* = L^0 \cup L^1 \cup L^2 \cup L^3 \cup \dots$$

Here it is clear that L^n is defined as L concatenated itself n-times with the special case $L^\circ = \{\wedge\}$. Now for given L we have $L^2 = \{a^n b^{n+1} a^m b^{m+1} : n, m \geq 0\}$ and

$$L^3 = \{a^n b^{n+1} a^m b^{m+1} a^p b^{p+1} : n, m, p \geq 0\}$$

Therefore, we have $L^* = L^0 \cup L^1 \cup L^2 \cup L^3 \cup \dots$

$$= \{\Lambda\} \cup \{a^n b^{n+1}\} \cup \{a^n b^{n+1} a^m b^{m+1}\} \cup \{a^n b^{n+1} a^m b^{m+1} a^p b^{p+1}\} \dots$$

i.e., $L^* = \{a^n b^{n+1}, a^n b^{n+1} a^m b^{m+1}, a^n b^{n+1} a^m b^{m+1} a^p b^{p+1}, \dots\}$

Hence, it is clear that all elements of L is in L^*, whereas all elements of L^* does not belong to L. For example $a^n b^{n+1} a^m b^{m+1}$ belongs to L^* but does not belongs to L. Therefore, L is a subset of L^* and hence $L \neq L^*$.

Example 25

Define the reverse of a string by the recursive rules.

Solution   For all $a \in \Sigma$ and $w \in \Sigma^*$, the reverse of a string can be defined recursively as $a^R = a$ and $(wa)^R = aw^R$.

Prove that $(uv)^R = v^R u^R$ for all $u, v \in \Sigma^+$.

Assume that $(uv)^R = v^R u^R$ for all $u, v \in \Sigma^+$ and all v of length n. Now, take a string of length $(n + 1)$, say $w = va$. Thus we have

$$(uw)^R = (uva)^R = a(uv)^R \;\; [\because (wa)^R = aw^R]$$

$$= av^R u^R = (va)^R u^R = w^R u^R$$

Thus, the assumption $(uv)^R = v^R u^R$; $u, v \in \Sigma^+$ is true for all strings of length $(n+1)$.

Example 26

Prove that $(w^R)^R = w$ for all $w \in \Sigma^*$.

Solution   Let $\Sigma = \{a, b\}$ and $w \in \Sigma^*$. Let us consider $w = abab$.

Now, $w^R = (abab)^R = baba$ and $(w^R)^R = (baba)^R = abab = w$. Therefore, it is clear that $(w^R)^R = w$.

REVIEW QUESTIONS

1. Define a set by method of recursion and write the different properties of sets.
2. Explain the partition and covering on a set A.
3. Define product of sets.
4. Explain the different properties of relation.
5. Define equivalence relation and explain closures of relation.
6. Write the different types of functions.
7. Describe regular, connected, digraph and complete graph.
8. Explain graph as a relation.
9. Explain the difference between full binary tree and complete binary tree.
10. State Pigeonhole principle.
11. What do you mean by alphabet and strings?

12. Define Kleene closure and language.

13. State the different properties of string.

14. What is the difference between induction hypothesis and induction step?

15. State the different properties of a tree.

PROBLEMS

1. Let A, B and C be three sets. Prove the following.

   (a) $\overline{(A \times B)} = \overline{A} \times \overline{B}$

   (b) $A \times (B - C) = (A \times B) - (A \times C)$

   (c) $A \Delta B = (A \cup B) - (A \cap B)$

2. If R be an equivalence relation defined in a set A, then R^{-1} is also an equivalence relation in the set A.

3. If R be a relation from A to B, where A and B be any two sets, then show that $(R^{-1})^{-1} = R$.

4. If R be a partial order relation on A, then R^{-1} is also a partial order relation on A.

5. Use mathematical induction to show that

   (a) $1 \cdot 2 + 2 \cdot 3 + 3 \cdot 4 + \ldots + n(n+1) = \dfrac{n(n+1)(n+2)}{3} \ldots$

   (b) $1 \cdot 2 \cdot 3 + 2 \cdot 3 \cdot 4 + 3 \cdot 4 \cdot 5 + \cdots \ldots + n(n+1)(n+2) = \dfrac{n(n+1)(n+2)(n+3)}{4}$

   (c) $n(n+1)(n+2)$ is divisible by 3.

6. Show that the relation $(x - y)$ is an even integer is an equivalence relation.

7. Let R be the relation in $A = \{1, 2, 3, 4, 5, 6\}$ defined by x and y are relative prime. Find the relation R.

8. Let N be the set of all natural numbers. R be a relation in N defined by $x\,Ry$ if and only if $x + 3y = 12$. Examine the relation for reflexive, symmetric and transitive.

9. Prove that the relation on the set of natural numbers N defined by xRy if and only if x divides y is reflexive, transitive but not symmetric.

10. Let R be a relation defined on the set $S = \{1, 2, 3, 4, 5\}$ by a rule (x, y) if $x^2 + y^2 \leq 16$. Find the reflexive, symmetric and transitive closures of R.

11. Let A be the set of non-zero integers and R be a relation on A defined by $(a, b)\, R\,(c, d)$ if and only if $ad = bc$. Show that R is an equivalence relation.

12. Let $A = R - \{3\}$ and $B = R - \{1\}$, where R is the set of real numbers. Let $f : A \to B$ defined by

   $f(x) = \dfrac{x-2}{x-3}, x \in A$. Show that f is bijective. Find the inverse function of f.

13. Determine whether the given functions are one-one, onto or bijective.

(a) $f: R^+ \to R^+$ defined by $f(x) = |x|$

(b) $f: R \to R$ defined by $f(x) = |x|$

(c) $f: I \to R$ defined by $f(x) = 2x + 5$

14. Let $f: R - \{0\} \to R - \{0\}$ defined by $f(x) = \dfrac{1}{x}$. Show that f is bijective and its inverse is given by

$f^{-1}(x) = \dfrac{1}{x}$.

15. Show that the function $f(x) = \dfrac{x}{x^2 + 1} : R \to R$ is neither one-one nor onto.

16. Construct (draw) the following graphs.

(a) 2 regular but not complete

(b) 3 regular but not complete

(c) 3 regular and complete

(d) 4 regular but not complete

(e) 5 regular but not complete

17. Construct a graph of order 5, where vertices have degrees 1, 2, 2, 3 and 4. What is the size of this graph?

18. Let G be the set of all graphs. Show that the relation "is isomorphic" is an equivalence relation on the set G.

19. Find the degree of every vertex for the following graphs.

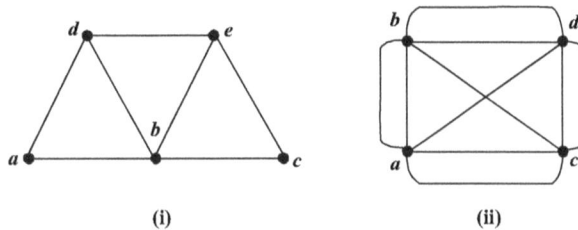

(i)   (ii)

20. Construct the binary tree for the following arithmetic expressions.

(a) $(A + B + C)(DE + F)$

(b) $(xy + 2x) + (3x + 2y)$

(c) $((x + y)(x - y))(2x + 4y)$

21. Let T be a tree of order p and size q having p_i vertices of degree i $(i = 1, 2, 3, ..)$. Let $\sum\limits_i p_i = p$ and

$\sum\limits_i ip_i = 2q = 2(p-1)$. Show that $p_1 = p_3 + 2p_4 + \cdots + 2$.

22. If T is a binary tree of height h and order p, then $(h+1) \le p \le 2^{(h+1)} - 1$

23. In a binary tree of height h, there are at most 2^{h-1} leaf nodes.

24. Let R be a relation in $A = \{1, 2, 3, 4, 5\}$. For the following relations, construct the digraphs. From the graphs identify the different relations for reflexive, symmetric and transitive.

   (a) $R = \{(1, 1), (1, 2), (2, 3), (4, 4), (4, 5), (2, 1), (5, 4)\}$

   (b) $R = \{(1, 1), (1, 2), (2, 2), (2, 3), (3, 3), (3, 4), (4, 4), (4, 5), (5, 5)\}$

   (c) $R = \{(1, 1), (1, 2), (1, 3), (2, 1), (2, 2), (3, 1), (3, 3), (3, 4), (4, 3), (4, 4), (4, 5), (5, 4), (5, 5)\}$

   (d) $R = \{(1, 4), (1, 2), (2, 3), (2, 1), (3, 4), (3, 2), (4, 1), (4, 3)\}$

   (e) $R = \{(1, 2), (2, 3), (1, 3), (2, 4), (4, 5), (3, 5), (1, 5), (1, 4), (2, 5)\}$

   (f) $R = \{(1, 1), (5, 2), (2, 3), (5, 3), (2, 2), (3, 3), (4, 4), (4, 1), (5, 5)\}$

25. Let $u = a^2 b^2 a^3 b^3$ and $w = abab$. Find the following strings and their length.

   (a) uw (b) wu (c) $uw^2 u$ (d) uwu

   (e) $wu^2 w$ (f) w^2 (g) u^2 (h) wuw

26. Given $\Sigma = \{a, b, c\}$. Find the Kleene closure on the alphabet Σ.

27. Check which of the following strings are palindromes?

   (a) $abababab$ (b) $gammag$ (c) $ab^2 ab^2 a$ (d) $a^2 b^2 c \ b^2 \ a^2$

28. Given a language $L = \{a^n b^{2n} : n \ge 0\}$ defined over some alphabet Σ. Obtain L^2 and L^R.

29. If $\Sigma = \{xx, xxx\}$, then obtain Σ^*.

30. Given $\Sigma = \{0, 1\}$. Obtain any two languages L_1 and L_2. Find $(L_1 \cup L_2), (L_1 \cap L_2)$, and $(L_1 - L_2)$,

31. Let $L = \{ab, aaba, aaa\}$. Identify which of the following strings are in L^*.

   (a) $abaaaaabaabab$ (b) $aaaaabaaabaab$

   (c) aabaaaababaaa   (d) baaaaabaababab

32. Given $\Sigma = \{0, 1\}$. Obtain the star closure Σ^* and the positive closure Σ^+.

33. Given $L = \{a^{n+1} b^n : n \ge 0\}$. Obtain L^*.

34. Let $A = \{aa, bb, ab\}$. Which of the following are not in A^*?

   (a) $aabbaaaabbab$ (b) $aabbbaabaa$

   (c) $bbbbbbbbaaab$ (d) $aabbaabbababbab$

35. Let $A = \{0, 1\}$. Give at least two strings that are members of and two strings that are not members of the following:

   (a) $a^* b^*$ (b) $A^* a A^* b$ (c) $(ab)^*$

36. Let $A = \{0, 1\}$. What are the elements of $P(A^2)$?

37. Show that for any two sets A and B, $|A \cup B| \le |A| + |B|$.

2

FINITE STATE AUTOMATA

2.0   INTRODUCTION

The theory of finite automata is the mathematical theory of a simple class of algorithms that are important in computer science and mathematics. It has primarily two applications in computer science. One is modelling of finite state systems and other is description of regular sets of finite words. It has many emerging applications such as computer algorithms and programs, verification and specification of protocols, designing of compilers, uses of regular expressions to professional programmers, searching for patterns in texts, natural language processing, combinatorial group theory and geometry, etc.

Algorithms are recipes that tell us how to solve problems. Mathematicians were interested initially in the dividing line between what was algorithmic and what was not. Certainly, this implied you could write a program to solve the problem, but it was observed that many of the programs would not run quickly or efficiently. Therefore, two approaches to classifying algorithms were developed. One is according to their running times and is known as complexity theory whereas the other is according to the types of memory used in implementing them is known as language theory. In language theory, the simplest algorithms are those that can be implemented by finite automata.

2.1   FINITE STATE MACHINE

First, we will discuss the general definition of automaton and later modify it to computer applications. A finite state machine (FSM) or finite automaton (automaton is singular of automata) is a machine that given an input, jumps through a series of states according to a transition function and tells the automaton which state to go next when a current state and a current symbol is given. That is, it is a model of behaviour composed of states, transitions and actions. The model of finite state machine is given in Figure 2.1. Finite state machines are widely used in modelling of application behaviour, design of hardware digital systems, software engineering and study of computation and languages. The characteristics of this model of finite state machine are given below.

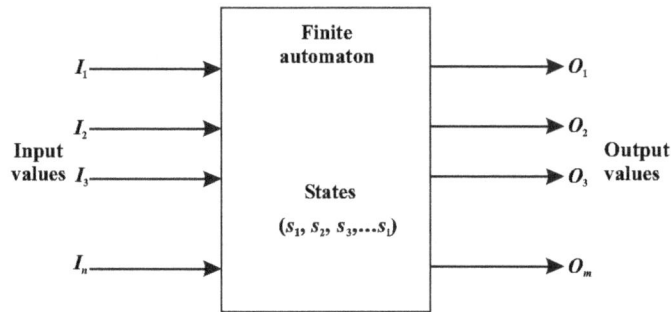

Figure 2.1   Model of finite state machine

Input   The input reads symbol by symbol, until it is completely consumed. Once the input is depleted, the automaton is said to have stopped. In the above model the input values are I_1, I_2, I_3, ..., I_n.

Output   The outputs of this model are O_1, O_2, O_3, ..., O_m. Each output can take finite number of fixed values from the output set O.

States   A state reflects the input changes from the system start to the present moment. That is it stores the information about the past. In the above model, at any instance of time the finite automaton will be in one of the states s_1, s_2, s_3, ..., s_l.

Transition   A transition indicates a state change and is described by a condition that would need to be fulfilled to enable the transition.

Action   An action is a description of an activity that is to be performed at a given moment. The different types of action are entry action, exit action, input action, and transition action. Entry action executes the action when entering the state whereas exit action executes the action when exiting the state. Input action executes the action dependent on present state and input conditions. Transition actions execute the action when performing a certain transition.

2.2   DETERMINISTIC FINITE AUTOMATA

In the theory of computation, a deterministic finite automaton (DFA) is a finite state machine in which each pair of state and input symbol has a deterministic next state. Analytically, a deterministic finite automaton is a five-tuple, $(S, \Sigma, \delta, s_0, A)$, consisting of the following.

S = Finite non-empty set of states

Σ = Finite non-empty set of inputs called the alphabet

δ = Transition function from $(S \times \Sigma) \to S$

s_0 = Starting state, which is the state in which the automaton is when no input has been processed yet, $s_0 \in S$

A = Set of final states or accepting states, $A \subseteq S$

2.2.1 Block Diagram of a DFA

The block diagram and the main components of finite automata are input tape, reading head and finite control. These are discussed in detail in Figure 2.2.

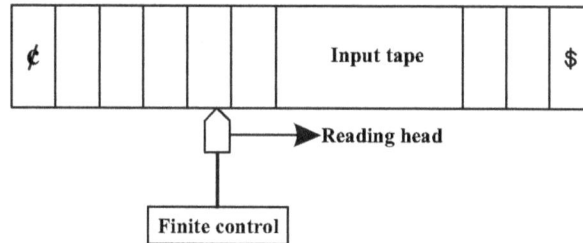

Figure 2.2 Block diagram of DFA

Input tape It is divided into blocks or squares. Each block contains a single symbol from the input alphabet å. The end blocks of the input tape-marked with ¢ and $. If the end markers are absent, then the tape is of infinite length. The symbol ¢ indicates the start of input tape whereas the symbol $ indicates the end of the tape. Every input string to be processed lies between these markers from left to right.

Reading head The reading head reads only one block at a time and jumps one block either to left or to right. We restrict the jump of read head only to the right side.

Finite control The symbol under the reading head is the input to the finite control. Suppose the input symbol is 1 and the present state of the machine is s_1; $(s_1 Î S)$ to give following outputs.

  i. The reading head moves to next block of input tape. If there is a null move, then the read head will remains at the same block.

  ii. The machine goes to next state $s_2 \in S$.

This can be explained with the following example. Let us consider the transition $\delta(s_1,10110) = \delta(s_2,0110)$. In this transition, s_1 is present state, the input symbol is 1 from the string 10110. When reading head reads symbol 1, then the machine goes to next state s_2 with the remaining string 0110. It can be cleared from the Figure 2.3a and b given below.

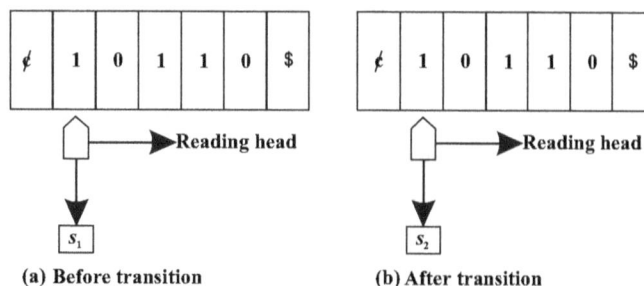

(a) Before transition   **(b) After transition**

Figure 2.3 Representation of transition

2.3   TRANSITION SYSTEM

A transition system is a finite directed labelled graph in which every vertex represents a state and each directed edge is a transition between two states. Each edge is labelled with input/output. The initial state is represented by a circle with an arrow towards it whereas the final state is represented by double concentric circles. The intermediate states are represented by simple circles.

These transition systems are used to graphically represent finite state machines. A transition system is also known as transition diagram, transition graph and state diagram. There are many forms of transition systems that differ slightly and have a different semantics. Consider the following transition system given below in which s_0 is the initial state, s_1 is the intermediate state whereas s_2 is the final state.

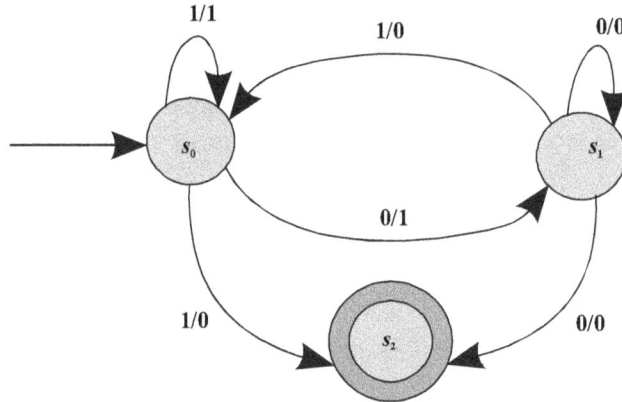

2.3.1   Transition Table

A state transition table is a table that desscribes the transition function δ of a finite automaton. This transition function governs what state (or states in the case of a nondeterministic finite automaton) the finite automaton will move to, given an input to the machine. A transition table can be derived from transition graph (state diagram) and vice versa. The transition table is also known as transition function table or state table. Transition function tables are typically two dimensional tables. There exists two common ways for arranging them.

   i.   The vertical (or horizontal) dimension indicates current states whereas horizontal (or vertical) dimension indicates events. The cells in the table contain next state if an event happens.

   ii. The vertical (or horizontal) dimension indicates current states whereas horizontal (or vertical) dimension indicates next states. The cells in the table contain the event which will lead to a particular next state.

Event	e_1	e_2	...	e_m
State				
s_1	s_i	—	...	—
s_2	—	—	...	s_j
...	...	...	...	...
s_n	—	s_k	...	—

Representation (i)

Next state	s_1	s_2	...	s_n
Current state				
s_1	E_i	—	...	—
s_2	—	—	...	E_j
...	...	...	...	...
s_n	—	E_k	...	—

Representation (ii)

Here s_i denotes the different states, E_i denotes the events whereas an illegal transition is represented through hyphen (-). An example of a transition table for a machine M together with the corresponding state diagram is given below.

Transition table

State	Input	
	0	**1**
→s_0	s_0	s_1
s_1	s_0	s_2
s_2	s_4	s_3
s_3	s_3	s_4
s_4	--	--

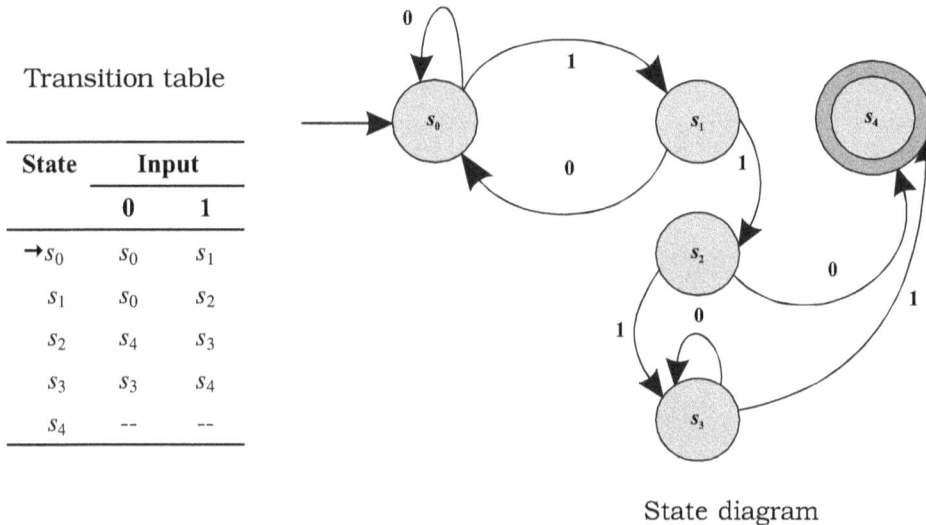

State diagram

Transformation to transition graph   It is possible to draw a transition graph from the transition table and vice versa. A sequence of steps is given below to construct a transition graph from a transition table.

Step 1   Draw the circles to represent the states given.

Step 2   Scan across the corresponding row and draw an arrow to the destination state(s) for each of the states given. There can be multiple arrows for an input character in case of nondeterministic finite automaton (NFA).

Step 3   Designate a state as the start state. The start state is given in the formal definition of the automaton.

Step 4   Designate one or more states as final or accepting state. It is also given in the formal definition.

2.3.2 Properties of Transition Functions

The important properties are as follows.

Property 1 In a finite automaton, $\delta(s, \Lambda) = s$. This means the state of the system can be changed only by giving an input symbol.

Property 2 For all strings w and input symbol a,

   i. $\delta(s, aw) = \delta(\delta(s, a), w)$

  ii. $\delta(s, wa) = \delta(\delta(s, w), a)$

It provides the state after the finite automaton reads the first symbol of a string "*aw*" and the state after the finite automaton reads a prefix "*w*" of the string "*wa*."

2.4 ACCEPTABILITY OF A STRING BY FINITE AUTOMATA

A string w is accepted by a finite automaton $M = (S, \Sigma, \delta, s_0, A)$ if $\delta(s_0, w) = s$ for some $s \in A$. This is basically the acceptability of a string by the accepting state. Consider the finite state machine M whose transition function δ is given in the following table.

State	Input	
	0	**1**
$\rightarrow$ s_0	s_1	s_2
s_1	s_2	s_3
s_2	s_0	s_3
s_3	s_0	s_4
s_4	s_2	s_1

Here, $S = \{s_0, s_1, s_2, s_3, s_4\}$; $\Sigma = \{0, 1\}$, $A = \{s_0\}$. Let us consider a string $w = 10011010$. Now we have

$$\delta(s_0, 10011010) = \delta(\delta(s_0, 1), 0011010) = \delta(s_2, 0011010)$$
$$= \delta(\delta(s_2, 0), 011010) = \delta(s_0, 011010)$$
$$= \delta(\delta(s_0, 0), 11010) = \delta(s_1, 11010)$$
$$= \delta(\delta(s_1, 1), 1010) = \delta(s_3, 1010)$$
$$= \delta(\delta(s_3, 1), 010) = \delta(s_4, 010)$$
$$= \delta(\delta(s_4, 0), 10) = \delta(s_2, 10)$$
$$= \delta(\delta(s_2, 1), 0) = \delta(s_3, 0)$$
$$= s_0$$

Hence, the string $w = 10011010$ is accepted by the finite state machine M with the transition function given above.

2.5   NONDETERMINISTIC FINITE AUTOMATA

In theory of computation, a nondeterministic finite automaton (NDFA / NFA) is a finite state machine in which states may or may not have a transition for each symbol in the alphabet. Sometimes the states of an automaton can have multiple transitions for a symbol.

A nondeterministic finite automaton (NFA) is a finite state machine where for each pair of state and input symbol there may be more than one possible next state. Consider the following transition graph.

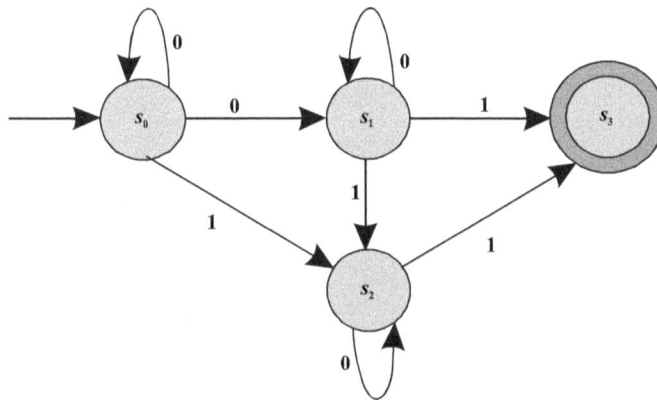

In the above transition, if the automaton is in a state $\{s_0\}$ and the input symbol is 0, then the next state will be either $\{s_0\}$ or $\{s_1\}$. Similarly, if the automaton is in a state $\{s_1\}$ and the input symbol is 1, then the next state will be either $\{s_2\}$ or $\{s_3\}$. Therefore, some moves of the machine cannot be determined uniquely by the input symbol and the present state. Thus the machines having such characteristics are called nondeterministic finite automata (NFA).

A nondeterministic finite automata (NFA) is a five-tuple $(S, \Sigma, \delta, s_0, A)$ consisting of the following.

S = Finite non-empty set of states

Σ = Finite non-empty set of inputs called alphabet

δ = Transition function from $(S \times \Sigma \to 2^S)$, where 2^S is the power set of S

s_0 = Initial state, $s_0 \in S$

A = Set of final states or accepting states, $A \subseteq S$.

2.5.1   Acceptability of a String by NFA

The nondeterministic finite automaton (NFA) accepts a string w if there exist at least one path from the initial state s_0 to a state in A (final state) labelled with the input string. In other words, a string w is accepted by NFA $M = (S, \Sigma, \delta, s_0, A)$ if $\delta(s_0, w)$ contains some final state.

Consider, for example, the nondeterministic finite automaton whose transition diagram is given below.

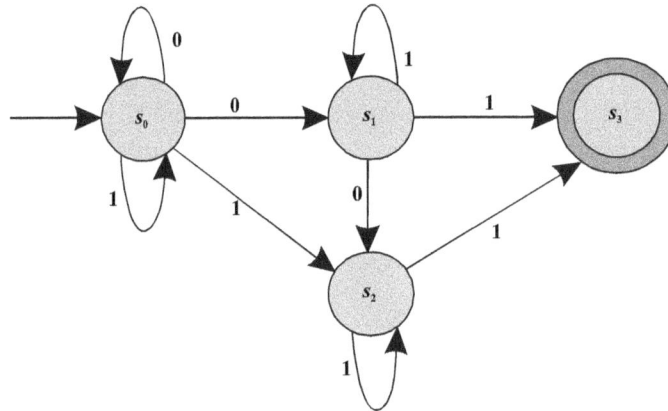

Here, $S = \{s_0, s_1, s_2, s_3\}$; $\Sigma = \{0, 1\}$, $A = \{s_3\}$. Let us consider the string $w = 01101$. The sequence of states for the input string w is given in the following figure.

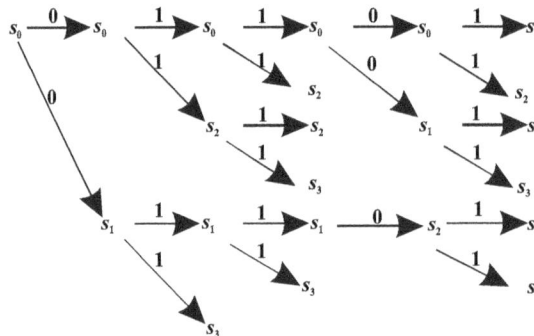

Therefore, $\delta(s_0, 01101) = \{s_0, s_1, s_2, s_3\}$. Since s_3 is an accepting state, the input string 01101 will be accepted by the nondeterministic finite automaton.

2.6   DIFFERENCE BETWEEN DFA AND NFA

The difference between the deterministic and nondeterministic automata is only the transition function δ. For DFA, the outcome is an element of S; whereas for NFA, the outcome is a subset of S. This indicates that, with a NFA is that it can potentially encounter a situation where it can make multiple choices. It has the ability to proceed from one state to another without consuming any input. The edges it follows in doing so are called ε edges.

Another way of thinking about a NFA is that whenever it has a choice, it makes copies of itself at that instant in time, and each selects one of the paths forward. Nondeterministic finite automata (NFA) are generally implemented by converting them to deterministic finite automata (DFA). In the worst case the generated DFA is exponentially bigger than the NFA.

2.7 EQUIVALENCE OF DFA AND NFA

In this section we try to find out the relation between deterministic finite automata (DFA) and nondeterministic finite automata (NFA). In contrast to the NFA, the DFA has no ε-transitions and for each (s, a); $s \in S$ and $a \in \Sigma$ there exists at most one successor state.

A DFA can simulate the behaviour of NFA by increasing the number of states, i.e., a DFA $(S, \Sigma, \delta, s_0, A)$ can be viewed as an NFA $(S, \Sigma, \delta', s_0, A)$ by defining $\delta'(s, a) = \{\delta(s, a)\}$. The theorem on equivalence of DFA and NFA is given below.

Theorem *If a language L is accepted by a nondeterministic finite automaton (NFA), then there exists a deterministic finite automaton (DFA) that accepts L.*

Proof Let $M = (S, \Sigma, \delta, s_0, A)$ be a nondeterministic finite automaton (NFA) accepting the language L. We construct a deterministic finite automaton (DFA) $M' = (S', \Sigma, \delta', s_0', A')$ where

   i. $S' = 2^S$ (any state in S' is represented by $[s_1, s_2, s_3 ..., s_i]$, where $s_1, s_2, s_3 ..., s_i \in S$);

   ii. $s_0' = [s_0]$

   iii. $A' =$ Set of all subsets of S containing an element of A.

Now, consider the construction of S', s_0' and A'. In case of nondeterministic finite automaton (NFA), M is initially at s_0. But M can reach any of the states in $\delta(s_0, a)$, on application of an input symbol, say a. Hence, to describe M' just after the application of the input symbol a, we require all the possible states that M can reach after application of a. Therefore, the states of M' are defined as subsets of S. Since M starts with initial state s_0, s_0' is defined as $[s_0]$. If the string w is accepted by M, then M reaches an accepting state on processing w. So, a final state in deterministic finite automata (DFA) M' is any subset of S containing some final state of M.

Now we can define transition function δ' of deterministic finite automata (DFA) M' as follows:

$$\delta'([s_1, s_2, s_3, ..., s_i], a) = \delta(s_1, a) \cup \delta(s_2, a) \cup \delta(s_3, a) \cup ... \cup \delta(s_i, a)$$

i.e.,
$$\delta'([s_1, s_2, s_3, ..., s_i], a) = [p_1, p_2, p_3, ..., p_j] \text{ if and only if}$$

$$\delta'(\{s_1, s_2, s_3, ..., s_i\}, a) = \{p_1, p_2, p_3, ..., p_j\}.$$

Now, we prove an auxiliary result:

$$\delta'(s_0', x) = [s_1, s_2, s_3, ..., s_i] \quad \text{if and only if} \quad \delta(s_0, x) = \{s_1, s_2, s_3, ..., s_i\} \forall x \in \Sigma^* \quad (1)$$

We prove the if part by method of induction. When $|x| = 0, \delta(s_0, \Lambda) = \{s_0\}$ and by definition of $\delta', \delta'(s_0', \Lambda) = \{s_0'\} = [s_0]$. So, the statement (1) is true for $|x| = 0$. Let us assume that the statement (1) is true for all strings 'y' with $|y| \leq n$. Now our aim is to show that the statement is true for all strings of length $(n + 1)$. Let x be a string of length $(n + 1)$. We can write $x = ya$, where $|y| = n$ and $a \in \Sigma$.

Let $\delta(s_0, y) = \{p_1, p_2, p_3, ..., p_i\}$ and $\delta(s_0, ya) = \{r_1, r_2, r_3, ..., r_k\}$

By induction hypothesis we have $\delta'(s_0', y) = [p_1, p_2, p_3, ... p_j]$ as $|y| \le n$.

Also, $\{r_1, r_2, r_3, \cdots, r_k\} = \delta(s_0, ya) = \delta(\delta(s_0, y)a) = (\delta\{p_1, p_2, p_3, \cdots, p_j\}, a)$

By definition of δ' we have $\delta'([p_1, p_2, p_3, \cdots, p_j], a) = \{r_1, r_2, r_3, \cdots, r_k\}$

Therefore, $\delta'(s_0', ya) = \delta'(\delta'(s_0', y), a) = \delta'([p_1, p_2, p_3, \cdots, p_j], a) = [r_1, r_2, r_3, \cdots, r_k]$

Hence, we have proved the statement (1) is true for $x = ya$. So, by method of induction, the statement is true for all strings x.

Now, $x \in T(M)$, the set of all input strings accepted by M, if and only if $\delta(s, x)$ contains a state of A. By equation (1) $\delta(s_0, x)$ contains a state of A if and only if $\delta'(s_0', x)$ is in A'. Thus, $x \in T(M)$ if and only if $x \in T(M')$. Therefore, M' accepts L.

2.7.1   Construction of DFA Equivalent to given NFA

The difficult part is in the construction of DFA M_2 equivalent to given NFA M_1 is the construction of transition function δ' for M_2. If δ for nondeterministic finite automata (NFA) M_1 is given in the form of a state table, the construction is easier. By definition, we have

$$\delta'([s_1, s_2, s_3, \cdots, s_k], a) = \bigcup_{i=1}^{k} \delta(s_i, a)$$

Consider the nondeterministic finite automata (NFA) $M_1 = (\{s_0, s_1\}, \{a, b\}, \delta, s_0, \{s_1\})$ where δ is given by its state table as below.

State	Input	
	a	*b*
$\rightarrow s_0$	s_1	s_0
s_1	s_0	s_0, s_1

Now, our aim is to construct a deterministic finite automaton (DFA) M_2 equivalent to M_1. For the deterministic finite automaton M_2 we have

   i. The states are subsets of $\{s_0, s_1\}$, i.e., ϕ, $[s_0]$, $[s_1]$, $[s_0, s_1]$;

  ii. The initial state is $[s_0]$

 iii. The accepting or final states are $[s_1]$ and $[s_0, s_1]$ as these are the only states containing s_1.

iv.   The transition function δ' is defined by the state table given below. In order to construct the state table we have to compute the following.

$\delta'(\phi, a) = \phi$; $\delta'(\phi, b) = \phi$

$\delta'([s_0], a) = \delta(s_0, a) = [s_1]$; $\delta'([s_0], b) = \delta(s_0, b) = [s_0]$;

$\delta'([s_1], a) = \delta(s_1, a) = [s_0]$; $\delta'([s_1], b) = \delta(s_1, b) = [s_0, s_1]$;

$\delta'([s_0, s_1], a) = \delta(s_0, a) \cup \delta(s_1, a) = [s_0, s_1]$; and

$\delta'([s_0, s_1], b) = \delta(s_0, b) \cup \delta(s_1, b) = [s_0, s_1]$

Therefore, we get the following table.

State	Input	
	a	b
ϕ	ϕ	ϕ
$[s_0]$	$[s_1]$	$[s_0]$
$[s_1]$	$[s_0]$	$[s_0, s_1]$
$[s_0, s_1]$	$[s_0, s_1]$	$[s_0, s_1]$

Note   It is clear that, when nondeterministic finite automaton (NFA) M_1 has n states, then the corresponding deterministic finite automaton (DFA) M_2 has 2^n states. However, it is not necessary to construct for all these 2^n states, but only for those states reachable from initial state $[s_0]$. This is because our aim is only in constructing M_2 accepting $T(M_1)$. Therefore, start the construction of δ' for $[s_0]$ and continue by considering only states appearing earlier under input columns and construct δ' for such states. We stop the process when no more new states appear under the input columns.

2.8   FINITE AUTOMATA WITH ε-MOVES

A nondeterministic finite automata with ε-moves is defined as five-tuple $(S, \Sigma, \delta, s_0, A)$ consisting of the following.

S = Finite non-empty set of states

Σ = Finite non-empty set of inputs called the alphabet

δ = Transition function from $(S \times (\Sigma \cup \varepsilon)) \rightarrow 2^S$, where 2^S is the power set of S

s_0 = Starting state; $s_0 \in S$

A = Set of final states or accepting states; $A \subseteq S$.

The only difference between nondeterministic finite automata and the nondeterministic finite automata with ε-moves is the transition function δ. In this case we extend our existing model of NFA to include transitions without giving any input, i.e., by providing empty input ε . This nondeterministic finite automaton accepts a string 'w' if there exists some path from the initial state 's_0' to a final state in A in which edges labelled ε may be included. Let us consider the following transition diagram.

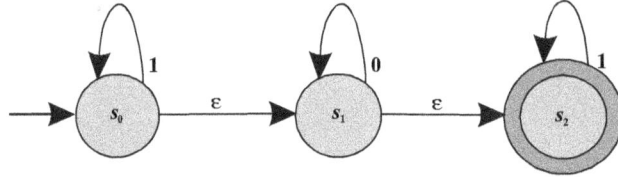

The nondeterministic finite automata whose transition diagram given above accepts the language consisting of any occurrence of 1's followed by any occurrence of 0's followed by any occurrence of 1's. For example, consider the string $w = 1110011$. This string is accepted by the following path.

$$s_0 \xrightarrow{1} s_0 \xrightarrow{1} s_0 \xrightarrow{1} s_0 \xrightarrow{\varepsilon} s_1 \xrightarrow{0} s_1 \xrightarrow{0} s_1 \xrightarrow{\varepsilon} s_2 \xrightarrow{1} s_2 \xrightarrow{1} s_2$$

For better understanding, state table for the above nondeterministic finite automata with ε-moves is given below in which if there is no transition in the transition graph, then we write ϕ in the corresponding place of transition table.

State	Input		
	0	**1**	ε
$\rightarrow s_0$	ϕ	$\{s_0\}$	$\{s_1\}$
s_1	$\{s_1\}$	ϕ	$\{s_2\}$
s_2	ϕ	$\{s_2\}$	ϕ

2.8.1 ε-Closure of a State

Assume that 's_0' be a state. ε-closure (s_0) is defined as the set of all states 'p' such that there is a path from s_0 to p labelled with ε. Generally, ε-closure (s_0) is denoted as $\hat{\delta}(s_0, \varepsilon)$. Let us consider P be a set of states. Then the ε-closure (P) is defined as

$$\varepsilon\text{-}closure\ (P) = \bigcup_{s \in P} \varepsilon\text{-}closure\ (s)$$

Now, we define ε-closure (s) as follows.

i. $\hat{\delta}(s, wa) = \varepsilon\text{-}closure(P)$ where $P = \{p : \text{for some } r \in \hat{\delta}(s, w);\ p \in \delta(r, a)\}$

ii. $\delta(R, a) = \bigcup_{s \in R} \delta(s, a)$ and

iii. $\hat{\delta}(R, a) = \bigcup_{s \in R} \hat{\delta}(s, a)$; where R is a set of states.

It is to be noted that $\hat{\delta}(s, a)$ and $\delta(s, a)$ are not necessarily equal, since $\hat{\delta}(s, a)$ includes all states that are reachable from 's' by paths labelled 'a' including paths labelled ε, whereas $\delta(s, a)$

includes only those states reachable from 's' labelled with 'a'. Consider the following nondeterministic finite automaton with ε-moves.

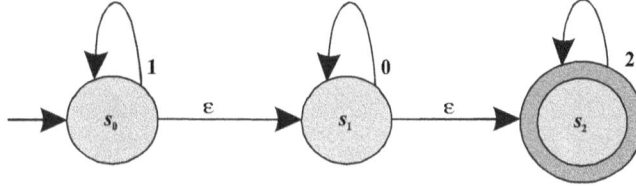

Therefore we have,

$$\hat{\delta}(s_0, \varepsilon) = \varepsilon\text{-}closure(s_0) = \{s_0, s_1, s_2\}$$

$$\hat{\delta}(s_0, 1) = \varepsilon\text{-}closure\ (\delta(\hat{\delta}(s_0, \varepsilon), 1)) = \varepsilon\text{-}closure\ (\delta(\{s_0, s_1, s_2\}, 1)$$

$$= \varepsilon\text{-}closure\ (\delta(s_0, 1) \cup \delta(s_1, 1) \cup \delta(s_2, 1))$$

$$= \varepsilon\text{-}closure\ (\{s_0\} \cup \varphi \cup \varphi)$$

$$= \varepsilon\text{-}closure\ (s_0) = \{s_0, s_1, s_2\}$$

$$\hat{\delta}(s_0, 0) = \varepsilon\text{-}closure\ (\delta(\hat{\delta}(s_0, \varepsilon), 0)) = \varepsilon\text{-}closure\ (\delta(\{s_0, s_1, s_2\}, 0)$$

$$= \varepsilon\text{-}closure\ (\delta(s_0, 0) \cup \delta(s_1, 0) \cup \delta(s_2, 0))$$

$$= \varepsilon\text{-}closure\ (\{s_1\} \cup \varphi \cup \varphi)$$

$$= \varepsilon\text{-}closure\ (s_1) = \{s_1, s_2\}$$

$$\hat{\delta}(s_0, 2) = \varepsilon\text{-}closure\ (\delta(\hat{\delta}(s_0, \varepsilon), 2)) = \varepsilon\text{-}closure\ (\delta(\{s_0, s_1, s_2\}, 2)$$

$$= \varepsilon\text{-}closure\ (\delta(s_0, 2) \cup \delta(s_1, 2) \cup \delta(s_2, 2))$$

$$= \varepsilon\text{-}closure\ (\{s_2\} \cup \varphi \cup \varphi)$$

$$= \varepsilon\text{-}closure\ (s_2) = \{s_2\}$$

Similarly, we have

$$\hat{\delta}(s_0, 01) = \varepsilon\text{-}closure\ (\delta(\hat{\delta}(s_0, 0), 1) = \varepsilon\text{-}closure\ (\delta(\{s_1, s_2\}, 1)$$

$$= \varepsilon\text{-}closure\ (\delta(s_1, 1) \cup \delta(s_2, 1))$$

$$= \varepsilon\text{-}closure\ (\varphi \cup \varphi) = \varphi$$

2.8.2 Construction of NFA Equivalent to NFA with ε-moves

In this section we will discuss how to construct an NFA if an NFA with ε-transitions is given. The construction of NFA from NFA with ε-moves is based on the following theorem.

Theorem   *If a language L is accepted by an NFA with ε-transitions, then there exists a NFA without ε-transitions that accepts L.*

Proof   Let $M = (S, \Sigma, \delta, s_0, A)$ be an NFA with ε-transition. Let us construct $M' = (S, \Sigma, \delta', s_0, A')$ where

$$A' = \begin{cases} A \cup \{s_0\}; & \text{if } \varepsilon\text{-closure}(s_0) \text{ contains a state of } A \\ A; & \text{otherwise} \end{cases}$$

and $\delta'(s, a) = \hat{\delta}(s, a)$ for $s \in S$ and $a \in \Sigma$.

Therefore, it is noted that M' has no ε-transitions. Hence, without loss of generality we may use δ' for $\hat{\delta}'$, but δ and $\hat{\delta}$ are not equal. We show by induction that $\delta'(s_0, w) = \hat{\delta}(s_0, w)$ for any string $w \in \Sigma^*$. Let us take $w = \varepsilon$. Therefore, we have $\delta'(s_0, \varepsilon) = \{s_0\}$ and $\hat{\delta}(s_0, \varepsilon) = \varepsilon$-closure (s_0). This implies that $\delta'(s_0, \varepsilon) \neq \hat{\delta}(s_0, \varepsilon)$. So, the statement $\delta'(s_0, w) = \hat{\delta}(s_0, w)$ is not true for the string having length zero.

Basis   Consider w be a string of length one. Let $w = a$. Therefore, by definition we have $\delta'(s_0, a) = \hat{\delta}(s_0, a)$ for $s_0 \in S$ and $a \in \Sigma$.

Induction   Assume that $\delta'(s_0, x) = \hat{\delta}(s_0, x)$ for any string x of length less than or equal to 'n'. Our aim is to show that the above statement is true for all strings of length $(n + 1)$. Let $w = xa$. Therefore, we have

$$\begin{aligned}
\delta'(s_0, w) &= \delta'(s_0, xa) \\
&= \delta'(\delta'(s_0, x), a) \\
&= \delta'(\hat{\delta}(s_0, x), a) \\
&= \delta'(P, a); \qquad \text{where} \quad P = \hat{\delta}(s_0, x) \\
&= \bigcup_{s \in P} \delta'(s, a) \\
&= \bigcup_{s \in P} \hat{\delta}(s, a) \\
&= \hat{\delta}(s_0, xa) = \hat{\delta}(s_0, w)
\end{aligned}$$

i.e., $\delta'(s_0, w) = \hat{\delta}(s_0, w)$

In order to complete the proof we shall show that $\delta'(s_0, x)$ contains a state of A' if and only if $\hat{\delta}(s_0, x)$ contains a state of A.

If $x = \varepsilon$, then $\delta'(s_0, \varepsilon) = \{s_0\} = \hat{\delta}(s_0, \varepsilon) = \varepsilon$-closure (s_0)

Also, it is clear that ε-closure (s_0) contains a state (possibly s_0) in A. If $x \neq \varepsilon$ then $x = wa$ for some symbol a. If $\hat{\delta}(s_0, x)$ contains a state of A, then surely $\delta'(s_0, x)$ contains the same state in A'. Conversely, if $\delta'(s_0, x)$ contains a state in A' other than s_0, then $\hat{\delta}(s_0, x)$ contains a state in A. If $\delta'(s_0, x)$ contains s_0 and s_0 is not in A, then $\hat{\delta}(s_0, x) = \varepsilon$-closure $\left(\delta\left(\hat{\delta}(s_0, w), a \right) \right)$, the state in ε-closure (s_0) and in A must be in $\hat{\delta}(s_0, x)$.

Consider a nondeterministic finite automaton with ε-moves as given below. Our aim is to construct a nondeterministic finite automaton without ε-moves.

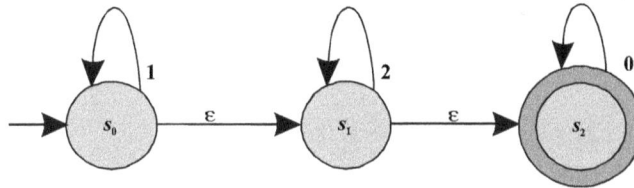

In the above nondeterministic finite automata with ε-moves we have $S = \{s_0, s_1, s_2\}$; $\Sigma = \{0, 1, 2, \varepsilon\}$; $A = \{s_2\}$ and the initial state is s_0.

Our aim is to construct a nondeterministic finite automaton without ε-transitions $M' = (S, \Sigma, \delta', s_0, A')$ such that

$$A' = \begin{cases} A \cup \{s_0\}; & \text{if } \varepsilon\text{-closure}(s_0) \text{ contains a state of } A \\ A; & \text{otherwise} \end{cases}$$

and $\delta'(s, a) = \hat{\delta}(s, a)$ for $s \in S$ and $a \in \Sigma$.

Now, we have

$$\delta'(s_0, 1) = \hat{\delta}(s_0, 1) = \varepsilon\text{-closure} \, (\delta \, (\, \hat{\delta}(s_0, \varepsilon), 1))$$
$$= \varepsilon\text{-closure} \, (\delta \, (\{s_0, s_1, s_2\}, 1))$$
$$= \varepsilon\text{-closure} \, (\delta \, (s_0, 1) \cup \delta \, (s_1, 1) \cup \delta \, (s_2, 1))$$
$$= \varepsilon\text{-closure} \, (\{s_0\} \cup \varphi \cup \varphi) = \varepsilon\text{-closure} \, (s_0)$$
$$= \{s_0, s_1, s_2\}$$

$$\delta'(s_0, 2) = \hat{\delta}(s_0, 2) = \varepsilon\text{-closure} \, (\delta \, (\, \hat{\delta}(s_0, \varepsilon), 2))$$

$$= \varepsilon\text{-closure} \, (\delta \, (\{s_0, s_1, s_2\}, 2))$$
$$= \varepsilon\text{-closure} \, (\delta \, (s_0, 2) \cup \delta \, (s_1, 2) \cup \delta \, (s_2, 2))$$
$$= \varepsilon\text{-closure} \, (\varphi \cup \{s_1\} \cup \varphi) = \varepsilon\text{-closure} \, (s_1)$$
$$= \{s_1, s_2\}$$

$$\delta'(s_0, 0) = \hat{\delta}(s_0, 0) = \varepsilon\text{-closure} \, (\delta \, (\, \hat{\delta}(s_0, \varepsilon), 0))$$
$$= \varepsilon\text{-closure} \, (\delta \, (\{s_0, s_1, s_2\}, 0))$$
$$= \varepsilon\text{-closure} \, (\delta \, (s_0, 0) \cup \delta \, (s_1, 0) \cup \delta \, (s_2, 0))$$
$$= \varepsilon\text{-closure} \, (\varphi \cup \varphi \cup \{s_2\}) = \varepsilon\text{-closure} \, (s_2)$$
$$= \{s_2\}$$

$$\delta'(s_1, 1) = \hat{\delta}(s_1, 1) = \varepsilon\text{-}closure\,(\delta\,(\,\hat{\delta}(s_1, \varepsilon), 1))$$
$$= \varepsilon\text{-}closure\,(\delta\,(\{s_1, s_2\}, 1))$$
$$= \varepsilon\text{-}closure\,(\delta\,(s_1, 1) \cup \delta\,(s_2, 1))$$
$$= \varepsilon\text{-}closure\,(\varphi \cup \varphi) = \varphi$$

$$\delta'(s_1, 2) = \hat{\delta}(s_1, 2) = \varepsilon\text{-}closure\,(\delta\,(\,\hat{\delta}(s_1, \varepsilon), 2))$$
$$= \varepsilon\text{-}closure\,(\delta\,(\{s_1, s_2\}, 2))$$
$$= \varepsilon\text{-}closure\,(\delta\,(s_1, 2) \cup \delta\,(s_2, 2))$$
$$= \varepsilon\text{-}closure\,(\{s_1\} \cup \varphi) = \varepsilon\text{-}closure\,(s_1)$$
$$= \{s_1, s_2\}$$

$$\delta'(s_1, 0) = \hat{\delta}(s_1, 0) = \varepsilon\text{-}closure\,(\delta\,(\,\hat{\delta}(s_1, \varepsilon), 0))$$
$$= \varepsilon\text{-}closure\,(\delta\,(\{s_1, s_2\}, 0))$$
$$= \varepsilon\text{-}closure\,(\delta\,(s_1, 0) \cup \delta\,(s_2, 0))$$
$$= \varepsilon\text{-}closure\,(\varphi \cup \{s_2\}) = \varepsilon\text{-}closure\,(s_2)$$
$$= \{s_2\}$$

$$\delta'(s_2, 1) = \hat{\delta}(s_2, 1) = \varepsilon\text{-}closure\,(\delta\,(\,\hat{\delta}(s_2, \varepsilon), 1))$$
$$= \varepsilon\text{-}closure\,(\delta\,(\{s_2\}, 1))$$
$$= \varepsilon\text{-}closure\,(\varphi) = \varphi$$

$$\delta'(s_2, 2) = \hat{\delta}(s_2, 2) = \varepsilon - closure\,(\delta\,(\,\hat{\delta}(s_2, \varepsilon), 2))$$
$$= \varepsilon\text{-}closure\,(\delta\,(\{s_2\}, 2))$$
$$= \varepsilon\text{-}closure\,(\varphi) = \varphi$$

$$\delta'(s_2, 0) = \hat{\delta}(s_2, 0) = \varepsilon\text{-}closure\,(\delta\,(\,\hat{\delta}(s_2, \varepsilon), 0))$$
$$= \varepsilon\text{-}closure\,(\delta\,(\{s_2\}, 0)) = \varepsilon\text{-}closure\,(\{\,s_2\,\}) = \{\,s_2\,\}$$

Therefore, the transition function δ' of M' is given in the following table.

States	Inputs		
	0	1	2
$\rightarrow s_0$	$\{\,s_2\,\}$	$\{\,s_0, s_1, s_2\,\}$	$\{\,s_1, s_2\,\}$
s_1	$\{\,s_2\,\}$	ϕ	$\{\,s_1, s_2\,\}$
s_2	$\{\,s_2\,\}$	ϕ	ϕ

The set of final states of $M' = A \cup \{s_0\} = \{s_0, s_2\}$ since $\varepsilon\text{-}closure\,(s_0) = \{s_0, s_1, s_2\}$.

Thus, the nondeterministic finite automata without ε-transitions is given below.

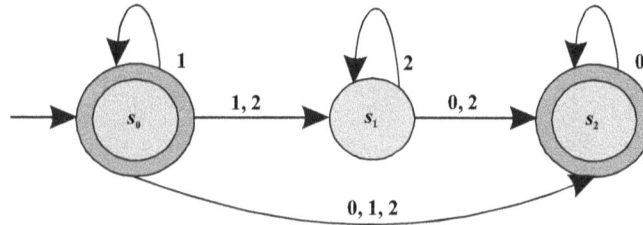

2.8.3 Construction of DFA Equivalent to NFA with ε-moves

If a language L is accepted by a nondeterministic finite automaton with ε-transitions, then there exists a deterministic finite automaton that accepts L. In order to construct a deterministic finite automaton from nondeterministic finite automaton with ε-transitions, first we have to construct a nondeterministic finite automaton equivalent to nondeterministic finite automaton with ε-moves and then a deterministic finite automaton can be constructed equivalent to the nondeterministic finite automaton.

Consider the following nondeterministic finite automaton with ε-moves.

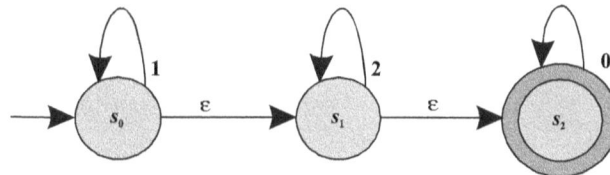

The nondeterministic finite automaton without ε-moves as discussed above is given below.

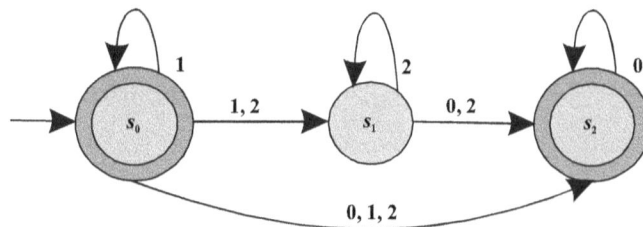

The deterministic finite automaton equivalent to the above nondeterministic finite automaton is given below.

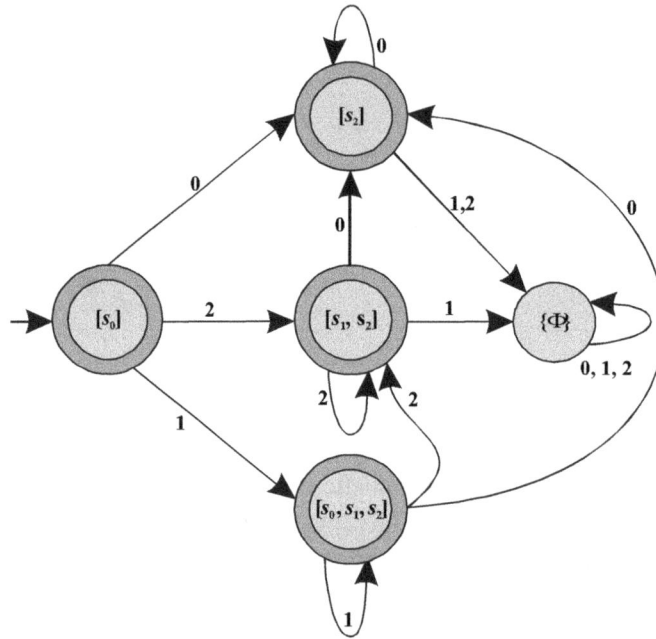

2.9   TWO-WAY FINITE AUTOMATA

We have discussed finite automaton as a control unit that reads a magnetic tape, moving one block right at each move. We added nondeterminism to this model, which allowed many copies of the control unit to read the magnetic tape simultaneously. Also we added ε-transitions, that allowed change of state without reading any input symbol from the tape. We also allowed the read head to move in either direction, i.e., left or right. This is called two-way finite automaton. The diagrammatic presentation is given here below (Figure 2.4).

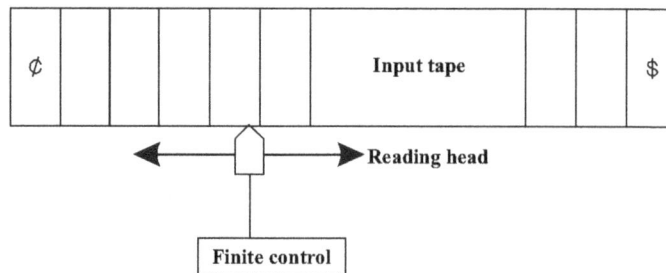

Figure 2.4   Block diagram of two-way finite automata

A two-way finite automata is defined as a five-tuple $(S, \Sigma, \delta, s_0, A)$, consisting of

S = Finite non-empty set of states

Σ = Finite non-empty set of inputs called the alphabet

δ = Transition function from $(S \times \Sigma) \rightarrow S \times \{L, R\}$

s_0 = Starting state; $s_0 \in S$.

A = Set of final states or accepting states; $A \subseteq S$.

Therefore, the only difference between two-way finite automaton and deterministic finite automaton is that of transition function. In two-way finite automaton, if it is in state s_1, reading input symbol a, ($a \in \Sigma$), it goes to next state s_2 and moves its read head either to left or right. This idea can be cleared from the given example.

S	$\times$	Σ	$\rightarrow$	S	$\times$	$\{L, R\}$
Present state (s_1)		**Current input** symbol (a)		**Next** state (s_2)		**Left or right** move of read head

Therefore, if δ $(s_1, a) = (s_2, R)$, then the read head in state s_1 scans the input symbol 'a', the two-way finite automaton moves to the next state s_2 by moving its read head one block right. Similarly, if δ $(s_1, a) = (s_2, L)$, then the two-way finite automaton enters the state s_2 by moving its read head one block left.

2.10 FINITE AUTOMATA WITH OUTPUTS

The limitation of finite automaton is that its output is limited to a binary output, i.e., they accept or do not accept the input string. This acceptability was decided on the basis of reachability of the final state by the initial state. Therefore, models in which the output is chosen from some other alphabet have been considered. It leads to two approaches. In the first case, the output may be associated with the state whereas in the second case the output is associated with the transition. The output associated with the current state alone is called a Moore machine whereas the output associated with the transition is called Mealy machine. The name Moore machine comes from that of their promoter E. F. Moore whereas the Mealy machine comes from that of their promoter G. H. Mealy.

2.10.1 Moore Machine

A Moore machine is defined as a six-tuple $(S, \Sigma, \Delta, \delta, \lambda, s_0)$ consisting of the following.

S = Finite non-empty set of states

Σ = Finite non-empty set of inputs called the alphabet

Δ = Output alphabet

δ = Transition function from $(S \times \Sigma) \rightarrow S$

λ = Output function from $S \rightarrow \Delta$

s_0 = Starting state $s_0 \in S$.

For example consider a Moore machine whose transition diagram and transition table is as follows:

Transition table of Moore machine

Present State	Next State		Output (z)
	$w = 0$	$w = 1$	
$\rightarrow s_0$	s_0	s_1	1
s_1	s_0	s_2	1
s_2	s_2	s_3	0
s_3	s_0	s_3	1

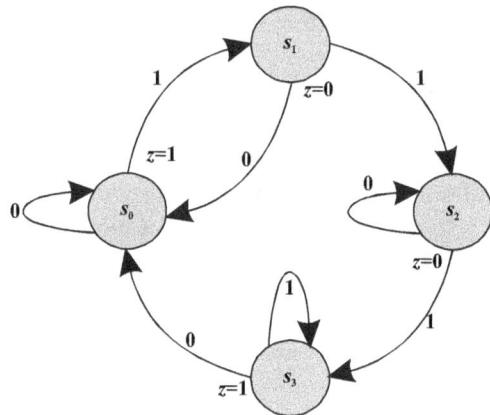

Transition diagram of Moore machine

2.10.2   Mealy Machine

A Mealy machine is a finite state machine where the outputs are obtained by the current state and the input. This implies that the transition diagram will include an output signal for each transition edge. For example, in moving from state s_1 to a state s_2 on input 0, the output might be 1. So, the edges are levelled 0/1. A Mealy machine is defined as a six-tuple $(S, \Sigma, \Delta, \delta, \lambda, s_0)$, consisting of the following.

S = Finite non-empty set of states

Σ = Finite non-empty set of inputs called the alphabet

Δ = Output alphabet

δ = Transition function from $(S \times \Sigma) \rightarrow S$

λ = Output function from $(S \times \Sigma) \rightarrow \Delta$

s_0 = Starting state $s_0 \in S$.

For example consider a Mealy machine whose transition diagram and transition table is given below.

Transition table of Mealy machine

Present State	Input ($w = 0$)		Input ($w = 1$)	
	State	Output	State	Output
$\rightarrow s_1$	s_1	0	s_2	0
s_2	s_1	1	s_2	1
s_3	s_1	1	s_2	1

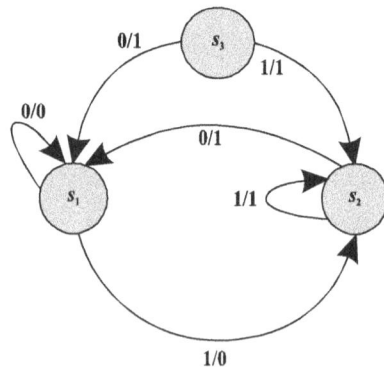

Transition diagram of Mealy machine

2.10.3   Construction of Mealy Machine from Moore Machine

Let M be a Moore or Mealy machine. Let us define $T_M(w)$, for input string w, to be the output produced by M. Let M_1 is a Moore machine and M_2 a Mealy machine. Therefore, $\left| T_{M_1}(w) \right|$ is one more than $\left| T_{M_2}(w) \right|$ for each string w. However, the acceptability of input string by a Moore machine is modified by neglecting the response of a Moore machine to input string Λ. Thus we say that Moore machine M_1 and Mealy machine M_2 are equivalent if for all inputs w, $b\,T_{M_2}(w) = T_{M_1}(w)$, where b is the output of Moore machine for its initial state.

Let $M_1 = (S, \Sigma, \Delta, \delta, \lambda, s_0)$ be a Moore machine. We construct a Mealy machine $M_2 = (S, \Sigma, \Delta, \delta, \lambda', s_0)$ such that

i. $\lambda'(s, a) = \lambda(\delta(s, a))$ for all states s and input symbol a.

ii. The transition function is same as that of Moore machine.

Consider the Moore machine $M_1 = (S, \Sigma, \Delta, \delta, \lambda, s_0)$ whose transition function δ is defined as below.

Present state	Next state		Output
	$a = 0$	$a = 1$	
$\rightarrow s_0$	s_2	s_3	1
s_1	s_1	s_2	0
s_2	s_3	s_0	1
s_3	s_0	s_1	0

Let the equivalent Mealy machine $M_2 = (S, \Sigma, \Delta, \delta, \lambda', s_0)$ such that $\lambda'(s, a) = \lambda(\delta(s, a))$. Therefore, we have

$$\lambda'(s_0, 0) = \lambda(\delta(s_0, 0)) = \lambda(s_2) = 1; \qquad \lambda'(s_0, 1) = \lambda(\delta(s_0, 1)) = \lambda(s_3) = 0$$
$$\lambda'(s_1, 0) = \lambda(\delta(s_1, 0)) = \lambda(s_1) = 0; \qquad \lambda'(s_1, 1) = \lambda(\delta(s_1, 1)) = \lambda(s_2) = 1$$
$$\lambda'(s_2, 0) = \lambda(\delta(s_2, 0)) = \lambda(s_3) = 0; \qquad \lambda'(s_2, 1) = \lambda(\delta(s_2, 1)) = \lambda(s_0) = 1$$
$$\lambda'(s_3, 0) = \lambda(\delta(s_3, 0)) = \lambda(s_0) = 1; \qquad \lambda'(s_3, 1) = \lambda(\delta(s_3, 1)) = \lambda(s_1) = 0$$

Thus the constructed Mealy machine is given by the following table.

Present state	Input ($w = 0$)		Input ($w = 1$)	
	State	Output	State	Output
$\rightarrow s_0$	s_2	1	s_3	0
s_1	s_1	0	s_2	1
s_2	s_3	0	s_0	1
s_3	s_0	1	s_1	0

2.10.4   Construction of Moore Machine from Mealy Machine

The output of a Moore machine depends only on the current state and does not depend on the current input. However, every Mealy machine is equivalent to a Moore machine whose state is the Cartesian product of the Mealy machines current and previous states.

Let $M_1 = (S, \Sigma, \Delta, \delta, \lambda, s_0)$ be a Mealy machine. We construct a Moore machine $M_2 = (S', \Sigma, \Delta, \delta', \lambda', s_0')$ where

i. $S' = S \times \Delta$

ii. $s_0' = [s_0, b_0]$, where b_0 is an arbitrarily selected member of Δ.

iii. $\delta'([s, b], a) = [\delta(s, a), \lambda(s, a)]$ and

iv. $\lambda'([s, b]) = b$

Therefore, it is clear that if M_1 enters states $s_0, s_1, s_2, \cdots, s_n$ on input $a_1, a_2, a_3, \cdots, a_n$ and results outputs $b_0, b_1, b_2, \cdots, b_n$ then M_2 enters states $[s_0, b_0]$, $[s_0, b_1], [s_1, b_0], [s_1, b_1], \cdots, [s_n, b_n]$ and emits outputs $b_0, b_1, b_2, \cdots, b_n$.

Consider the Mealy machine M_1 given by following transition table. Our aim is to construct an equivalent Moore machine.

Present state	Input (0)		Input (1)	
	State	Output	State	Output
$\rightarrow s_0$	s_2	1	s_1	0
s_1	s_0	1	s_3	1
s_2	s_1	0	s_0	1
s_3	s_3	1	s_2	0

Here we have, $S = \{s_0, s_1, s_2, s_3\}$; $\Sigma = \{0, 1\}$; $\Delta = \{0, 1\}$. Let the equivalent Moore machine be $M_2 = (S', \Sigma, \Delta, \delta', \lambda', s_0')$. Therefore, the states of Moore machine are $[s_0, 0], [s_0, 1]$, $[s_1, 0], [s_1, 1]$, $[s_2, 0], [s_2, 1]$, $[s_3, 0]$, and $[s_3, 1]$. On choosing $b_0 = 0$ the initial state is given as $[s_0, 0]$. However, we can choose $[s_0, 1]$ as also the initial state on considering $b_0 = 1$.

Now, $\lambda'([s_0, 1]) = 1;$ $\lambda'([s_0, 0]) = 0;$ $\lambda'([s_1, 0]) = 0;$ $\lambda'([s_1, 1]) = 1$

$\lambda'([s_2, 1]) = 1;$ $\lambda'([s_2, 0]) = 0;$ $\lambda'([s_3, 0]) = 0;$ $\lambda'([s_3, 1]) = 1$

Again we have

$$\delta'([s_0, 0], 0) = [\delta(s_0, 0), \lambda(s_0, 0)] = [s_2, 1]; \; \delta'([s_0, 0], 1) = [\delta(s_0, 1), \lambda(s_0, 1)] = [s_1, 0]$$
$$\delta'([s_0, 1], 0) = [\delta(s_0, 0), \lambda(s_0, 0)] = [s_2, 1]; \; \delta'([s_0, 1], 1) = [\delta(s_0, 1), \lambda(s_0, 1)] = [s_1, 0]$$

Similarly, we get the following

$$\delta'([s_1, 0], 0) = [s_0, 1] = \delta'([s_1, 1], 0); \qquad \delta'([s_1, 0], 1) = [s_3, 1] = \delta'([s_1, 1], 1);$$
$$\delta'([s_2, 0], 0) = [s_1, 0] = \delta'([s_2, 1], 0); \qquad \delta'([s_2, 1], 1) = [s_0, 1] = \delta'([s_2, 1], 1);$$
$$\delta'([s_3, 0], 0) = [s_3, 1] = \delta'([s_3, 1], 0); \qquad \delta'([s_3, 0], 1) = [s_2, 0] = \delta'([s_3, 1], 1);$$

The state graph of Moore machine constructed from Mealy machine is given below.

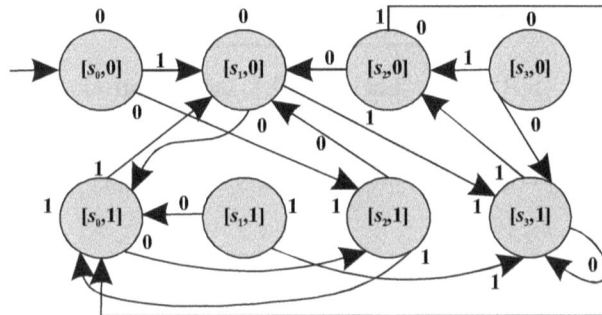

Therefore, the transition table of the Moore machine is given below.

Present state	Next state		Output
	Input (0)	Input (1)	
$[s_0, 0]$	$[s_2, 1]$	$[s_1, 0]$	0
$[s_0, 1]$	$[s_2, 1]$	$[s_1, 0]$	1
$[s_1, 0]$	$[s_0, 1]$	$[s_3, 1]$	0
$[s_1, 1]$	$[s_0, 1]$	$[s_3, 1]$	1
$[s_2, 0]$	$[s_1, 0]$	$[s_0, 1]$	0
$[s_2, 1]$	$[s_1, 0]$	$[s_0, 1]$	1
$[s_3, 0]$	$[s_3, 1]$	$[s_2, 0]$	0
$[s_3, 1]$	$[s_3, 1]$	$[s_2, 0]$	1

2.10.5   Construction of Moore Machine from Finite Automata

A Moore machine can be constructed from finite automaton by introducing output alphabet $\Delta = \{0,1\}$ and defining the output function $\lambda(s)$, such that

$$\lambda(s) = \begin{cases} 1 & \text{if } s \in A \\ 0 & \text{if } s \notin A \end{cases}$$

Consider the finite automaton given by the following table, where s_0 is the initial state and s_2 is the final state. Our aim is to construct a Moore machine.

State	Next state	
	$a = 0$	$a = 1$
$\rightarrow s_0$	s_2	s_1
s_1	s_3	s_0
s_2	s_0	s_3
s_3	s_1	s_2

Here, $S = \{s_0, s_1, s_2, s_3\}$; $A = \{s_2\}$. Let $\Delta = \{0,1\}$ be the introduced alphabet. The value of output is defined by output function $\lambda(s)$. Therefore, we get

$$\lambda(s_0) = 0 = \lambda(s_1) = \lambda(s_3) \quad [\because s_0, s_1, s_3 \notin A] \quad \text{and}$$
$$\lambda(s_2) = 1 \qquad\qquad\qquad [\because s_2 \in A]$$

Therefore, the transition table and transition graph of the Moore machine equivalent to the given finite automaton is given below.

State	Next state		Output
	$a = 0$	$a = 1$	
$\rightarrow s_0$	s_2	s_1	0
s_1	s_3	s_0	0
s_2	s_0	s_3	1
s_3	s_1	s_2	0

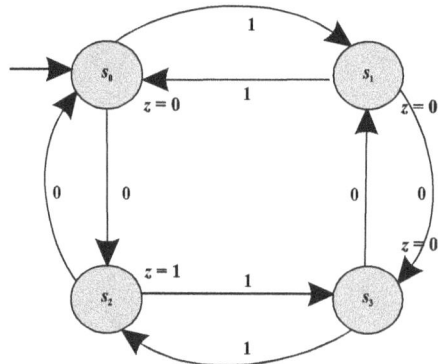

Note A Mealy machine can also be constructed from finite automaton. In such cases, a Moore machine is to be constructed from finite automaton and the Mealy machine can be constructed from Moore machine. Therefore, it is clear that finite automaton, Moore machine and Mealy machine are all equivalent.

2.11   MINIMIZATION OF FINITE AUTOMATA

A deterministic finite automata (DFA) can recognize a unique language but the converse is not true. It indicates that for a given language L there exists more than one deterministic finite automaton that accepts it. These deterministic finite automata are called equivalent automaton. Two or more equivalent automata will differ only in the number of possible states. So, it is desirable to reduce the number of states as it requires less space and is easily processed.

In this section our aim is to construct an automaton M_1 with minimum number of states equivalent to given finite automaton M. Two states s_1 and s_2 are equivalent if both $\delta(s_1, x)$ and $\delta(s_2, x)$ are final states, or both of them are non-final states for all input string $x \in \Sigma^*$. Two states s_1 and s_2 are k-equivalent if both $\delta(s_1, x)$ and $\delta(s_2, x)$ are final states, or both of them are non-final states for all input string $x \in \Sigma^*$ of length k or less. So, it is clear that any two final states are 0-equivalent and also any two non-final states are 0-equivalent.

2.11.1   Construction of Minimum State Automaton

In this section we will discuss regarding the construction of minimum state automaton. The following steps are used to construct a minimum state automaton for a given finite automaton.

Step 1   Construction of π_0—by applying the definition of 0-equivalence we get $\pi_0 = (S_1^0, S_2^0)$, where S_1^0 is the set of all final states and $S_2^0 = (S - S_1^0)$ is the set of all non-final states.

Step 2   Construction of π_{k+1} from π_k—suppose S_i^k is any subset in π_k. If s_1 and s_2 are in S_i^k, then they are $(k+1)$-equivalent provided $\delta(s_1, a)$ and $\delta(s_2, a)$ are k-equivalent. Determine whether $\delta(s_1, a)$ and $\delta(s_2, a)$ are in the same equivalence class of π_k for every $a \in \Sigma$. If so, s_1 and s_2 are $(k+1)$-equivalent. Therefore, S_i^k is further divided into $(k+1)$-equivalence classes. Repeat this process for every S_i^k in π_k to find all the elements of π_{k+1}.

Step 3   Now we construct π_n for $n = 1, 2, 3, \cdots$ until $\pi_n = \pi_{n+1}$.

Step 4   The states obtained in step 3 are the equivalence classes. Obtain the transition table by replacing a state 's' by the corresponding equivalence class $[s]$.

Note   It is to be noted that the construction of $\pi_0, \pi_1, \pi_2, \cdots$ is easier when the transition table is given. It is clear that $\pi_0 = \{S_1^0, S_2^0\}$, where $S_1^0 = A$ and $S_2^0 = (S - A)$. The subsets of π_1 are obtained by partitioning subsets of π_0 further. If s_1 and s_2 are in S_1^0, consider the states in each a-column, $a \in \Sigma$, corresponding to s_1 and s_2. If they are in the same subset of π_0, then s_1 and s_2 are said to be 1-equivalent. If the states under same a-column, $a \in \Sigma$ are different, then s_1 and s_2 are said to be not 1-equivalent. Generally $(k+1)$ equivalent states are obtained by applying the above concept for s_1 and s_2 in S_i^k.

Let us consider the following finite automaton

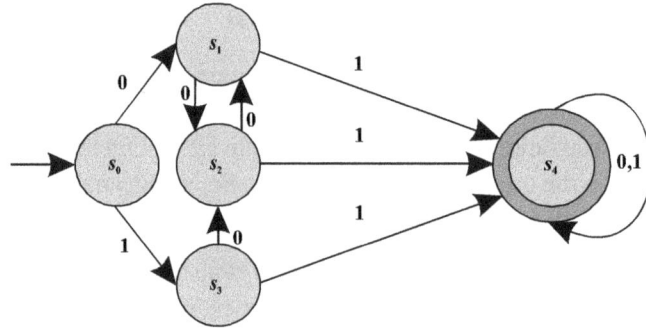

Our aim is to construct a minimum state finite automaton. In order to construct the minimum state automaton, first of all we have to get the transition table. The transition table for the above figure is given below.

States	Inputs	
	0	1
$\rightarrow s_0$	s_1	s_3
s_1	s_2	s_4
s_2	s_1	s_4
s_3	s_2	s_4
s_4	s_4	s_4

Now, we have $S_1^0 = A = \{s_4\}$ and $S_2^0 = S - A = \{s_0, s_1, s_2, s_3\}$, Therefore, $\pi_0 = \{S_1^0, S_2^0\} = \{\{s_4\}, \{s_0, s_1, s_2, s_3\}\}$

Since $S_1^0 = A = \{s_4\}$ has only one state, so it cannot be partitioned further. Hence, $S_1^1 = \{s_4\}$. Now, consider s_0 and s_1 of S_2^0. From the table it is easy to see that under 0-column, the states corresponding to s_0 and s_1 are s_1 and s_2 that are in S_2^0 whereas under 1-column, the states corresponding to s_0 and s_1 are s_3 and s_4 with $s_3 \in S_2^0$ and $s_4 \in S_1^0$. Therefore, s_0 and s_1 are not 1-equivalent. Similarly, it can be shown that s_0 is not 1-equivalent to s_2 and s_3.

Again consider s_1 and s_2 of S_2^0. From the state table, it is clear that under 0-column, the states corresponding to s_1 and s_2 are s_2 and s_1 that are in S_2^0. Similarly, under 1-column, the states corresponding to s_1 and s_2 are s_4 and s_4 with $s_4 \in S_1^0$. Therefore s_1 and s_2 are 1-equivalent. Similarly, it can be shown that s_1 and s_3 are 1-equivalent. Therefore, we get

$$\pi_1 = \{\{s_4\}, \{s_0\}, \{s_1, s_2, s_3\}\} = \{S_1^1, S_2^1, S_3^1\} \text{ (say), in which}$$

$$S_1^1 = \{s_4\}, S_2^1 = \{s_0\} \text{ and } S_3^1 = \{s_1, s_2, s_3\}.$$

Now, $\{s_4\}$ and $\{s_0\}$ are also in π_2 as they can not be partitioned further. Consider, s_1 and s_2 of S_3^1. The states under 0-column are s_2 and s_1 that are in S_3^1. In the other end, states under 1-column are s_4 and s_4 with $s_4 \in S_1^1$. So s_1 and s_2 are 2-equivalent. It can also be shown that s_1 and s_3 are 2-equivalent. Therefore, we get

$$\pi_2 = \{\{s_4\}, \{s_0\}, \{s_1, s_2, s_3\}\} = \pi_1$$

Hence, the minimum state automaton is given as $M_1 = \{S', \Sigma, \delta', s_0', A'\}$, in which $S' = \{[s_4], [s_0], [s_1, s_2, s_3]\}, \Sigma = \{0, 1\}, s_0' = [s_0]$, $A' = [s_4]$ and δ' is given by the following table.

States	Input	
	0	**1**
$[s_0]$	$[s_1, s_2, s_3]$	$[s_1, s_2, s_3]$
$[s_4]$	$[s_4]$	$[s_4]$
$[s_1, s_2, s_3]$	$[s_1, s_2, s_3]$	$[s_4]$

The transition diagram for minimum state automaton is given below.

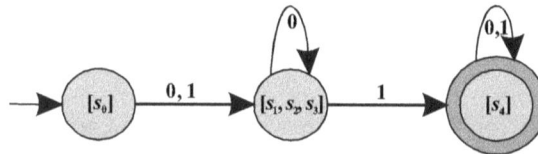

SOLVED EXAMPLES

Example 1

Consider the following transition diagram.

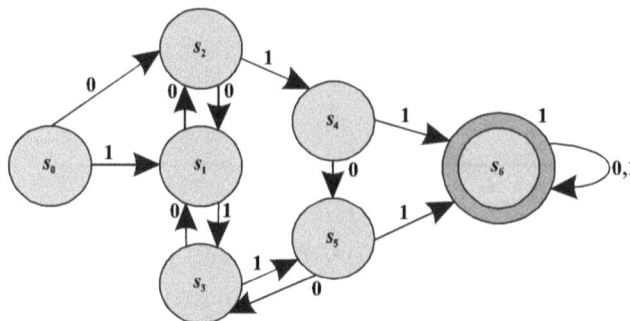

Determine the initial state, final state, intermediate states and input alphabet for the given transition diagram.

Solution   In the above transition diagram we have the initial state as $\{s_0\}$ final state (A) as $\{s_6\}$ the set of states $(S) = \{s_0, s_1, s_2, s_3, s_4, s_5, s_6\}$ whereas the intermediate states are $\{s_1, s_2, s_3, s_4, s_5\}$. The input alphabet $\Sigma = \{0, 1\}$.

Example 2

Consider the deterministic finite automaton (DFA) whose transition diagram is given below. Show that the string $w = 0201201$ is accepted by the DFA.

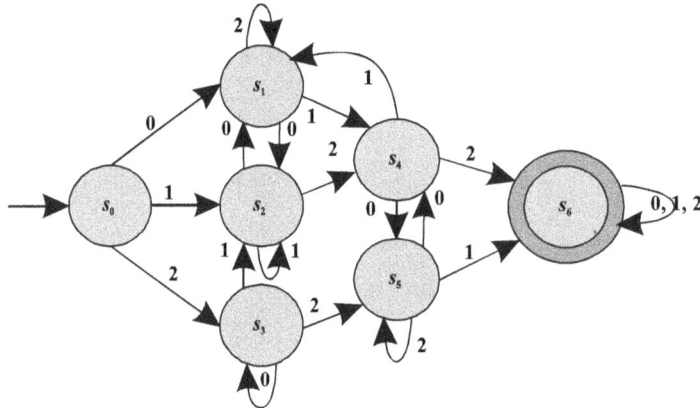

Solution   The transition table of the above DFA is given below. Here, we have $S = \{s_0, s_1, s_2, s_3, s_4, s_5, s_6\}$; $A = \{s_6\}$; $\Sigma = \{0, 1, 2\}$. Given that $w = 0201201$.

State	Next state		
	0	**1**	**2**
$\to s_0$	s_1	s_2	s_3
s_1	s_2	s_4	s_1
s_2	s_1	s_2	s_4
s_3	s_3	s_2	s_5
s_4	s_5	s_1	s_6
s_5	s_4	s_6	s_5
s_6	s_6	s_6	s_6

Therefore, $\delta(s_0, 0201201) = \delta(\delta(s_0, 0)201201)$

$$= \delta(s_1, 201201) = \delta(\delta(s_1, 2)01201)$$
$$= \delta(s_1, 01201) = \delta(\delta(s_1, 0)1201)$$
$$= \delta(s_2, 1201) = \delta(\delta(s_2, 1)201)$$
$$= \delta(s_2, 201) = \delta((\delta(s_2, 2)01)$$

$$= \delta(s_4, 01) = \delta(\delta(s_4, 0)1)$$
$$= \delta(s_5, 1) = s_6$$

Thus, we get $\delta(s_0, 0201201) = s_6$ with $s_6 \in A$. Therefore, the string w is accepted by the deterministic finite automata.

Example 3

Find the language accepted by the following deterministic finite automata, whose transition diagram is given below.

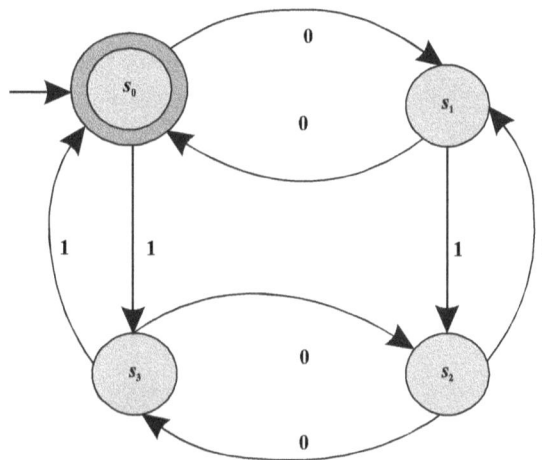

Solution   In the above transition diagram, we have $S = \{s_0, s_1, s_2, s_3\}$; $\Sigma = \{0, 1\}$; $A = \{s_0\}$ and the initial state $= \{s_0\}$. The transition table is given below.

Present state	Next state	
	0	1
$\rightarrow s_0$	s_1	s_3
s_1	s_0	s_2
s_2	s_3	s_1
s_3	s_2	s_0

Let us consider $w_1 = 11$; $w_2 = 00$; $w_3 = 0101$; $w_4 = 1101$. Now, we have

$$\delta(s_0, w_1) = \delta(s_0, 11) = \delta(\delta(s_0, 1)1)$$
$$= \delta(s_3, 1) = s_0 \in A$$
$$\delta(s_0, w_2) = \delta(s_0, 00) = \delta(\delta(s_0, 0)0)$$
$$= \delta(s_1, 0) = s_0 \in A$$

$$\delta(s_0, w_3) = \delta(s_0, 0101) = \delta(\delta(s_0, 0)101)$$
$$= \delta(s_1, 101) = \delta(\delta(s_1, 1)01)$$
$$= \delta(s_2, 01) = \delta(\delta(s_2, 0)1)$$
$$= \delta(s_3, 1) = s_0 \in A \quad \text{and}$$
$$\delta(s_0, w_4) = \delta(s_0, 1101) = \delta(\delta(s_0, 1)101)$$
$$= \delta(s_3, 101) = \delta(\delta(s_3, 1)01)$$
$$= \delta(s_0, 01) = \delta(\delta(s_0, 0)1)$$
$$= \delta(s_1, 1) = s_2 \notin A$$

Therefore, it is clear that the strings accepted by the given finite automata are w_1, w_2 and w_3. Hence, the language accepted by the given automaton is the set of strings with even number of ones and zeros.

Example 4

Consider the following deterministic finite automaton. Check the acceptability of the strings (a) *aabab* and (b) *bbaba*.

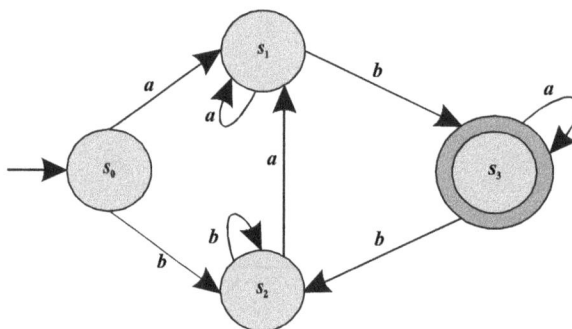

Solution Let $M = (S, \Sigma, \delta, s_0, A)$ be the deterministic finite automaton. Here, $S = \{s_0, s_1, s_2, s_3\}$; $\Sigma = \{a, b\}$; $A = \{s_3\}$. The transition function δ is defined as

Present state	Next state	
	a	b
$\rightarrow s_0$	s_1	s_2
s_1	s_1	s_3
s_2	s_1	s_2
s_3	s_3	s_2

(a) Let us take $w = aabab$. Therefore, we have

$$\delta(s_0, aabab) = \delta(\delta(s_0, a)abab)$$
$$= \delta(s_1, abab) = \delta(\delta(s_1, a)bab)$$
$$= \delta(s_1, bab) = \delta(\delta(s_1, b)ab)$$
$$= \delta(s_3, ab) = \delta(\delta(s_3, a)b)$$
$$= \delta(s_3, b) = s_2 \notin A$$

(b) Let us take $w = bbaba$. Therefore, we have

$$\delta(s_0, bbaba) = \delta(\delta(s_0, b)baba)$$
$$= \delta(s_2, baba) = \delta(\delta(s_2, b)aba)$$
$$= \delta(s_2, aba) = \delta(\delta(s_2, a)ba)$$
$$= \delta(s_1, ba) = \delta(\delta(s_1, b)a)$$
$$= \delta(s_3, a) = s_3 \in A$$

So, the string $w = aabab$ is not accepted by the given deterministic finite automata whereas, the string $w = bbaba$ is not accepted by the given deterministic finite automata.

Example 5

Show that the string $w = bacb$ is accepted by the machine whose transition table is given below. Construct the transition diagram. It is given that the accepting state is s_3.

States	Inputs		
	a	b	c
$\to s_0$	$\{s_0\}$	$\{s_0, s_1\}$	$\{s_1, s_2\}$
s_1	$\{s_1, s_2\}$	$\{s_1\}$	$\{s_1, s_4\}$
s_2	$\{s_1, s_2\}$	$\{s_3\}$	$\{s_2\}$
s_3	ϕ	ϕ	ϕ
s_4	$\{s_4\}$	$\{s_3, s_4\}$	$\{s_3, s_4\}$

Solution The transition diagram is given below.

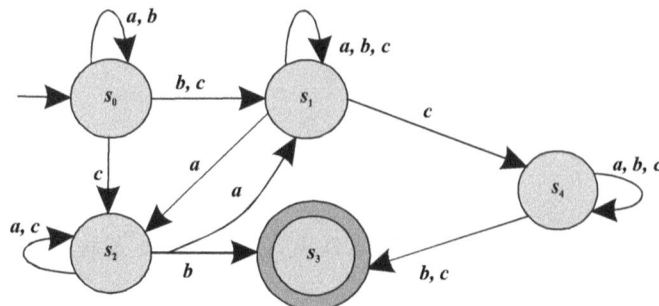

Here, $S = \{s_0, s_1, s_2, s_3, s_4\}$; $A = \{s_3\}$; $\Sigma = \{a, b, c\}$. Given that $w = bacb$. Therefore, we have

$$
\begin{aligned}
\delta(s_0, bacb) &= \delta(\delta(s_0, b)acb) \\
&= \delta(\{s_0, s_1\}, acb) \\
&= \delta(s_0, acb) \cup \delta(s_1, acb) \quad\quad\quad (1)
\end{aligned}
$$

$$
\begin{aligned}
\delta(s_0, acb) &= \delta(\delta(s_0, a)cb) \\
&= \delta(s_0, cb) = \delta(\delta(s_0, c)b) \\
&= \delta(\{s_1, s_2\}, b) \\
&= \delta(s_1, b) \cup \delta(s_2, b) = \{s_1, s_3\}
\end{aligned}
$$

$$
\begin{aligned}
\delta(s_1, acb) &= \delta(\delta(s_1, a)cb) \\
&= \delta(\{s_1, s_2\}, cb) \\
&= \delta(s_1, cb) \cup \delta(s_2, cb) \quad\quad\quad (2)
\end{aligned}
$$

$$
\begin{aligned}
\delta(s_1, cb) &= \delta(\delta(s_1, c)b) \\
&= \delta(\{s_1, s_4\}, b) \\
&= \delta(s_1, b) \cup \delta(s_4, b) = \{s_1, s_3, s_4\}
\end{aligned}
$$

$$
\begin{aligned}
\delta(s_2, cb) &= \delta(\delta(s_2, c)b) \\
&= \delta(s_2, b) = \{s_3\}
\end{aligned}
$$

Therefore, from equation (2) we get $\delta(s_1, acb) = \{s_1, s_3, s_4\}$. Hence, equation (1) reduces to

$$\delta(s_0, bacb) = \{s_1, s_3\} \cup \{s_1, s_3, s_4\} = \{s_1, s_3, s_4\}.$$

Therefore, it is clear that $\delta(s_0, bacb)$ contains the final state s_3 and so the string $w = bacb$ is accepted by the given machine.

Example 6

Construct a deterministic finite automata equivalent to the nondeterministic finite automata $M = (\{s_0, s_1\}, \{0,1\}, \delta, s_0, \{s_1\})$ where the transition function δ is described in the following table.

State	Input	
	0	1
$\rightarrow s_0$	$\{s_1\}$	$\{s_0, s_1\}$
s_1	$\{s_0, s_1\}$	–

Solution   Given NFA $M = (\{s_0, s_1\}, \{0,1\}, \delta, s_0, \{s_1\})$, where δ is defined as

State	Input	
	0	**1**
$\rightarrow s_0$	$\{s_1\}$	$\{s_0, s_1\}$
s_1	$\{s_0, s_1\}$	$-$

Our aim is to construct an equivalent deterministic finite automaton. Let the equivalent deterministic finite automaton is $M_1 = (S', \Sigma, \delta', s'_0, A')$. For the deterministic finite automata M_1 we have

i. The state S' are subset of $\{s_0, s_1\}$, i.e., $\phi, [s_0], [s_1]$ and $[s_0, s_1]$.

ii. The initial state $s'_0 = [s_0]$

iii. The final states are $[s_1]$ and $[s_0, s_1]$

iv. The transition function is defined by the state table given below. In order to construct the state transition table, we have to compute the followings

$$\delta'(\phi, 0) = \phi = \delta'(\phi, 1)$$
$$\delta'([s_0], 0) = \delta(s_0, 0) = [s_1]$$
$$\delta'([s_0], 1) = \delta(s_0, 1) = [s_0, s_1]$$
$$\delta'([s_1], 0) = \delta(s_1, 0) = [s_0, s_1]$$
$$\delta'([s_1], 1) = \delta(s_1, 1) = \phi$$
$$\delta'([s_0, s_1], 0) = \delta(s_0, 0) \cup \delta(s_1, 0) = [s_0, s_1]$$
$$\delta'([s_0, s_1], 1) = \delta(s_0, 1) \cup \delta(s_1, 1) = [s_0, s_1]$$

Therefore, the transition table is given as

State	Input	
	0	**1**
$\rightarrow [s_0]$	$[s_1]$	$[s_0, s_1]$
$[s_1]$	$[s_0, s_1]$	$[\phi]$
$[s_0, s_1]$	$[s_0, s_1]$	$[s_0, s_1]$

The transition diagram of the equivalent deterministic finite automata is given below.

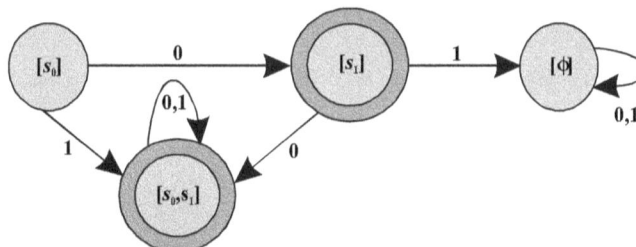

Example 7

Find a deterministic acceptor equivalent to $M = (\{s_0, s_1, s_2, s_3\},\ \{a, b\},\ \delta, s_0, \{s_2\})$ where δ is given in the following table.

State	Input	
	a	b
$\rightarrow s_0$	$\{s_1\}$	$\{s_1, s_2\}$
s_1	$\{s_0\}$	$\{s_2, s_3\}$
s_2	$\{s_3\}$	$-$
s_3	$\{s_1, s_2\}$	$\{s_0\}$

Solution   Let the deterministic automaton equivalent to M is $M_1 = (S', \{a, b\}, \delta', [s_0], A')$ defined as follows, where $A' = \{[s_2], [s_0, s_2],\ [s_1, s_2], [s_2, s_3],\ [s_0, s_2, s_3],\ [s_0, s_1, s_2],\ [s_1, s_2, s_3],\ [s_0, s_1, s_2, s_3]\}$. We start the construction by considering $[s_0]$ first. Therefore, we have

$$\delta'([s_0], a) = \delta(s_0, a) = [s_1]$$
$$\delta'([s_0], b) = \delta(s_0, b) = [s_1, s_2]$$
$$\delta'([s_1], a) = \delta(s_1, a) = [s_0]$$
$$\delta'([s_1], b) = \delta(s_1, b) = [s_2, s_3]$$
$$\delta'([s_1, s_2], a) = \delta(s_1, a) \cup \delta(s_2, a) = [s_0, s_3]$$
$$\delta'([s_1, s_2], b) = \delta(s_1, b) \cup \delta(s_2, b) = [s_2, s_3]$$
$$\delta'([s_2, s_3], a) = \delta(s_2, a) \cup \delta(s_3, a) = [s_1, s_2, s_3]$$
$$\delta'([s_2, s_3], b) = \delta(s_2, b) \cup \delta(s_3, b) = [s_0]$$
$$\delta'([s_0, s_3], a) = \delta(s_0, a) \cup \delta(s_3, a) = [s_1, s_2]$$
$$\delta'([s_0, s_3], b) = \delta(s_0, b) \cup \delta(s_3, b) = [s_0, s_1, s_2]$$
$$\delta'([s_1, s_2, s_3], a) = \delta(s_1, a) \cup \delta(s_2, a) \cup \delta(s_3, a) = [s_0, s_1, s_2, s_3]$$
$$\delta'([s_1, s_2, s_3], b) = \delta(s_1, b) \cup \delta(s_2, b) \cup \delta(s_3, b) = [s_0, s_2, s_3]$$
$$\delta'([s_0, s_2, s_3], a) = \delta(s_0, a) \cup \delta(s_2, a) \cup \delta(s_3, a) = [s_1, s_2, s_3]$$
$$\delta'([s_0, s_2, s_3], b) = \delta(s_0, b) \cup \delta(s_2, b) \cup \delta(s_3, b) = [s_0, s_1, s_2]$$
$$\delta'([s_0, s_1, s_2], a) = \delta(s_0, a) \cup \delta(s_1, a) \cup \delta(s_2, a) = [s_0, s_1, s_3]$$
$$\delta'([s_0, s_1, s_2], b) = \delta(s_0, b) \cup \delta(s_1, b) \cup \delta(s_2, b) = [s_1, s_2, s_3]$$

$\delta'([s_0, s_1, s_3], a) = \delta(s_0, a) \cup \delta(s_1, a) \cup \delta(s_3, a) = [s_0, s_1, s_2]$

$\delta'([s_0, s_1, s_3], b) = \delta(s_0, b) \cup \delta(s_1, b) \cup \delta(s_3, b) = [s_0, s_1, s_2, s_3]$

$\delta'([s_0, s_1, s_2, s_3], a) = \delta(s_0, a) \cup \delta(s_1, a) \cup \delta(s_2, a) \cup \delta(s_3, a) = [s_0, s_1, s_2, s_3]$

$\delta'([s_0, s_1, s_2, s_3], b) = \delta(s_0, b) \cup \delta(s_1, b) \cup \delta(s_2, b) \cup \delta(s_3, b) = [s_0, s_1, s_2, s_3]$

Now, we do not get any new states and so, we terminate the construction of δ'. The transition table is given below.

States	Input	
	a	*b*
$\rightarrow [s_0]$	$[s_1]$	$[s_1, s_2]$
$[s_1]$	$[s_0]$	$[s_2, s_3]$
$[s_1, s_2]$	$[s_0, s_3]$	$[s_2, s_3]$
$[s_2, s_3]$	$[s_1, s_2, s_3]$	$[s_0]$
$[s_0, s_3]$	$[s_1, s_2]$	$[s_0, s_1, s_2]$
$[s_1, s_2, s_3]$	$[s_0, s_1, s_2, s_3]$	$[s_0, s_2, s_3]$
$[s_0, s_2, s_3]$	$[s_1, s_2, s_3]$	$[s_0, s_1, s_2]$
$[s_0, s_1, s_2]$	$[s_0, s_1, s_3]$	$[s_1, s_2, s_3]$
$[s_0, s_1, s_3]$	$[s_0, s_1, s_2]$	$[s_0, s_1, s_2, s_3]$
$[s_0, s_1, s_2, s_3]$	$[s_0, s_1, s_2, s_3]$	$[s_0, s_1, s_2, s_3]$

Example 8

Construct a state diagram of the deterministic finite automata equivalent to the following nondeterministic finite automata.

Solution   In the above diagram the nondeterministic finite automata is defined as $M = (\{s_0, s_1, s_2\}, \{a, b\}, \delta, s_0, \{s_2\})$. Here $S = \{s_0, s_1, s_2\}$; $\Sigma = \{a, b\}$ and $A = \{s_2\}$. The transition function δ is defined as below.

States	Input	
	a	*b*
$\rightarrow s_0$	$\{s_0, s_1\}$	$\{s_2\}$
s_1	$\{s_0\}$	$\{s_1\}$
s_2	$--$	$\{s_0, s_1\}$

Let the equivalent deterministic finite automata M_1 be defined as $M_1 = (S', \Sigma, \delta', [s_0], A')$, where $A' = \{[s_2], [s_0, s_2], [s_1, s_2], [s_0, s_1, s_2]\}$. We start construction of transition function by considering $[s_0]$ first. Therefore, we have

$$\delta'([s_0], a) = \delta(s_0, a) = [s_0, s_1]$$
$$\delta'([s_0], b) = \delta(s_0, b) = [s_2]$$
$$\delta'([s_2], a) = \delta(s_2, a) = \phi$$
$$\delta'([s_2], b) = \delta(s_2, b) = [s_0, s_1]$$
$$\delta'([s_0, s_1], a) = \delta(s_0, a) \cup \delta(s_1, a) = [s_0, s_1]$$
$$\delta'([s_0, s_1], b) = \delta(s_0, b) \cup \delta(s_1, b) = [s_1, s_2]$$
$$\delta'([s_1, s_2], a) = \delta(s_1, a) \cup \delta(s_2, a) = [s_0]$$
$$\delta'([s_1, s_2], b) = \delta(s_1, b) \cup \delta(s_2, b) = [s_0, s_1]$$

Therefore, the state diagram of the deterministic finite automata is given below, where ϕ is called as a dummy state, trap state or dead state.

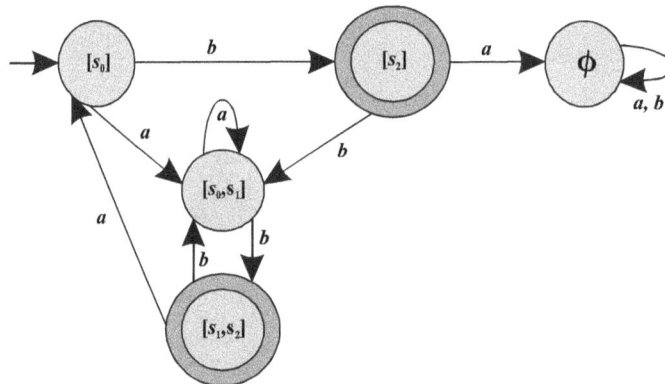

Example 9

Construct a nondeterministic finite automaton for the following nondeterministic finite automata with ε-moves.

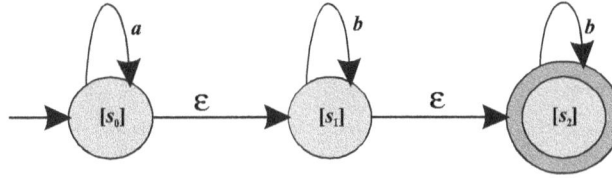

Solution From the transition diagram given above, the nondeterministic finite automata with ε-moves M is described as $M = (S,\Sigma,\delta,s_0,A)$ where $S = \{s_0, s_1, s_2\}$ $\Sigma = \{a, b, \varepsilon\}$ and $A = \{s_2\}$. Our aim is to construct a nondeterministic finite automaton M_1 defined as $M_1 = (S,\Sigma',\delta',s_0,A')$ where $\Sigma' = \{a, b\}$. The transition function δ' and A' is defined as below.

$$\delta'(s,a) = \hat{\delta}(s,a) \text{ for } s \in S \text{ and } a \in \Sigma$$

$$A' = \begin{cases} A \cup \{s_0\}; & \text{if } \varepsilon\text{-}closure\,(s_0) \text{ contains a state of } A \\ A; & \text{otherwise} \end{cases}$$

From the transition diagram, it is clear that

$$\varepsilon\text{-}closure\,(s_0) = \{s_0,s_1,s_2\}$$
$$\varepsilon\text{-}closure\,(s_1) = \{s_1,s_2\}$$
$$\varepsilon\text{-}closure\,(s_2) = \{s_2\}$$

Since, $\varepsilon\text{-}closure\,(s_0) = \{s_0,s_1,s_2\}$ and $s_2 \in A$, we get $A' = A \cup \{s_0\} = \{s_0,s_2\}$. Now,

$$\begin{aligned}
\delta'(s_0,a) &= \hat{\delta}(s_0,a) = \varepsilon\text{-}closure\,(\delta(\hat{\delta}(s_0,\varepsilon),a)) \\
&= \varepsilon\text{-}closure\,(\delta(\{s_0,s_1,s_2\},a)) \\
&= \varepsilon\text{-}closure\,(\delta(s_0,a)\cup\delta(s_1,a)\cup\delta(s_2,a)) \\
&= \varepsilon\text{-}closure\,(\{s_0\}\cup\phi\cup\phi) \\
&= \varepsilon\text{-}closure\,(\{s_0\}) = \{s_0,s_1,s_2\} \\
\delta'(s_0,b) &= \hat{\delta}(s_0,b) = \varepsilon\text{-}closure\,(\delta(\hat{\delta}(s_0,\varepsilon),b)) \\
&= \varepsilon\text{-}closure\,(\delta(\{s_0,s_1,s_2\},b)) \\
&= \varepsilon\text{-}closure\,(\delta(s_0,b)\cup\delta(s_1,b)\cup\delta(s_2,b)) \\
&= \varepsilon\text{-}closure\,(\phi\cup\{s_1\}\cup\{s_2\}) \\
&= \varepsilon\text{-}closure\,(\{s_1,s_2\}) = \{s_1,s_2\} \\
\delta'(s_1,a) &= \hat{\delta}(s_1,a) = \varepsilon\text{-}closure\,(\delta(\hat{\delta}(s_1,\varepsilon),a)) \\
&= \varepsilon\text{-}closure\,(\{s_1,s_2\},a)) \\
&= \varepsilon\text{-}closure\,(\delta(s_1,a)\cup\delta(s_2,a)) \\
&= \varepsilon\text{-}closure\,(\phi\cup\phi) = \phi
\end{aligned}$$

$$\delta'(s_1,b) = \hat{\delta}(s_1,b) = \varepsilon\text{-}closure\,(\delta(\hat{\delta}(s_1,\varepsilon),b))$$
$$= \varepsilon\text{-}closure\,(\delta(\{s_1,s_2\},b))$$
$$= \varepsilon\text{-}closure\,(\delta(s_1,b)\cup\delta(s_2,b))$$
$$= \varepsilon\text{-}closure\,(\{s_1\}\cup\{s_2\}) = \{s_1,s_2\}$$
$$\delta'(s_2,a) = \hat{\delta}(s_2,a) = \varepsilon\text{-}closure\,(\delta(\hat{\delta}(s_2,\varepsilon),a))$$
$$= \varepsilon\text{-}closure\,(\delta(\{s_2\},a))$$
$$= \varepsilon\text{-}closure\,(\phi) = \phi$$
$$\delta'(s_2,b) = \hat{\delta}(s_2,b) = \varepsilon\text{-}closure\,(\delta(\hat{\delta}(s_2,\varepsilon),b))$$
$$= \varepsilon\text{-}closure\,(\delta(\{s_2\},b))$$
$$= \varepsilon\text{-}closure\,(\{s_2\}) = \{s_2\}$$

Therefore, the transition function δ' of M_1 is given in the following table.

States	Input	
	a	*b*
$\rightarrow s_0$	$\{s_0,s_1,s_2\}$	$\{s_1,s_2\}$
s_1	ϕ	$\{s_1,s_2\}$
s_2	ϕ	$\{s_2\}$

Thus, the nondeterministic finite automata without ε-transition is given below.

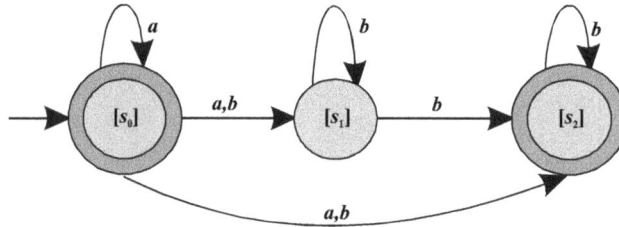

Example 10

Construct a nondeterministic finite automaton for the given machine.

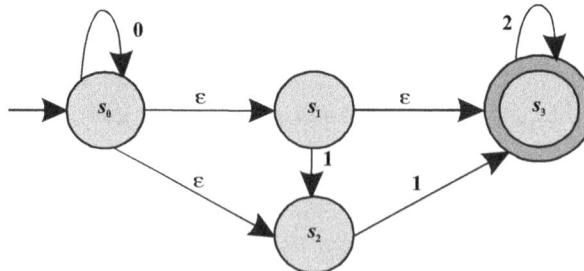

Solution   In the above machine $S = \{s_0, s_1, s_2, s_3\}$; $\Sigma = \{0, 1, 2, \varepsilon\}$; $A = \{s_3\}$; and initial state $= [s_0]$. Our aim is to construct a nondeterministic finite automaton. From the transition diagram it is clear that

$$\hat{\delta}(s_0, \varepsilon) = \varepsilon\text{-}closure \ (s_0) = \{s_0, s_1, s_2, s_3\}$$

$$\hat{\delta}(s_1, \varepsilon) = \varepsilon\text{-}closure \ (s_1) = \{s_1, s_2, s_3\}$$

$$\hat{\delta}(s_2, \varepsilon) = \varepsilon\text{-}closure \ (s_2) = \{s_2, s_3\}$$

$$\hat{\delta}(s_3, \varepsilon) = \varepsilon\text{-}closure \ (s_3) = \{s_3\}$$

Let us define the nondeterministic finite automata $M = (S, \Sigma', \delta', s_0, A')$ where $S = \{s_0, s_1, s_2, s_3\}$; $\Sigma' = \{0, 1, 2\}$. The final state A' is defined as $A' = \{s_0, s_3\}$ since $A' = A \cup \{s_0\}$. The transition function δ' is defined as $\delta'(s, a) = \hat{\delta}(s, a)$ for $s \in S$ and $a \in \Sigma$. Therefore, we get

$$\delta'(s_0, 0) = \hat{\delta}(s_0, 0) = \varepsilon\text{-}closure \ (\delta(\hat{\delta}(s_0, \varepsilon)0)) \ ?$$
$$= \varepsilon\text{-}closure \ (\delta(\{s_0, s_1, s_2, s_3\}, 0))$$
$$= \varepsilon\text{-}closure \ (\delta(s_0, 0) \cup \delta(s_1, 0) \cup \delta(s_2, 0) \cup \delta(s_3, 0))$$
$$= \varepsilon\text{-}closure \ (s_0) = \{s_0, s_1, s_2, s_3\}$$

$$\delta'(s_0, 1) = \hat{\delta}(s_0, 1) = \varepsilon\text{-}closure \ (\delta(\hat{\delta}(s_0, \varepsilon)1))$$
$$= \varepsilon\text{-}closure \ (\delta(\{s_0, s_1, s_2, s_3\}, 1)) = \{s_2, s_3\}$$

$$\delta'(s_0, 2) = \hat{\delta}(s_0, 2) = \varepsilon\text{-}closure \ (\delta(\hat{\delta}(s_0, \varepsilon)2))$$
$$= \varepsilon\text{-}closure \ (\delta(\{s_0, s_1, s_2, s_3\}, 2)) = \{s_3\}$$

$$\delta'(s_1, 0) = \hat{\delta}(s_1, 0) = \varepsilon\text{-}closure \ (\delta(\hat{\delta}(s_1, \varepsilon)0))$$
$$= \varepsilon\text{-}closure \ (\delta(\{s_1, s_2, s_3\}, 0)) = \phi$$

$$\delta'(s_1, 1) = \hat{\delta}(s_1, 1) = \varepsilon\text{-}closure \ (\delta(\hat{\delta}(s_1, \varepsilon)1))$$
$$= \varepsilon\text{-}closure \ (\delta(\{s_1, s_2, s_3\}, 1))$$
$$= \{s_2, s_3\}$$

$$\delta'(s_1, 2) = \hat{\delta}(s_1, 2) = \varepsilon\text{-}closure \ (\delta(\hat{\delta}(s_1, \varepsilon)2))$$
$$= \varepsilon\text{-}closure \ (\delta(\{s_1, s_2, s_3\}, 2)) = \{s_3\}$$

$$\delta'(s_2, 0) = \hat{\delta}(s_2, 0) = \varepsilon\text{-}closure \ (\delta(\hat{\delta}(s_2, \varepsilon)0))$$
$$= \varepsilon\text{-}closure \ (\delta(\{s_2, s_3\}, 0)) = \phi$$

$$\delta'(s_2, 1) = \hat{\delta}(s_2, 1) = \varepsilon\text{-}closure \ (\delta(\hat{\delta}(s_2, \varepsilon)1))$$
$$= \varepsilon\text{-}closure \ (\delta(\{s_2, s_3\}, 1)) = \{s_3\}$$

$$\delta'(s_2, 2) = \hat{\delta}(s_2, 2) = \varepsilon\text{-}closure \ (\delta(\hat{\delta}(s_2, \varepsilon)2))$$
$$= \varepsilon\text{-}closure \ (\delta(\{s_2, s_3\}, 2)) = \{s_3\}$$

$$\delta'(s_3, 0) = \hat{\delta}(s_3, 0) = \varepsilon\text{-}closure \ (\delta(\hat{\delta}(s_3, \varepsilon)0))$$
$$= \varepsilon\text{-}closure \ (\delta(\{s_3\}, 0)) = \phi$$

$$\delta'(s_3,1) = \hat{\delta}(s_3,1) = \varepsilon\text{-}closure\,(\delta(\hat{\delta}(s_3,\varepsilon)1))$$
$$= \varepsilon\text{-}closure\,(\delta(\{s_3\},1)) = \phi$$
$$\delta'(s_3,2) = \hat{\delta}(s_3,2) = \varepsilon\text{-}closure\,(\delta(\hat{\delta}(s_3,\varepsilon)2))$$
$$= \varepsilon\text{-}closure\,(\delta(\{s_3\},2)) = \{s_3\}$$

Therefore, the transition table is given below.

States	Input		
	0	**1**	**2**
$\to s_0$	$\{s_0,s_1,s_2,s_3\}$	$\{s_2,s_3\}$	$\{s_3\}$
s_1	ϕ	$\{s_2,s_3\}$	$\{s_3\}$
s_2	ϕ	$\{s_3\}$	$\{s_3\}$
s_3	ϕ	ϕ	$\{s_3\}$

Example 11

Construct the minimum state automaton equivalent to the transition diagram given below.

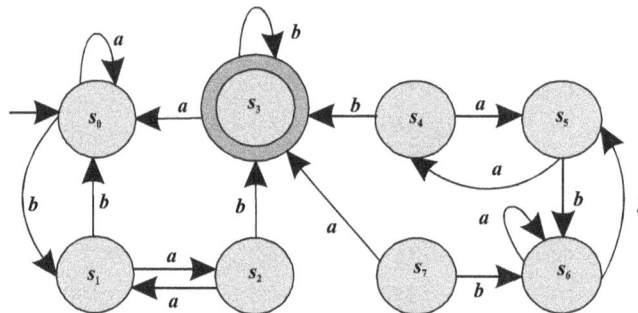

Solution   The transition table for the above diagram is given below.

States	Input	
	a	**b**
$\to s_0$	s_0	s_1
s_1	s_2	s_0
s_2	s_1	s_3
s_3	s_0	s_3
s_4	s_5	s_3
s_5	s_4	s_6
s_6	s_6	s_5
s_7	s_3	s_6

Now, we have $S_1^0 = A = \{s_3\}$ and $S_2^0 = \{s_0, s_1, s_2, s_4, s_5, s_6, s_7\}$. Therefore, we get $\pi_0 = \{S_1^0, S_2^0\} = \{\{s_3\}, \{s_0, s_1, s_2, s_4, s_5, s_6, s_7\}\}$.

Since, $S_1^0 = \{s_3\}$ has only one state, so it can not get further partition. Thus, $S_1^1 = \{s_3\}$. Now, consider s_0, s_1, of S_2^0. From the transition table it is clear that under a-column, the states corresponding to s_0, s_1 are s_0, s_2 respectively, which are in S_2^0. Again the states under b-column corresponding to s_0, s_1 are s_1, s_0 respectively, which are in S_2^0. Therefore, s_0, s_1 are 1-equivalent. Similarly, s_0, s_5 and s_0, s_6 are 1-equivalent. Therefore, we get $S_2^1 = \{s_0, s_1, s_5, s_6\}$. Now, consider s_2, s_4 of S_2^0. It is clear from the transition table that under a-column, the states corresponding to s_2, s_4 are s_1 and s_5 respectively, which are in S_2^1 whereas under b-column, the corresponding states are s_3 and s_3 that are in S_1^1. Therefore, s_2 and s_4 are 1-equivalent. Again under a-column the states corresponding to s_2 and s_7 are s_1 and s_3 with $s_1 \in S_2^1$ and $s_3 \in S_1^1$. So, s_2 and s_7 are not 1-equivalent. Hence, we get $S_3^1 = \{s_2, s_4\}$ and $S_4^1 = \{s_7\}$. Therefore,

$$\pi_1 = \{\{s_3\}, \{s_0, s_1, s_5, s_6\}, \{s_2, s_4\}, \{s_7\}\}$$

Consider s_1 and s_5 of S_2^1. From the transition table, it is clear that under a-column, the states corresponding to s_1, s_5 are s_2 and s_4 respectively that are in S_3^1. Again, the states under b-column corresponding to s_1, s_5 are s_0 and s_6 respectively that are in S_2^1. So, s_1, s_5 are 2-equivalent. Similarly, it can be shown that s_0, s_6 are 2-equivalent. Therefore, $S_2^2 = \{\{s_1, s_5\}, \{s_0, s_6\}\}$. Also s_2, s_4 are 2-equivalent. So $S_3^2 = \{s_2, s_4\}$. As $S_1^1 = \{s_3\}$ and $S_4^1 = \{s_7\}$ contains only one state, so we have $S_1^2 = \{s_3\}$ and $S_4^2 = \{s_7\}$. Therefore, we get

$$\pi_2 = \{\{s_3\}, \{s_0, s_6\}, \{s_1, s_5\}, \{s_2, s_4\}, \{s_7\}\}$$

Similarly, it can be shown that s_0, s_6; s_1, s_5 and s_2, s_4 are 3-equivalent. Therefore,

$$\pi_3 = \{\{s_3\}, \{s_0, s_6\}, \{s_1, s_5\}, \{s_2, s_4\}, \{s_7\}\} = \pi_2$$

Hence, the minimum state automaton is given as $M_1 = (S', \Sigma, \delta', s_0', A')$, where $S' = \{[s_3], [s_0, s_6], [s_1, s_5], [s_2, s_4], [s_7]\}$; $\Sigma = \{a, b\}$; $s_0' = [s_0, s_6]$ and $A' = [s_3]$. The transition function δ' is given in the following table.

Present State	Input	
	a	*b*
$\rightarrow [s_0, s_6]$	$[s_0, s_6]$	$[s_1, s_5]$
$[s_1, s_5]$	$[s_2, s_4]$	$[s_0, s_6]$
$[s_2, s_4]$	$[s_1, s_5]$	$[s_3]$
$[s_3]$	$[s_0, s_6]$	$[s_3]$
$[s_7]$	$[s_3]$	$[s_0, s_6]$

The transition diagram is given below.

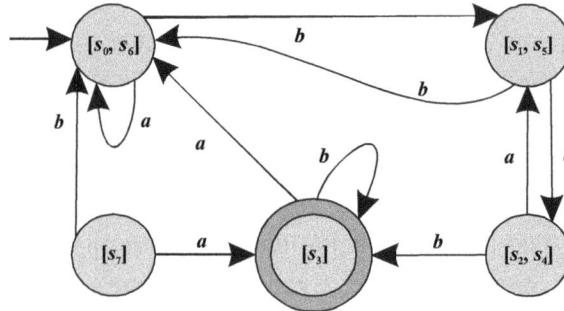

Construct the minimum state automaton equivalent to a given automaton M whose state table is given below. It is given that s_6 is the final state.

States	Input	
	0	**1**
$\to s_0$	s_0	s_3
s_1	s_2	s_5
s_2	s_3	s_4
s_3	s_0	s_5
s_4	s_0	s_6
s_5	s_1	s_4
s_6	s_1	s_3

Solution   From the given transition table we have $S_1^0 = A = \{s_6\}$ and $S_2^0 = S - A = \{s_0, s_1, s_2, s_3, s_4, s_5\}$. Therefore, we have

$$\pi_0 = \{S_1^0, S_2^0\} = \{\{s_6\}, \{s_0, s_1, s_2, s_3, s_4, s_5\}\}$$

Since $S_1^0 = \{s_6\}$ has only one state, so $S_1^1 = \{s_6\}$. Now consider s_0 and s_1 of S_2^0. From the state table it is clear that under 0-column, the state corresponding to s_0 and s_1 are s_0 and s_2 respectively, which are belongs to S_2^0. On the other hand states under 1-column coresponding to s_0 and s_1 are s_3 and s_5 repectively, which are also belongs to S_2^0. Therefore, s_0 is 1-equivalent to s_1. Similarly, it can be shown that s_0 is 1-equivalent to s_2, s_3 and s_5 but not to s_4. Therefore, $S_2^1 = \{s_0, s_1, s_2, s_3, s_5\}$ and $S_3^1 = \{s_4\}$. Hence, we get

$$\pi_1 = \{S_1^1, S_2^1, S_3^1\} = \{\{s_6\}, \{s_0, s_1, s_2, s_3, s_5\}, \{s_4\}\}$$

Now consider s_0 and s_1 of S_2^1. From the state table it is clear that the states under 0-column corresponding to s_0 and s_1 are s_0 and s_2 respectively, which are belonging to S_2^1. On the other hand states under 1-column corresponding to s_0 and s_1 are s_3 and s_5 respectively, which are also belonging to S_2^1. Therefore, s_0 is 2-equivalent to s_1. Proceeding in this manner it can be shown that s_0 is 2-equivalent to s_3 but not to s_2 and s_5. Hence we get, $S_2^2 = \{\{s_0, s_1, s_3\}, \{s_2, s_5\}\}$. Again, $S_1^1 = \{s_6\}$ and $S_3^1 = \{s_4\}$ each contains only one state, so we must have $S_1^2 = \{s_6\}$ and $S_3^2 = \{s_4\}$. Therefore, we have

$$\pi_2 = \{\{s_6\}, \{s_0, s_1, s_3\}, \{s_4\}, \{s_2, s_5\}\} = \{S_1^2, S_2^2, S_3^2, S_4^2\} \quad \text{(Say)}$$

Since $S_1^2 = \{s_6\}$ contains only state, we have $S_1^3 = \{s_6\}$. Similarly, we have $S_3^3 = \{s_4\}$. Now consider s_0 and s_1 of S_2^2. From the state table it is clear that the states under 0-column corresponding to s_0 and s_1 are s_0 and s_2 respectively with $s_0 \in S_2^2$ and $s_2 \in S_4^2$. Therefore, s_0 is not 3-equivalent to s_1. Similarly, it can be shown that s_0 is not 3-equivalent to s_3. Again, it can be shown that s_1 is not 3-equivalent to s_3. Also, it can be shown that s_2 is 3-equivalent to s_5. So we have $S_2^3 = \{\{s_0\}, \{s_1\}, \{s_3\}\}$ and $S_4^3 = \{s_2, s_5\}$. Therefore, we get

$$\pi_3 = \{\{s_6\}, \{\{s_0\}, \{s_1\}, \{s_3\}\}, \{s_4\}, \{s_2, s_5\}\} = \{S_1^3, S_2^3, S_3^3, S_4^3\}$$

i.e., $\quad \pi_3 = \{\{s_6\}, \{s_0\}, \{s_4\}, \{s_2, s_5\}, \{s_1\}, \{s_3\}\} = \{S_1^3, S_2^3, S_3^3, S_4^3, S_5^3, S_6^3\} \quad \text{(Say)}$

Now consider s_2 and s_5 of S_4^3. From the state table it is clear that under 0-column, the states corresponding to s_2 and s_5 are s_3 and s_1 respectively with $s_3 \in S_6^3$ and $s_1 \in S_5^3$. Therefore, s_2 is not 4-equivalent to s_5. Hence, we get

$$\pi_4 = \{\{s_6\}, \{s_0\}, \{s_4\}, \{s_2\}, \{s_5\}, \{s_1\}, \{s_3\}\} = S$$

In this case we have $\pi_4 = S$. Therefore, the minimum state automaton is the given automaton itself. The tansition diagram is given below.

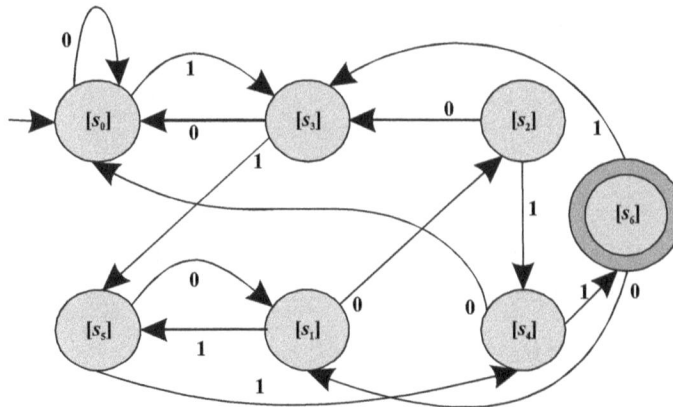

Example 13

Construct a Moore machine equivalent to the finite automata $M = (\{s_1, s_2, s_3, s_4, s_5\}, \{0,1\}, \delta, s_1, \{s_5\})$ whose transition function is given as $\delta(s_1,0) = s_2$; $\delta(s_1,1) = s_4$; $\delta(s_2,0) = s_5$; $\delta(s_2,1) = s_3$; $\delta(s_3,0) = s_3$; $\delta(s_3,1) = s_3$; $\delta(s_4,0) = s_5$; $\delta(s_4,1) = s_2$; $\delta(s_5,0) = s_5$ and $\delta(s_5,1) = s_3$

Solution   Here, $S = \{s_1, s_2, s_3, s_4, s_5\}$; $A = \{s_5\}$. Let $\Delta = \{0,1\}$ be the introduced alphabet. The output is defined by output function $\lambda(s)$. Therefore, we get

$$\lambda(s_1) = \lambda(s_2) = \lambda(s_3) = \lambda(s_4) = 0 \quad [\because s_1, s_2, s_3, s_4 \notin A] \text{ and}$$

$$\lambda(s_5) = 1 \qquad\qquad [\because s_5 \in A]$$

Therefore, the transition table of the equivalent constructed Moore machine for the given transition function is given below.

State	Next state		Output
	0	**1**	
$\rightarrow s_1$	s_2	s_4	0
s_2	s_5	s_3	0
s_3	s_3	s_3	0
s_4	s_5	s_2	0
s_5	s_5	s_3	1

Example 14

Construct a Moore machine equivalent to the finite automaton *M* whose transition diagram is given below.

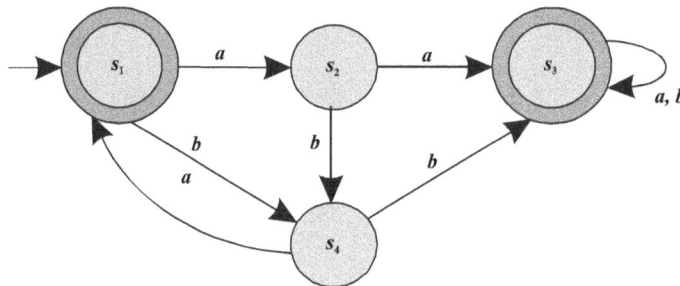

Solution   Here, $S = \{s_1, s_2, s_3, s_4\}$; $A = \{s_1, s_3\}$. Let the introduced alphabet be $\Delta = \{0,1\}$. The output is defined by output function $\lambda(s)$. Therefore, we get

$$\lambda(s_2) = 0 = \lambda(s_4) \quad [\because s_2, s_4 \notin A] \text{ and}$$

$$\lambda(s_1) = 1 = \lambda(s_3) \qquad [\because s_1, s_3 \in A]$$

Therefore, the transition table of equivalent constructed Moore machine for the given transition diagram is given below.

State	Next state		Output
	0	**1**	
$\rightarrow s_1$	s_2	s_4	1
s_2	s_3	s_1	0
s_3	s_3	s_3	1
s_4	s_1	s_3	0

Example 15

Construct a Mealy machine equivalent to the Moore machine *M* given in the following transition table.

State	Next state		Output
	0	**1**	
$\rightarrow s_0$	s_1	s_2	1
s_1	s_3	s_2	0
s_2	s_2	s_1	1
s_3	s_0	s_3	1

Solution Consider the Moore machine $M = (S, \Sigma, \Delta, \delta, \lambda, s_0)$ whose transition function δ is defined in the above table. Let the equivalent Mealy machine $M_1 = (S, \Sigma, \Delta, \delta, \lambda', s_0)$ such that $\lambda'(s,a) = \lambda(\delta(s,a))$ for $a \in \Sigma$. Therefore, we get

$$\lambda'(s_0, 0) = \lambda(\delta(s_0, 0)) = \lambda(s_1) = 0; \quad \lambda'(s_0, 1) = \lambda(\delta(s_0, 1)) = \lambda(s_2) = 1$$
$$\lambda'(s_1, 0) = \lambda(\delta(s_1, 0)) = \lambda(s_3) = 1; \quad \lambda'(s_1, 1) = \lambda(\delta(s_1, 1)) = \lambda(s_2) = 1$$
$$\lambda'(s_2, 0) = \lambda(\delta(s_2, 0)) = \lambda(s_2) = 1; \quad \lambda'(s_2, 1) = \lambda(\delta(s_2, 1)) = \lambda(s_1) = 0$$
$$\lambda'(s_3, 0) = \lambda(\delta(s_3, 0)) = \lambda(s_0) = 1; \quad \lambda'(s_3, 1) = \lambda(\delta(s_3, 1)) = \lambda(s_3) = 1$$

Therefore, the transition table of the equivalent Mealy machine is given below.

Present state	Input (0)		Input (1)	
	State	**Output**	**State**	**Output**
$\rightarrow s_0$	s_1	0	s_2	1
s_1	s_3	1	s_2	1
s_2	s_2	1	s_1	0
s_3	s_0	1	s_3	1

Example 16

Construct a Mealy machine equivalent to the Moore machine $M = (\{s_0, s_1, s_2, s_3\},$ $\{a, b\}, \{0,1\}, \delta, \lambda, s_0)$ whose transition function is given in the following table.

Present state	Next state		Output
	a	b	
$\rightarrow s_0$	s_3	s_1	0
s_1	s_1	s_2	1
s_2	s_2	s_3	0
s_3	s_3	s_0	0

Solution   Consider the Moore machine $M = (S, \Sigma, \Delta, \delta, \lambda, s_0)$ whose transition function δ is defined in the above table. Let the equivalent Mealy machine $M_1 = (S, \Sigma, \Delta, \delta, \lambda', s_0)$ such that $\lambda'(s,a) = \lambda(\delta(s,a))$ for $a \in \Sigma$. Therefore, we get

$$\lambda'(s_0, a) = \lambda(\delta(s_0, a)) = \lambda(s_3) = 0; \quad \lambda'(s_0, b) = \lambda(\delta(s_0, b)) = \lambda(s_1) = 1$$
$$\lambda'(s_1, a) = \lambda(\delta(s_1, a)) = \lambda(s_1) = 1; \quad \lambda'(s_1, b) = \lambda(\delta(s_1, b)) = \lambda(s_2) = 0$$
$$\lambda'(s_2, a) = \lambda(\delta(s_2, a)) = \lambda(s_2) = 0; \quad \lambda'(s_2, b) = \lambda(\delta(s_2, b)) = \lambda(s_3) = 0$$
$$\lambda'(s_3, a) = \lambda(\delta(s_3, a)) = \lambda(s_3) = 0; \quad \lambda'(s_3, b) = \lambda(\delta(s_3, b)) = \lambda(s_0) = 0$$

Therefore, the transition table of the equivalent Mealy machine is given below.

Present state	Input (0)		Input (1)	
	State	Output	State	Output
$\rightarrow s_0$	s_3	0	s_1	1
s_1	s_1	1	s_2	0
s_2	s_2	0	s_3	0
s_3	s_3	0	s_0	0

Example 17

Given $M = (\{s_1, s_2, s_3\}, \{0,1\}, \delta, s_1, \{s_3\})$, where the transition function is given by $\delta(s_1,0) = \{s_2, s_3\}$; $\delta(s_1,1) = \{s_1\}$; $\delta(s_2,0) = \{s_1, s_2\}$; $\delta(s_2,1) = \phi$; $\delta(s_3,0) = \{s_2\}$ and $\delta(s_3,1) = \{s_1, s_2\}$. Construct a deterministic finite automaton.

Solution   Given nondeterministic finite automata $M = (\{s_1, s_2, s_3\}, \{0,1\}, \delta, s_1, \{s_3\})$, where the transition function δ is defined as $\delta(s_1,0) = \{s_2, s_3\}$; $\delta(s_1,1) = \{s_1\}$; $\delta(s_2,0) = \{s_1, s_2\}$; $\delta(s_2,1) = \phi$; $\delta(s_3,0) = \{s_2\}$ and $\delta(s_3,1) = \{s_1, s_2\}$.

Let the equivalent deterministic finite automata be $M_1 = (S', \Sigma, \delta', s_1', A')$ where

i.   The state S' are subsets of $\{s_1, s_2, s_3\}$.

ii. The initial state is given as $s_1' = [s_1]$

iii. The final states are $[s_3], [s_1, s_3], [s_2, s_3]$ and $[s_1, s_2, s_3]$

iv. The transition function δ' is defined below.

$$\delta'([s_1], 0) = \delta(s_1, 0) = [s_2, s_3]$$
$$\delta'([s_1], 1) = \delta(s_1, 1) = [s_1]$$
$$\delta'([s_2, s_3], 0) = \delta(s_2, 0) \cup \delta(s_3, 0) = [s_1, s_2]$$
$$\delta'([s_2, s_3], 1) = \delta(s_2, 1) \cup \delta(s_3, 1) = [s_1, s_2]$$
$$\delta'([s_1, s_2], 0) = \delta(s_1, 0) \cup \delta(s_2, 0) = [s_1, s_2, s_3]$$
$$\delta'([s_1, s_2], 1) = \delta(s_1, 1) \cup \delta(s_2, 1) = [s_1]$$
$$\delta'([s_1, s_2, s_3], 0) = \delta(s_1, 0) \cup \delta(s_2, 0) \cup \delta(s_3, 0) = [s_1, s_2, s_3]$$
$$\delta'([s_1, s_2, s_3], 1) = \delta(s_1, 1) \cup \delta(s_2, 1) \cup \delta(s_3, 1) = [s_1, s_2]$$

The transition table of deterministic finite automata is given below.

State	Input	
	0	1
$\rightarrow [s_1]$	$[s_2, s_3]$	$[s_1]$
$[s_2, s_3]$	$[s_1, s_2]$	$[s_1, s_2]$
$[s_1, s_2]$	$[s_1, s_2, s_3]$	$[s_1]$
$[s_1, s_2, s_3]$	$[s_1, s_2, s_3]$	$[s_1, s_2]$

The transition diagram of the equivalent deterministic finite automata of the given nondeterministic finite automata M is described below, where the final states are $[s_2, s_3]$ and $[s_1, s_2, s_3]$.

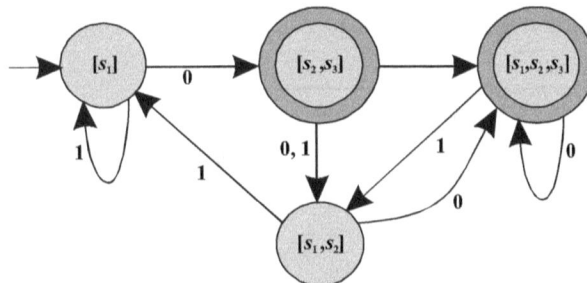

Example 18

Construct a Moore machine equivalent to the Mealy machine $M = (\{s_1, s_2, s_3, s_4\}, \{0,1\}, \{0,1\}, \delta, \lambda, s_1)$, whose transition function δ is given in the following table.

Present state	Input (0)		Input (1)	
	State	Output	State	Output
$\rightarrow s_1$	s_1	1	s_2	0
s_2	s_4	1	s_4	1
s_3	s_2	1	s_3	1
s_4	s_3	0	s_1	1

Solution   Here, $S = \{s_1, s_2, s_3, s_4\}; \Sigma = \{0,1\}$ and $\Delta = \{0,1\}$. Let the equivalent Moore machine be $M_1 = (S', \Sigma, \Delta, \delta', \lambda', s_1')$ such that

i. $S' = \{[s_1,0], [s_1,1], [s_2,0], [s_2,1], [s_3,0], [s_3,1], [s_4,0], [s_4,1]\}$.

ii.   Initial state $s_1' = [s_1,0]$. However, we can choose $[s_1,1]$ as also the initial state.

iii.   The output function λ' is defined below.

$\lambda'[s_1,0] = 0;$ $\lambda'[s_1,1] = 1;$ $\lambda'[s_2,0] = 0;$ $\lambda'[s_2,1] = 1;$
$\lambda'[s_3,0] = 0;$ $\lambda'[s_3,1] = 1;$ $\lambda'[s_4,0] = 0;$ $\lambda'[s_4,1] = 1.$

iv.   The transition function δ' is defined below.

$\delta'([s_1,0],0) = [\delta(s_1,0), \lambda(s_1,0)] = [s_1,1]; \; \delta'([s_1,0],1) = [\delta(s_1,1), \lambda(s_1,1)] = [s_2,0]$
$\delta'([s_1,1],0) = [\delta(s_1,0), \lambda(s_1,0)] = [s_1,1]; \; \delta'([s_1,1],1) = [\delta(s_1,1), \lambda(s_1,1)] = [s_2,0]$

Similarly, we will get

$$\delta'([s_2,0],0) = [s_4,1] = \delta'([s_2,1],0); \; \delta'([s_2,0],1) = [s_4,1] = \delta'([s_2,1],1)$$
$$\delta'([s_3,0],0) = [s_2,1] = \delta'([s_3,1],0); \; \delta'([s_3,0],1) = [s_3,1] = \delta'([s_3,1],1)$$
$$\delta'([s_4,0],0) = [s_3,0] = \delta'([s_4,1],0); \; \delta'([s_4,0],1) = [s_1,1] = \delta'([s_4,1],1)$$

Therefore, the transition diagram of the equivalent Moore machine is defined below.

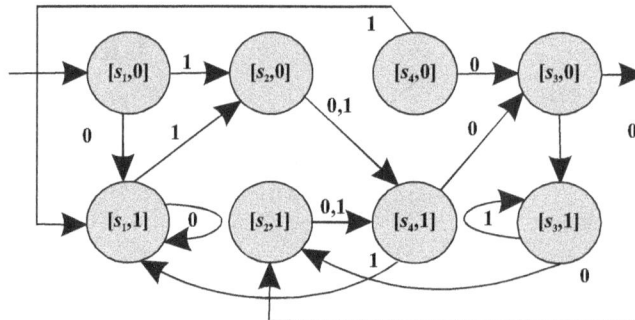

The transition table of the equivalent Moore machine is given below.

Present state	Next state		Output
	Input (0)	Input (1)	
$\rightarrow [s_1,0]$	$[s_1,1]$	$[s_2,0]$	0
$[s_1,1]$	$[s_1,1]$	$[s_2,0]$	1
$[s_2,0]$	$[s_4,1]$	$[s_4,1]$	0
$[s_2,1]$	$[s_4,1]$	$[s_4,1]$	1
$[s_3,0]$	$[s_2,1]$	$[s_3,1]$	0
$[s_3,1]$	$[s_2,1]$	$[s_3,1]$	1
$[s_4,0]$	$[s_3,0]$	$[s_1,1]$	0
$[s_4,1]$	$[s_3,0]$	$[s_1,1]$	1

Example 19

Construct the minimum state automaton equivalent to a given automaton M whose transition table is given below, where s_2 is the final state.

Present state	Next state	
	Input (a)	Input (b)
$\rightarrow s_0$	s_1	s_5
s_1	s_6	s_2
s_2	s_0	s_2
s_3	s_2	s_6
s_4	s_1	s_6
s_5	s_2	s_6
s_6	s_6	s_4
s_7	s_6	s_2

Solution Here, $S_1^0 = \{s_2\} = A$ is the final state and $S_2^0 = \{s_0, s_1, s_3, s_4, s_5, s_6, s_7\}$ is the set of non-final states. Therefore, we have $\pi_0 = (S_1^0, S_2^0) = \{\{s_2\}, \{s_0, s_1, s_3, s_4, s_5, s_6, s_7\}\}$

Since, $S_1^0 = \{s_2\}$ contains only one state, so $S_1^1 = \{s_2\}$. Now, consider s_0 and s_1 of S_2^0. The states under a-column corresponding to s_0 and s_1 are s_1 and s_6 respectively with $s_1, s_6 \in S_2^0$ whereas the states under b-column corresponding to s_0 and s_1 are s_5 and s_2 respectively with $s_5 \in S_2^0$ and $s_2 \in S_1^0$.

Therefore, s_0 is not 1-equivalent to s_1. Similarly, it can be shown that s_0 is not 1-equivalent to s_3, s_5 and s_7. Consider s_0 and s_4 of S_2^0. The states under a-column corresponding to s_0 and s_4 are s_1 and s_1 respectively with $s_1 \in S_2^0$ whereas the states under b-column corresponding to s_0 and s_4 are s_5 and s_6 respectively with $s_5, s_6 \in S_2^0$. Therefore, s_0 is 1-equivalent to s_4. Similarly, it can be shown that s_0 is 1-equivalent to s_6. Therefore, we have $S_2^1 = \{\{s_0, s_4, s_6\}, \{s_1, s_3, s_5, s_7\}\}$. Therefore, we get

$$\pi_1 = \{\{s_2\}, \{s_0, s_4, s_6\}, \{s_1, s_3, s_5, s_7\}\} = (S_1^1, S_2^1, S_3^1) \text{ (Say)}$$

Since, $S_1^1 = \{s_2\}$ contains only one state, so $S_1^2 = \{s_2\}$. Consider s_0 and s_4 of S_2^1. The states under a-column corresponding to s_0 and s_4 are s_1 and s_1 respectively with $s_1 \in S_3^1$ whereas the states under b-column corresponding to s_0 and s_4 are s_5 and s_6 respectively with $s_5 \in S_3^1$ and $s_6 \in S_2^1$. So s_0 is not 2-equivalent to s_4. Similarly it can be shown that s_0 is not 2-equivalent to s_6. Also it can be shown that s_4 is not 2-equivalent to s_6. Therefore, we have $S_2^2 = \{\{s_0\}, \{s_4\}, \{s_6\}\}$. Again, on considering s_1 and s_3 of S_3^1. The states under a-column corresponding to s_1 and s_3 are s_6 and s_2 respectively with $s_6 \in S_2^1$ and $s_2 \in S_1^1$. So, s_1 is not 2-equivalent to s_3. Similarly, it can be shown that s_1 is not 2-equivalent to s_5. Also, it can be shown that s_1 is 2-equivalent to s_7. Therefore, we get $S_3^2 = \{\{s_1, s_7\}, \{s_3, s_5\}\}$. Thus, we have

$$\pi_2 = \{\{s_2\}, \{s_0\}, \{s_4\}, \{s_6\}, \{s_1, s_7\}, \{s_3, s_5\}\}$$

Proceeding in this manner it can be shown that $\pi_2 = \{\{s_2\}, \{s_0\}, \{s_4\}, \{s_6\}, \{s_1, s_7\}, \{s_3, s_5\}\} = \pi_3$

Therefore, the minimum state automaton is given as $M_1 = (S', \Sigma, \delta', s_0', A')$ such that $S' = ([s_2], [s_0], [s_4], [s_6], [s_1, s_7], [s_3, s_5]); \Sigma = \{a, b\}; s_0' = [s_0]; A' = [s_2]$ and δ' is described in the following state table.

Present state	Next state	
	a	**b**
$\rightarrow [s_0]$	$[s_1, s_7]$	$[s_3, s_5]$
$[s_2]$	$[s_0]$	$[s_2]$
$[s_4]$	$[s_1, s_7]$	$[s_6]$
$[s_6]$	$[s_6]$	$[s_4]$
$[s_1, s_7]$	$[s_6]$	$[s_2]$
$[s_3, s_5]$	$[s_2]$	$[s_6]$

The transition diagram of the minimum state automaton is as follows:

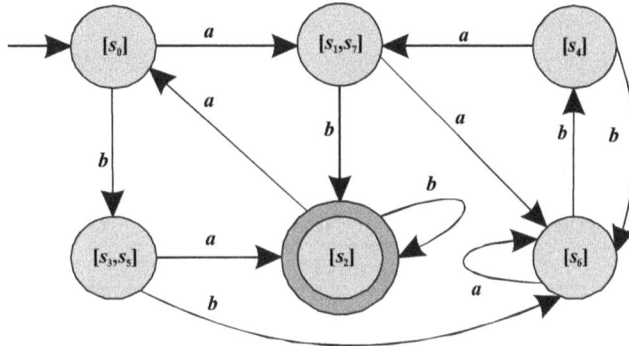

Example 20

Construct a Moore machine equivalent to the Mealy machine whose transition function is given in the following table.

Present state	Input (a)		Input (b)	
	State	**Output**	**State**	**Output**
$\rightarrow s_1$	s_3	a	s_2	a
s_2	s_1	b	s_4	a
s_3	s_2	b	s_1	b
s_4	s_4	b	s_3	a

Solution   Here, $S = \{s_1, s_2, s_3, s_4\}; \Sigma = \{a, b\}$ and $\Delta = \{a, b\}$. Let the equivalent Moore machine be $M_1 = (S', \Sigma, \Delta, \delta', \lambda', s_1')$ such that

i. $S' = \{[s_1, a], [s_1, b], [s_2, a], [s_2, b], [s_3, a], [s_3, b], [s_4, a], [s_4, b]\}$

ii. $s_1' = [s_1, a]$. However, we can choose $[s_1, b]$ as also the initial state.

iii. The output function λ' is defined below.

$\lambda'[s_1, a] = a; \quad \lambda'[s_1, b] = b; \quad \lambda'[s_2, a] = a; \quad \lambda'[s_2, b] = b;$
$\lambda'[s_3, a] = a; \quad \lambda'[s_3, b] = b; \quad \lambda'[s_4, a] = a; \quad \lambda'[s_4, b] = b.$

iv. The transition function δ' is defined below.

$$\delta'([s_1, a], a) = [\delta(s_1, a), \lambda(s_1, a)] = [s_3, a];$$
$$\delta'([s_1, a], b) = [\delta(s_1, b), \lambda(s_1, b)] = [s_2, a]$$
$$\delta'([s_1, b], a) = [\delta(s_1, a), \lambda(s_1, a)] = [s_3, a];$$
$$\delta'([s_1, b], b) = [\delta(s_1, b), \lambda(s_1, b)] = [s_2, a]$$

Similarly, we will get

$$\delta'([s_2,a],a)=[s_1,b]=\delta'([s_2,b],a); \qquad \delta'([s_2,a],b)=[s_4,a]=\delta'([s_2,b],b)$$

$$\delta'([s_3,a],a)=[s_2,b]=\delta'([s_3,b],a); \qquad \delta'([s_3,a],b)=[s_1,b]=\delta'([s_3,b],b)$$

$$\delta'([s_4,a],a)=[s_4,b]=\delta'([s_4,b],a); \qquad \delta'([s_4,a],b)=[s_3,a]=\delta'([s_4,b],b)$$

Therefore, the state table of the equivalent Moore machine is given below.

Present state	Next state		Output
	Input (a)	Input (b)	
$\rightarrow [s_1,a]$	$[s_3,a]$	$[s_2,a]$	a
$[s_1,b]$	$[s_3,a]$	$[s_2,a]$	b
$[s_2,a]$	$[s_1,b]$	$[s_4,a]$	a
$[s_2,b]$	$[s_1,b]$	$[s_4,a]$	b
$[s_3,a]$	$[s_2,b]$	$[s_1,b]$	a
$[s_3,b]$	$[s_2,b]$	$[s_1,b]$	b
$[s_4,a]$	$[s_4,b]$	$[s_3,a]$	a
$[s_4,b]$	$[s_4,b]$	$[s_3,a]$	b

The transition diagram of the moore machine is given below.

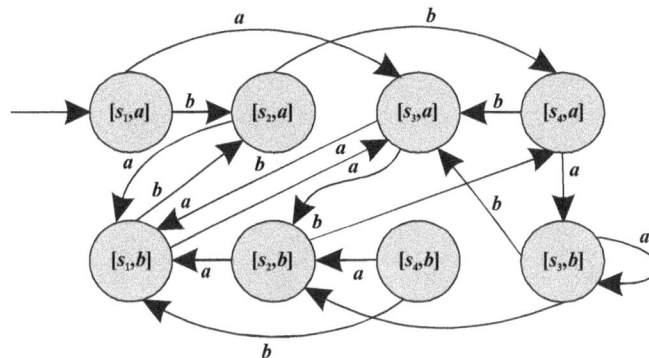

REVIEW QUESTIONS

1. Define finite state machine and give an account of its different characteristics.

2. Define a deterministic finite automaton and explain its different components with a neat diagram.

3. Explain transition diagram and transition table with an example.

4. Prove that if a language L is accepted by a nondeterministic finite automaton, then there exists a deterministic finite automaton that accepts L.

5. Define a nondeterministic finite automaton and explain how it is different from deterministic finite automaton.

6. State and prove the properties of transition functions.

7. Define finite automata with ε-moves and explain ε-closure of a state by considering an example.

8. Write a note on two-way finite automata.

9. Explain Moore machine with an example. Also explain the construction of Moore machine from Mealy machine.

10. Define Mealy machine and explain the construction of Mealy machine from Moore machine.

11. Explain the construction of Moore machine from finite automata.

12. What do you mean by minimization of finite automata and write the different steps used to construct a minimum state automaton.

13. Explain the construction of Mealy machine from finite automaton.

PROBLEMS

1. Show that the string $w = 02102112$ is accepted by the following deterministic finite automata whose transition graph is given below.

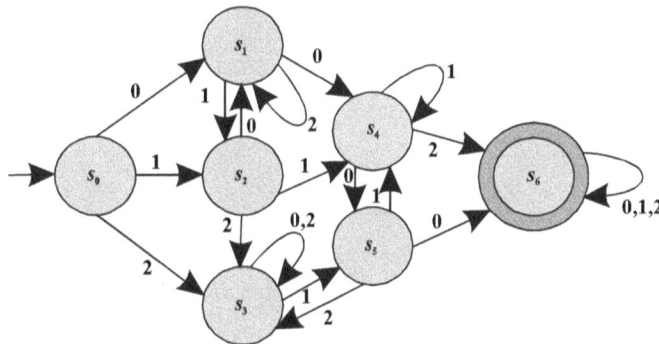

2. Determine the initial state, final state, intermediate state and input alphabet for the following transition diagrams.

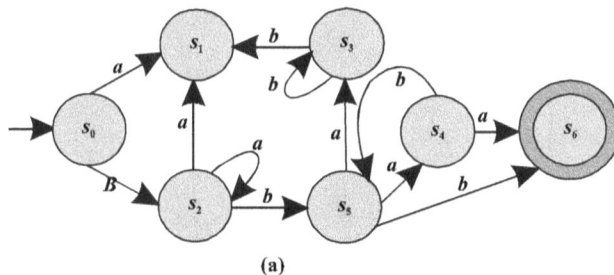

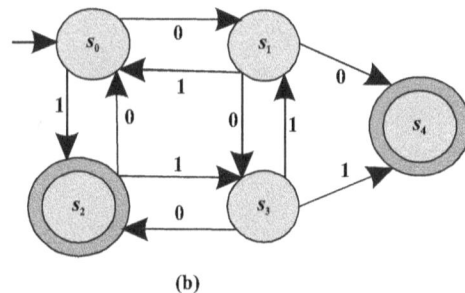

(a)   (b)

3. Consider the deterministic finite automata whose transition graph is given below. Check the acceptability of the following strings by the given deterministic finite automata.

   (a) $w = aaab$ (b) $w = baaba$ (c) $w = aabbbaa$

   (d) $w = bbabb$ (e) $w = bbabba$ (f) $w = bbbaab$

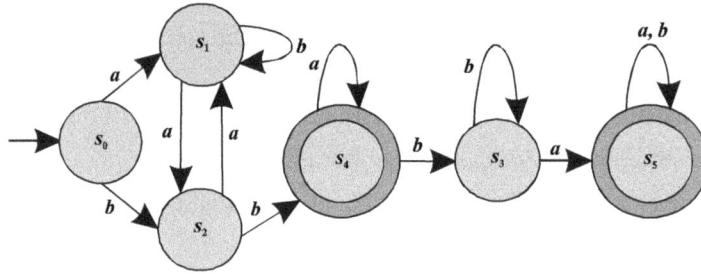

4. Find the language accepted by the following deterministic finite automata, whose transition diagrams are given below.

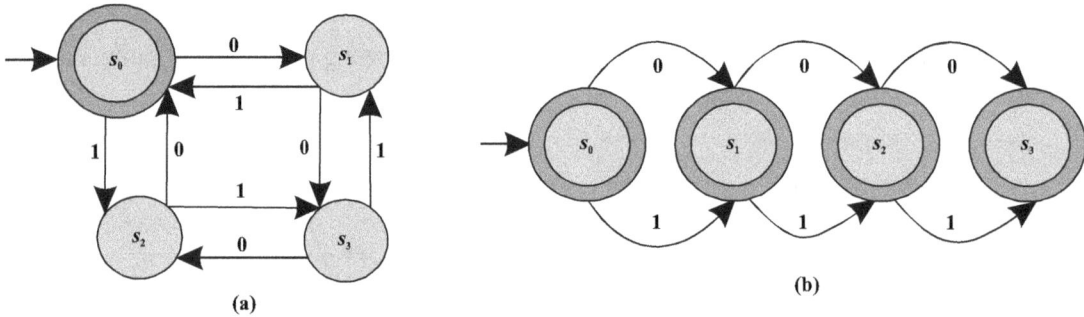

   (a)

   (b)

5. Show that the string $w = 00102012$ is accepted by the machine whose transition table is given below, where s_4 and s_5 are final states. Also, construct the transition graph.

Present state	Next state		
	0	**1**	**2**
$\rightarrow s_0$	$\{s_1\}$	$\{s_2\}$	$\{\cdots\}$
s_1	$\{s_2\}$	$\{s_1, s_4\}$	$\{s_4\}$
s_2	$\{s_4\}$	$\{s_1\}$	$\{s_3\}$
s_3	$\{s_2, s_3\}$	$\{s_3\}$	$\{s_5\}$
s_4	$\{\cdots\}$	$\{\cdots\}$	$\{s_5\}$
s_5	$\{\cdots\}$	$\{s_4\}$	$\{\cdots\}$

6. Find the language accepted by the following nondeterministic finite automata whose transition graph is given below.

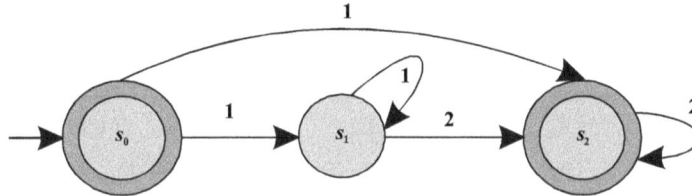

7. Construct the deterministic finite automata equivalent to the nondeterministic finite automata whose transition graph is given below.

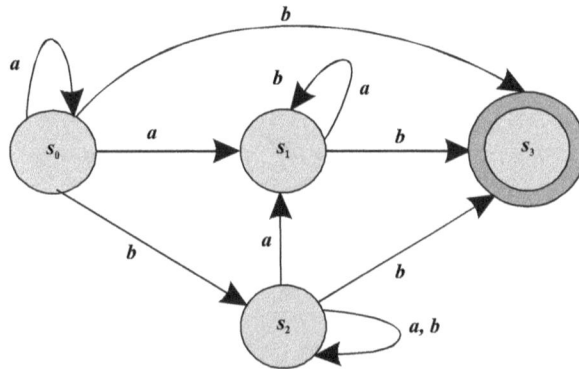

8. Construct a finite automaton accepting all strings over $\{a, b\}$ ending with *aba* or *aaba*.

9. Construct a finite automaton accepting all strings over $\Sigma = \{0, 1\}$ having odd number of 0's.

10. Construct a deterministic finite automaton equivalent to the nondeterministic finite automaton $M = (\{s_0, s_1, s_2\}, \{0,1\}, \delta, s_0, \{s_2\})$, where the transition function δ is described in the following table.

Present state	Next state	
	0	**1**
$\to s_0$	$\{s_0, s_1\}$	$\{s_2\}$
s_1	$\{s_0\}$	$\{s_1, s_2\}$
s_2	$\cdots$	$\{s_2\}$

11. Construct a nondeterministic finite automaton accepting the set of all strings over $\Sigma = \{0, 1\}$ ending in 101. Construct a deterministic finite automaton equivalent to the above nondeterministic finite automaton.

12. Construct a nondeterministic finite automaton accepting $\{01, 10\}$. Find a deterministic finite automaton accepting the same set.

13. Find a deterministic finite automaton equivalent to the nondeterministic finite automaton $M = (\{s_0, s_1, s_2, s_3\}, \{a, b, c\}, \delta, s_0, \{s_2\})$, where the transition function is given in the following table.

Present state	Next state		
	a	b	c
$\rightarrow s_0$	$\{s_0, s_1\}$	$\{s_1\}$	$\{s_1\}$
s_1	...	$\{s_1, s_2, s_3\}$	$\{s_0, s_1\}$
s_2	$\{s_2, s_3\}$	...	$\{s_0\}$
s_3	$\{s_1\}$	$\{s_3\}$	$\{s_2\}$

14. Construct a nondeterministic finite automaton for the following finite automaton with ε-moves.

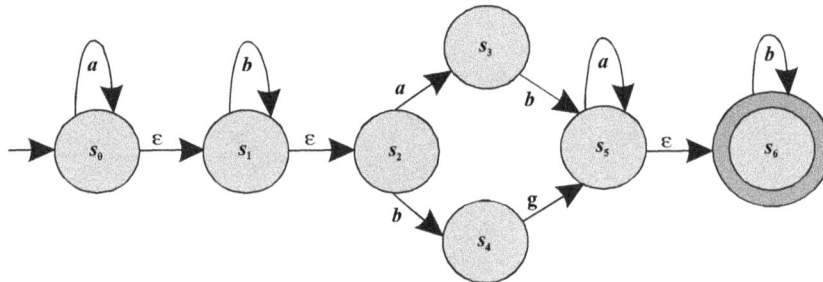

15. Construct a deterministic finite automaton for the given nondeterministic finite automata with ε-moves.

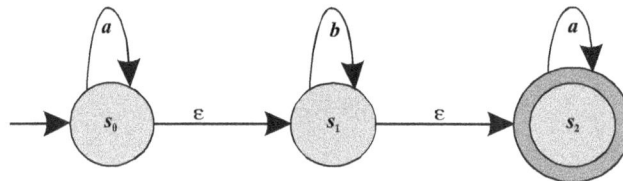

16. Construct a Moore machine equivalent to the finite automaton M whose transition graph is given below.

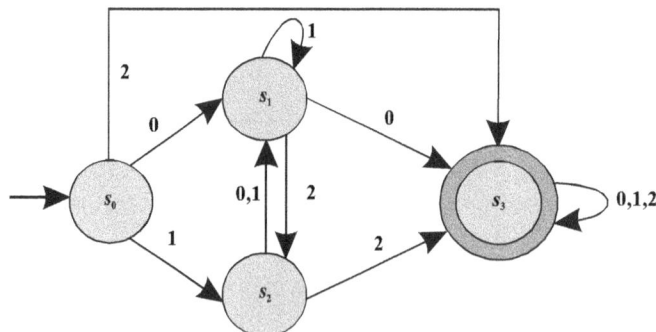

17. Obtain the minimum state automaton for the following finite automaton M whose state table is given below. It is given that s_3 is the final state.

Present state	Inputs	
	0	1
$\rightarrow s_0$	s_1	s_5
s_1	s_6	s_2
s_2	s_0	s_1
s_3	s_2	s_6
s_4	s_5	s_1
s_5	s_2	s_6
s_6	s_4	s_6
s_7	s_6	s_2

18. Convert the following Mealy machine whose transition table is given below into equivalent Moore machine.

Present state	Input (0)		Input (1)	
	State	Output	State	Output
$\rightarrow s_1$	s_3	1	s_2	1
s_2	s_1	0	s_4	1
s_3	s_4	1	s_1	0
s_4	s_2	0	s_3	1

19. Assume the finite automaton M whose transition diagram is given below. Construct a Moore machine that is equivalent to the given finite automaton M.

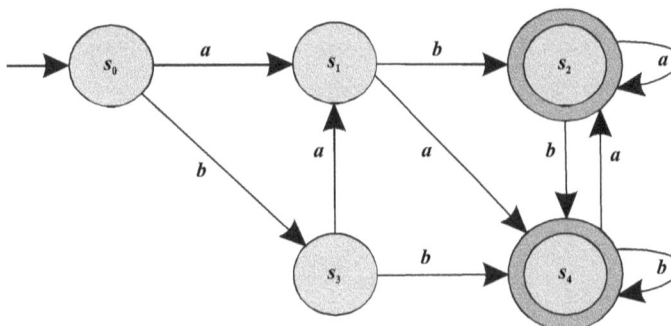

20. Prove that for every nondeterministic finite automaton there is an equivalent nondeterministic finite automata that has only one final state.

21. Can a deterministic finite automaton have an infinite computation? Justify your answer.

22. Find the minimum state automaton equivalent to the transition graph given below.

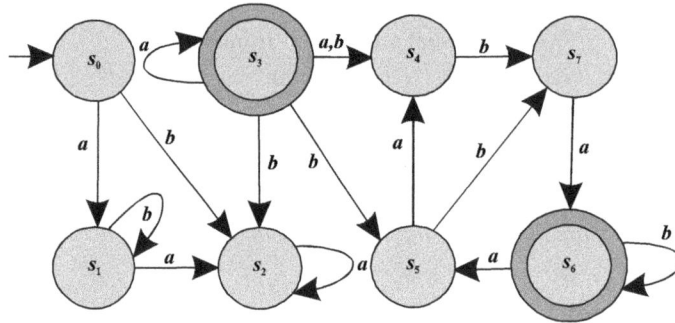

23. Find the equivalent Moore machine for the following Mealy machine whose transition table is given below.

Present state	Input (0)		Input (1)	
	State	Output	State	Output
$\rightarrow s_1$	s_1	0	s_2	0
s_2	s_4	0	s_4	0
s_3	s_2	0	s_3	1
s_4	s_3	1	s_1	1

24. Find the equivalent Mealy machine for the following Moore machine M whose transition table is given below.

Present state	Inputs		Output
	a	b	
$\rightarrow s_0$	s_1	s_2	1
s_1	s_3	s_2	0
s_2	s_2	s_3	1
s_3	s_0	s_1	1
s_4	s_3	s_4	0

25. Draw a deterministic finite automaton for the following languages over $\{0, 1\}$.

   (a) All strings with even number of zeros and even number of ones.

   (b) All strings of length at most seven.

   (c) All strings with odd number of ones and odd number zeros.

26. Convert the following Moore machine whose transition table is given below into an equivalent Mealy machine.

Present state	Inputs		Output
	a	b	
$\rightarrow s_0$	s_1	s_0	1
s_1	s_1	s_2	1
s_2	s_1	s_2	0

27. Consider the finite automaton M whose transition graph is given below. Construct a minimum state automaton equivalent to M.

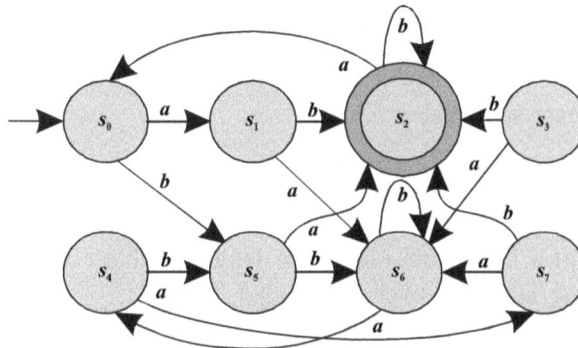

28. Given $M = (\{s_1, s_2, s_3\}, \{0,1\}, \delta, s_1, \{s_2\})$ where the transition function δ is given by $\delta(s_1, 0) = \{s_2\}$; $\delta(s_1, 1) = \{s_2, s_3\}$; $\delta(s_2, 0) = \{s_1, s_3\}$; $\delta(s_2, 1) = \phi$; $\delta(s_3, 0) = \{s_2\}$ and $\delta(s_3, 1) = \{s_1, s_2\}$. Construct an equivalent deterministic finite automaton.

29. Is an alphabet a language? Justify your answer.

30. Every nondeterministic finite automaton is a deterministic finite automaton. If yes, justify your answer.

31. Construct a deterministic finite automaton over $\{0, 1\}$ that accepts all strings w starts with a 0 and has odd length or starts with a 1 and has even length.

32. Construct a deterministic finite automaton over $\{0, 1\}$ that accepts all strings w that contains an odd number of 0's.

33. Construct a deterministic finite automaton over $\{0, 1\}$ that accepts all strings w that contains an even number of 1's or odd number of 0's.

34. Write a DFA that takes a binary string as input and accepts the string only if the last two digits are the same (either 00 or 11).

35. Write a DFA that takes a binary string as input and accepts it only if it does not end with two consecutive 0s.

36. Write a DFA that recognizes all binary strings that are made of alternating 0s and 1s (For example 010101, 10101010, etc.).

37. Write a DFA that takes a binary string as input and accepts the string only if it contains at least two 1s or at most three 0s.

38. Suppose a DFA has n states. Suppose it accepts a string of length $m > n$. Then prove that w contains at least one substring, which repeats more than once.

39. Write a DFA that takes a binary number as input and accepts it only if it is a multiple of 3.

40. Write a C program to implement a DFA for the above problems 34, 35 and 36.

41. Write a Mealy machine that takes a binary string as input and outputs the same string with a delay of 1 bit. The first output bit will always be 0. For example, if the given input is 11011, the output will be 01101.

42. Convert the above Mealy machine to Moore machine.

43. Give state diagram of a DFA recognizing the language that begins with a 1 and ends with a 0 over an alphabet $\Sigma = \{0,1\}$.

44. Give state diagram of a DFA recognizing the language that contains at least three 1s over an alphabet $\Sigma = \{0,1\}$.

45. Give state diagram of a DFA recognizing the language that contains substrings 0101, i.e., $z = w_1 0101 w_2$ for some w_1 and w_2 over an alphabet $\Sigma = \{0,1\}$.

46. Give state diagram of a NFA with three states recognizing the language that ends with 00 over an alphabet $\Sigma = \{0,1\}$.

47. Give state diagram of a NFA with five states recognizing the language that contains substrings 0101, i.e., $z = w_1 0101 w_2$ for some w_1 and w_2 over an alphabet $\Sigma = \{0,1\}$.

3

GRAMMAR AND
CHOMSKY CLASSIFICATION

3.0   INTRODUCTION

The study of formal languages is an area with a number of applications in computability theory and complexity theory. In 1950s linguists were trying to define a formal grammar to describe English. It was Noam Chomsky who gave a formal language theory in 1956. Though it was not useful for natural languages but it was most useful for computer languages. In this chapter we discuss the grammatical definitions of regular language, context-free language, context-sensitive language and recursively enumerable language. The details about these languages are studied in the subsequent chapters. These concepts of grammars are defined as potential models of natural languages. The four classes of languages are often called the Chomsky hierarchy, named after Noam Chomsky.

3.1   GRAMMAR

A grammar basically consists of a set of rules for transforming strings. A string in a language is generated by applying the set of rules any number of times, in any order starting with a string consisting of only a single start symbol. The set of all the strings that can be generated in this manner is known as a language. One particular string in the language can be generated by a particular sequence of legal choices taken during this rewriting process. If there is atleast one string in the language that can be generated in multiple ways, then the grammar is said to be ambiguous.

For example, consider the start symbol Q, the alphabet a and b with the rules

(i) $Q_0 \rightarrow aQ_0 b$ and   (ii) $Q_0 \rightarrow ab$

In order to obtain strings we start with the start symbol Q, and can choose a rule to apply to it. Also from the rules it is clear that the least length string is ab. If we start with the rule (i) we obtain the string aQb. Again, if we apply rule (ii) then we obtain the string $aabb$ whereas we obtain $aaQbb$ if we apply rule (i) again. This process can be repeated at until all occurrences of Q are removed, and only symbols from the alphabet remain. For example, if we choose the string $aaabbb,$ then more briefly we can write the derivation using symbols as $Q \rightarrow aQb \rightarrow aaQbb \rightarrow aaabbb.$

The language L of the above grammar is the set of all the strings that can be generated using this process and is given as $L = \{ab, aabb, aaabbb, \ldots\}$.

Therefore, the classic formalization of generative grammars G proposed by Noam Chomsky consists of the following components:

1. A non-empty finite set V_N of non-terminal symbols or variables.
2. A non-empty finite set T of terminal symbols that is disjoint from V_N.
3. A finite set P of production rules or productions, each rule of the form $\alpha \to \beta$ where $\alpha, \beta \in (V_N \cup T)^*$. That is, each production rule maps from one string of symbols α to another β, where the first string α contains at least one non-terminal symbol from V_N.
4. A distinguished symbol $Q_0 \in V_N$ is known as the start symbol.

Analytically, a grammar G is defined as the ordered qudrapule $G = (V_N, T, P, Q_o)$, where the notations are defined above.

Note We have the following observations regarding the productions of a grammar.

1. Reverse substitution of productions is not permitted. It indicates that, if $Q \to Q_1 Q_2$ is a production, then we can replace Q by $Q_1 Q_2$, but we cannot replace $Q_1 Q_2$ by Q.
2. Inverse operation of productions is not permitted. It indicates that, if $Q \to Q_1 Q_2$ is a production, then it is not necessary that $Q_1 Q_2 \to Q$ is a production.

3.2   CHOMSKY CLASSIFICATION

Noam Chomsky first formalized generative grammars and classified them into different types known as the Chomsky hierarchy. These classes of languages are defined as potential models of natural languages. The main difference between these types of grammars is due to strict production rules and can express fewer formal languages. The important types among these grammars are context-free grammars and regular grammars. The languages corresponding to such grammars are known as context-free languages and regular languages. These two restricted types of grammars are most often used because parsers for them can be efficiently implemented. For example, all regular languages can be recognized by a finite state machine, and for useful subsets of context-free grammars there are well-known algorithms to generate efficient LL parsers and LR parsers to recognize the corresponding languages those grammars generate.

3.2.1   Formal Grammar

Formal grammars consists of a finite set of terminal symbols, a finite set of variables or non-terminals, a set of production rules, and a start symbol. A production rule may be applied to a word by replacing the left-hand side by the right-hand side. In general, we apply a sequence of production rules to get a word. These sequence of rule applications is known as a derivation. Such a grammar defines the formal languages of all words consisting solely of terminal symbols that can be reached by a derivation from the start symbol. Broadly we state, a formal language L over an alphabet Σ is just a subset of Kleene star Σ^*, that is, a set of words over that alphabet. The adjective format is usually omitted as we do not deal with natural languages while studying computer science and mathematics.

Formal language rarely concerns itself with some particular languages except as examples, but is mainly concerned with the study of various types of formalisms to describe languages. For example, a language can be given as, those strings generated by some formal grammar; those strings described by a particular regular expression; those strings accepted by some automaton; those strings for which some decision procedure produces the answer.

For example, we can simply enumerate all well-formed words as finite language such as $L = \{a, abb, baa, cba\}$. However, even over a finite alphabet such as $\Sigma = \{a, b\}$ there are infinitely many words: "a", "ab", "ba", "aa", "bb", "aab"$\cdots$. Therefore, formal languages are typically infinite, and defining an infinite formal language is not as simple as writing $L = \{a, abb, baa, cba\}$.

3.2.2 The Hierarchy

In this section we discuss in brief the Chomsky hierarchy that consists of the four classification levels type 0, type 1, type 2 and type 3. The first classification (type 0) is the most general category in the hierarchy; the next category (type 1) is a subset of that one, and so on, down to the most restrictive category.

1. *Type-0 grammars* These are unrestricted grammars that include all formal grammars recognized by a Turing machine. These languages are otherwise known as recursively enumerable languages. It is to be noted that this is different from the recursive languages which can be decided by an always-halting Turing machine.

2. *Type-1 grammars* These are also known as context-sensitive grammars. They generate exactly context-sensitive languages that can be recognized by a linear-bounded automaton or non-deterministic Turing machine. These grammars have productions of the form $\alpha B\beta \to \alpha\gamma\beta$, where $B \in V_N$ is a non-terminal and α, β and γ are strings of terminals and non-terminals, i.e., $\alpha, \beta, \gamma \in (V_N \cup T)^*$. The strings α and β may be empty whereas γ must be non-empty. Also, the production $Q \to \wedge$ is allowed provided Q does not appear on the right side of any rule.

3. *Type-2 grammars* These are also known as context-free grammars and are used to derive a context-free language. These languages are exactly all languages that can be recognized by a non-deterministic pushdown automaton. These are defined by productions of the form $Q \to \alpha$ with $Q \in V_N$ a non-terminal and α a string of terminals and non-terminals, i.e., $\alpha \in (V_N \cup T)^*$. This is an important class of languages as it is the theoretical basis for the syntax of most programming languages.

4. *Type-3 grammars* This is the most specific grammar and is the smallest subset called as regular grammars. These are used to generate the regular languages. These languages are exactly all languages that can be recognized by a finite state automaton. Also, this family of formal languages can be obtained by regular expressions. Such a grammar restricts its productions to a single non-terminal on the left-hand side and a right-hand side consisting of a single terminal, possibly followed or preceded by a single non-terminal but not both. Analytically, the productions are of the form $Q \to a$ or $Q \to aQ_1$, where $a \in T$ and $Q, Q_1 \in V_N$. The production $Q \to \wedge$ is also allowed here if Q does not appear on the right side of any rule. This is also an important class of languages as these are used to define search patterns and the lexical structure of programming languages.

The following Figure 3.1 describes the types of languages and their corresponding automata. It also describes the relation among four types of languages, and the relation among automata. From Figure 3.1 it is clear that, every regular language is context-free, every context-free language is context-sensitive and every context-sensitive language is recursively enumerable. It implies that the function carried out by a finite automaton can also be carried out by a pushdown automaton. Similarly, the function carried out by a pushdown automaton can be carried out by a linear-bounded automaton and the function carried out by a linear-bounded automaton can also be carried out by a Turing machine. These are all proper inclusions, meaning that there exist recursively enumerable languages which are not context-sensitive, context-sensitive languages which are not context-free and context-free languages which are not regular.

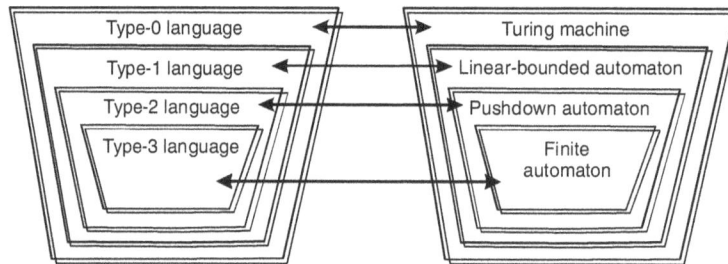

Figure 3.1   Languages and corresponding machines

The following table summarizes each of Chomsky's four types of grammars, the class of language it generates, the type of automaton that recognizes it, and the form its productions must have.

Table 3.1   Chomsky hierarchy of grammar and languages

Grammar	Languages	Automaton	Productions
Type-0 grammar	Recursively enumerable	Turing machine	No restrictions
Type-1 grammar	Context-sensitive	Linear-bounded	$\alpha B \beta \to \alpha \gamma \beta$
Type-2 grammar	Context-free	Pushdown	$Q \to \alpha$
Type-3 grammar	Regular	Finite state	$Q \to \alpha$ $Q \to aQ_1$

3.3   LANGUAGES AND THEIR RELATION

In this section we will study the relationship among the different class of languages classified by Noam Chomsky. In order to get the relationship we denote L_{RE} as the family of recursively enumerable languages, L_{CSL} as the family of context-sensitive languages, L_{CFL} as the family of context-free languages, and L_{RL} as the family of regular languages. To obtain the relationship, we state the following properties of languages.

Properties 1 According to the definition of languages, regular language is a subset of context-free language, context-sensitive language is a subset of recursively enumerable language and context-free language is a subset of recursively enumerable language. Symbolically, $L_{RL} \subseteq L_{CFL}$, $L_{CSL} \subseteq L_{RE}$ and $L_{CFL} \subseteq L_{RE}$.

Properties 2 The context-free language is a subset of context-sensitive language that is not immediate. This is not immediate because we follow production $Q \to \wedge$ in context-free grammar even when $Q \neq Q_0$, the start symbol, but not in context-sensitive grammars.

A context-free grammar G with productions of the form $Q \to \wedge$ is equivalent to a context-free grammar G_1 that has no productions of the form $Q \to \wedge$. Also, when grammar G_1 has $Q_0 \to \wedge$ and Q_0 does not appear in right-hand side of any production of G_1, then G_1 is context-sensitive grammar. This proves that $L_{CFL} \subseteq L_{CSL}$.

Properties 3 On the basis of above properties 1 and 2, the inclusion relation among L_{RL}, L_{CFL}, L_{CSL} and L_{RE} is given as $L_{RL} \subseteq L_{CFL} \subseteq L_{CSL} \subseteq L_{RE}$. This relation also holds for proper inclusion. Therefore, $L_{RL} \subset L_{CFL} \subset L_{CSL} \subset L_{RE}$.

3.3.1 Operations on Languages

Certain operations on languages are common in general. This includes the standard set operations, such as union, intersection, complement and concatenation. A class of languages is closed under a particular operation when the operation, applied to languages in the class, always produces a language in the same class again. In this section we study the closer properties of languages over union and concatenation that holds for all class of languages. The closer properties of some remained operations are discussed in the subsequent chapters. Before proving the closer properties of languages over union and concatenation we first state and prove an important theorem.

Theorem *Let G be a type-0 grammar. Then there exists an equivalent grammar G_1 in which each production is either of the form $\alpha \to \beta$ or of the form $Q \to a$, where α, β are strings of variables, $Q \in V_N$ and $a \in T$. The grammar G_1 is of type 1, type 2 or type 3 according as G is of type 1, type 2 or type 3 respectively.*

Proof Let G be a type-0 grammar. We construct the equivalent grammar G_1 as follows: Consider a production $\alpha \to \beta$ in G, where α or β has some terminals. In such case introduce a new variable Q_a for every terminal a appearing in α or β. This leads to the new set of productions $\alpha' \to \beta'$ and $Q_a \to a$ for every terminal a appearing in α or β. The productions for G_1 are the new set of productions obtained through the above construction.

Therefore, it is clear that G_1 contains the variables of G together with the new variables of type Q_a along with the new set of productions obtained through the above construction. The terminals and the start symbol of G_1 remains same as that of G. So, G_1 satisfies the required conditions of G and hence $L(G) = L(G_1)$.

Theorem   *Each of the classes $L_{RL}, L_{CFL}, L_{CSL}, L_{RE}$ is closed under union.*

Proof   Let L_1 and L_2 be two languages of the same type i, $i = 0, 1, 2, 3$. Hence, by immediate preceding theorem we get grammars $G_1 = (V'_N, T_1, P_1, Q'_0)$ and $G_2 = (V''_N, T_2, P_2, Q''_0)$ of type i generating L_1 and L_2 respectively. Therefore, any production in G_1 or G_2 is either of the form $\alpha \to \beta$ where α, β contain only variables or of the form $Q_a \to a$ where Q_a is a variable and a is a terminal. Without loss of generality, we can further assume that V'_N and V''_N are disjoint. This can be achieved by renaming the variables of V''_N if they appear in V'_N, i.e., $V'_N \cap V''_N = \phi$.

Now, we define a new grammar $G = (V_N, T, P, Q_0)$, where $V_N = V'_N \cup V''_N \cup \{Q_0\}$, where Q_0 is the new start symbol, i.e., $Q_0 \notin V'_N \cup V''_N$, $T = T_1 \cup T_2$ and $P = P_1 \cup P_2 \cup \{Q_0 \to Q'_0 | Q''_0\}$.

Now our claim is $L(G) = L_1 \cup L_2$, i.e., $(L_1 \cup L_2) \subseteq L(G)$ and $L(G) \subseteq (L_1 \cup L_2)$. We prove $(L_1 \cup L_2) \subseteq L(G)$ as follows: If $w \in L_1 \cup L_2$, then either $Q'_0 \xrightarrow[G_1]{*} w$ or $Q''_0 \xrightarrow[G_2]{*} w$. Therefore, we have $Q_0 \xrightarrow[G]{} Q'_0 \xrightarrow[G]{} w$ or $Q_0 \xrightarrow[G]{} Q''_0 \xrightarrow[G]{} w$. Hence, it is clear that $w \in L(G)$, i.e., $(L_1 \cup L_2) \subseteq L(G)$.

In order to prove the second part $L(G) \subseteq (L_1 \cup L_2)$, consider a derivation of w. In fact in grammar G, the first step should be $Q_0 \to Q'_0$ or $Q_0 \to Q''_0$. If the first step is $Q_0 \to Q'_0$, then in the subsequent steps Q'_0 is changed according to the productions P_1 and involve only variables of V'_N. Therefore, $Q_0 \xrightarrow[G_1]{*} w$. Similarly, if the first step is $Q_0 \to Q''_0$, then $Q_0 \xrightarrow[G_2]{} Q''_0 \xrightarrow[G_2]{*} w$. Therefore, $L(G) \subseteq (L_1 \cup L_2)$ and hence $L(G) = (L_1 \cup L_2)$.

Again, $L(G)$ is of type 0 or type 2 according as L_1 and L_2 are of type 0 or type 2. If $\wedge \notin (L_1 \cup L_2)$, then $L(G)$ is of type 1 or type 3 according as L_1 and L_2 are of type 1 or type 3. If $\wedge \in L_1$, then we define the grammar $G = (V_N, T, P, Q_1)$, such that $V_N = V'_N \cup V''_N \cup \{Q_0, Q_1\}$, where Q_1 is a new start symbol, i.e., $Q_1 \notin (V'_N \cup V''_N \cup Q_0)$, $T = T_1 \cup T_2$ and $P = P_1 \cup P_2 \cup \{Q_1 \to Q_0, Q_0 \to Q'_0 | Q''_0\}$. Therefore, $L(G)$ is of type 1 or type 3 according as L_1 and L_2 are of type 1 or type 3. Similar argument also holds when $\wedge \in L_2$.

Theorem   *Each of the classes $L_{RL}, L_{CFL}, L_{CSL}, L_{RE}$ is closed under concatenation.*

Proof   Let L_1 and L_2 be two languages of the same type i, $i = 0, 1, 2, 3$. Hence, by the first theorem of this section 3.3.1 we get grammars $G_1 = (V'_N, T_1, P_1, Q'_0)$ and $G_2 = (V''_N, T_2, P_2, Q''_0)$ of type i generating L_1 and L_2 respectively. Therefore, any production in G_1 or G_2 is either of the form

$\alpha \rightarrow \beta$, where α, β contain only variables or of the form $Q_a \rightarrow a$, where Q_a is a variable and a is a terminal. Without loss of generality, we can further assume that V_N' and V_N'' are disjoint. This can be achieved by renaming the variables of V_N'' if they appear in V_N', i.e., $V_N' \cap V_N'' = \phi$.

Now, we define a new grammar $G = (V_N, T, P, Q_0)$, where $V_N = V_N' \cup V_N'' \cup \{Q_0\}$, where Q_0 is a newly introduced start symbol, i.e., $Q_0 \notin V_N' \cup V_N''$, $T = T_1 \cup T_2$ and $P = P_1 \cup P_2 \cup \{Q_0 \rightarrow Q_0' Q_0''\}$.

Now our claim is $L(G) = L_1 L_2$, i.e., $L_1 L_2 \subseteq L(G)$ and $L(G) \subseteq L_1 L_2$. We prove $L_1 L_2 \subseteq L(G)$ as follows: If $w = w_1 w_2 \in L_1 L_2$, then we have $Q_0' \xrightarrow[G_1]{*} w_1$ and $Q_0'' \xrightarrow[G_2]{*} w_2$. Therefore, we get $Q_0 \xrightarrow[G]{} Q_0' Q_0'' \xrightarrow[G]{} w_1 w_2 = w$. Hence, it is clear that $w \in L(G)$, i.e., $L_1 L_2 \subseteq L(G)$.

Now we prove $L(G) \subseteq L_1 L_2$ as follows: If $w \in L(G)$, then definitely the first step in the derivation of w is $Q_0 \rightarrow Q_0' Q_0''$. Again, as $V_N' \cap V_N'' = \phi$ and productions in G_1 or G_2 involve only variables, so definitely $w = w_1 w_2$ where $Q_0' \rightarrow w_1$ and $Q_0'' \rightarrow w_2$. Therefore, we get $w = w_1 w_2 \in L_1 L_2$ and so $L(G) \subseteq L_1 L_2$. Hence, we get $L_1 L_2 \subseteq L(G)$ and $L(G) \subseteq L_1 L_2$. Therefore, $L(G) = L_1 L_2$.

Again, $L(G)$ is of type 0 or type 2 according as L_1 and L_2 are of type 0 or type 2. If $\wedge \notin L_1 \cup L_2$, then above construction is sufficient when L_1 and L_2 are of type 1 or type 3.

Assume that L_1 and L_2 are of type 1 or type 3 with $\wedge \in L_1$ or $\wedge \in L_2$. Let us define $L_1' = L_1 - \{\wedge\}$ and $L_2' = L_2 - \{\wedge\}$. Then we define $L_1 L_2$ as follows:

$$L_1 L_2 = \begin{cases} L_1' L_2' \cup L_2' & \text{if } \wedge \text{ is in } L_1 \text{ but not in } L_2 \\ L_1' L_2' \cup L_1' & \text{if } \wedge \text{ is in } L_2 \text{ but not in } L_1 \\ L_1' L_2' \cup L_1' \cup L_2' \cup \{\wedge\} & \text{if } \wedge \text{ is in } L_1 \text{ and also in } L_2 \end{cases}$$

From the immediate previous theorem we know that L_{RL} and L_{CSL} are closed under union. Therefore, it is clear that $L_1 L_2$ is of type 3 or type 1 according as L_1 and L_2 are of type 3 or type 1.

Note The families of languages L_{RL}, L_{CSL} and L_{RE} is closed under intersection whereas L_{CFL} is not closed under intersection. But the intersection of a L_{CFL} and L_{RL} is context-free.

3.3.2 Tabular Presentation of Closure Properties of Languages

At present it is very difficult to establish property under all operations. This is because lack of properties of families of languages under consideration. Thus, we state them in tabular form the properties without proof. Some of them we prove in the subsequent chapters.

Operation	Meaning	Regular	CFL	CSL	RE
Union	$\{w \mid w \in L_1 \lor w \in L_2\}$	Yes	Yes	Yes	Yes
Concatenation	$\{w_1 w_2 \mid w_1 \in L_1 \land w_2 \in L_2\}$	Yes	Yes	Yes	Yes
Intersection	$\{w \mid w \in L_1 \land w \in L_2\}$	Yes	No	Yes	Yes
Complement	$\{w \mid w \notin L_1\}$	Yes	No	Yes	No
Kleene star	$\{\land\} \cup \{w_1 w_2 \mid w_1 \in L_1 \land w_2 \in L_1^*\}$	Yes	Yes	Yes	Yes
Homomorphism	$h : h(a)$ contains single string for each a.	Yes	Yes	Yes	Yes
Substitution	$f : \Sigma \to \Delta^*$ defined by $f(a) = Ra$	Yes	Yes	Yes	Yes
Reverse	$\{w^R \mid w \in L_1\}$	Yes	Yes	Yes	Yes

3.3.3 Recursive and Recursively Enumerable Sets

A set Q is said to be recursive if there is an algorithm to determine whether or not a given element belongs to set Q. An algorithm is a procedure that terminates after a finite number of steps for any valid input. A procedure for solving a problem is a finite sequence of instructions that can be carried out on valid input data. A set Q is said to be recursively enumerable if there exists a recursive function that can eventually generate any element in Q. Therefore, it is clear that any recursive set is also recursively enumerable.

Theorem *Every context-sensitive language, L_{CSL} is recursive.*

Proof Given a context-sensitive grammar, $G = (V_N, T, P, Q_0)$ and a word w in Σ^* of length n, we can determine whether w is in $L(G)$ as follows.

Construct a graph whose vertices are the strings in $(V_N \cup T)^*$ of length n or less. We put an edge from α to β if $\alpha \Rightarrow \beta$. Then the paths in the transition graph correspond to derivations in G, and w is in $L(G)$ if and only if there is a path from the vertex for Q_0 to the vertex for w. We can use any number of path-finding algorithms to decide whether such a path exists or not.

SOLVED EXAMPLES

Example 1

Define rules for constructing a formal language L over the alphabet $\Sigma = \{0, 1, 2, 3, 4, 5, 6, 7, 8, 9, +, =\}$

Solution The following rules are defined for constructing a formal language L over the alphabet $\Sigma = \{0, 1, 2, 3, 4, 5, 6, 7, 8, 9, +, =\}$.

1. Every non-empty string that does not contain + or = and does not start with 0 is in the language L.

2. The string 0 is in L.

3. A string containing = is in L if and only if there is exactly one =, and it separates two strings in L.

4. A string containing + is in L if and only if every + in the string separates two valid strings in L.

5. No string is in L other than those implied by the previous rules.

Under these rules, the string "$25 + 34 = 223$" is in L, but the string "$= 245 + =$" is not. This formal language expresses natural numbers.

Example 2

Define the string reversal in formal language.

Solution The string reversal in formal language is defined as follows:

1. If $\wedge$ be the empty string, then $\wedge^R = \wedge$.

2. For each non-empty string $w = x_1 x_2 \cdots x_n$ over some alphabet Σ, the reversal is defined as $w^R = x_n x_{n-1} \cdots x_1$.

3. For a formal language L, $L^R = \{ w^R \mid w \in L \}$.

Example 3

Explain formal languages are typically infinite.

Solution For finite languages one can simply enumerate all well-formed words. For example, we can define a finite language L as just $L = \{$"a", "b", "ab", "aba"$\}$. However, even over a finite non-empty alphabet such as $\Sigma = \{a,b\}$ there are infinitely many words: "a", "ab", "baa", "$abbb$", "$aaaab$", $\cdots$. Therefore, formal languages are typically infinite. Also, it is clear that defining an infinite formal language is not as simple as writing $L = \{$"a", "b", "ab", "aba"$\}$.

Example 4

Derive the language for the grammar G having productions $Q_0 \to aQ_0 bb$ and $Q_0 \to \wedge$.

Solution Let the grammar be $G = (V_N, T, P, Q_0)$. So, we have $V_N = \{Q_0\}$, $T = \{a,b\}$ and $P = \{Q_0 \to aQ_0 bb, Q_0 \to \wedge\}$. From the productions it is clear that the least length string generated by the grammar G is empty string. The next string generated by the grammar is 'abb'. It is generated by substituting '$\wedge$' on place of non-terminal Q_0 in the right-hand side of the production $Q_0 \to aQ_0 bb$. On substituting '$aQ_0 bb$' n-times on place of non-terminal Q_0 in the right-hand side of the production $Q_0 \to aQ_0 bb$ we will get

$$Q_0 \to aaa \cdots aQ_0 bb \cdots bbbbbb \text{, i.e., } Q_0 \to a^n Q_0 b^{2n}.$$

On substituting $Q_0 \to \wedge$ we get finally as $Q_0 \to a^n b^{2n}$. Since the least length string is $\wedge$ the smallest value of n is equal to 0. Therefore, the language $L(G)$ derived by the grammar G is given as:

$$L(G) = \{a^n b^{2n} \mid n \geq 0\}$$

Example 5

Find the language $L(G)$ generated by the grammar $G = (\{Q_0\}, \{a,b\}, P, Q_0)$, where production P is defined as $Q_0 \to a$, $Q_0 \to b$, $Q_0 \to aQ_0$ and $Q_0 \to bQ_0$.

Solution   Given that $G = (\{Q_0\}, \{a,b\}, P, Q_0)$ with productions $Q_0 \to a$, $Q_0 \to b$, $Q_0 \to aQ_0$ and $Q_0 \to bQ_0$.

Therefore, it is clear that the smallest string generated by the grammar G is either 'a' or 'b'. In the right-hand side of productions $Q_0 \to aQ_0$ and $Q_0 \to bQ_0$, Q_0 can be replaced by a, b, aQ_0 or bQ_0 in any order. Thus all strings having any combination of a's and b's can be generated except the null string. So, we get

$$L(G) = \{a,b\}^* - \{\wedge\}$$

Example 6

Let $G = (\{Q_0, A, B\}, \{a,b\}, P, Q_0)$, where the set of productions P consists of $Q_0 \to aABa$, $B \to Aab$ and $aA \to baa$. Test whether $w = babaaaba$ is in $L(G)$.

Solution   Given that $G = (\{Q_0, A, B\}, \{a,b\}, P, Q_0)$, where productions P consists of $Q_0 \to aABa$, $B \to Aab$ and $aA \to baa$. At each step of derivation we apply a suitable production from the set of productions to get $w = babaaaba$. So, we have

$$
\begin{aligned}
Q_0 &\to aABa \\
&\to baaBa & [aA \to baa] \\
&\to baaAaba & [B \to Aab] \\
&\to babaaaba & [aA \to baa]
\end{aligned}
$$

Therefore, $w = babaaaba \in L(G)$

Example 7

Find the language derived by the grammar $G = (\{Q_0, A, B\}, \{a\}, P, Q_0)$, where production P is defined as $Q_0 \to AB, A \to BB$ and $B \to AA$.

Solution   Given productions are $Q_0 \to AB, A \to BB$ and $B \to AA$. It is clear that there is no terminal on the right-hand side of the productions. So, we get $L(G) = \phi$.

Example 8

Derive the language for the grammar G having productions $Q_0 \to 0Q_0A2, Q_0 \to 012, 2A \to A2$ and $1A \to 11$.

Solution   Let the grammar be $G = (V_N, T, P, Q_0)$. So, we have $V_N = \{Q_0, A\}$, $T = \{0, 1, 2\}$ and $P = \{Q_0 \rightarrow 0Q_0A2, Q_0 \rightarrow 012, 2A \rightarrow A2, 1A \rightarrow 11\}$. From the productions it is clear that the least length string generated by the grammar G is '012', i.e., $012 \in L(G)$. On substituting '$0Q_0A2$' in place of Q_0, $(n-1)$ times we get

$$Q_0 \rightarrow 0Q_0A2 \rightarrow 00Q_0A2A2 \rightarrow 00\cdots0Q_0A2\cdots A2A2$$
$$\rightarrow 0^{n-1}012(A2)^{n-1} \qquad [Q_0 \rightarrow 012]$$
$$\rightarrow 0^n12(A2)^{n-1}$$
$$\rightarrow 0^n1A^{n-1}2^n \qquad [2A \rightarrow A2]$$
$$\rightarrow 0^n11A^{n-2}2^n \qquad [1A \rightarrow 11]$$
$$\rightarrow 0^n111A^{n-3}2^n \qquad [1A \rightarrow 11]$$
$$\rightarrow 0^n1^n A^{n-n}2^n \qquad [1A \rightarrow 11]$$
$$\rightarrow 0^n1^n2^n$$

Therefore, we get $L(G) = \{0^n1^n2^n \mid n \geq 1\}$.

Example 9

Let G be the grammar having productions $Q_0 \rightarrow aQ_0 \mid aA \mid a$, $A \rightarrow ab$ and $A \rightarrow aAb$. Obtain the language generated by the grammar G.

Solution   Consider the A productions $A \rightarrow aAb$ and $A \rightarrow ab$. The smallest string generated by A productions is 'ab'. On substituting 'aAb' $(n-1)$ times on place of non-terminal A in the right-hand side of the production $A \rightarrow aAb$, we get

$$A \rightarrow aaa\cdots aAbbb\cdots b, \text{ i.e., } A \rightarrow a^{n-1}Ab^{n-1}.$$

Finally substituting $A \rightarrow ab$ we will get $A \rightarrow a^nb^n, n \geq 1$.

On considering productions $Q_0 \rightarrow aA$ and $A \rightarrow a^nb^n$, we get $Q_0 \rightarrow aa^nb^n = a^{n+1}b$, $n \geq 1$. Similarly, on considering productions $Q_0 \rightarrow aQ_0 \mid a$, we can generate $a^n, n \geq 1$. It indicates that an arbitrary number of a's to be introduced at the beginning. Again on considering $Q_0 \rightarrow aQ_0 \mid aA$, we can generate $Q_0 \rightarrow a^nQ_0 \rightarrow a^naA \rightarrow a^{2n+1}b^n$, $Q_0 \rightarrow a^{n-1}Q_0 \rightarrow a^{2n}b$, $Q_0 \rightarrow a^{n-1}Q_0 \rightarrow a^{3n}b^n$, etc. It is observed that in each case the number of occurrence a's is more than number of occurrence b's. Thus, we get

$$L(G) = \{a^mb^n \mid m > n \geq 0\}$$

Example 10

Construct a grammar generating the language $L(G) = \{0^n10^n \mid n \geq 1\}$.

Solution   Given that $L(G) = \{0^n10^n \mid n \geq 1\}$. Our aim is to construct a grammar for the above language. It is clear that the smallest string in the language is '010'. The next string in the language is '$0^2 10^2$'. Therefore, we formally define the grammar $G = (V_N, T, P, Q_0)$ as $V_N = \{Q_0\}$, $T = \{0,1\}$, and $P = \{Q_0 \to 0 Q_0 0, Q_0 \to 010\}$.

Example 11

Let $G = (\{Q_0, A\}, \{0,1\}, P, Q_0)$, where the set of productions P consists of $Q_0 \to 0A \mid 1Q_0 \mid 0 \mid 1$ and $A \to 1A \mid 1Q_0 \mid 1$. Test whether $w = 01010$ is in $L(G)$.

Solution   Given that $G = (\{Q_0, A\}, \{0,1\}, P, Q_0)$, where productions P consists of $Q_0 \to 0A \mid 1Q_0 \mid 0 \mid 1$ and $A \to 1A \mid 1Q_0 \mid 1$. At each step of derivation we apply a suitable production to get $w = 01010$. Therefore, we have

$$Q_0 \to 0A$$
$$\to 01Q_0 \qquad [A \to 1Q_0]$$
$$\to 010A \qquad [Q_0 \to 0A]$$
$$\to 0101Q_0 \qquad [A \to 1Q_0]$$
$$\to 01010 \qquad [Q_0 \to 0]$$

Therefore, $w = 01010 \in L(G)$.

Example 12

Find a grammar generating the language $L(G) = \{ ww^R \mid w \in \{0,1\}^* \}$.

Solution   It is clear that any string in the language $L(G) = \{ ww^R \mid w \in \{0,1\}^* \}$ is generated by recursion as follows: (a) $\land \in L$; (b) if $x \in L$, then $wxw^R \in L$. Therefore, we formally define the grammar as $G = (V_N, T, P, Q_0)$ such that $V_N = \{Q_0\}$, $T = \{0,1\}$ with the productions $P = \{ Q_0 \to 0 Q_0 0, \ Q_0 \to 1 Q_0 1, \ Q_0 \to \land \}$.

Example 13

Let $L(G) = \{ a^n b^n c^m \mid n \geq 1, m \geq 0 \}$. Construct a grammar generating the language $L(G)$.

Solution   Given that $L(G) = \{ a^n b^n c^m \mid n \geq 1, m \geq 0 \}$. Therefore, the language can be expressed as $L = L_1 L_2$, where $L_1 = \{ a^n b^n \mid n \geq 1 \}$ and $L_2 = \{ c^m \mid m \geq 0 \}$.

We construct $L_1 = \{ a^n b^n \mid n \geq 1 \}$ by recursion as follows. The productions generating L_1 is given as $A \to aAb, \ A \to ab$. Similarly, we construct $L_2 = \{ c^m \mid m \geq 0 \}$ with the set of productions $B \to cB, B \to \land$. Therefore, we formally define the grammar as $G = (V_N, T, P, Q_0)$ such that $V_N = \{Q_0, A, B\}$, $T = \{a,b,c\}$ with the set of productions $P = \{Q_0 \to AB, \ A \to aAb, \ A \to ab, B \to cB, \ B \to \land \}$.

Example 14

Construct a grammar that generates all even integers up to 98.

Solution   Our aim is to construct all even integers up to 98. Therefore, it is clear that the grammar should generate all one and two digit even numbers.

We define the grammar G as $G = (V_N, T, P, Q_0)$, where $V_N = \{Q_0, A, B\}$, $T = \{0, 1, 2, 3, 4, 5, 6, 7, 8, 9\}$ with the productions $Q_0 \rightarrow 0|2|4|6|8$, $Q_0 \rightarrow AB$, $A \rightarrow 1|2|3|4|5|6|7|8|9$ and $B \rightarrow 0|2|4|6|8$.

The production $Q_0 \rightarrow 0|2|4|6|8$ generates all one digit even numbers. The production $Q_0 \rightarrow AB$ together with $A \rightarrow 1|2|3|4|5|6|7|8|9$ and $B \rightarrow 0|2|4|6|8$ generates all remaining two digit even numbers.

Example 15

Find a grammar G for the language $L(G) = \{w \in \{a,b\}^* \,|\, n_a(w) \neq n_b(w)\}$, where $n_a(w)$ denotes the number of occurrence of a's in string w and $n_b(w)$ denotes the number of occurrence of b's in string w.

Solution   Given language is $L(G) = \{w \in \{a,b\}^* \,|\, n_a(w) \neq n_b(w)\}$, where $n_a(w)$ denotes the number of occurrence of a's in string w and $n_b(w)$ denotes the number of occurrence of b's in string w. Therefore, it will generate two different languages L_0 and L_1 such that $L(G) = L_0(G) \cup L_1(G)$, where $L_0(G) = \{w \in \{a,b\}^* \,|\, n_a(w) > n_b(w)\}$ and $L_1(G) = \{w \in \{a,b\}^* \,|\, n_a(w) < n_b(w)\}$.

Clearly, $a \in L_0$ and for any $w \in L_0$, both wa and aw are in L_0. This leads to the productions $Q_0' \rightarrow a|\, aQ_0'|\, Q_0'a$, where Q_0' is the initial symbol of L_0.

Now, we need to introduce b's to our strings. On introducing one b to a string of L_0 may not always produce a string of L_0. However, the concatenation of two strings in L_0 always produces a string with at least two more a's than b's. Therefore, introducing a single b will still produce a string of L_0. We can introduce it at the left position, at the right position or in between of the two strings. This leads to the productions $Q_0' \rightarrow bQ_0'Q_0'|\, Q_0'Q_0'b|\, Q_0'bQ_0'$. Therefore, the grammar for $L_0(G)$ is given as $G_0 = (\{Q_0'\}, \{a,b\}, P_1, Q_0')$, where $P_1 = \{Q_0' \rightarrow bQ_0'Q_0'|\, Q_0'Q_0'b|\, Q_0'bQ_0'|\, aQ_0'|\, Q_0'a|\, a\}$.

Similarly, we can construct a grammar for $L_1(G) = \{w \in \{a,b\}^* \,|\, n_a(w) < n_b(w)\}$ and is given as $G_1 = (\{Q_0''\}, \{a,b\}, P_2, Q_0'')$, where the productions P_2 is defined as $P_2 = \{Q_0'' \rightarrow aQ_0''Q_0''|\, Q_0''Q_0''a|\, Q_0''aQ_0''|\, bQ_0''|\, Q_0''b|\, b\}$.

Therefore, the grammar for the language $L(G)$ is given as $G = (V_N, T, P, Q_0)$, where $V_N = \{Q_0, Q_0', Q_0''\}$, $T = \{a,b\}$ and $P = P_1 \cup P_2 \cup \{Q_0 \rightarrow Q_0'|\, Q_0''\}$.

Example 16

Test whether $w = 001100$ is in the language $L(G)$ generated by the grammar $G = (\{Q_0, A, B\}, \{0,1\}, P, Q_0)$, where the set of productions P consists of $Q_0 \to 0Q_0BA$, $Q_0 \to 01A$, $AB \to BA$, $1B \to 11$, $1A \to 10$ and $0A \to 00$.

Solution   Given that $G = (\{Q_0, A, B\}, \{0,1\}, P, Q_0)$, where productions P consists of $Q_0 \to 0Q_0BA$, $Q_0 \to 01A$, $AB \to BA$, $1B \to 11$, $1A \to 10$ and $0A \to 00$. At each step of derivation we apply a suitable production which is likely to derive $w = 001100$. Therefore, we start with the initial production $Q_0 \to 0Q_0BA$ and proceed as follows:

$$Q_0 \to 0Q_0BA$$
$$\to 001ABA \qquad [Q_0 \to 01A]$$
$$\to 001BAA \qquad [AB \to BA]$$
$$\to 00 11AA \qquad [1B \to 11]$$
$$\to 001 10A \qquad [1A \to 10]$$
$$\to 001100 \qquad [0A \to 00]$$

Therefore, $w = 001100 \in L(G)$.

REVIEW QUESTIONS

1. Explain in detail the Chomsky hierarchy.

2. State the properties of languages and their relation.

3. Define formal grammar. Also give examples to define formal grammar.

4. Explain the Chomsky hierarchy in tabular form.

5. Explain with suitable examples the difference between type-2 and type-3 language.

6. Explain with suitable examples the difference between type-3 and type-4 language.

7. Show that a context-sensitive language is recursive.

8. Show that the family of type-3 languages is closed under union.

9. Show that the family of type-3 languages is closed under concatenation.

10. Show that the family of type-2 languages is closed under union.

11. Write a note on recursively enumerable sets.

PROBLEMS

1. Construct a grammar G generating the language L of all palindromes over $\{0, 1\}$.

2. Find a grammar generating the language $L(G) = \{w2w^R \mid w \in \{0,1\}^*\}$.

3. Construct a grammar for the language $L(G) = \{ww \mid w \in \{0,1\}^*\}$.

4. Obtain a grammar G for the language $L(G) = \{w \in \{a,b\}^* \mid n_a(w) = n_b(w)\}$, where $n_a(w)$ denotes the number of occurrence of a's in string w.

5. Compute $L(G)$ if G has the following productions.

   (a) $Q_0 \to aAa$, $A \to aAa$, $A \to b$

   (b) $Q_0 \to aQ_0b$, $Q_0 \to b$

   (c) $Q_0 \to aQ_0b$, $Q_0 \to aa$

6. Construct a grammar for the following languages.

   (a) $L(G) = \{b^n c^m b^n \mid n, m \geq 1\}$

   (b) $L(G) = \{ww^R \mid w \in \{a, b\}^+\}$

   (c) $L(G) = \{a^m b^n \mid m > n \geq 1\}$

   (d) $L(G) = \{a^n b^{n+m} c^m \mid m \geq 1, n \geq 0\}$

7. Obtain a grammar for the languages given below.

   (a) $L(G) = \{w \in \{a,b\}^+ \mid n_a(w) = n_b(w) + 1\}$

   (b) $L(G) = \{w \in \{a,b\}^+ \mid n_a(w) = 2n_b(w)\}$

   (c) $L(G) = \{w \in \{a,b\}^+ \mid n_a(w) < n_b(w)\}$

   Here, $n_a(w)$ and $n_b(w)$ are number of a's and b's in a string w respectively.

8. Derive the language for the grammar $G = (\{Q_0, A\}, \{a, b\}, P, Q_0)$ having productions $Q_0 \to aAbb$, and $A \to Q_0 \mid abbb$. Show that the string $w = a^3 b^9$ is accepted by the given grammar.

9. Obtain a grammar generating the language $L(G)$, where $L(G)$ is given below.

   (a) $L(G) = \{0^n 1^{3n} 2^n \mid n \geq 1\}$

   (b) $L(G) = \{0^3 1^n 2^n \mid n \geq 0\}$

10. Find the language $L(G)$ generated by the grammar $G = (V_N, T, P, Q_0)$, where $V_N = \{Q_0, A, B, E\}$, $T = \{a, b, c\}$ and productions P is defined as $Q_0 \to AB$, $A \to aEb$, $E \to A \mid b$ and $B \to cB \mid \wedge$.

11. Let G be the grammar having productions $Q_0 \to aQ_0cc$ and $Q_0 \to b$. Obtain the language generated by the grammar G.

12. Let $G = (\{Q_0, A\}, \{0, 1, 2\}, P, Q_0)$, where the set of productions P consists of $Q_0 \to 0Q_0A2$, $Q_0 \to 012$, $2A \to A2$ and $1A \to 11$. Test whether $w = 0^3 1^3 2^3$ is in $L(G)$.

13. Construct a grammar that generates all even integers up to 998.

14. Derive the language for the grammar G having productions $Q_0 \to AB$, $A \to aB$ and $B \to bA$, where Q_0 is the start symbol.

15. Obtain the language generated by following grammars:

    (a) $Q_0 \to aQ_0b \mid aA \mid a \mid bB \mid b$, $A \to aA \mid a$, $B \to bB \mid b$

    (b) $Q_0 \to 0Q_0BA \mid 01A$, $AB \to BA$, $1B \to 11$, $1A \to 10$, $0A \to 00$

16. Construct the grammar, accepting each of the following sets.

    (a) $\{a^n b^m a^m b^n \mid m, n \geq 1\}$

    (b) $\{a^n b^n \mid n \geq 1\} \cup \{b^m a^m \mid m \geq 1\}$

    (c) $\{a^n b^{3n+1} \mid n \geq 0\}$

17. Test whether $w = 0011111$ is in the language $L(G)$ generated by the grammar $G = (\{Q_0, A, B\}, \{0, 1\}, P, Q_0)$, where the set of productions P consists of $Q_0 \to 0Q_0 1 \mid 0A \mid 1B \mid 0 \mid 1$, $A \to 0A \mid 0$, and $B \to 1B \mid 1$.

18. Let us consider the grammar $G = (\{Q_0, A\}, \{0, 1\}, P, Q_0)$, where P consists of $Q_0 \to 0Q_0 1 \mid 0A1$, and $A \to 1A \mid 1$. Test whether $w = 00111$ is in $L(G)$.

19. Construct a grammar to generate $L_1 \cup L_2$, where the language L_1 and L_2 is defined as $L_1 = \{(01)^n \mid n \geq 1\}$ and $L_2 = \{(10)^n \mid n \geq 1\}$.

20. Derive the language for the grammar G having productions $Q_0 \to aQ_0 b \mid aAb$ and $A \to 1A0 \mid 10$, where Q_0 is the start symbol.

4

REGULAR LANGUAGES AND EXPRESSIONS

4.0   INTRODUCTION

Languages are likely to be infinite but must be defined in some finite way. In general, these languages are described in two different ways. One way the strings in the language can be generated from simpler strings using string operations, whereas on the other way specify an algorithmic procedure for determining whether a given string is in the language.

In this book, we consider the regular languages that have a number of applications in computer science. It is the simplest language that can be generated from one element language by using certain standard operations a finite number of times. Therefore, they are also recognized by finite automaton. In this chapter we will discuss regular languages, regular expressions, regular sets and grammar with pumping lemma.

4.1   REGULAR LANGUAGES

Regular languages are those languages that are described by regular expressions and can be accepted by a deterministic finite automata, nondeterministic finite automata, regular grammar and read-only Turing machine. A regular language may be defined inductively as follows.

(a) ϕ is a regular language.

(b)   For each $a \in \Sigma$, $\{a\}$ is a regular language.

(c)   If $L_1, L_2, L_3, \cdots, L_n; n \geq 2$ are regular languages, then $(L_1 \cup L_2 \cup L_3 \cup \cdots \cup L_n)$ is also a regular language.

(d)   If $L_1, L_2, L_3, \cdots, L_n; n \geq 2$ are regular languages, then $L_1 L_2 L_3 \cdots L_n$ is also a regular language.

(e)   If L is a regular language, then the star closure or Kleene star L^* is also a regular language.

(f)   Nothing else is a regular language unless its construction follows from (a) to (e) defined above.

For example, $L = \{0, 01\}$ is a regular language defined over the alphabet $\Sigma = \{0, 1\}$. From the definition it is clear that $\{0\}$ and $\{1\}$ are regular languages and so $\{01\}$. Again, $\{0\}$ and $\{01\}$ are regular languages and so $\{0\} \cup \{01\} = \{0, 01\}$.

Theorem   *If L_1, L_2 are two regular languages, then $(L_1 + L_2), L_1L_2, L_1^*$ and L_2^* are also regular languages.*

Proof   Let the regular languages L_1 and L_2 be described by the regular expression r_1 and r_2 respectively. It indicates that the regular expression $(r_1 + r_2)$ defines the language $(L_1 + L_2)$. Similarly, r_1r_2 defines the language L_1L_2 whereas r_1^* and r_2^* defines the language L_1^* and L_2^* respectively.

Therefore, $(L_1 + L_2), L_1L_2, L_1^*$ and L_2^* are also regular languages. It is to be noted that the above proof uses the fact that the regular languages L_1 and L_2 must be described by the regular expressions.

Theorem   *If L be a regular language, then the complement of L, i.e., L' is also a regular language.*

Proof   Let L be a regular language. Let us denote the complement of L as L'. We know a regular language L can be accepted by finite automaton. It indicates that some of the states of this finite automaton are final states whereas the remaining are intermediate states. Let us reverse the status of each state. If it was a final state, make it to an intermediate state and if it was an intermediate state, make it to a final state. Hence, it is clear that if an input string is accepted by an intermediate state (non-final state), it now accepts a final state and vice versa. It indicates that the later machine accepts all input strings that were not accepted by the original finite automaton.

So, the converted machine accepts the complement of the language L, i.e., L'. Therefore, the complement of a regular language is also regular.

Theorem   *If L_1 and L_2 are two regular languages, then $(L_1 \cap L_2)$ is also a regular language.*

Proof   Let L_1 and L_2 be two regular languages. Therefore, the complement of L_1 and L_2, i.e., L_1' and L_2' are also regular languages. So, $(L_1' + L_2')$ is also a regular language. Therefore, the complement of $(L_1' + L_2')$, i.e., $(L_1' + L_2')'$ is also a regular language. By De Morgan's law we have $(L_1' + L_2')' = (L_1 \cap L_2)$. It indicates that $(L_1 \cap L_2)$ is also a regular language.

4.2   REGULAR EXPRESSIONS

The languages accepted by finite automaton are called regular expressions. In this section, we define regular expressions and prove that the class of languages that are accepted by finite automata is the class of languages describable by regular expressions. In general, regular expressions are useful for representing certain sets of strings algebraically. Now we give a recursive definition of regular expression.

Let Σ be an alphabet. The regular expressions over Σ are defined recursively as below.

(*a*) ϕ is a regular expression.

(*b*) $\wedge$ or ε is a regular expression.

(*c*) If $a \in \Sigma$, then a is a regular expression. We use in boldface of a symbol when the symbol is a part of regular expression.

(*d*) If r is a regular expression, then (r) is also a regular expression.

(*e*) If *r* and *s* are two regular expressions, then their union (*r* + *s*) is also a regular expression.

(*f*) If *r* and *s* are two regular expressions, then their concatenation (*rs*) is   also a regular expression.

(*g*) If *r* is a regular expression, then the closure of *r*, i.e., r^* is also a regular expression.

Note　Inside a parenthesis, we have the hierarchy of operations as follows: closure, concatenation and union. It indicates that, we perform closure first, and then concatenation followed by union while evaluating a regular expression.

4.2.1　Identities for Regular Expressions

The identities for regular expressions are given below. These are generally used for simplifying regular expressions. Let us assume *r*, *s* and *t* be three different regular expressions. Therefore, the identities are:

(*a*)　$\phi + r = r$

(*b*)　$\phi r = r\phi = \phi$

(*c*)　$\wedge r = r \wedge = r$

(*d*)　$\wedge^* = \wedge$ *and* $\phi^* = \wedge$

(*e*)　$rr^* = r^* r$

(*f*)　$r + r = r$

(*g*)　$(r^*)^* = r^*$

(*h*)　$r^* r^* = r^*$

(*i*)　$\wedge + rr^* = r^* = \wedge + r^* r$

(*j*)　$(rs)^* r = r(sr)^*$

(*k*)　$(r + s)t = rt + st$ and $t(r + s) = tr + ts$

(*l*)　$(r + s)^* = (r^* s^*)^* = (r^* + s^*)^*$

4.2.2　Arden's Theorem

Let *r* and *s* be two regular expressions defined over Σ. If *r* does not contain $\wedge$, then the equation $t = s + tr$ has a unique solution $t = sr^*$

Proof　Let *r* and *s* be two regular expressions defined over Σ and assume that *r* does not contain $\wedge$. On taking $t = sr^*$ we have

$$s + tr = s + (sr^*)r = s(\wedge + r^* r)$$

$$= sr^* \qquad \text{[By identity (i)]}$$

$$= t$$

i.e., $t = sr^*$ is a solution of the equation $t = s + tr$.

Now our aim is to show that the solution is unique. On replacing t by $s + tr$ on the right-hand side, we get

$$s + tr = s + (s + tr)r$$
$$= s + sr + tr^2$$
$$= s + sr + (s + tr)r^2$$
$$= s + sr + sr^2 + tr^3$$
$$= s + sr + sr^2 + \cdots + sr^i + tr^{i+1}$$
$$= s(\wedge + r + r^2 + \cdots + r^i) + tr^{i+1}$$

i.e., $t = s(\wedge + r + r^2 + \cdots + r^i) + tr^{i+1}$ for $i \geq 0$ \hfill (1)

Now we show that any solution of $t = s + tr$ is equivalent to sr^*. Let t satisfies $t = s + tr$, then it also satisfies equation (1). Let 'a' be a string of length i in the set t, then definitely 'a' belongs to $s(\wedge + r + r^2 + \cdots + r^i) + tr^{i+1}$. As we know that r does not contain $\wedge$, so tr^{i+1} has no string of length i and hence the string 'a' is not in the string tr^{i+1}. It indicates that 'a' belongs to the set $s(\wedge + r + r^2 + \cdots + r^i)$, i.e., sr^*.

Now consider a string 'a' in sr^*. Then definitely 'a' is in the set sr^k for $k \geq 0$. It implies that, 'a' is in right-hand side of equation (1). Therefore, 'a' is in left-hand side of equation (1). So, we can say that t and sr^* represent the same set of strings.

4.2.3   Application of Arden's Theorem

Arden's theorem is applicable to determine the regular expression represented by a transition diagram. The following assumptions are made regarding transition diagrams.

(a)   The transition diagram does not have any $\wedge$ move.

(b)   It has only one initial state, say s_1.

(c)   Its finite set of states are $s_1, s_2, s_3, \cdots, s_n$.

(d)   If s_i is the final state, then r_i is the regular expression representing the set of strings accepted by the transition diagram.

(e)   The regular expression representing the set of labels of edges from s_i to s_j is r_{ij}. If there is no path from s_i to s_j, then $r_{ij} = \phi$.

Therefore, we can get the following set of equations in terms of $s_1, s_2, s_3, \cdots, s_n$.

$$s_1 = s_1 r_{11} + s_2 r_{21} + s_3 r_{31} + \cdots + s_n r_{n1} + \wedge$$
$$s_2 = s_1 r_{12} + s_2 r_{22} + s_3 r_{32} + \cdots + s_n r_{n2}$$
$$s_3 = s_1 r_{13} + s_2 r_{23} + s_3 r_{33} + \cdots + s_n r_{n3}$$
$$\cdots \quad \cdots \quad \cdots \quad \cdots \quad \cdots \quad \cdots \quad \cdots$$
$$s_n = s_1 r_{1n} + s_2 r_{2n} + s_3 r_{3n} + \cdots + s_n r_{nn}$$

By applying substitutions and Arden's theorem repeatedly we can express final state s_i in terms of r_{ij}'s. If there is more than one final state in the transition diagram, then the resultant set of strings recognized by the transition diagram can be obtained by taking the union of all s_i's (final states).

Consider the following transition diagram

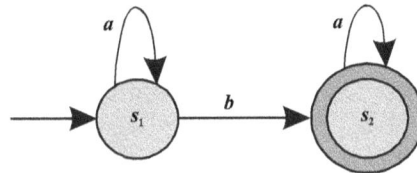

Here, $S = \{s_1, s_2\}; \Sigma = \{a, b\}$. Therefore, we have two equations: one for s_1 and another for s_2. Thus, we have the following equations.

$$s_1 = s_1 a + \wedge \tag{1}$$
$$s_2 = s_1 b + s_2 a \tag{2}$$

By using Arden's theorem, we have from equation (1)

$$s_1 = \wedge a^* = a^*$$

Therefore, equation (2) reduces to

$$s_2 = a^* b + s_2 a \qquad [\because s_1 = a^*]$$

Again, by applying Arden's theorem we have $s_2 = (a^* b) a^*$.

Thus, the regular expression corresponding to the transition diagram is given as $a^* b a^*$.

4.3   FINITE AUTOMATA AND REGULAR EXPRESSIONS

Our aim is to showing that the languages accepted by finite automaton are the languages denoted by regular expressions. In this section we establish the relation between regular expressions and finite automata.

4.3.1   Transition System and Regular Expressions

In this subsection we proceed to show that there exists a nondeterministic finite automaton (NFA) with ε-moves for every regular expression.

Theorem   *Let r be a regular expression. Then there exists an NFA with ε-moves that accepts L(r).*

Proof   We prove this theorem by mathematical induction on the number of characters in the regular expression r. Let the number of characters in r be 1. Therefore, r must be $\wedge$, ϕ or a for some $a \in \Sigma$. The nondeterministic finite automaton that will recognize these regular expressions are given in the following figures.

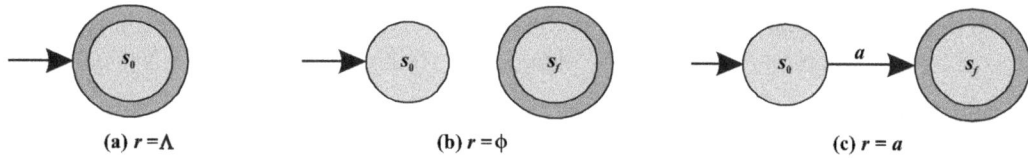

(a) $r = \Lambda$ (b) $r = \phi$ (c) $r = a$

The finite automaton for basic step is given above. Assume that the theorem is true for regular expressions with characters n or less. Our aim is to show that the theorem is true for $(n+1)$ characters. Let r have n characters. Hence there arises three cases, i.e., $r = s + t$; $r = st$ or $r = s^*$ where s and t are regular expressions each having n characters or less.

Case 1   Let us consider $r = s + t$. Thus there exists nondeterministic automata $M_1 = (S_1, \Sigma_1, \delta_1, s_1, \{f_1\})$ and $M_2 = (S_2, \Sigma_2, \delta_2, s_2, \{f_2\})$ such that $L(M_1) = s$ and $L(M_2) = t$. Since we may rename states of a nondeterministic finite automaton at all, we may assume that S_1 and S_2 are disjoint. Let s_0 be the newly introduced initial state whereas f_0 be the newly introduced final state. Construct M such that $M = (S, \Sigma, \delta, s_0, \{f_0\})$, where $S = S_1 \cup S_2 \cup \{s_0, f_0\}$, $\Sigma = \Sigma_1 \cup \Sigma_2$ and the transition function δ is defined as

i. $\delta(s_0, \varepsilon) = \{s_1, s_2\}$
ii. $\delta(s, a) = \delta_1(s, a)$ for $s \in S_1 - \{f_1\}$; $a \in \Sigma_1 \cup \{\varepsilon\}$
iii. $\delta(s, a) = \delta_2(s, a)$ for $s \in S_2 - \{f_2\}$; $a \in \Sigma_2 \cup \{\varepsilon\}$
iv. $\delta(f_1, \varepsilon) = \{f_0\} = \delta(f_2, \varepsilon)$

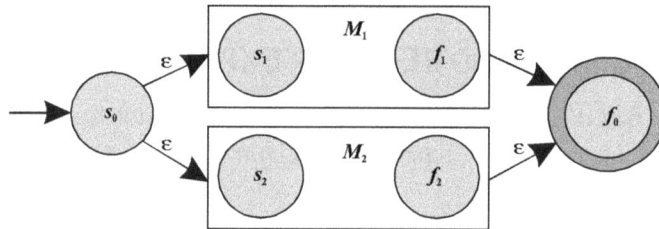

Therefore, all moves of M_1 and M_2 are present in M. The construction of M is given above. From the diagram it is clear that any path of M from s_0 to f_0 must start by going to either s_1 or s_2 on ε. If the path goes to s_1, then it may follow any path in M_1 to reach f_1 and then go to f_0 on ε. Similarly, if the path goes to s_2, then it may follow any path in M_2 to reach f_2 and then go to f_0 on ε. Thus, it is clear that $L(M) = L(M_1) + L(M_2)$.

Case 2   Let us consider $r = st$. Thus, there exists nondeterministic automata $M_1 = (S_1, \Sigma_1, \delta_1, s_1, \{f_1\})$ and $M_2 = (S_2, \Sigma_2, \delta_2, s_2, \{f_2\})$ such that $L(M_1) = s$ and $L(M_2) = t$. Construct $M = (S, \Sigma, \delta, s_1, \{f_2\})$ such that $S = S_1 \cup S_2$, $\Sigma = \Sigma_1 \cup \Sigma_2$ and the transition function δ is defined as

i. $\delta(s,a) = \delta_1(s,a)$ for $s \in S_1 - \{f_1\}; a \in \Sigma_1 \cup \{\varepsilon\}$

ii. $\delta(f_1, \varepsilon) = \{s_2\}$

iii. $\delta(s,a) = \delta_2(s,a)$ for $s \in S_2 - \{f_2\}; a \in \Sigma_2 \cup \{\varepsilon\}$

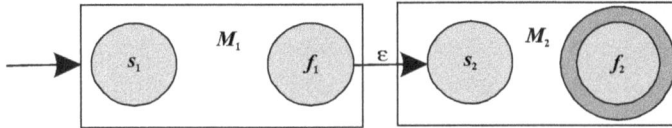

Therefore, all moves of M_1 and M_2 are present in M. The construction of M is given above. It is clear from the figure that every path in M from s_1 to f_2 is a path labelled by some string w_1 from s_1 to f_1 followed by the edge from f_1 to s_2 labelled ε and then further followed by a path from s_2 to f_2 by some string w_2. Therefore,

$$L(M) = \{w_1 w_2 \mid w_1 \in L(M_1) \text{ and } w_2 \in L(M_2)$$

i.e., $L(M) = L(M_1)L(M_2)$

Case 3   Let us consider $r = s^*$. Let $M_1 = (S_1, \Sigma_1, \delta_1, s_1, \{f_1\})$ such that $L(M_1) = s$. Let s_0 be the newly introduced initial state whereas f_0 be the newly introduced final state. Now construct $M = (S, \Sigma_1, \delta, s_0, \{f_0\})$ such that $S = S_1 \cup \{s_0, f_0\}$ and the transition function δ is defined as

i. $\delta(s_0, \varepsilon) = \{s_1, f_0\} = \delta(f_1, \varepsilon)$

ii. $\delta(s,a) = \delta_1(s,a)$ for $s \in S_1 - \{f_1\}; a \in \Sigma_1 \cup \{\varepsilon\}$

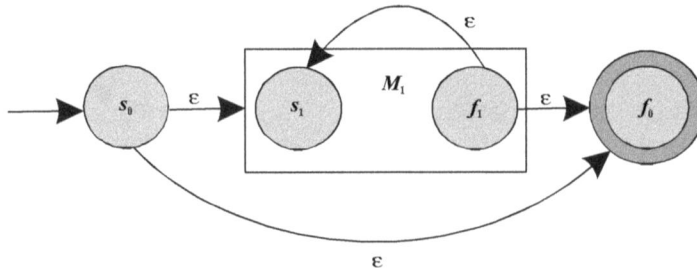

The construction of M is given above. It is clear from the figure that any path from s_0 to f_0 consists either of a path from s_0 to f_0 labelled ε or a path from s_0 to s_1 on ε followed by some path from s_1 to f_1 then back to s_1 labelled ε, each labelled by a string $L(M_1)$, followed by a path from s_1 to f_1 on a string in $L(M_1)$ and then to f_0 labelled on ε. Therefore, $L(M) = (L(M_1))^*$.

Consider a regular expression $a^*b + ab$. The above expression can be written as $r = a^*b + ab = s + t$, where $s = a^*b$ and $t = ab$. Now $t = ab$ can be written as $t = t_1 t_2$, where $t_1 = a$ and $t_2 = b$. The transition diagrams of $t_1 = a$, $t_2 = b$ are simple and are given below.

Therefore, the transition diagram of $t = t_1 t_2$ is given as

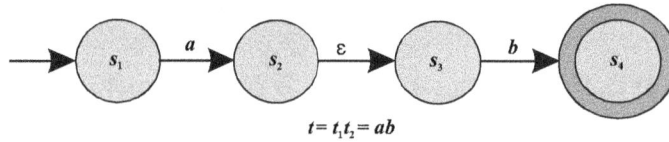

$$t = t_1 t_2 = ab$$

Similarly, $s = a^* b$ can be written as $s = r_1 r_2$, where $r_1 = a^*$ and $r_2 = b$. Again, $r_1 = a^*$ can be written as $r_1 = r_3^*$, where $r_3 = a$. The construction of $r_2 = b$ and $r_3 = a$ are simple and are given below.

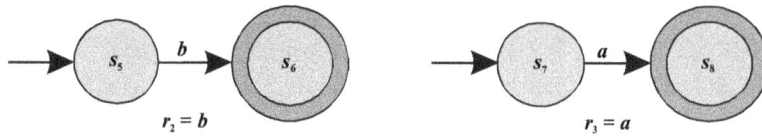

Therefore, the construction of $r_1 = r_3^*$ is given below.

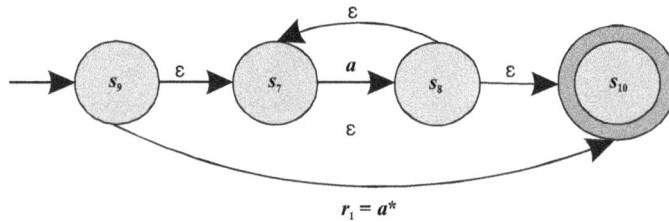

$$r_1 = a^*$$

Thus, the construction of $s = r_1 r_2$ is given below.

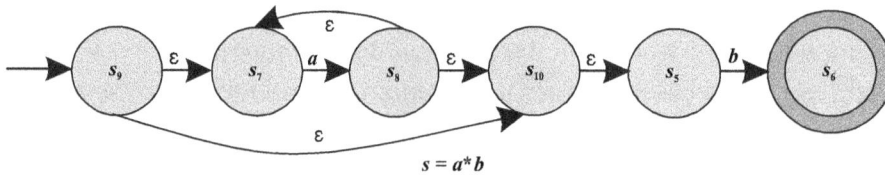

$$s = a^* b$$

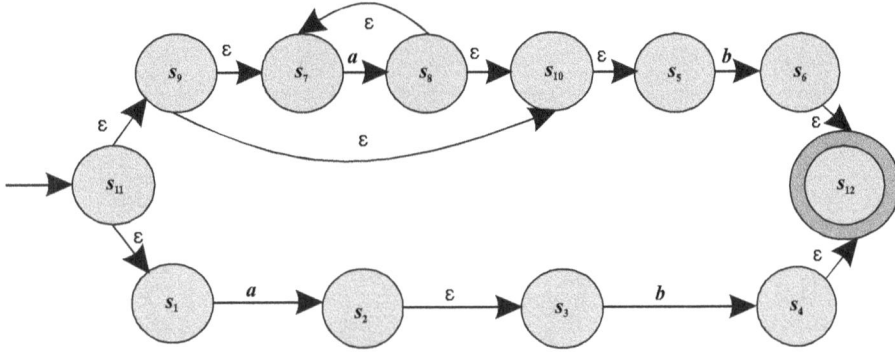

Finally, the transition diagram of $r = s + t$ is given above, where s_{11} is the initial state and s_{12} is the final state.

Theorem   *If L is accepted by deterministic finite automata, then L is denoted by a regular expression.*

Proof   Let L be the set accepted by the deterministic finite automata $M = (S, \Sigma, \delta, s_1, A)$, where $S = \{s_1, s_2, s_3, \cdots, s_n\}$. Let R_{ij}^k be the set of strings that takes the deterministic finite automata M from state s_i to state s_j without going through any state numbered higher than k. It indicates that, if $x \in R_{ij}^k$, then $\delta(s_i, x) = s_j$ and if $\delta(s_i, y) = s_l$ for any y that is of the form $y x_i \cdots x_n$, other than x or ε, then $l \le k$.

Note that R_{ij}^n denotes all strings that take s_i to s_j since there is no state numbered greater than n. Let us define R_{ij}^k recursively as follows:

$$R_{ij}^k = R_{ik}^{k-1}(R_{kk}^{k-1})^* R_{kj}^{k-1} \cup R_{ij}^{k-1} \tag{1}$$

$$R_{ij}^0 = \begin{cases} \{a \mid \delta(s_i, a) = s_j\} & \text{if} \quad i \ne j \\ \{a \mid \delta(s_i, a) = s_j\} \cup \{\varepsilon\} & \text{if} \quad i = j \end{cases}$$

The above definition of R_{ij}^k means that the inputs that cause M to move from s_i to s_j without passing through a state higher than s_k are either in

(a) R_{ij}^{k-1} (i.e., they never pass through a state as high as s_k) or

(b) Composed of a string in R_{ik}^{k-1} that takes M to s_k for the first time followed by zero or more strings in R_{kk}^{k-1} (that take M from s_k back to s_k without passing through s_k or a higher numbered state) followed by a string R_{kj}^{k-1} (that takes M from state s_k to s_j).

Now, we must show that for each i, j and k, there exists a regular expression r_{ij}^k denoting the language R_{ij}^k. We will prove this by the method of induction on k.

Basis When $k = 0$, we have $R_{ij}^0 = \{a\}$ or ε. It indicates that r_{ij}^0 can be written as $(a_1 + a_2 + \cdots + a_p)$ if $i \neq j$ or $(a_1 + a_2 + \cdots + a_p + \varepsilon)$ if $i = j$, where $\{a_1, a_2, \cdots, a_p\}$ is the set of symbols a such that $\delta(s_i, a) = s_j$. If there exists no such a's, then $r_{ij}^0 = \phi$ if $i \neq j$ or $r_{ij}^0 = \varepsilon$ if $i = j$.

Induction By the induction hypothesis, for each l and m there exists a regular expression r_{lm}^{k-1} such that $L(r_{lm}^{k-1}) = R_{lm}^{k-1}$. Hence by equation (1) there exists a regular expression r_{ij}^k such that

$$r_{ij}^k = r_{ik}^{k-1}(r_{kk}^{k-1})^* r_{kj}^{k-1} + r_{ij}^{k-1}$$

In order to complete the proof, we have only to observe that $L(M) = \bigcup_{s_j \in A} R_{1j}^n$. Since R_{1j}^n denotes the set of all strings that take s_1 to s_j. Therefore, $L(M)$ is denoted by the regular expression

$$r_{1j_1}^n + r_{1j_2}^n + r_{1j_3}^n + \cdots + r_{1j_p}^n, \text{ where } A = \{j_1, j_2, \cdots, j_p\}.$$

Consider the deterministic finite automata whose transition diagram is given below.

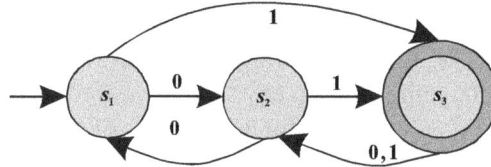

Let us take $M = (S, \Sigma, \delta, s_1, A)$ be a deterministic finite automaton with $S = \{s_1, s_2, s_3\}$ and $A = \{s_3\}$. Therefore, the required regular expression is r_{13}^3. Therefore, we have

$$r_{13}^3 = r_{13}^2 (r_{33}^2)^* r_{33}^2 + r_{13}^2 \tag{2}$$

Now,

$$r_{11}^0 = \varepsilon; \quad r_{12}^0 = 0 \quad [\because \delta(s_1, 0) = s_2]; \quad r_{13}^0 = 1 \quad [\because \delta(s_1, 1) = s_3].$$

Similarly, we get

$$r_{21}^0 = 0; \quad r_{22}^0 = \varepsilon; \quad r_{23}^0 = 1; \quad r_{31}^0 = \phi; \quad r_{33}^0 = \varepsilon \text{ and}$$

$$r_{32}^0 = (0 + 1) \quad [\because \delta(s_3, 0) = s_2; \delta(s_3, 1) = s_2]$$

Again, $r_{12}^1 = r_{11}^0 (r_{11}^0)^* r_{12}^0 + r_{12}^0 = \varepsilon(\varepsilon)^* 0 + 0 = 0$

$$r_{13}^1 = r_{11}^0 (r_{11}^0)^* r_{13}^0 + r_{13}^0 = \varepsilon(\varepsilon)^* 1 + 1 = 1$$

$$r_{22}^1 = r_{21}^0 (r_{11}^0)^* r_{12}^0 + r_{22}^0 = 0(\varepsilon)^* 0 + \varepsilon = (00 + \varepsilon)$$

$$r_{23}^1 = r_{21}^0 (r_{11}^0)^* r_{13}^0 + r_{23}^0 = 0(\varepsilon)^* 1 + 1 = (01 + 1)$$

$$r_{32}^1 = r_{31}^0 (r_{11}^0)^* r_{12}^0 + r_{32}^0 = \phi(\varepsilon)^* 0 + (0 + 1) = (0 + 1)$$

$$r_{33}^1 = r_{31}^0 (r_{11}^0)^* r_{13}^0 + r_{33}^0 = \phi(\varepsilon)^* 1 + \varepsilon = \varepsilon$$

Similarly, we have

$$r_{33}^2 = r_{32}^1 (r_{22}^1)^* r_{23}^1 + r_{33}^1$$
$$= (0+1)(00+\varepsilon)^* (01+1) + \varepsilon$$
$$= (0+1)(00)^* (0+\varepsilon)1 + \varepsilon$$
$$= (0+1)0^*1 + \varepsilon$$
$$r_{13}^2 = r_{12}^1 (r_{22}^1)^* r_{23}^1 + r_{13}^1$$
$$= 0(00+\varepsilon)^* (01+1) + 1$$
$$= 0(00)^* (0+\varepsilon)1 + 1$$
$$= 00^*1 + 1 = (00^* + \varepsilon)1 = 0^*1$$

Therefore, from equation (2) the regular expression is given as

$$r_{13}^3 = r_{13}^2 (r_{33}^2)^* r_{33}^2 + r_{13}^2$$
$$= 0^*1((0+1)0^*1+\varepsilon)^* ((0+1)0^*1+\varepsilon) + 0^*1$$
$$= 0^*1((0+1)0^*1)^* + 0^*1$$
$$= 0^*1(((0+1)0^*1)^* + \varepsilon) = 0^*1((0+1)0^*1)^*$$

i.e., $r_{13}^3 = 0^*1((0+1)0^*1)^*$

4.3.2   Transition Diagrams with ∧ Moves

The transition systems that are associated with null symbol (∧) are considered as transition systems with ∧ moves. These transitions generally occur when no input is applied. However it is possible to find a transition system without ∧ moves from a transition system with ∧ moves. The following steps are used to get a transition system without ∧ moves from a transition system with ∧ moves. Suppose we want to eliminate a ∧ move from a state s_i to state s_j, then we proceed as follows:

1. Find all transitions starting from the state s_j.
2. Draw all these transitions starting from s_i, without changing their transition labels. It indicates that, if $\delta(s_j, a) = s_k$ for an input symbol a, then draw a transition such that $\delta(s_i, a) = s_k$.
3. If s_i is an initial state, then make s_j also an initial state.
4. If s_j is a final state, then make s_i also the final state.

Consider a finite automaton with ∧-moves given in the following diagram. Our aim is to remove the ∧-moves from the given finite automaton.

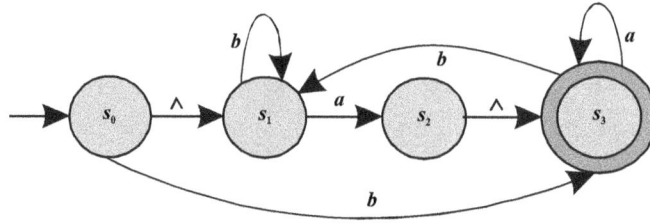

It is clear from the state graph that, there are two $\wedge$-moves: one from s_0 to s_1 and another from s_2 to s_3. First we remove the $\wedge$-move from s_0 to s_1. From the system it is clear that $\delta(s_1, a) = s_2$ and $\delta(s_1, b) = s_1$. Therefore, $\delta(s_0, a) = s_2$ and $\delta(s_0, b) = s_1$. Again, s_0 is an initial state, so s_1 is made an initial state. Thus, we get the following state graph.

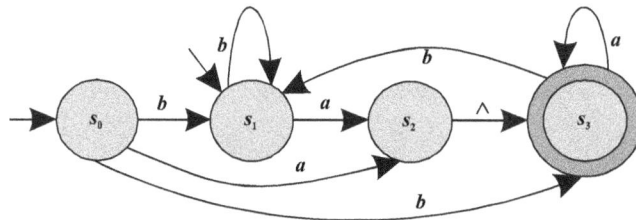

Now, we have to remove the $\wedge$-move from s_2 to s_3. From the transition system it is clear that $\delta(s_3, a) = s_3$ and $\delta(s_3, b) = s_1$. Therefore, $\delta(s_2, a) = s_3$ and $\delta(s_2, b) = s_1$. Again, s_3 is a final state, so we made s_2 also a final state. Thus, the transition system without $\wedge$-moves is given below.

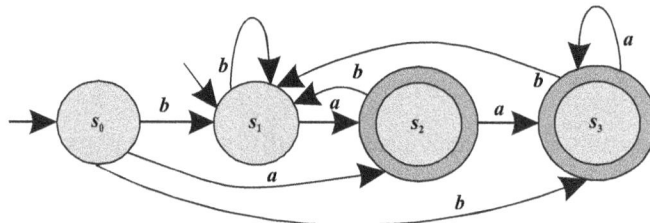

4.4   PUMPING LEMMA

In this section we prove a basic result, called the pumping lemma. It is made up of two words: pumping and lemma. The word pumping means to generate many input strings from a given string whereas the word lemma refers to an intermediate result in a proof. It is a powerful tool for proving that certain languages are not regular.

Statement   Let $M = (S, \Sigma, \delta, s_0, A)$ be a finite automaton with n-states. Let L be the regular set accepted by M. Let $z \in L$ and $|z| \geq m$. If $m \geq n$, then there exists u, v, y such that $z = uvy$, with $v \neq \wedge$ and $uv^i y \in L$ for each positive integer i, i.e., $i \geq 0$.

Proof   Let $M = (S, \Sigma, \delta, s_0, A)$ be a finite automaton with n-states. Assume that L be the regular set accepted by M. Let us take $z = a_1 a_2 a_3 \cdots a_m$ with $m \geq n$ and $\delta(s_0, a_1 a_2 a_3 \cdots a_i) = s_i$ for

$i = 1, 2, 3, \cdots, m$ with the sequence of states $S_1 = \{s_0, s_1, s_2, \cdots, s_m\}$ having path value $z = a_1 a_2 a_3 \cdots a_m$. As there are only n-states in the finite automaton M, so at least two states in S_1 must coincide. Therefore, there exists j and k, $0 \le j \le k \le n$ such that $s_j = s_k$. Therefore, the string $z = a_1 a_2 a_3 \cdots a_m$ can be decomposed into three substrings $u = a_1 a_2 \cdots a_j$; $v = a_{j+1} a_{j+2} \cdots a_k$ and $y = a_{k+1} a_{k+2} \cdots a_m$. As $k \le n, |uv| \le n$ and $z = uvy$, the path labelled $z = a_1 a_2 a_3 \cdots a_m$ in the transition diagram of M is given below.

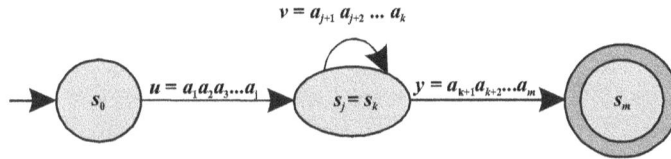

The automaton M starts from the initial state s_0 and reaches $s_j = s_k$ after applying the string u. On applying the string v, it comes back to $s_j = s_k$. It implies that, after applying v^i for each $i \ge 0$, the automaton remains on the same state $s_j = s_k$. On applying z, it reaches the final state s_m. Therefore, $uv^i y \in L$. As every state in S_1 is obtained by applying an input symbol, $v \ne \wedge$.

4.4.1   Application of Pumping Lemma

The pumping lemma is extremely useful to prove that certain sets are not regular. The following steps are used for proving that a given set is not regular. Here, we use method of contradiction to prove certain languages are not regular.

(a) Assume that the given language L is regular. Let the number of states in the corresponding finite automaton be n.

(b) Choose a string z such that $|z| \ge n$. Using pumping lemma express $z = uvy$ with $|uv| \le n$ and $|v| \ge 0$.

(c) Find a suitable integer i such that $uv^i y \notin L$. It leads to a contradiction. Therefore, our assumption is wrong and hence L is not regular.

For example consider the language L as $L = \{0^n 1^n \mid n \ge 1\}$. Assume that L is regular and let n be the number of states in finite automaton. Let $z = 0^n 1^n$. Therefore, $|z| = 2n > n$. By pumping lemma we write $z = uvy$ with $|uv| \le n$ and $|v| \ne 0$. Our aim is to find out at least one i such that $uv^i y \notin L$ for getting a contradiction. Therefore, it is essential to identify the string v. Here the string v can be of the following types:

(a) v has only 0's, i.e., $v = 0^k$ for some $k \ge 1$

(b) v has only 1's, i.e., $v = 1^l$ for some $l \ge 1$

(c) v has both 0's and 1's, i.e., $v = 0^k 1^l$ for some $k, l \ge 1$

Case (a)   In this case, $z = uvy = 0^{n-k} 0^k 1^n$. Thus we have $u = 0^{n-k}$ and $y = 1^n$. On taking $i = 0$ we have $uv^i y = uy = 0^{n-k} 1^n$. As $k \ge 1$ $(n - k) \ne n$. Therefore, $uv^i y \notin L$.

Case (*b*)   In this case, $z = uvy = 0^n 1^l 1^{n-l}$. Thus we have $u = 0^n$ and $y = 1^{n-l}$. On taking $i = 0$ we have $uv^i y = uy = 0^n 1^{n-l}$. As $l \geq 1$, $(n - l) \neq n$. Therefore, $uv^i y \notin L$.

Case (*c*)   In this case, $z = uvy = 0^{n-k} 0^k 1^l 1^{n-l}$. Thus we have $u = 0^{n-k}$ and $y = 1^{n-l}$. On taking $i = 2$ we have $uv^i y = uv^2 y = 0^{n-k} 0^k 1^l 0^k 1^l 1^{n-l}$ and it is not in the form of $0^n 1^n$. Thus, $uv^i y \notin L$.

Therefore, we get a contradiction for all cases. Hence, our assumption is wrong. Thus the given language $L = \{0^n 1^n \mid n \geq 1\}$ is not regular.

4.5   REGULAR SETS AND REGULAR GRAMMAR

A regular set is a set accepted by a deterministic finite automaton. In this section, we will discuss closure properties of regular sets, decision algorithms for regular sets and finally we give the relation between regular sets and regular grammar.

4.5.1   Closure Properties of Regular Sets

In Chapter 3, we have discussed closure properties in general. Here, we discuss the closure properties of regular sets under union, concatenation, Kleene star, complementation, intersection, substitution and homomorphism.

Theorem   *The regular sets are closed under union, concatenation and Kleene closure.*

Proof   Assume that r and s are two regular expressions denoting regular sets R and S respectively. Therefore, by definition of regular expression $(r + s)$ denotes the regular set $(R \cup S)$. So, $(R \cup S)$ is also a regular set.

Similarly, rs denotes the regular set RS. So, RS is also a regular set. Again r^* is a regular expression denoting the regular set R^* and hence R^* is also a regular set.

Theorem   *The class of regular sets is closed under complementation. In other words, if R is a regular set and $R \subseteq \Sigma^*$, then $\Sigma^* - R$ is a regular set.*

Proof   Let R be the regular set for the deterministic finite automaton $M = (S, \Sigma_1, \delta, s_0, A)$ and let $R \subseteq \Sigma^*$. Hence, there arises two cases, either $\Sigma_1 = \Sigma$ or $\Sigma_1 \neq \Sigma$. If $\Sigma_1 = \Sigma$, then there is no problem. If $\Sigma_1 \neq \Sigma$, then there arises two cases, i.e., either symbols in Σ_1 not in Σ, or symbols in Σ not in Σ_1. If there are symbols in Σ_1 that are not in Σ, we may delete all transitions of M on symbols that does not belongs to Σ. If there are symbols in Σ that are not in Σ_1, then introduce a dead state '*d*' into M such that

$$\delta(d, a) = d \quad \forall \, a \in \Sigma \quad \text{and}$$
$$\delta(s, a) = d \quad \forall \, s \in S, a \in \Sigma - \Sigma_1$$

Now, in order to accept $\Sigma^* - R$, construct another deterministic finite automaton $M_1 = (S, \Sigma, \delta, s_0, A')$ such that $A' = S - A$. It implies that a final state of M is a non-final state of M_1 and vice versa. The transition diagrams of M and M_1 remains same except for the final states.

Now, $z \in T(M_1)$, i.e., M_1 accepts a string z, if and only if $\delta(s_0, z) \in A'$, i.e., $z \notin R$. This implies that $z \in \Sigma^* - R$. Therefore, $T(M_1) = \Sigma^* - R$.

Theorem *The regular sets are closed under intersection, i.e., if R and S are two regular sets defined over Σ, then $(R \cap S)$ is also a regular set over Σ.*

Proof   Let R and S be two regular sets defined over Σ. Our aim is to show that $(R \cap S)$ is also a regular set over Σ. By De Morgan's law we have

$$(R \cap S) = \overline{(\overline{R} \cup \overline{S})} = \Sigma^* - (\overline{R} \cup \overline{S})$$

We know that if R is a regular set, then $\overline{R} = \Sigma^* - R$ is also a regular set. Similarly, we get $\overline{S}$ as a regular set. Now, $\overline{R}$ and $\overline{S}$ are regular sets and so their union, i.e., $(\overline{R} \cup \overline{S})$ is also regular. It implies that $\Sigma^* - (\overline{R} \cup \overline{S})$ is regular. Therefore, $(R \cap S)$ is regular.

Theorem *The class of regular sets is closed under substitution.*

Proof   Let R be a regular set such that $R \subseteq \Sigma^*$ and for each $a \in \Sigma$, let $R_a \subseteq \Delta^*$ be a regular set for some alphabet Δ.

Let $f : \Sigma \to \Delta^*$ be the substitution defined by $f(a) = R_a$. Select regular expressions denoting R and R_a. Replace each occurrence of the symbol *a* in the regular expression for R by the regular expression for R_a. In order to prove the resulting regular expression denotes $f(R)$, observe that the substitution of a union, product or closure is the union, product or closure of the substitution.

The proof will be complete by applying mathematical induction on the number of operators in the regular expression.

For example, consider $f(0) = b^*$ and $f(1) = a$. This implies that $f(0)$ is the language of all strings of b's whereas $f(1)$ is the language $\{a\}$. Then, $f(101)$ is the regular set ab^*a. If $L = 0^*(0+1)1^*$, then $f(L) = (b^*)^*(b^* + a)a^*$.

Theorem *The class of regular sets is closed under homomorphism and inverse homomorphism.*

Proof   Homomorphism h is a substitution such that $h(a)$ contains a single string for each a. As homomorphism is a substitution and regular sets are closed under substitution, so regular sets is closed under homomorphism. Now our aim is to show that regular sets are closed under inverse homomorphism.

To show closure under inverse homomorphism, it is useful to define the inverse homomorphic image of language L. It is defined as

$$h^{-1}(L) = \{x \mid h(x) \in L\}$$

Similarly, for a string z we have

$$h^{-1}(z) = \{x \mid h(x) = z\}$$

Let $M = (S, \Sigma, \delta, s_0, A)$ be a deterministic finite automata (DFA) accepting the language L and let h be a homomorphism from Δ to Σ^*. In order to accept $h^{-1}(L)$ by reading symbol $a \in \Delta$ and simulating M on $h(a)$, let us construct a deterministic finite automata (DFA) $M_1 = (S, \Sigma, \delta', s_0, A)$ such that $\delta'(s, a) = \delta(s, h(a))$ for $s \in S$ and $a \in \Delta$. It is to be noted that $h(a)$ may be a long string or ε, but δ is defined on all strings by extension. By method of induction on $|x|$ it can be shown that

$$\delta'(s, x) = \delta(s, h(x))$$

This implies that, M_1 accepts x if and only if M accepts $h(x)$, i.e., $L(M_1) = h^{-1}(L(M))$.

4.5.2 Decision Algorithms for Regular Sets

It is important to have algorithms to answer various questions related to regular sets such as: is a given language empty, finite or infinite?; is one regular set equivalent to another? Before we discuss the existence of algorithms we must know about decision algorithm. A decision algorithm is a finite set of instructions that, if followed, accomplishes a particular task. It has the following characteristics.

1. Each instruction in the algorithm must be clear and unambiguous.

2. For every possible input, if we trace out the algorithm, then the algorithm must terminate after a finite number of steps.

3. Every time the output must be same for the same input.

Algorithms to determine whether a regular set is empty, finite or infinite may be based on the following theorem.

Theorem *The set of sentences accepted by a finite automaton M with n-states is:*

(a) nonempty if and only if the finite automaton accepts a sentence of length less than n.

(b) infinite if and only if the automaton accepts some sentence of length l, where $n \leq l \leq 2n$.

Thus there is an algorithm to determine whether a finite automaton accepts zero, finite or an infinite number of sentences.

Proof (*a*) Suppose finite automaton M accepts a non-empty set. Let z be a string as short as any other string accepted. Therefore, by pumping lemma $|z| < n$. If the string z were the shortest and $|z| \geq n$, then by pumping lemma $z = uvy$, and uy is a shorter word in the language. It shows that whether language accepted by finite automaton, $L(M)$ is empty or not. Thus, there is a procedure guaranteed to halt.

(*b*) If z is in $L(M)$ and $n \leq l \leq 2n$, then by pumping lemma there exists u, v, y such that $z = uvy$ and for all $i, uv^i y \in L$. It indicates that $L(M)$ is infinite.

Conversely, if the language accepted by finite automaton M, i.e., $L(M)$ is infinite, then there exists $z \in L(M)$ such that $|z| \geq n$. Our aim is to show that $|z| \leq 2n$. Suppose on the contrary consider the length of string z is at least $2n$, but as short as any string in $L(M)$ whose length is greater than or equal to $2n$. Therefore, by pumping lemma $z = uvy$ with $1 \leq |v| \leq n$ and $uy \in L(M)$. This implies that either z was not a shortest string of length $2n$ or more, or $n \leq |uy| \leq (2n-1)$. It leads to a contradiction in either case. Therefore, $n \leq |z| < 2n$. It decides whether $L(M)$ is infinite. For example if any string of length between n and $(2n-1)$ is in $L(M)$, then there is a procedure that is guaranteed to halt.

Note A finite automata is said to be empty if there exists no path from the initial state to the final state. A diagrammatic representation is given below.

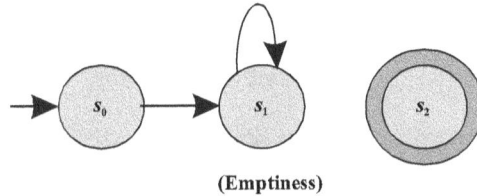

(Emptiness)

A finite automaton can also be treated as a directed graph. If the finite automaton contains no cycles, then it is said to be finite whereas the finite automaton containing any cycle is termed as infinite. A diagrammatic representation is given below.

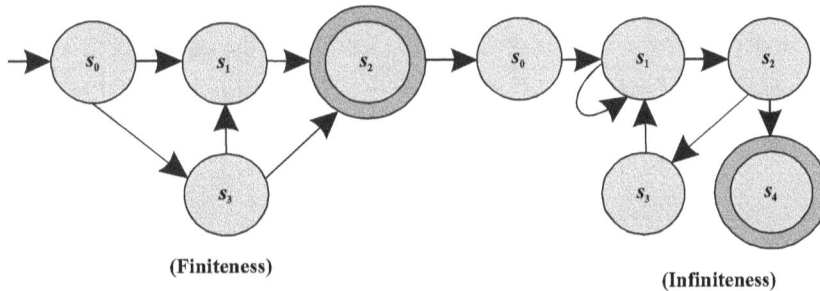

(Finiteness)

(Infiniteness)

4.5.3 Myhill–Nerode Theorem

We extend our idea by giving another characterization of the regular sets and regular languages on an alphabet Σ. We begin with some definitions.

Definition Let $L \subseteq \Sigma^*$, where Σ is an alphabet. For strings $x, y \in \Sigma^*$, we write $x \equiv_L y$ to mean that for each $z \in \Sigma^*$ we have $xz \in L$ if and only if $yz \in L$. It indicates that $x \equiv_L y$ if and only if for every $z \in \Sigma^*$, either both or neither of xz and yz is in L. It is clear that $\equiv_L$ has the following properties.

(a) $x \equiv_L x$.
(b) If $x \equiv_L y$, then $y \equiv_L x$.
(c) If $x \equiv_L y$ and $y \equiv_L z$, then $x \equiv_L z$.

Therefore, the relation $\equiv_L$ is an equivalence relation. It is also obvious that, if $x \equiv_L y$, then for all $z \in \Sigma^*$, $xz \equiv_L yz$.

Definition (Right invariant) Let $M = (S, \Sigma, \delta, s_0, A)$ be a deterministic finite automata. An equivalence relation $\equiv$ such that $x \equiv y$ implies $xz \equiv yz$ is called right invariant with respect to concatenation. It is obvious that every finite automaton induces a right invariant equivalence relation, defined as $\equiv_M$ such that $x \equiv_M y \Rightarrow xz \equiv_M yz$ for all $z \in \Sigma^*$, on its set of input strings.

Definition (Spanning set)   Let $L \subseteq \Sigma^*$, where Σ is an alphabet. Let $Q \subseteq \Sigma^*$. Then Q is said to be a spanning set for L if

   (a) Q is finite and

   (b) for every $x \in \Sigma^*$, there is a $y \in Q$ such that $x \equiv_L y$.

Theorem *The following three statements are equivalent.*

   (a) *The set $L \subseteq \Sigma^*$ is accepted by some deterministic finite automaton.*

   (b) *L is the union of some of the equivalence classes of a right invariant (with respect to concatenation) equivalence relation of finite index.*

   (c) *Let the equivalence relation $\equiv_L$ be defined by, $x \equiv_L y$ if and only if $z \in \Sigma^*$, $xz \in L$ exactly when $yz \in L$. Then the equivalence relation $\equiv_L$ is of finite index.*

Proof $(a) \rightarrow (b)$ Let $M = (S, \Sigma, \delta, s_0, A)$ be a deterministic finite automaton. Assume that L is accepted by some deterministic finite automaton M. Let $\equiv_M$ be the equivalence relation $x \equiv_M y$ if and only if $\delta(s_0, x) = \delta(s_0, y)$. Again, for any $z \in \Sigma^*$ we have

$$\delta(s_0, xz) = \delta(\delta(s_0, x), z)$$
$$= \delta(\delta(s_0, y), z) = \delta(s_0, yz); \qquad [\because \delta(s_0, x) = \delta(s_0, y)]$$
$$\text{i.e.,} \delta(s_0, xz) = \delta(s_0, yz)$$

It indicates that $\equiv_M$ is right invariant. Also, the index $\equiv_M$ is finite, since the index is at most the number of states of S. Furthermore, L is the union of those equivalence classes that include a string x such that $\delta(s_0, x)$ is in the set of final states A, i.e., the equivalence classes corresponding to final states.

$(b) \rightarrow (c)$ We prove that any equivalence relation $\equiv$ satisfying (b) is a refinement of $\equiv_L$, i.e., every equivalence class of $\equiv$ is entirely contained in some equivalence class of $\equiv_L$. Therefore, the index of equivalence relation $\equiv_L$ cannot be greater than the index of $\equiv$. So, the index of $\equiv_L$ is finite.

Assume that $x \equiv y$. Then, we have $xz \equiv yz$ for each $z \in \Sigma^*$, since $\equiv$ is right invariant. It indicates that yz is in L if and only if xz is in L. Therefore, $x \equiv_L y$, and hence the equivalence class of x in $\equiv$ is contained in the equivalence class of x in $\equiv_L$. So, we conclude that each equivalence class of $\equiv$ is contained within some equivalence class of $\equiv_L$.

$(c) \rightarrow (a)$ First we have to prove that $\equiv_L$ is right invariant. Suppose $x \equiv_L y$, and let $z \in \Sigma^*$. Since $x \equiv_L y$, we know by definition of $\equiv_L$ that for any w, xw is in L exactly when yw is in L. Let $w = vz$. It indicates that xvz is in L exactly when yvz is in L. Thus, the equivalence relation $\equiv_L$ is right invariant.

Let S' be the finite set of equivalence classes of $\equiv_L$ and let $[x]$ be the element of S' containing x. Let us define $\delta'([x], a) = [xa]$. Since $\equiv_L$ is right invariant the definition is consistent. If we choose y instead of x from equivalence class $[x]$, then we would have $\delta'([x], a) = [ya]$. But $x \equiv_L y$, so xz is in L exactly when yz is in L. In particular, if $z = aw$, xaw is in L exactly when yaw is in L. Thus, $xa \equiv_L ya$ and $[xa] = [ya]$.

Let $s_0' = [\varepsilon]$ and let $A' = \{[x] | x \text{ is in } L\}$. Therefore, the finite automaton $M' = (S', \Sigma, \delta', s_0', A')$ accepts L, since $\delta'(s_0', x) = [x]$, and thus x is in $L(M')$ if and only if $[x]$ is in A'.

4.5.4   Construction of Regular Grammar for a DFA

Let $M = (S, \Sigma, \delta, s_0, A)$ be a deterministic finite automaton. Let us define $T(M)$ as the path values concatenated from the initial state to the final state. So, for the grammar G, the productions should correspond to the transitions and there should be a provision for terminating the derivation once a transition terminating at some final state is achieved. Now, we construct the grammar G as $G = (V_N, T, P, Q_0)$ where,

V_N = Finite non-empty set of variables $\{Q_0, Q_1, Q_2, \cdots, Q_n\}$

T = Finite non-empty set of terminals

Q_0 = Special variable called the start symbol

P = Set of production rules.

The production rules P is defined as

(a) $Q_i \rightarrow aQ_j$ is in P if $\delta(s_i, a) = s_j$; $s_j \notin A$

(b) $Q_i \rightarrow aQ_j$ and $Q_i \rightarrow a$ are in P if $\delta(s_i, a) = s_j$; $s_j \in A$

Theorem *If G is the regular grammar and T(M) is the set of strings accepted by deterministic finite automaton M, then L(G) = T(M).*

Proof   Given $T(M)$ is the set of strings accepted by deterministic finite automaton $M = (S, \Sigma, \delta, s_0, A)$. We construct the grammar G as $G = (V_N, T, P, Q_0)$, where the productions P is defined as above.

Therefore, $Q_0 \rightarrow a_1 Q_1$ if and only if $\delta(s_0, a_1) = s_1$

$$Q_1 \rightarrow a_2 Q_2 \quad \text{if and only if} \quad \delta(s_1, a_2) = s_2$$
$$Q_2 \rightarrow a_3 Q_3 \quad \text{if and only if} \quad \delta(s_2, a_3) = s_3$$
$$\cdots \quad \cdots \quad \cdots \quad \cdots \quad \cdots \quad \cdots \quad \cdots \quad \cdots$$
$$\cdots \quad \cdots \quad \cdots \quad \cdots \quad \cdots \quad \cdots \quad \cdots \quad \cdots$$
$$Q_{k-2} \rightarrow a_{k-1} Q_{k-1} \quad \text{if and only if} \quad \delta(s_{k-2}, a_{k-1}) = s_{k-1}$$

with $\delta(s_{k-1}, a_k)$ belongs to some final state in A. Let it be s_k. Therefore, we have $Q_{k-1} \rightarrow a_k s_k$ and $Q_{k-1} \rightarrow a_k$. On combining these results we get

$$Q_0 \rightarrow a_1 Q_1 \rightarrow a_1 a_2 Q_2 \rightarrow a_1 a_2 a_3 Q_3$$

$$\cdots \quad \cdots \quad \cdots$$

$$\rightarrow a_1 a_2 a_3 \cdots a_{k-1} Q_{k-1} \rightarrow a_1 a_2 a_3 \cdots a_{k-1} a_k$$

This proves that $z = a_1 a_2 a_3 \cdots a_{k-1} a_k \in L(G)$ if and only if $\delta(s_0, a_1 a_2 a_3 \cdots a_{k-1} a_k) \in A$ i.e., $z \in T(M)$.

Consider the following deterministic finite automaton

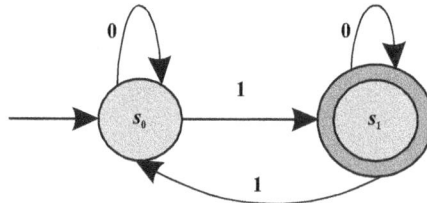

Here $S = \{s_0, s_1\}$; $\Sigma = \{0,1\}$; $A = \{s_1\}$. Thus, the regular grammar G corresponding to the above deterministic finite automata is given by the following productions.

$$Q_0 \rightarrow 0 Q_0 \qquad\qquad [\because \delta(s_0, 0) = s_0 \notin A]$$

$$Q_0 \rightarrow 1 Q_1 \text{ and } Q_0 \rightarrow 1 \qquad [\because \delta(s_0, 1) = s_1 \in A]$$

$$Q_1 \rightarrow 0 Q_1 \text{ and } Q_1 \rightarrow 0 \qquad [\because \delta(s_1, 0) = s_1 \in A]$$

$$Q_1 \rightarrow 1 Q_0 \qquad\qquad [\because \delta(s_1, 1) = s_0 \notin A]$$

Therefore, the regular grammar G is defined as (V_N, T, P, Q_0) such that $V_N = \{Q_0, Q_1\}$; $T = \{0,1\}$ with the productions P as $Q_0 \rightarrow 0 Q_0$; $Q_0 \rightarrow 1 Q_1$; $Q_0 \rightarrow 1$; $Q_1 \rightarrow 0 Q_1$; $Q_1 \rightarrow 0$ and $Q_1 \rightarrow 1 Q_0$, where Q_0 is the start symbol.

4.5.5   Construction of Regular Expression for a Regular Grammar

Let $G = (V_N, T, P, Q_0)$ be the regular grammar. A regular grammar consists of the productions P in the form of

$$Q_0 \rightarrow \alpha Q_0 | \alpha B | \wedge; \quad B \rightarrow \beta C; \quad C \rightarrow \gamma C | \beta, \text{ etc., where } \alpha, \beta, \gamma \in T^*.$$

Our aim is to construct the regular expression. The following rules are adopted to construct a regular expression for a given regular grammar.

(*a*) If the production is in the form of $C \rightarrow \gamma C | \beta$, then the equivalent regular expression is given as $C \equiv \gamma^* \beta$; $\beta, \gamma \in T^*$.

(*b*) If the production is in the form of $B \rightarrow \beta C$, then the equivalent regular expression is given as $B \equiv \beta \gamma^* \beta$; $\beta, \gamma \in T^*$.

(*c*) If the production is in the form of $Q_0 \rightarrow \alpha Q_0 | \alpha B | \wedge$, then the equivalent regular expression is given as

$$Q_0 \to \alpha^* \big| \alpha\beta\gamma^*\beta \qquad \text{or} \qquad Q_0 \equiv \alpha^* + \alpha\beta\gamma^*\beta$$

In this case, the productions are treated separately. The regular expression described by the start symbol is the required regular expression.

Consider the following productions representing the regular grammar G

$$Q_0 \to aQ_1 \big| bQ_2; \quad Q_1 \to bQ_1 \big| c; \quad Q_2 \to b \big| cQ_2$$

The regular expression corresponding to $Q_2 \to b \big| cQ_2$ is given as $Q_2 \equiv c^* b$.

The regular expression corresponding to $Q_1 \to bQ_1 \big| c$ is given as $Q_1 \equiv b^* c$.

The regular expression corresponding to $Q_0 \to aQ_1 \big| bQ_2$ is given as

$$Q_0 \equiv ab^* c \big| bc^* b \qquad \text{or} \quad Q_0 \equiv ab^* c + bc^* b$$

Therefore, the required regular expression is $ab^* c + bc^* b$.

4.5.6 Construction of Transition System for a Regular Grammar

Let $G = (V_N, T, P, Q_0)$ be the regular grammar, where $V_N = \{Q_0, Q_1, Q_2, \cdots, Q_n\}$. Our aim is to construct a transition system M where

(a) States correspond to variables V_N.

(b) Initial state corresponds to Q_0.

(c) Transitions in M correspond to the productions P.

If the production applied in any derivation is in the form of $Q_i \to a$, then the corresponding transition terminates at a new state that is unique final state. We define the transition system M as $M = (S, \Sigma, \delta, s_0, A)$, where $S = \{s_0, s_1, s_2, \cdots, s_n, s_f\}$; $A = \{s_f\}$ and δ is defined as follows:

(a) For each production of the form $Q_i \to aQ_j$, there exists a transition from state s_i to s_j on input a.

(b) For each production of the form $Q_k \to a$, there exists a transition from the state s_k to the final state s_f on input a.

Therefore it is clear that, the total number of states in M is one greater than total number of variables in G.

For example, consider the grammar G as:

$$G = (\{Q_0, Q_1\}, \{a, b\}, \{Q_0 \to bQ_1, Q_1 \to aQ_1 \big| b \big| bQ_0\}, Q_0)$$

Here, $V_N = \{Q_0, Q_1\}; T = \{a, b\}; P = \{Q_0 \to bQ_1, Q_1 \to aQ_1 \big| b \big| bQ_0\}$ and initial variable Q_0. Suppose that the transition system M to be constructed is defined as $M = (S, \Sigma, \delta, s_0, s_f)$, where $S = \{s_0, s_1, s_f\}$. From S, it is clear that the states s_0, s_1 correspond to the variables Q_0 and Q_1

respectively whereas s_f is the newly introduced final state. In order to find the transitions, we have to consider the productions P. Therefore, we have the following transitions. The production:

(a) $Q_0 \rightarrow bQ_1$ represents the transition $\delta(s_0, b) = s_1$.

(b) $Q_1 \rightarrow aQ_1$ represents the transition $\delta(s_1, a) = s_1$.

(c) $Q_1 \rightarrow b$ represents the transition $\delta(s_1, b) = s_f$.

(d) $Q_1 \rightarrow bQ_0$ represents the transition $\delta(s_1, b) = s_0$.

Therefore, the transition system M is given below.

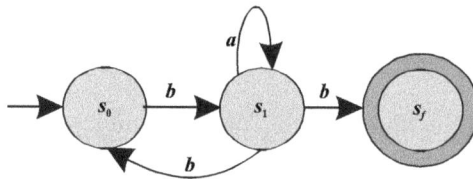

4.6 EQUIVALENCE OF TWO FINITE AUTOMATA

In chapter 2 we have studied minimization of finite automata. It implies that, there exist at least two automata that accept the same language. As a result equivalence of finite automata is developed. In this section we discuss the equivalence of finite automata. Before explaining any method, first we state the equivalence of finite automata. Two finite automata M_1 and M_2 defined over Σ are said to be equivalent if they accept the same set of strings over Σ. Two finite automata M_1 and M_2 are said to be not equivalent if there exists at least one string z of Σ^* such that one automaton reaches a final state whereas the other automaton reaches a non-final state. Here, we discuss a method, called as comparison method, to test the equivalence of two finite automata over Σ.

4.6.1 Comparison Method

Suppose M_1 and M_2 are two finite automata defined over Σ. Here, we construct a comparison table consisting of $(n + 1)$ columns, where n is the number of input symbols. The first column represents (s, s'), where $s \in M_1$ and $s' \in M_2$. If (s, s') appears in some row of first column, then the corresponding entry in a-column is (s_a, s'_a), where $s_a \in M_1$ and $s'_a \in M_2$, that are reachable from s and s' on application of a.

The comparison process starts from (s_0, s'_0) where s_0 and s'_0 are initial states of M_1 and M_2 respectively. The first element in the second column is (s_a, s'_a), where s_a and s'_a are reachable from s_0 and s'_0 respectively on application of a. The process is repeated by considering the pairs obtained in the second and subsequent columns that are not present in the first column. Therefore, there arise two cases:

Case 1 If we find a pair (s, s') such that s is a final state of M_1 whereas s' is a non-final state of M_2 or vice versa, then the process terminates. In this case we conclude that M_1 and M_2 are not equivalent.

Case 2 Here the construction is terminated when no new pair appears in the second and subsequent columns that are not in the first column. In this case we conclude that M_1 and M_2 are equivalent.

For example, consider the following two deterministic finite automata M_1 and M_2 whose transition diagram is given below defined over $\Sigma = \{a,b\}$. Our aim is to show that finite automaton M_1 is equivalent to the finite automaton M_2.

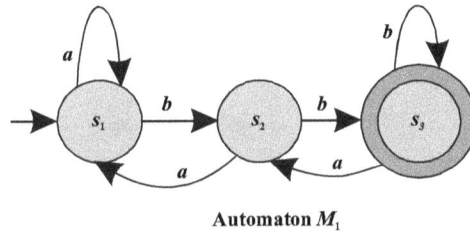

Automaton M_1

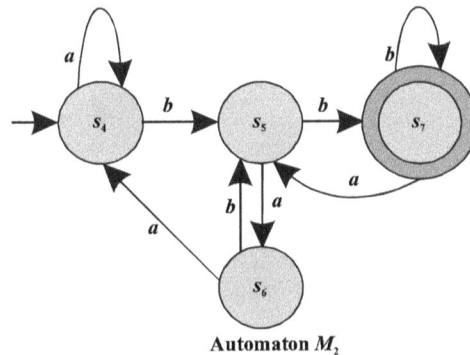

Automaton M_2

From the transition diagram it is clear that the initial states in M_1 and M_2 are s_1 and s_4 respectively. Thus the first element of the first column in the comparison table is (s_1,s_4). The first element in the second column is (s_1,s_4), since s_1 and s_4 are a-reachable from the initial states s_1 and s_4 respectively. The complete comparison table is given below.

(s,s')	(s_a,s'_a)	(s_b,s'_b)
(s_1,s_4)	(s_1,s_4)	(s_2,s_5)
(s_2,s_5)	(s_1,s_6)	(s_3,s_7)
(s_1,s_6)	(s_1,s_4)	(s_2,s_5)
(s_3,s_7)	(s_2,s_5)	(s_3,s_7)

Here, the construction is terminated as no new pair appears in the second and third column that is not present in the first column. We also do not get a pair (s,s') such that s is a final state of M_1 and s' is a non-final state of M_2 or vice versa. Therefore, we conclude that M_1 and M_2 are equivalent.

SOLVED EXAMPLES

Example 1

Show that the automata M_1 and M_2 given below are not equivalent.

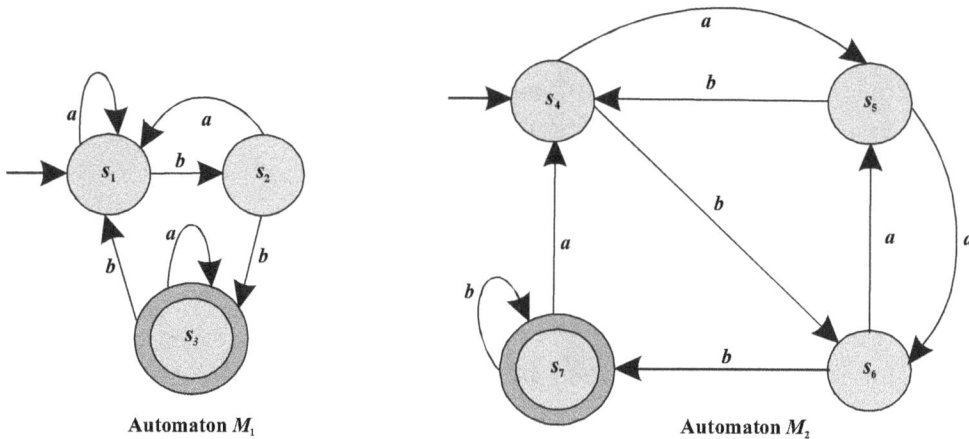

Automaton M_1 Automaton M_2

Solution   From the transition diagram given above the initial states in M_1 and M_2 are s_1 and s_4 respectively. Hence the first element of the first column in the comparison table is (s_1, s_4). It is clear that (s_1, s_5) is a-reachable from (s_1, s_4). The comparison table is given below.

(s, s')	(s_a, s'_a)	(s_b, s'_b)
(s_1, s_4)	(s_1, s_5)	(s_2, s_6)
(s_1, s_5)	(s_1, s_6)	(s_2, s_4)
(s_2, s_6)	(s_1, s_5)	(s_3, s_7)
(s_1, s_6)	(s_1, s_5)	(s_2, s_7)

From the table it is clear that (s_2, s_7) is b-reachable from (s_1, s_6). As s_2 is a non-final state in M_1 and s_7 is a final state of M_2, we see that M_1 and M_2 are not equivalent.

Example 2

Let $L_1 = \{11, 00\}; L_2 = \{1100, 101\}$. Compute $L_1 L_2, L_1^*$ and L_1^+.

Solution   Given that $L_1 = \{11, 00\}; L_2 = \{1100, 101\}$

Therefore, $L_1 L_2 = \{111100, 11101, 001100, 00101\}$

The Kleene closure L_1^* and the positive closure L_1^+ are defined as below.

$$L_1^* = \bigcup_{i=0}^{\infty} L_1^i = L_1^0 \cup L_1^1 \cup L_1^2 \cup \cdots$$
$$= \{\wedge\} \cup \{11,00\} \cup \{1111,0000,1100,0011\} \cup \cdots$$
$$= \{\wedge, 11, 00, 1111, 0000, 1100, 0011, \cdots\}$$

$$L_1^+ = \bigcup_{i=1}^{\infty} L_1^i = L_1^1 \cup L_1^2 \cup \cdots$$
$$= \{11,00\} \cup \{1111,0000,1100,0011\} \cup \cdots$$
$$= \{11, 00, 1111, 0000, 1100, 0011, \cdots\}$$

Example 3

Construct a regular expression corresponding to the following transition diagram.

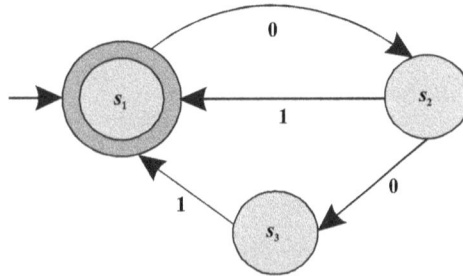

Solution Here, $S = \{s_1, s_2, s_3\}$ and the final state is s_1. From the given transition graph we obtain the following equations for s_1, s_2 and s_3. Therefore,

$$s_1 = s_2 1 + s_3 1 + \wedge \tag{1}$$

$$s_2 = s_1 0 \tag{2}$$

$$s_3 = s_2 0 \tag{3}$$

From equations (2) and (3) we have

$$s_3 = s_2 0 = s_1 00 \tag{4}$$

On putting $s_3 = s_1 00$ and $s_2 = s_1 0$ in equation (1) we get

$$s_1 = s_1 01 + s_1 001 + \wedge$$
$$= s_1 (01 + 001) + \wedge$$

By applying Arden's theorem we get $s_1 = \wedge (01+001)^* = (01+001)^*$. Therefore, the regular expression corresponding to the transition diagram given above is $(01+001)^*$.

Example 4

Find a regular expression for the following transition graph.

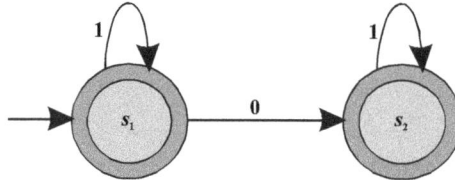

Solution   Here, $S = \{s_1, s_2\}$; $\Sigma = \{0, 1\}$ and the final states are s_1 and s_2. From the transition graph given above, we obtain the following equations for s_1 and s_2.

Therefore, we get

$$s_1 = s_1 1 + \wedge \qquad (1)$$

$$s_2 = s_1 0 + s_2 1 \qquad (2)$$

By applying Arden's theorem to equation (1) we have $s_1 = \wedge 1^* = 1^*$.

Therefore, equation (2) reduces to

$$s_2 = s_1 0 + s_2 1$$

$$= 1^* 0 + s_2 1 \qquad (3)$$

On applying Arden's theorem again to equation (3) we get $s_2 = (1^* 0)1^*$. As the state graph contains two final states s_1 and s_2, so the required regular expression is $(s_1 + s_2)$. Therefore,

$$(s_1 + s_2) = 1^* + 1^* 01^* = 1^* (\wedge + 01^*)$$

Hence, the regular expression corresponding to the state graph is $1^* (\wedge + 01^*)$

Example 5

Construct a regular expression for the following transition diagram.

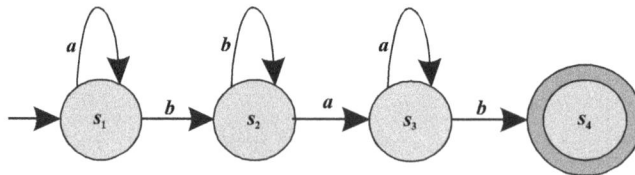

Solution   Here, $S = \{s_1, s_2, s_3, s_4\}$; $\Sigma = \{a, b\}$ with the final state s_4. From the transition graph given above, we obtain the following equations for s_1, s_2, s_3 and s_4. Therefore, we get

$$s_1 = s_1 a + \wedge \tag{1}$$

$$s_2 = s_1 b + s_2 b \tag{2}$$

$$s_3 = s_2 a + s_3 a \tag{3}$$

$$s_4 = s_3 b \tag{4}$$

By applying Arden's theorem to equation (1), i.e., $s_1 = s_1 a + \wedge$ we get $s_1 = \wedge a^* = a^*$.

Therefore, equation (2), i.e., $s_2 = s_1 b + s_2 b$ reduces to $s_2 = a^* b + s_2 b$. By applying Arden's theorem we get

$$s_2 = (a^* b) b^* = a^* b b^*$$

Therefore, equation (3), i.e., $s_3 = s_2 a + s_3 a$ reduces to $s_3 = a^* b b^* a + s_3 a$. By applying Arden's theorem we get

$$s_3 = (a^* b b^* a) a^* = a^* b b^* a a^*$$

So, the required regular expression corresponding to the transition diagram is given as $a^* b b^* a a^*$.

Example 6

Construct a nondeterministic finite automaton with ε moves for the regular expression $10^* + 1$.

Solution   Given regular expression is $10^* + 1$. By precedence rules this expression can be written as $1(0)^* + 1$. Therefore, it is in the form of $(s + t)$, where $s = (1(0)^*)$ and $t = 1$. The transition diagram of $t = 1$ is easy and is given below.

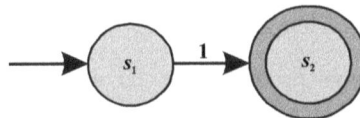

Now, s can be expressed as $r_1 r_2$, where $r_1 = 1$ and $r_2 = 0^*$. Again transition diagram of $r_1 = 1$ is easy and is given as

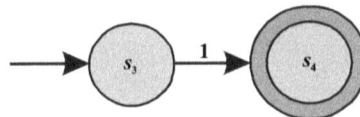

Similarly, the construction of $r_2 = 0^*$ is given as

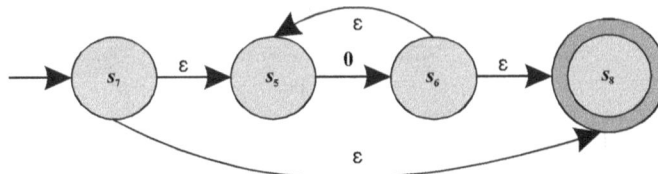

Therefore, the construction of $s = 1(0^*) = r_1 r_2$ is given as

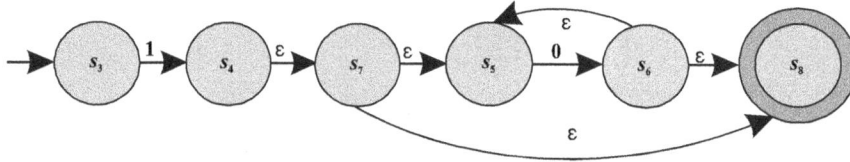

Therefore, the final construction of $r = (s + t)$ is given as below.

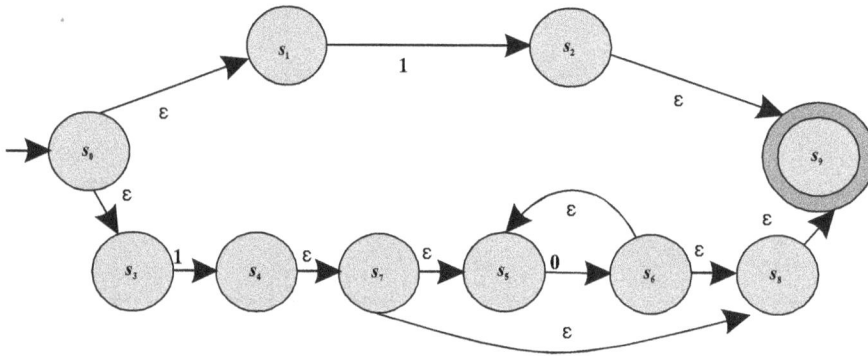

Example 7

Construct finite automata equivalent to the regular expression $10 + (0+11)0^*1$.

Solution   Given regular expression is $r = 10 + (0+11)0^*1$.

This can be expressed as $r = s + t$, where $s = 10$ and $t = (0+11)0^*1$. Now, $s = 10$ can be expressed as $s = r_1 r_2$ with $r_1 = 1$ and $r_2 = 0$. So, the transition diagrams are given below.

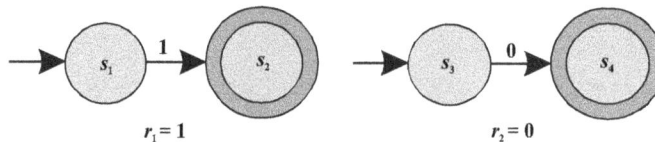

Therefore, the construction of $s = r_1 r_2 = 10$ is given as

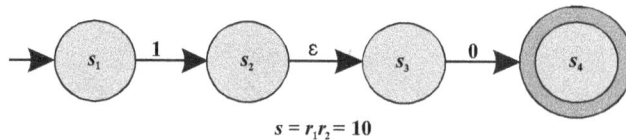

Similarly, $t = (0+11)0^*1$ can be expressed as $t = t_1 t_2 t_3$, where $t_1 = (0+11)$; $t_2 = 0^*$ and $t_3 = 1$. The transition diagram for $t_3 = 1$ and $t_2 = 0^*$ are easy and is given below.

$t_3 = 1$

$t_2 = 0*$

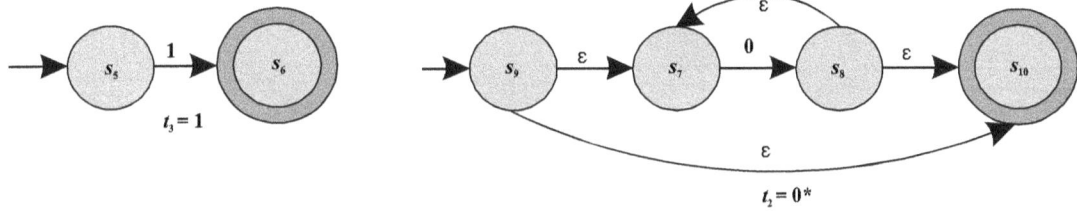

Now, $t_1 = (0 + 11)$ can be expressed as $(t_5 + t_6)$, where $t_5 = 0$ and $t_6 = 11$. Again t_6 can be expressed as $t_6 = 11 = t_7 t_8$ with $t_7 = 1$ and $t_8 = 1$. The construction of $t_5 = 0$, $t_7 = 1$ and $t_8 = 1$ are given below.

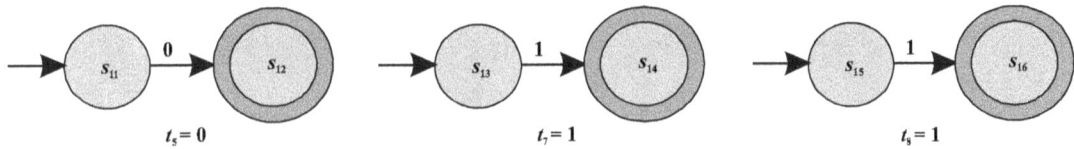

$t_5 = 0$

$t_7 = 1$

$t_8 = 1$

Therefore, the construction of $t_6 = 11 = t_7 t_8$ is given below.

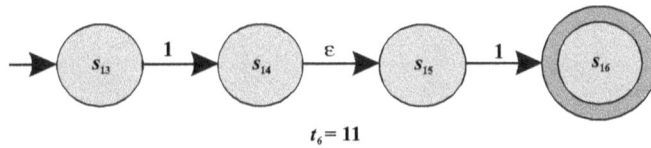

$t_6 = 11$

Similarly, the construction of $t_1 = (t_5 + t_6)$ is given below.

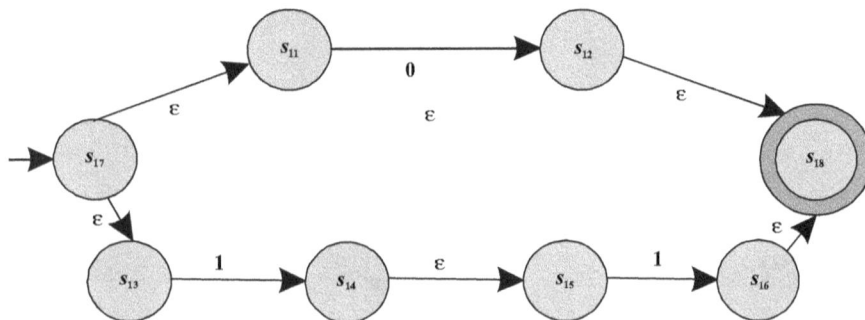

Therefore, the construction of $t = t_1 t_2 t_3$ is given below.

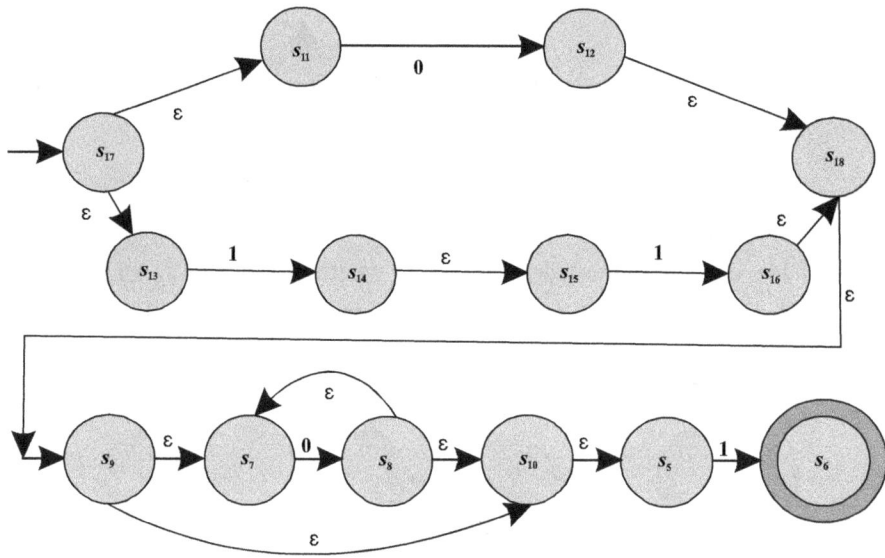

Finally, the construction of $r = s + t$ is given below.

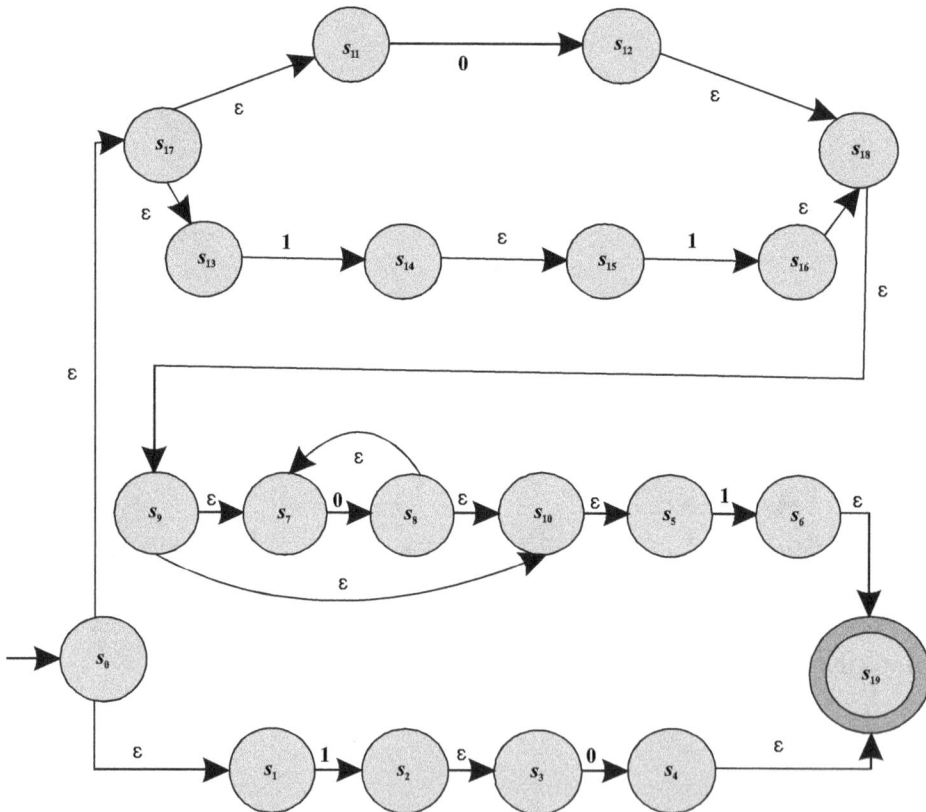

Example 8

Construct a transition diagram for the regular expression $(ba^*)(b+a)^*(a+ba)$.

Solution   Given regular expression is $r = (ba^*)(b+a)^*(a+ba)$. The above expression can be written as $r = stu$, where $s = ba^*$, $t = (b+a)^*$ and $u = (a+ba)$. Now, $s = ba^*$ can be written as $s = s_1 s_2$ with $s_1 = b$ and $s_2 = a^*$. The transition diagram of s_1 and s_2 is given below.

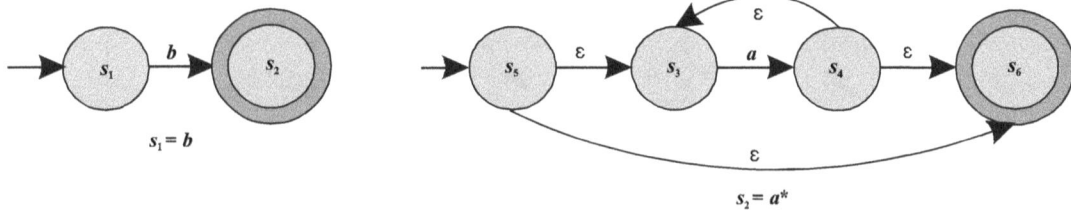

$s_1 = b$

$s_2 = a^*$

Therefore, the transition diagram of $s = s_1 s_2$ is given below.

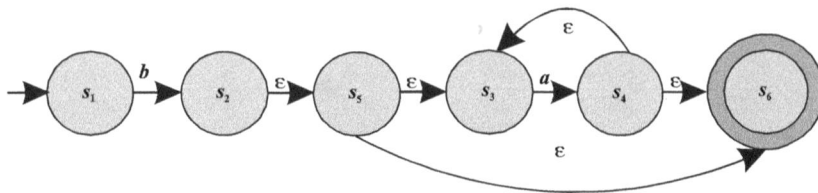

Now, $t = (b+a)^*$ can be expressed as $t = t_1^*$ with $t_1 = (b+a)$. Again $t_1 = (b+a)$ can be expressed as $t_1 = t_2 + t_3$, where $t_2 = b$ and $t_3 = a$. The transition diagram of $t_2 = b$ and $t_3 = a$ are given below.

$t_2 = b$

$t_3 = a$

Therefore, the transition diagrams of $t_1 = t_2 + t_3$ and $t = t_1^*$ is given below.

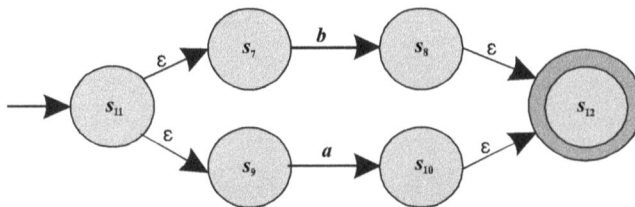

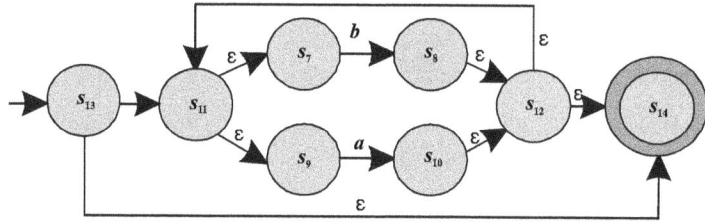

Similarly, the transition diagram of $u = (a + ba)$ is given below.

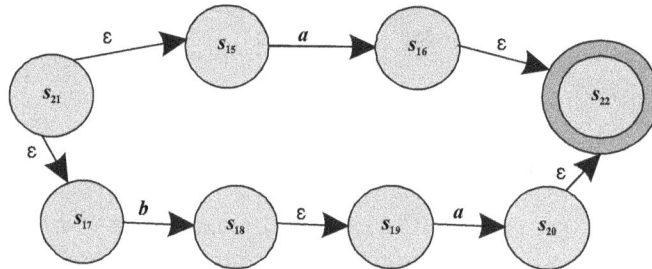

Finally, the transition diagram of $r = stu$ is given below.

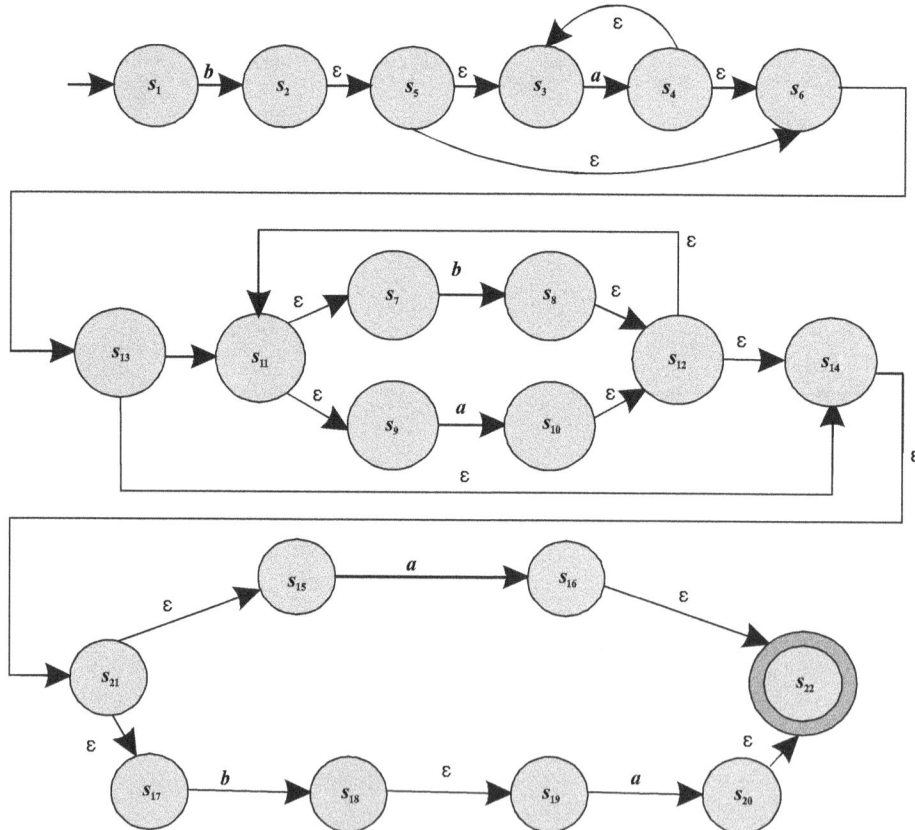

Example 9

Obtain the regular expression for the following deterministic finite automaton whose transition diagram is given below.

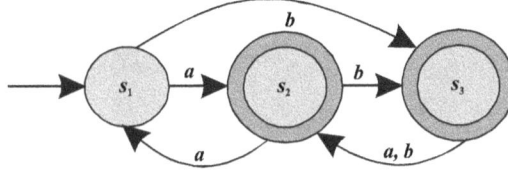

Solution   Let $M = (S, \Sigma, \delta, s_1, A)$ be a deterministic finite automata. Here, we have $S = \{s_1, s_2, s_3\}; \Sigma = \{a, b\}$ and $A = \{s_2, s_3\}$. Therefore, the required regular expression is given as $r_{12}^3 + r_{13}^3$. In order to get ($r_{12}^3 + r_{13}^3$), we have to find out r_{ij}^k for all i, j and $k = 0, 1, 2$. Therefore, we get

$$r_{11}^0 = \varepsilon$$

$$r_{13}^0 = b \quad [\because \delta(s_1, b) = s_3]$$

$$r_{22}^0 = \varepsilon$$

$$r_{31}^0 = \phi$$

$$r_{12}^0 = a \quad [\because \delta(s_1, a) = s_2]$$

$$r_{21}^0 = a \quad [\because \delta(s_2, a) = s_1]$$

$$r_{23}^0 = b \quad [\because \delta(s_2, b) = s_3]$$

$$r_{32}^0 = a + b \quad [\because \delta(s_3, a) = s_2 \text{ and } \delta(s_3, b) = s_2]$$

Similarly, we get $r_{11}^1 = r_{11}^0 (r_{11}^0)^* r_{11}^0 + r_{11}^0 = \varepsilon(\varepsilon)^* \varepsilon + \varepsilon = \varepsilon$ and the following:

$$r_{12}^1 = r_{11}^0 (r_{11}^0)^* r_{12}^0 + r_{12}^0 = \varepsilon(\varepsilon)^* a + a = a; \quad r_{13}^1 = r_{11}^0 (r_{11}^0)^* r_{13}^0 + r_{13}^0 = \varepsilon(\varepsilon)^* b + b = b;$$

$$r_{21}^1 = r_{21}^0 (r_{11}^0)^* r_{11}^0 + r_{21}^0 = a(\varepsilon)^* \varepsilon + a = a; \quad r_{22}^1 = r_{21}^0 (r_{11}^0)^* r_{12}^0 + r_{22}^0 = (aa + \varepsilon);$$

$$r_{23}^1 = r_{21}^0 (r_{11}^0)^* r_{13}^0 + r_{23}^0 = (ab + b); \quad r_{31}^1 = r_{31}^0 (r_{11}^0)^* r_{11}^0 + r_{31}^0 = \phi(\varepsilon)^* \varepsilon + \phi = \phi;$$

$$r_{32}^1 = r_{31}^0 (r_{11}^0)^* r_{12}^0 + r_{32}^0 = (a + b); \quad r_{33}^1 = r_{31}^0 (r_{11}^0)^* r_{13}^0 + r_{33}^0 = \phi(\varepsilon)^* b + \varepsilon = \varepsilon;$$

Therefore, we have the following computations:

$$r_{11}^2 = r_{12}^1 (r_{22}^1)^* r_{21}^1 + r_{11}^1 = a(aa + \varepsilon)^* a + \varepsilon = a(aa)^* a + \varepsilon = (aa)^*$$

$$r_{12}^2 = r_{12}^1 (r_{22}^1)^* r_{22}^1 + r_{12}^1 = a(aa + \varepsilon)^* (aa + \varepsilon) + a = a(aa + \varepsilon)^* + a = a(aa)^*$$

$$r_{13}^2 = r_{12}^1 (r_{22}^1)^* r_{23}^1 + r_{13}^1 = a(aa + \varepsilon)^* (ab + b) + b = (aa^* + \varepsilon)b = a^* b$$

$$r_{21}^2 = r_{22}^1 (r_{22}^1)^* r_{21}^1 + r_{21}^1 = (aa + \varepsilon)(aa + \varepsilon)^* a + a = (aa)^* a + a = (aa)^* a$$

$$r_{22}^2 = r_{22}^1 (r_{22}^1)^* r_{22}^1 + r_{22}^1 = (aa + \varepsilon)(aa + \varepsilon)^* (aa + \varepsilon) + (aa + \varepsilon)$$

$$= ((aa)^* + \varepsilon)(aa + \varepsilon) = (aa)^*$$

$$r_{23}^2 = r_{22}^1 (r_{22}^1)^* r_{23}^1 + r_{23}^1 = (aa + \varepsilon)(aa + \varepsilon)^* (ab + b) + (ab + b)$$

$$= (aa)^* (ab + b) + (ab + b) = a^* b$$

$$r_{31}^2 = r_{32}^1 (r_{22}^1)^* r_{21}^1 + r_{31}^1 = (a+b)(aa+\varepsilon)^* a + \phi$$
$$= (a+b)(aa)^* a = a^* b$$
$$r_{32}^2 = r_{32}^1 (r_{22}^1)^* r_{22}^1 + r_{32}^1 = (a+b)(aa+\varepsilon)^* (aa+\varepsilon) + (a+b)$$
$$= (a+b)(aa)^* (aa+\varepsilon) + (a+b)$$
$$= (a+b)(aa)^* + (a+b) = (a+b)(aa)^*$$
$$r_{33}^2 = r_{32}^1 (r_{22}^1)^* r_{23}^1 + r_{33}^1 = (a+b)(aa+\varepsilon)^* (ab+b) + \varepsilon$$
$$= (a+b)(aa)^* (a+\varepsilon)b + \varepsilon$$
$$= (a+b)a^* b + \varepsilon$$

Finally, we have

$$r_{12}^3 = r_{13}^2 (r_{33}^2)^* r_{32}^2 + r_{12}^2 = a^* b((a+b)a^* b + \varepsilon)^* (a+b)(aa)^* + a(aa)^*$$
$$= a^* b((a+b)a^* b)^* (a+b)(aa)^* + a(aa)^*$$
$$r_{13}^3 = r_{13}^2 (r_{33}^2)^* r_{33}^2 + r_{13}^2 = a^* b((a+b)a^* b + \varepsilon)^* ((a+b)a^* b + \varepsilon) + a^* b$$
$$= a^* b((a+b)a^* b)^* + a^* b = a^* b((a+b)a^* b)^*$$

Therefore, the regular expression is given as

$$r_{12}^3 + r_{13}^3 = a^* b((a+b)a^* b)^* (a+b)(aa)^* + a(aa)^* + a^* b((a+b)a^* b)^*$$
$$= a^* b((a+b)a^* b)^* (\varepsilon + (a+b)(aa)^*) + a(aa)^*$$

Example 10

Find the regular expression for the following deterministic finite automaton whose transition diagram is given below.

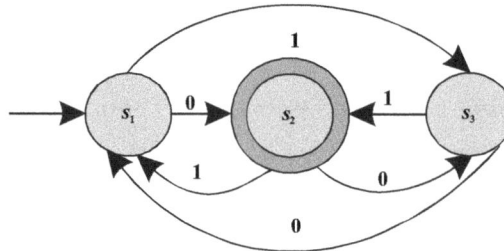

Solution   Let $M = (S, \Sigma, \delta, s_1, A)$ be a deterministic finite automata whose transition function is given above. Here, we have $S = \{s_1, s_2, s_3\}$; $\Sigma = \{0,1\}$ and $A = \{s_2\}$. Therefore, the required regular expression r_{12}^3 is given as

$$r_{12}^3 = r_{13}^2 (r_{33}^2)^* r_{32}^2 + r_{12}^2 \tag{1}$$

In order to find r_{12}^3, we have to compute r_{ij}^k for all i, j and $k = 0, 1, 2$. Therefore, for $k = 0$ we have

$r_{11}^0 = \varepsilon$ $\qquad\qquad\qquad\qquad\qquad$ $r_{12}^0 = 0$ $\qquad$ $[\because \delta(s_1, 0) = s_2]$

$r_{13}^0 = 1$ $\qquad$ $[\because \delta(s_1, 1) = s_3]$ $\qquad$ $r_{21}^0 = 1$ $\qquad$ $[\because \delta(s_2, 1) = s_1]$

$r_{22}^0 = \varepsilon$ $\qquad\qquad\qquad\qquad\qquad$ $r_{23}^0 = 0$ $\qquad$ $[\because \delta(s_2, 0) = s_3]$

$r_{31}^0 = 0$ $\qquad$ $[\because \delta(s_3, 0) = s_1]$ $\qquad$ $r_{32}^0 = 1$ $\qquad$ $[\because \delta(s_3, 1) = s_2]$

Similarly, we get for $k = 1$ the following.

$r_{12}^1 = r_{11}^0 (r_{11}^0)^* r_{12}^0 + r_{12}^0 = 0;$

$r_{22}^1 = r_{21}^0 (r_{11}^0)^* r_{12}^0 + r_{22}^0 = (10 + \varepsilon);$

$r_{32}^1 = r_{31}^0 (r_{11}^0)^* r_{12}^0 + r_{32}^0 = (00 + 1);$

$r_{13}^1 = r_{11}^0 (r_{11}^0)^* r_{13}^0 + r_{13}^0 = 1;$

$r_{23}^1 = r_{21}^0 (r_{11}^0)^* r_{13}^0 + r_{23}^0 = (11 + 0);$

$r_{33}^1 = r_{31}^0 (r_{11}^0)^* r_{13}^0 + r_{33}^0 = (01 + \varepsilon);$

Similarly, we get for $k = 2$ the following.

$$r_{12}^2 = r_{12}^1 (r_{22}^1)^* r_{22}^1 + r_{12}^1 = 0(10 + \varepsilon)^* (10 + \varepsilon) + 0$$
$$= 0(10 + \varepsilon)^* + 0 = 0((10 + \varepsilon)^* + \varepsilon)$$
$$= 0(10 + \varepsilon)^*$$

$$r_{13}^2 = r_{12}^1 (r_{22}^1)^* r_{23}^1 + r_{13}^1 = 0(10 + \varepsilon)^* (11 + 0) + 1$$

$$r_{32}^2 = r_{32}^1 (r_{22}^1)^* r_{22}^1 + r_{32}^1 = (00 + 1)(10 + \varepsilon)^* (10 + \varepsilon) + (00 + 1)$$
$$= (00 + 1)(10 + \varepsilon)^* + (00 + 1)$$
$$= (00 + 1)((10 + \varepsilon)^* + \varepsilon) = (00 + 1)(10 + \varepsilon)^*$$

$$r_{33}^2 = r_{32}^1 (r_{22}^1)^* r_{23}^1 + r_{33}^1 = (00 + 1)(10 + \varepsilon)^* (11 + 0) + (01 + \varepsilon)$$

Finally, from equation (1) the regular expression is given as

$$r_{12}^3 = r_{13}^2 (r_{33}^2)^* r_{32}^2 + r_{12}^2$$
$$= (0(10 + \varepsilon)^* (11 + 0) + 1)((00 + 1)(10 + \varepsilon)^* (11 + 0) + (01 + \varepsilon))^*$$
$$(00 + 1)(10 + \varepsilon)^* + 0(10 + \varepsilon)^*$$

Example 11

Consider the following two deterministic finite automata M_1 and M_2 over $\{0, 1\}$ given below. Determine whether M_1 and M_2 are equivalent.

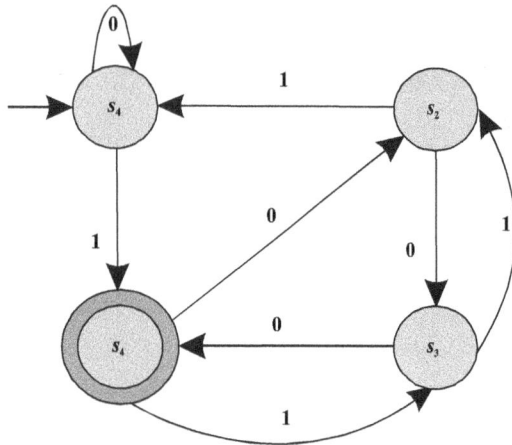

Automaton M_1

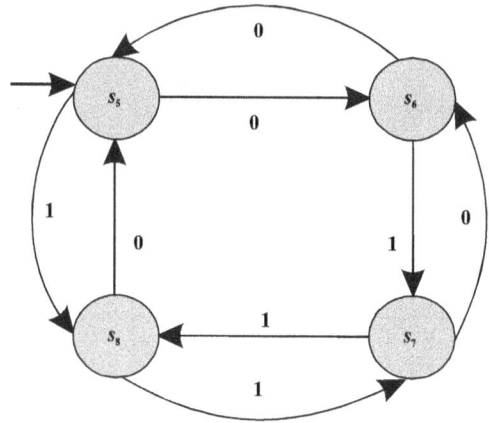

Automaton M_2

Solution   From the transition diagram given above, it is clear that the initial state in M_1 and M_2 is s_1 and s_5 respectively. Hence the first element in first column of the comparison table is (s_1, s_5). From the diagram it is clear that (s_1, s_6) is 0-reachable from (s_1, s_5) whereas (s_4, s_8) is 1-reachable from (s_1, s_5). The comparison table is given below.

(s, s')	(s_0, s_0')	(s_1, s_1')
(s_1, s_5)	(s_1, s_6)	(s_4, s_8)
(s_1, s_6)	(s_1, s_5)	(s_4, s_7)

From the table it is clear that s_4 and s_7 are 1-reachable from s_1 and s_6 respectively. Also it is clear that s_4 is a non-final state of M_1 whereas s_7 is a final state of M_2. Therefore, the process terminates and hence M_1 and M_2 are not equivalent.

Example 12

Show that the following automata M_1 and M_2 whose transition diagram is given below are equivalent.

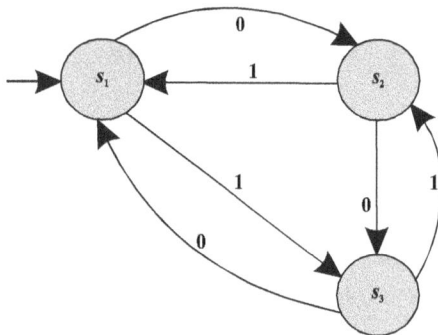

Automaton M_1

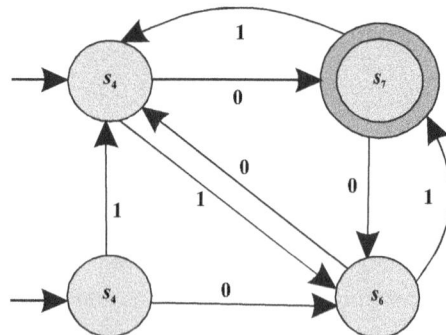

Automaton M_2

Solution   From the state diagram given above it is clear that the initial states in M_1 and M_2 are s_1 and s_4 respectively. Hence the first element in first column of the comparison table is (s_1, s_4). From the state diagram it is clear that (s_3, s_5) is 1-reachable from (s_1, s_4) whereas (s_2, s_6) is 0-reachable from (s_1, s_4). Similarly, we get

(s, s')	(s_0, s_0')	(s_1, s_1')
(s_1, s_4)	(s_2, s_6)	(s_3, s_5)
(s_2, s_6)	(s_3, s_5)	(s_1, s_7)
(s_3, s_5)	(s_1, s_7)	(s_2, s_6)
(s_1, s_7)	(s_2, s_6)	(s_3, s_5)

From the table it is clear that no new pair appears in the second and third column that are not present in the first column and hence the construction is terminated. From the table it is also clear that there is no such pair whose first element belongs to a final state of M_1 and second element belongs to a non-final state of M_2 or vice versa. Therefore, M_1 and M_2 are equivalent.

Example 13

Construct a regular expression for the following state diagram.

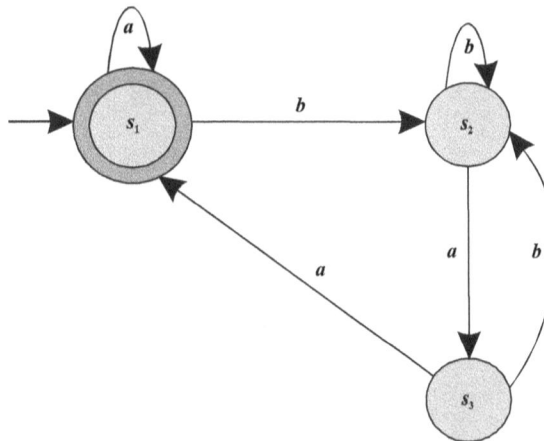

Solution   From the diagram it is clear that, it has only one final state without any moves. Therefore, we get the following set of equations:

$$s_1 = s_1 a + s_3 a + \wedge \tag{1}$$

$$s_2 = s_1 b + s_2 b + s_3 b \tag{2}$$

$$s_3 = s_2 a \tag{3}$$

From equation (2) we get

$$s_2 = s_1 b + s_2 b + (s_2 a)b \qquad [\because s_3 = s_2 a]$$
$$= s_1 b + s_2(b + ab)$$

Therefore, by applying Arden's theorem we get

$$s_2 = s_1 b(b + ab)^* \qquad (4)$$

Now, from equation (1) we have

$$s_1 = s_1 a + s_3 a + \wedge$$
$$= s_1 a + (s_2 a)a + \wedge \qquad [\text{By (2)}]$$
$$= s_1 a + s_2 aa + \wedge$$
$$= s_1 a + s_1 b(b + ab)^* aa + \wedge$$
$$= s_1(a + b(b + ab)^* aa) + \wedge$$

Therefore, by Arden's theorem we get

$$s_1 = \wedge(a + b(b + ab)^* aa)^*$$
$$\text{i.e., } s_1 = (a + b(b + ab)^* aa)^*$$

As the transition diagram contains only one final state, the regular expression corresponding to the given state diagram is $(a + b(b + ab)^* aa)^*$.

Example 14

Construct a finite automaton recognizing $L(G)$, where G is the grammar $Q_0 \to aQ_0 | bQ_1 | b$ and $Q_1 \to aQ_1 | bQ_0 | a$.

Solution   Given that G is a grammar having productions

$$Q_0 \to aQ_0 | bQ_1 | b \text{ and } Q_1 \to aQ_1 | bQ_0 | a.$$

Our aim is to construct a transition system. Let the transition system be $M = (S, \Sigma, \delta, s_0, s_f)$, where $S = \{s_0, s_1, s_f\}$. The states s_0 and s_1 correspond to the variables Q_0 and Q_1 whereas s_f is the newly introduced final state. Now in order to find the transitions, we have to consider the productions P. The production:

(a) $Q_0 \to aQ_0$ represents the transition $\delta(s_0, a) = s_0$.

(b) $Q_0 \to bQ_1$ represents the transition $\delta(s_0, b) = s_1$.

(c) $Q_0 \to b$ represents the transition $\delta(s_0, b) = s_f$.

(d) $Q_1 \to aQ_1$ represents the transition $\delta(s_1, a) = s_1$.

(e) $Q_1 \to bQ_0$ represents the transition $\delta(s_1, b) = s_0$.

(f) $Q_1 \to a$ represents the transition $\delta(s_1, a) = s_f$.

Therefore, the finite automaton recognizing $L(G)$ is given as

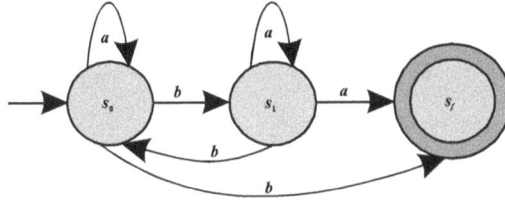

Example 15

Construct a finite automaton recognizing $L(G)$, where the productions P of the grammar G are $Q \to aQ|bA|bB;\ A \to bA|aB|b$ and $B \to bA|a$.

Solution   Given that G is a grammar having productions

$$Q \to aQ|bA|bB;\ A \to bA|aB|b \text{ and } B \to bA|a.$$

Construct a transition system M as $M = (S, \Sigma, \delta, s_0, s_f)$, where $S = \{s_0, s_1, s_2, s_f\}$. The states s_0, s_1 and s_2 correspond to the variables Q, A and B respectively whereas s_f is a newly introduced final state. The transition system δ is defined in terms of transition table as below.

State	Input	
	a	b
$\to s_0$	s_0	$\{s_1, s_2\}$
s_1	s_2	$\{s_1, s_f\}$
s_2	s_f	s_1

Therefore, the finite automaton M recognizing $L(G)$ is given below.

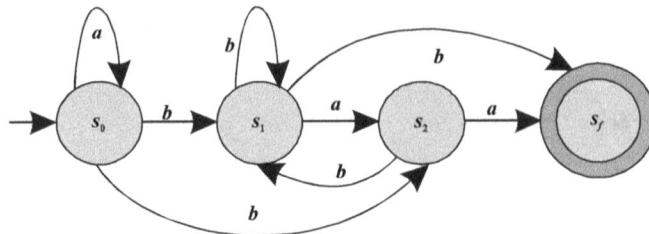

Example 16

Construct a transition diagram without Λ moves for the given transition graph with Λ moves.

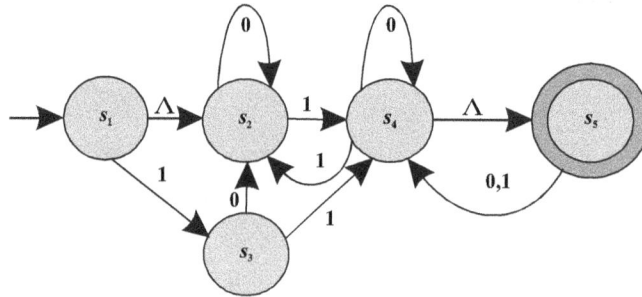

Solution   From the given diagram it is clear that there exists two Λ moves one from s_1 to s_2 whereas another from s_4 to s_5. Our aim is to remove these Λ moves. First we remove the Λ move between s_1 to s_2. From the state diagram it is clear that $\delta(s_2, 0) = s_2$ and $\delta(s_2, 1) = s_4$. Therefore, $\delta(s_1, 0) = s_2$ and $\delta(s_1, 1) = s_4$. Again, s_1 is an initial state, so s_2 is made an initial state. Hence, we get the following state graph.

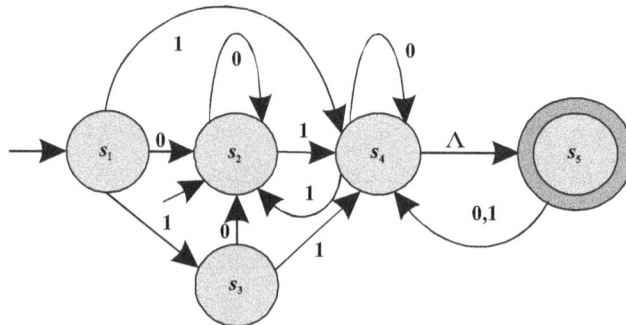

Now we remove the Λ move from s_4 to s_5. It is clear that, $\delta(s_5, 0) = s_4$ and $\delta(s_5, 1) = s_4$. Therefore, $\delta(s_4, 0) = s_4$ and $\delta(s_4, 1) = s_4$. Again, s_5 is a final state, so s_4 is made a final state. Therefore, we get the following state diagram.

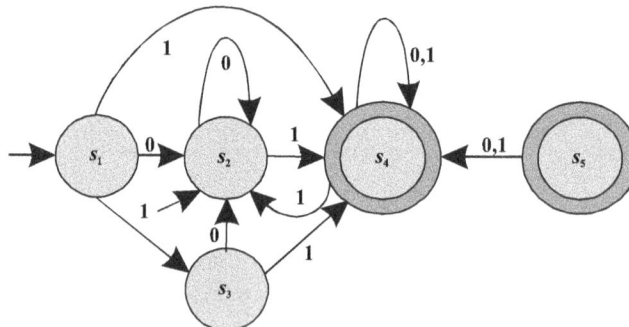

Example 17

Obtain an equivalent automaton without Λ moves for the finite automaton given below.

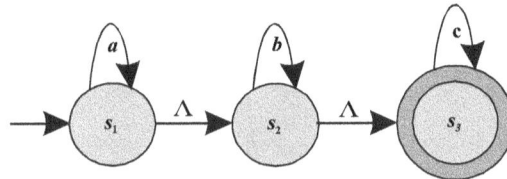

Solution   Here there exists two Λ moves one from s_1 to s_2 whereas another from s_2 to s_3. Our aim is to remove these Λ moves. First we remove the Λ move between s_1 to s_2. From the state diagram it is clear that $\delta(s_2,b)=s_2$ and $\delta(s_2,\wedge)=s_3$. Therefore, we have $\delta(s_1,b)=s_2$ and $\delta(s_1,\wedge)=s_3$. Again, s_1 is an initial state, so s_2 is made an initial state. Now we have to remove the Λ transition from s_2 to s_3. From the transition graph it is clear that $\delta(s_3,c)=s_3$ and hence we have the transition $\delta(s_2,c)=s_3$. Also it is clear that s_3 is a final state and so s_2 is made as a final state. Therefore, we get the following state diagram.

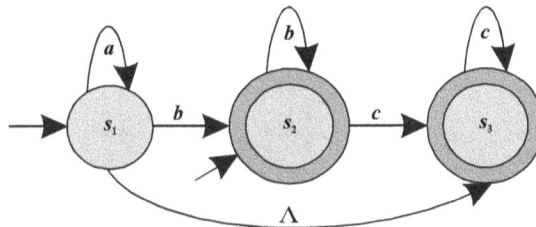

Finally, we have to remove the Λ move between s_1 to s_3. Now, $\delta(s_1,c)=s_3$ as $\delta(s_3,c)=s_3$. Again s_3 is a final state, so s_1 is made as a final state. On the other hand, s_1 is an initial state and so s_3 is made as an initial state. Therefore, the constructed automaton is given below.

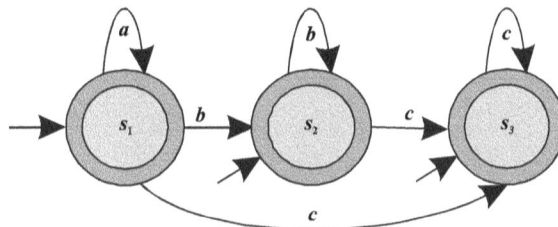

Example 18

Show that the language $L = \{0^p \mid p \text{ is a prime}\}$ is not regular.

Solution   Consider the language $L = \{0^p \mid p \text{ is a prime}\}$. Let us assume that L is regular. Let n be the number of states in the finite automaton accepting the language L.

Let us assume that p be a prime number greater than n. Let, $z = 0^p$. Thus, $|z| = p > n$. Therefore, by pumping lemma, we can write $z = uvy$, with $|uv| \leq n$ and $|v| > 0$. So, $v = 0^m$ for some m such that $1 \leq m \leq n$, i.e., $|v| = m$.

Let us choose $i = p + 1$. Therefore,

$$|uv^i w| = |uvy| + |v^{i-1}|$$
$$= p + (i-1)m$$
$$= p + pm \qquad [\because i = p + 1]$$
$$= p(m+1)$$

But it is clear that $p(m+1)$ is divisible by $(m+1)$ and hence it is not a prime number. Therefore, $uv^i y \notin L$. This leads to a contradiction. So our assumption is wrong. Thus, $L = \{0^p \mid p \text{ is a prime}\}$ is not regular.

Example 19

Show that the language $L = \left\{ a^{i^2} \mid i \geq 1 \right\}$ is not regular.

Solution   Consider the set $L = \left\{ a^{i^2} \mid i \geq 1 \right\}$ is regular. Let n be the number of states in the finite automaton accepting the language L.

Let $z = a^{n^2}$. Therefore, $|z| = \left| a^{n^2} \right| = n^2 > n$. Hence, by pumping lemma we can write $z = uvy$, with $|uv| \leq n$ and $|v| > 0$. Let us choose $i = 2$. Therefore,

$$|uv^2 y| = |u| + 2|v| + |y|$$
$$> |u| + |v| + |y|$$
$$> |uvy| = |z| = n^2$$

i.e., $|uv^2 y| > n^2 \qquad (1)$

Again we have

$$|uv^2 y| = |uvy| + |v|$$
$$= |z| + |v|$$
$$\leq n^2 + n \qquad [\because |v| < |uv| \leq n]$$
$$< n^2 + 2n + 1 = (n+1)^2$$

i.e., $\left|uv^2y\right| < (n+1)^2$ (2)

Therefore, from equations (1) and (2) we get $n^2 < \left|uv^2y\right| < (n+1)^2$. Hence, it is clear that $\left|uv^2y\right|$ strictly lies between n^2 and $(n+1)^2$ but not equal to anyone of them. So, $uv^iy \notin L$ and hence L is not regular.

Example 20

Show that $L = \left\{a^nb^{2n} \mid n > 0\right\}$ is not regular.

Solution   Consider the language $L = \left\{a^nb^{2n} \mid n > 0\right\}$. Assume that L is regular. Let n be the number of states in the finite automaton accepting the language L.

Let $z = a^nb^{2n}$. Therefore, $|z| = n + 2n = 3n > n$. Hence, by pumping lemma we can write $z = uvy$, with $|uv| \le n$ and $|v| > 0$. Therefore, v can be expressed in any one of the following forms:

(*a*) v has only *a*'s, i.e., $v = a^k$ for some $k \ge 1$.

(*b*) v has only *b*'s, i.e., for some $l \ge 1$.

(*c*) v has both *a*'s and *b*'s, i.e., $v = a^kb^l$ for some $k, l \ge 1$.

Case (*a*)   In this case $z = uvy = a^{n-k}a^kb^{2n}$. On taking $i = 0$, we have

$$uv^iy = uy = a^{n-k}b^{2n} \notin L \qquad [\because (n-k) \ne n]$$

Case (*b*)   In this case $z = uvy = a^nb^lb^{2n-l}$. On taking $i = 0$, we have

$$uv^iy = uy = a^nb^{2n-l} \notin L \qquad [\because (2n-l) \ne 2n]$$

Case (*c*)   In this case $z = uvy = a^{n-k}a^kb^lb^{2n-l}$. On taking $i = 0$, we have

$$uv^iy = uy = a^{n-k}b^{2n-l} \notin L \qquad [\because (2n-l) \ne 2(n-k)]$$

Therefore, our assumption is wrong. Thus the given language $L = \left\{a^nb^{2n} \mid n > 0\right\}$ is not regular.

Example 21

Prove that $L = \{0^m1^n \mid \text{GCD }(m,n) = 1\}$ is not regular.

Solution   Consider $L = \{0^m1^n \mid \text{GCD }(m,n) = 1\}$. Assume that L is regular. Let n be the number of states in the finite automaton accepting the language L.

Let $z = 0^m 1^n$. Thus, $|z| = m + n > n$. So, by pumping lemma we have $z = uvy, |uv| \leq n$. So, v can be expressed in any one of the following forms:

(a) v has only 0's, i.e., $v = 0^k$ for some $k \geq 1$.

(b) v has only 1's, i.e., $v = 1^l$ for some $l \geq 1$.

(c) v has both 0's and 1's, i.e., $v = 0^k 1^l$ for some $k, l \geq 1$.

Case (a)   In this case $z = uvy = 0^{m-k} 0^k 1^n$. On taking $i = 0$, we have

$$uv^i y = uy = 0^{m-k} 1^n \notin L$$

This is because GCD $(m - k, n)$ may not be equal to 1. For example, consider $m = 27$ and $n = 16$. Therefore, GCD $(m, n) = 1$. On taking $k = 1$ we get

$$\text{GCD} (m - k, n) = \text{GCD} (26, 16) = 2 \neq 1.$$

Thus, GCD $(m - k, n)$ may not always be equal to 1.

Case (b)   In this case $z = uvy = 0^m 1^l 1^{n-l}$. On taking $i = 0$, we have

$$uv^i y = uy = 0^m 1^{n-l} \notin L$$

This is because GCD $(m, n - l)$ may not be equal to 1. For example, consider $m = 27$ and $n = 16$. Therefore, GCD $(m, n) = 1$. On taking $l = 1$ we get GCD $(m, n - l) = $ GCD $(27, 15) = 3 \neq 1$. Thus, GCD $(m, n - 1)$ may not always be equal to 1.

Case (c)   In this case $z = uvy = 0^{m-k} 0^k 1^l 1^{n-l}$. On taking $i = 0$, we have.

$$uv^i y = uy = 0^{m-k} 1^{n-l} \notin L$$

This is because GCD $(m - k, n - l)$ may not equal to 1. For example, consider $m = 27$ and $n = 16$. Therefore, GCD$(m, n) = 1$. On taking $k = l = 5$ we get GCD $(m - k, n - l) = $ GCD $(22, 11) = 11 \neq 1$.

Thus, GCD $(m - k, n - l)$ may not always be equal to 1. Thus, it is a contradiction and hence $L = \{0^m 1^n | \text{GCD} (m, n) = 1\}$ is not regular.

Example 22

Let $h(0) = bb$ and $h(1) = bab$. If L_1 is $(01)^*$ and L_2 is $(ba + ab)^* b$, then compute $h(010), h(L_1)$ and $h^{-1}(L_2)$.

Solution   Given that $h(0) = bb, h(1) = bab, L_1 = (01)^*$ and $L_2 = (ba + ab)^* b$.

Therefore, $h(010) = bbbabbb$ and $h(L_1) = (bbbab)^*$.

Now, we have to compute $h^{-1}(L_2)$. It is to be noted that, a string in L_2 that begins with 'a' cannot be $h(x)$ for any string x of 0's and 1's because $h(0)$ and $h(1)$ each begin with a 'b'. Therefore,

$h^{-1}(w)$ is nonempty and if $w \in L_2$, then 'w' must begin with 'b'. Therefore, there exist two cases: $w = b$ or $w = baw'$ for some w' in $(ba + ab)^* b$.

If $w = b$, then $h^{-1}(w)$ is surely empty. If $w = baw'$ for some w' in $(ba + ab)^* b$, then we conclude that every word in $h^{-1}(w)$ must begin with a 1 and since $h(1) = bab$, w' must begin with 'b'. If $w' = b$, we have $w = bab$ and $h^{-1}(w) = \{1\}$. If $w' \neq b$, then $w' = baw''$ and hence $w = babaw''$. But no string x in $(0+1)^*$ has $h(x)$ beginning with $baba$. Therefore, we conclude that $h^{-1}(w)$ is empty in this case. Thus the only string in L_2 which has an inverse image under 'h' is bab and so $h^{-1}(L_2) = \{1\}$.

Example 23

Construct a regular grammar G generating the regular expression $a^* b(a + b)^*$.

Solution Given that the regular expression $a^* b(a + b)^*$. In order to construct a regular grammar, first we have to construct a deterministic finite automaton for the given regular expression. The constructed deterministic finite automaton is given below.

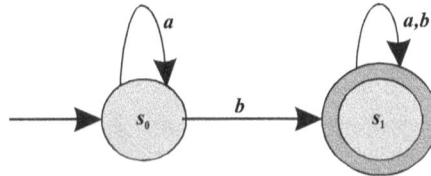

The regular grammar G corresponding to the above deterministic finite automaton is given by the following productions.

$$Q_0 \rightarrow aQ_0 \qquad\qquad [\because \delta(s_0, a) = s_0 \notin A]$$
$$Q_0 \rightarrow bQ_1 \text{ and } Q_0 \rightarrow b \qquad\qquad [\because \delta(s_0, b) = s_1 \in A]$$
$$1Q_1 \rightarrow aQ_1 \text{ and } Q_1 \rightarrow a \qquad\qquad [\because \delta(s_1, a) = s_1 \in A]$$
$$Q_1 \rightarrow bQ_1 \text{ and } Q_1 \rightarrow b \qquad\qquad [\because \delta(s_1, b) = s_1 \in A]$$

Therefore, the regular grammar G is defined as (V_N, T, P, Q_0) such that $V_N = \{Q_0, Q_1\}$; $T = \{a, b\}$; Q_0 is the start symbol and the productions P is defined as $P = (Q_0 \rightarrow aQ_0 | bQ_1 | b, Q_1 \rightarrow aQ_1 | bQ_1 | b | a)$.

Example 24

Construct a regular grammar G generating the regular set represented by $\{a, aa, aaa, \ldots\}$.

Solution The regular expression corresponding to the regular set is $\{a, aa, aaa, \ldots\}$ is a^+ or aa^*. The deterministic finite automaton equivalent to the regular expression aa^* is given below.

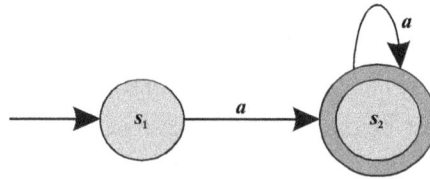

The regular grammar G corresponding to the above deterministic finite automaton is given by the following productions.

$$Q_1 \to aQ_2 \text{ and } Q_1 \to a \qquad [\because \delta(s_1, a) = s_2 \in A]$$
$$Q_2 \to aQ_2 \text{ and } Q_2 \to a \qquad [\because \delta(s_2, a) = s_2 \in A]$$

Therefore, the regular grammar $G = (V_N, T, P, Q_0)$ is defined as $V_N = \{Q_1, Q_2\}$; $T = \{a\}$; and Q_1 is the start symbol with the productions P as defined above.

Example 25

Find the regular expression corresponding to the regular grammar G having productions $Q \to aQ|B, B \to bB|a$.

Solution   Given that the productions are $Q \to aQ|B, B \to bB|a$.

The regular expression corresponding to $B \to bB|a$ is given as $B = b^*a$. Therefore, $Q \to aQ|B$ reduces to $Q \to aQ|b^*a$. Thus, we get $Q = a^*b^*a$. So, the regular expression corresponding to the grammar G is given as a^*b^*a.

Example 26

Consider the regular grammar G having productions $Q \to aQ|A$; $A \to bB|a$; $B \to bB|a$. Find the equivalent regular expression corresponding to regular grammar G.

Solution   Given that the productions of the regular grammar G are

$$Q \to aQ|A; \ A \to bB|a; \ \ B \to bB|a$$

The regular expression corresponding to $B \to bB|a$ is given as $B = b^*a$.

The regular expression corresponding to $A \to bB|a$ is given as $A = bb^*a + a$.

The regular expression corresponding to the production $Q \to aQ|A$ is given as $Q = aQ|(bb^*a + a) = a^*(bb^*a + a)$. Therefore, the required regular expression is given as $a^*(bb^*a + a)$.

Example 27

Prove that the language L of palindromes defined over Σ is not regular.

Solution   First of all, we have to define the language L of palindromes. Let Σ be any alphabet. Therefore, L can be defined as follows:

(a) $\wedge \in L$

(b)   For any $a \in \Sigma, a \in L$.

(c)   For any $z \in L, a \in \Sigma; aza \in L$.

(d)   No string is in L, unless it can be derived by using (a) (b) and (c).

Now we have to show that L is not regular. Assume that the language L of palindromes defined over Σ is regular. Let n be the number of states in the finite automaton accepting the language L. Let us consider $\Sigma = \{0,1\}$.

Let $z = 0^n 10^n$. Then we have $|z| = n+1+n = 2n+1 > n$. Therefore, by pumping lemma we can write $z = uvy$, with $|uv| \le n$ and $|v| > 0$. Therefore, v can be expressed as 0^k for some $k > 0$. Thus,

$$z = uvy = 0^{n-k}(0^k)10^n$$

On taking $i = 0$ we have $uv^i y = uy = 0^{n-k}10^n \notin L$. Therefore, it is a contradiction. So our assumption is wrong. Thus the language L of palindromes defined over Σ is not regular.

REVIEW QUESTIONS

1.   What do you mean by regular language? State the different properties of regular language.

2.   Give a recursive definition of regular expression.

3.   State and prove Arden's theorem.

4.   Prove that there exists a nondeterministic finite automaton with ε-moves that accepts $L(r)$, where r be a regular language.

5.   Show that if L is accepted by deterministic finite automata, then L is denoted by a regular expression.

6.   What do you mean by transition diagram with $\wedge$ moves? Write the steps that are used to construct a transition diagram without $\wedge$ moves.

7.   State pumping lemma and its applications.

8.   Explain a regular set. State the closure properties of regular sets.

9.   Explain the decision algorithms for regular sets.

10. When two finite automata are said to be equivalent? Explain comparison method with a suitable example.

11. State and prove Myhill–Nerode theorem.

PROBLEMS

1. Consider the following transition diagram and construct a regular expression for it by using Arden's theorem.

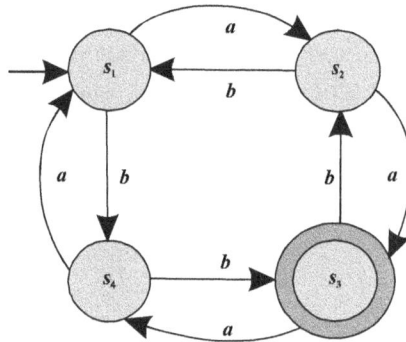

2. Find the regular expressions for the following transition systems using Arden's theorem.

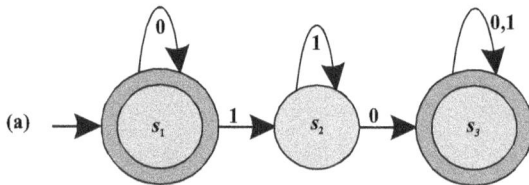

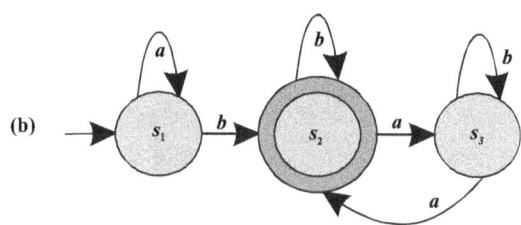

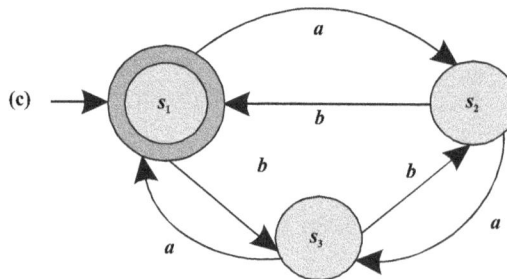

3. Find regular expressions for the transition diagrams given in problem 2 by using the following relations.

$$R_{ij}^k = R_{ik}^{k-1}(R_{kk}^{k-1})^* R_{kj}^{k-1} \cup R_{ij}^{k-1} \quad \text{and}$$

$$R_{ij}^0 = \begin{cases} \{a \mid \delta(s_i, a) = s_j\} & \text{if } i \neq j \\ \{a \mid \delta(s_i, a) = s_j\} \cup \{\varepsilon\} & \text{if } i = j \end{cases}$$

4. Construct nondeterministic finite automata with ε transition for the following regular expressions.

   (a) $01^* + 10^* + 001$

   (b) $(a+b)^* ab + a^*$

   (c) $(ab)^* + aa^* + b$

   (d) $(101 + 01)^* 1 + 01$

   (e) $((a+b)c)^* + ba^* + c$

   (f) $((1+0)(0+1))^* + 01$

   (g) $(ab + bc + ca)^* + a^* b + bc$

   (h) $(1+0)(0+1)(00+11) + 0^* 1$

5. Show that the following automata M_1 and M_2 given below are not equivalent. Use comparison method to justify your answer.

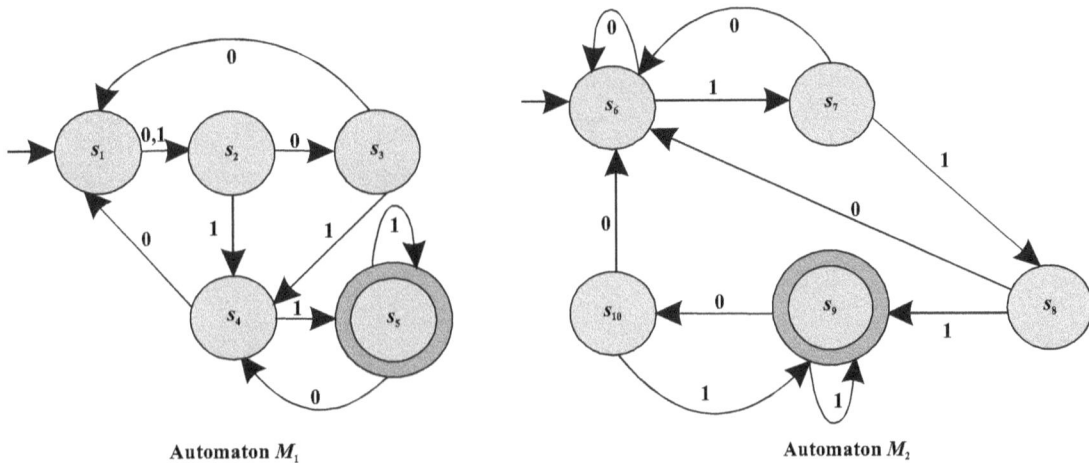

Automaton M_1

Automaton M_2

6. Let $L_1 = \{aab, bba\}$ and $L_2 = \{ab, ba\}$. Compute $L_1 L_2$, L_1^*, L_2^*, L_1^+ and L_2^+.

7. Find a nondeterministic finite automaton for the regular expression $(ab + ba)^*$.

8. Construct a finite automaton equivalent to the regular expression $11 + 00 + 10^* 1$.

9. Construct a transition diagram for the regular expression $((0+1)^* + 11^* + 01)1$.

10. Construct a finite automaton recognizing $L(G)$, where G is the grammar $Q_0 \rightarrow abQ_0 \mid bQ_1 \mid a$ and $Q_1 \rightarrow abQ_1 \mid aQ_0 \mid b$.

11. The productions P of the grammar G are $Q \rightarrow aA \mid bbA \mid bB$; $A \rightarrow aA \mid ab$ and $B \rightarrow bA \mid a$. Construct a finite automaton recognizing the language $L(G)$.

12. Consider the transition system M_1 and M_2 given below. Show that they are equivalent.

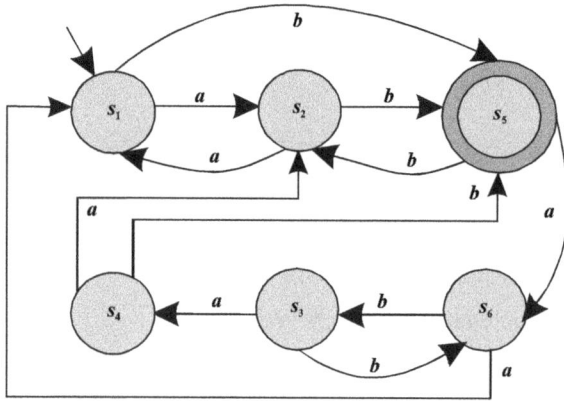

Automaton M_1

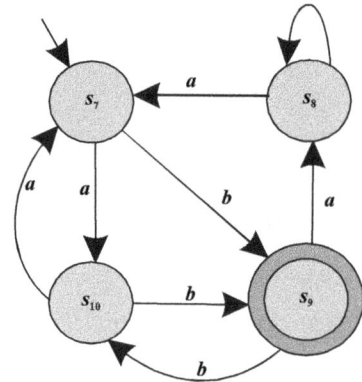

Automaton M_2

13. Find the regular expression for the deterministic finite automaton whose transition diagram is given below by using the following relation.

$$R_{ij}^k = R_{ik}^{k-1}(R_{kk}^{k-1})^* R_{kj}^{k-1} \bigcup R_{ij}^{k-1} \quad \text{and}$$

$$R_{ij}^0 = \begin{cases} \{a | \delta(s_i, a) = s_j\} & \text{if } i \neq j \\ \{a | \delta(s_i, a) = s_j\} \cup \{\varepsilon\} & \text{if } i = j \end{cases}$$

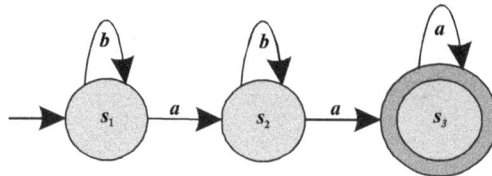

14. Construct a transition diagram without Λ moves for the transition diagram given below.

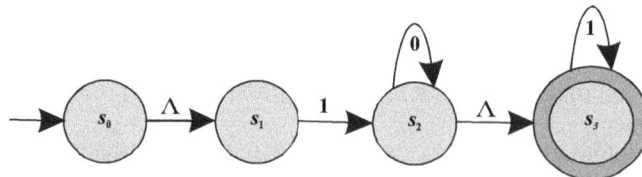

15. Consider the transition diagrams given below. Construct a transition diagram without $\wedge$ moves for each of the given diagram.

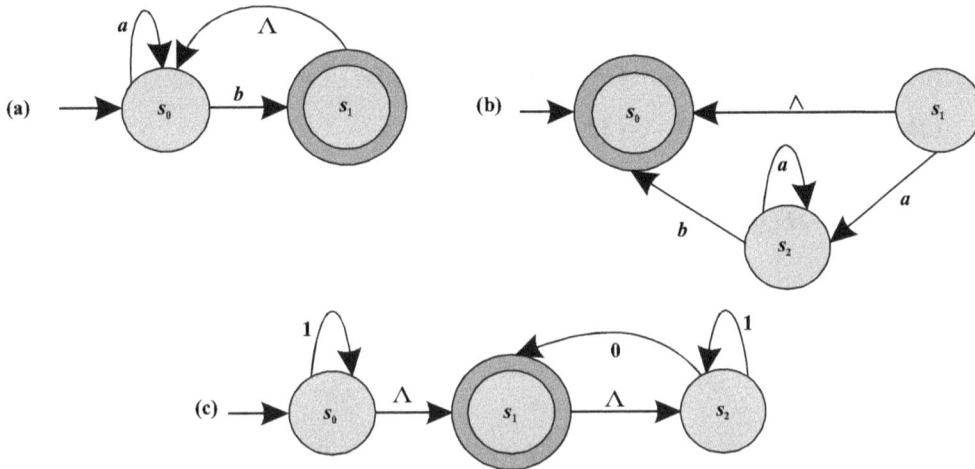

16. Show that the language $L = \{a^n b^n \mid n > 0\}$ is not regular by using pumping lemma.

17. Use pumping lemma to show $L = \{a^n b^m \mid 0 < n < m\}$ is not regular.

18. Construct a deterministic finite automata equivalent to the grammar $Q_0 \rightarrow aQ_0 \mid bQ_0 \mid bA; A \rightarrow aB; B \rightarrow aA \mid bC$ and $C \rightarrow b \mid \wedge$.

19. Show by using pumping lemma that $L = \{ww \mid w \in \{0,1\}^*\}$ is not regular.

20. Which of the following languages are regular sets? Prove your answer by using pumping lemma.

   (a) $L = \{a^{2n} \mid n \geq 1\}$

   (b) $L = \{0^n 1^m 0^{n+m} \mid m \geq 1 \text{ and } n \geq 1\}$

   (c) $L = \{xwx^R \mid x, w \in (0+1)^+\}$

21. Construct a regular grammar G for the transition diagram given below.

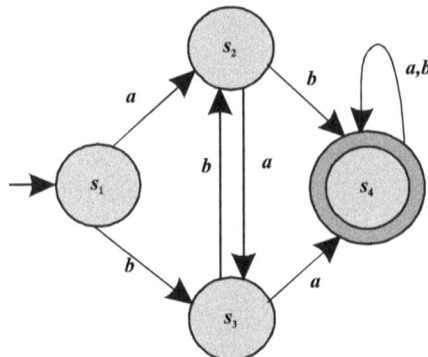

22. Construct a regular grammar G generating the regular expression $ab + ab^*$.

23. Find the regular grammar G generating the regular expression $(a + b)^* abb$.

24. Construct a transition system corresponding to the regular expression given below.

 (a) $(ab + c^*)^* b + ac$ (b) $0 + 11 + 22 + 01^* 2^*$

25. Construct a regular expression representing the regular grammar G corresponding to the productions $Q_0 \rightarrow bQ_0 | B$ and $B \rightarrow aB | b$.

26. Find regular expressions that represent the following sets:

 (a) Set of all strings over $\{a, b\}$ having at most one pair of a's or at most one pair of b's.

 (b) Set of all strings over $\{0, 1\}$ having at least two occurrences of 1 between any two occurrences of 0.

 (c) Set of all strings over $\{a, b\}$ beginning with aaa and ending with b.

 (d) Set of all strings over $\{0, 1\}$ beginning with 000.

27. Consider the transition diagrams given below. Find out the set of strings over $\Sigma = \{0,1\}$ recognized by the given transition systems.

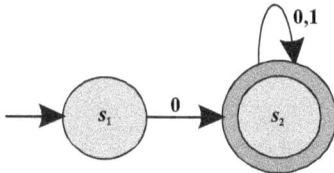

(a)

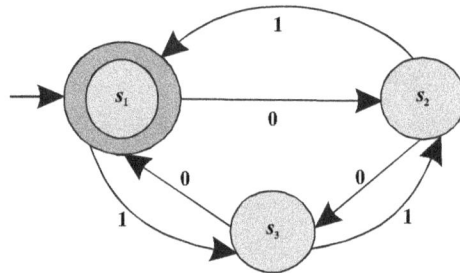

(b)

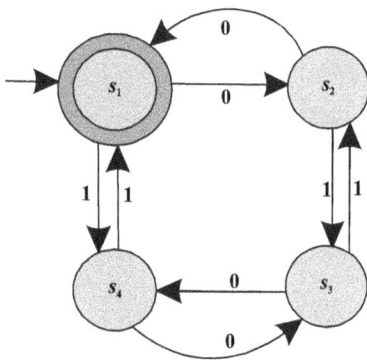

(c)

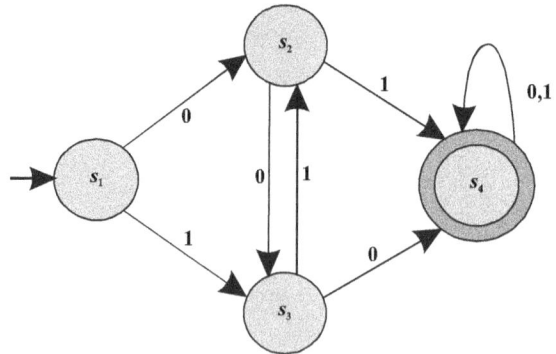

(d)

28. Represent the following sets by regular expression.

   (a) $\{a, aa, aaa, aaaa, \cdots\}$

   (b) $\{a, b, ab, ba, aa, bb, \cdots\}$

   (c) $\{a, b, c\}$

   (d) $\{a^m \,|\, m \text{ is divisible by 2 or 4 or 5}\}$

   (e) $\{a^{2n+1} \,|\, n > 0\}$

   (f) $\{\wedge, a, aa, aaa, \cdots\}$

29. Construct a finite automaton accepting all strings over the set $\{a, b\}$ ending with *aba* or *aabb*.

30. Let r, s and t be three regular expressions. Prove the following identities. It is given that $r = s$ means $L(r) = L(s)$.

   (a) $r + s = s + r$ (b) $(r + s) + t = r + (s + t)$

   (c) $(rs)t = r(st)$ (d) $r(s + t) = rs + rt$

   (e) $(r + s)t = rt + st$ (f) $(r^*)^* = r^*$

   (g) $\phi^* = \wedge$ (h) $(r^* s^*)^* = (r + s)^*$

31. Show by using pumping lemma that the set $L = \{a^m b^n c^p \,|\, m, n, p \geq 0\}$ is not regular.

32. Prove that the set $L = \{a^n b^{2n} \,|\, n > 0\}$ is not regular by using pumping lemma.

33. Let R be a regular set for the deterministic finite automaton whose transition diagram is given below. Construct a transition diagram that accepts the complement of R, i.e., $\Sigma^* - R$.

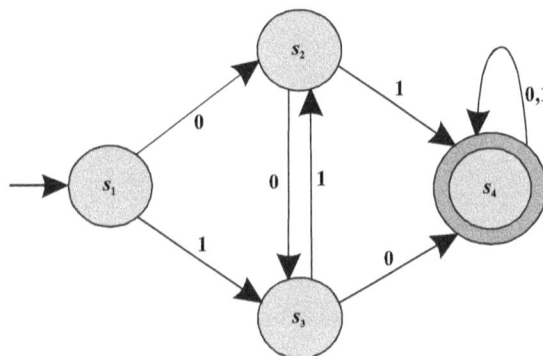

34. Prove that the following.

   (a) $(0 + 1^*)^* = (0 + 1)^*$

   (b) $(aaa^*)^* = (aa + aaa)^*$

5

CONTEXT-FREE LANGUAGES

5.0 INTRODUCTION

In the previous chapter, we studied that all languages are not regular. One does not need to look very far for non-regular languages while languages are effective in describing certain simple patterns. It is observed that, the language therefore describes a simple kind of nested structure as found in programming languages by indicating that some properties of programming languages require some thing beyond regular languages. In order to overcome this problem we must enlarge the family of languages and it leads to context-free languages and grammar.

The study of context-free grammar is mainly due to Chomsky (1956). We begin this chapter by defining context-free grammars that are based on a more complete understanding of the structure of the words belonging to the language, and context-free languages. We define derivation trees, ambiguity in context-free grammars and give procedures for simplifying context-free grammars. The two normal forms: Chomsky normal form (CNF) and Greibach normal form (GNF) are dealt with. The chapter is concluded after proving pumping lemma and giving some decision algorithms.

5.1 CONTEXT-FREE GRAMMAR

A context-free grammar (CFG) is defined as four-tuple (V_N, T, P, Q_0), where V_N and T are finite set of variables and terminals respectively such that $(V_N \cap T) = \phi$. The finite set of productions P is the kernel of grammars and language specification of the form $Q \rightarrow \alpha$, where Q is a variable and α is a string of symbols from $(V_N \cup T)^*$. Finally Q_0 is denoted as a special variable called the start symbol.

We use the following conventions regarding grammars:

(a) The upper-case letters $A, B, C, D, E, Q_1, Q_2, Q_3, \cdots$ denote variables with Q_0 as the start symbol unless otherwise stated.

(b) The lower-case letters $a, b, c, d, e, \ldots$ and digits denote terminals.

(c) The lower-case letters x, y, z, w, u and v denote strings of terminals.

(*d*) The capital letters X, Y and Z denote symbols that may be either terminals or variables.

(*e*) The lower-case Greek letters α, β, γ denote strings of variables and terminals, i.e., α, β, γ are elements of $(V_N \cup T)^*$.

(*f*) $X^0 = \wedge$ for any symbol $X \in (V_N \cup T)$.

Note

1. If $Q \to a_1, Q \to a_2, \cdots, Q \to a_k$ are the productions for the variable Q of some grammar G, then the productions may be expressed as $Q \to a_1|a_2|a_3|\cdots|a_k$, where the vertical line (|) represents "or".

2. If $Q \to Q_1Q_2$ is a production, then we can replace Q by Q_1Q_2, but at the same time we cannot replace Q_1Q_2 by Q. It implies that reverse substitution is not permitted.

3. If $Q \to Q_1Q_2$ is a production, then it is not necessary that $Q_1Q_2 \to Q$ is a production. It implies that inversion operation is not permitted.

5.2 CONTEXT-FREE LANGUAGES

We now formally define the language. Generally language is generated by a grammar $G = (V_N, T, P, Q_0)$. If the grammar is context-free, then the generated languages are called context-free languages (CFL). The different applications of context-free languages are parser design and describing block structure in programming languages. An example of context-free languages is the syntax of programming languages. However, it is to be noted that not all syntactic aspects of programming languages are captured by the context-free grammar. For example, the fact that a variable has to be declared before we use and the type correctness of expressions are not captured by context-free grammar. In general, we use tree structures to visualize derivations in context-free languages called as derivation trees. In the next section we discuss in detail about derivation trees.

For example, consider a grammar $G = (V_N, T, P, Q_0)$, where $V_N = \{Q_0\}$, $T = \{0,1\}$ and $P = \{Q_0 \to 0Q_01, Q_0 \to 01\}$. Here there exists only one variable Q_0 and two terminals 0 and 1. On using first production repeatedly we have

$$Q_0 \to 0Q_01 \to 00Q_011 \to 0^2 Q_0 1^2$$

After applying first production $(n-1)$ times, we get

$$Q_0 \to 0^{n-1} Q_0 1^{n-1}$$
$$\to 0^{n-1} 011^{n-1} \quad [\because Q_0 \to 01]$$
$$\to 0^n 1^n$$

Therefore, the language generated by the grammar G is $L(G) = \{0^n 1^n \mid n \geq 1\}$.

5.2.1 Derivation Tree

The string generated by a context-free grammar $G = (V_N, T, P, Q_0)$ can be represented by using trees. A derivation tree is an ordered tree where nodes are labelled with the left side of productions whereas the children of a node represent its corresponding right side. A derivation tree is otherwise known as parse tree, generation tree, syntax tree or production tree. Now we give a rigorous definition of a derivation tree.

A tree is a derivation tree for a context-free grammar $G = (V_N, T, P, Q_0)$ if it has the following characteristics.

(a) Every vertex has a label which is a variable, terminal or $\wedge$. It implies that every vertex has a label, which is a symbol of $(V_N \cup T \cup \{\wedge\})$.

(b) The root is labelled as Q_0, where Q_0 is the start symbol.

(c) The label of an interior vertex is always a variable. It implies that if a vertex A is interior, then $A \in V_N$.

(d) If a vertex has label $A \in V_N$ and its children are labelled $a_1, a_2, \cdots, a_n$ from left to right, then the set of productions P must have a production of the form $A \rightarrow a_1 a_2 a_3 \cdots a_n$.

(e) If a vertex has label $\wedge$, then the vertex is a leaf and is the only child of its parent.

(f) Every leaf has a label from $T \cup \{\wedge\}$.

Consider the grammar $G = (V_N, T, P, Q_0)$, where $V_N = \{Q_0, A, B\}$; $T = \{a, b\}$ and P consists of productions $Q_0 \rightarrow aAB|b$; $A \rightarrow Q_0 B|a$ and $B \rightarrow A|b$. We draw a derivation tree, with circles instead of points for the vertices. The vertices are numbered for reference.

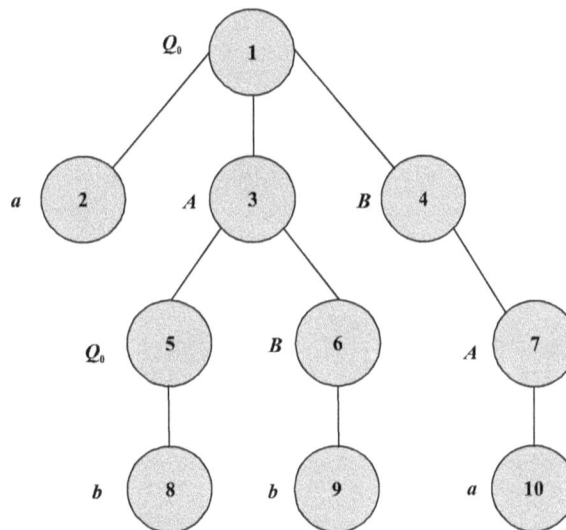

Derivation tree

There is no need to draw arrow on the edges because the direction of productions is always downward. From the tree, it is clear that the interior vertices are 1, 3, 4, 5, 6 and 7. Vertex 1 has label Q_0 with children labelled a, A and B from left to right. It leads to the production $Q_0 \rightarrow aAB$. Similarly, vertex 3 labelled with A has children labelled Q_0 and B from left to right. It gives the production $A \rightarrow Q_0 B$. Likewise we will get the productions $B \rightarrow A, Q_0 \rightarrow b, B \rightarrow b$ and $A \rightarrow a$ from the vertices 4, 5, 6 and 7 respectively. Also, it is clear that the derivation tree yield the string "*abba*".

5.2.2 Basic Terminologies

There are certain terms that are used frequently in this chapter. Here we discuss the basic terminologies.

Yield   The string of symbols obtained by reading the leaves of the derivation tree from left to right omitting any $\wedge$'s encountered is termed as the yield of the derivation tree. For example, the above derivation tree yield the string '*abba*'.

Subtree   A subtree of a derivation tree has the following characteristics:

(*a*) The root of subtree is a particular vertex A of derivation tree, where $A \in V_N$.

(*b*) The vertices of subtree are descendents of A, the edges connecting them, and their labels.

For example, consider the derivation tree and a subtree of this derivation tree are given below.

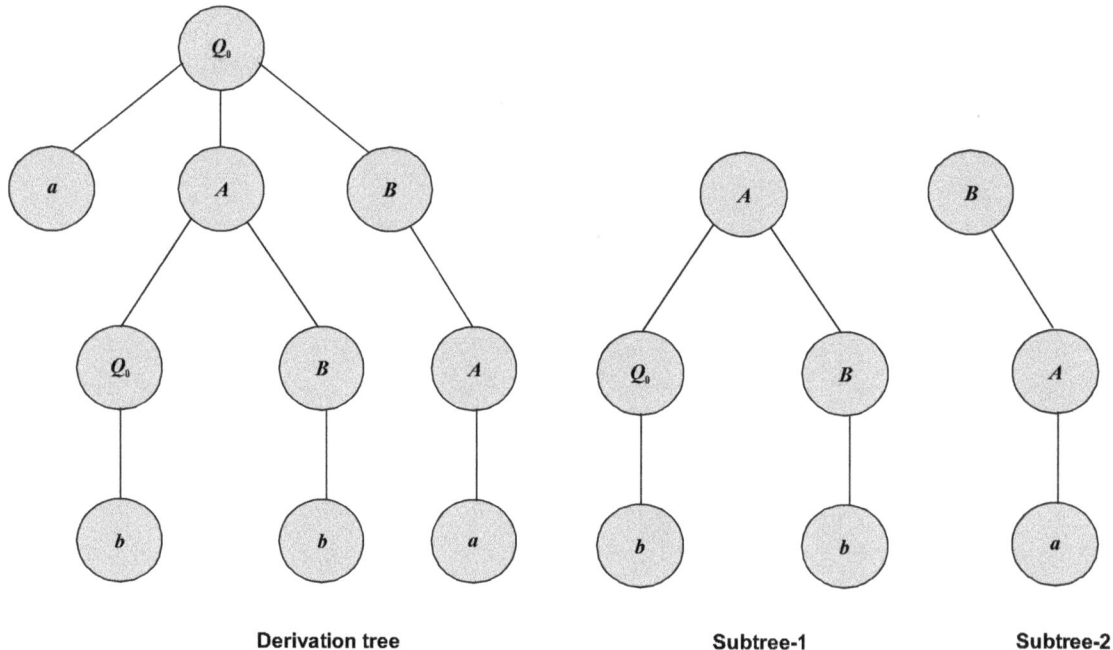

Derivation tree   Subtree-1   Subtree-2

A-tree   A subtree looks just like a derivation tree except that the label of the root may not be the start symbol Q_0 of the grammar G if and only if Q_0 does not appear in the right-hand side of any production. If the root of the subtree is labelled with A, then it is known as A-tree. For example, the above subtree-1 is called A-tree whereas subtree-2 is called as B-tree.

Partial derivation tree   A tree that has properties (a), (c), (d) and (e) defined earlier but in which the property (b) does not necessarily hold and the property (f) is replaced by "every leaf has a label from $V_N \cup T \cup \{\wedge\}$ is said to be a partial derivation tree.

Interior node   A node in a derivation tree whose label is a variable or non terminal is known as an interior node. If A is a node and $A \in V_N$, then A is an interior node.

Leaf node   A node in a derivation tree whose label is a terminal is known as a leaf node. If b is a node and $b \in T$, then b is a leaf node.

Sentential form   A string of terminals and variables α is called a sentential form if $Q_0 \overset{*}{\Rightarrow} \alpha$, where $Q_0 \in V_N$ and $\alpha \in (V_N \cup T)^*$.

Relation between derivation tree and derivations   Let $G = (V_N, T, P, Q_0)$ be a context-free grammar and then $Q_0 \overset{*}{\Rightarrow} \alpha$ if and only if there is a derivation tree in grammar G with yield α.

Proof   Let $G = (V_N, T, P, Q_0)$ be a context-free grammar. First of all we shall prove that for any $A \in V_N$, $A \overset{*}{\Rightarrow} \alpha$ if and only if there is an A-tree with yield α. Once it is proved, the above relation follows by assuming that $A = Q_0$.

Assume that, α be the yield of an A-tree T. We show $A \overset{*}{\Rightarrow} \alpha$ by mathematical induction on the number of interior nodes in tree T. If there is only one interior vertex, then the remaining vertices are leaves and are children of the root A. In this case the tree must look like as the following figure where $X_1 X_2 \cdots X_n$ must be α.

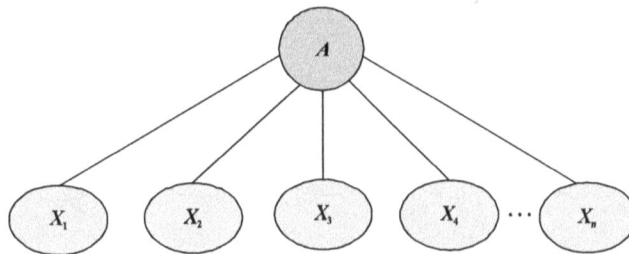

Therefore, by definition of derivation tree we get $A \to X_1 X_2 \cdots X_n = \alpha$ and $A \to \alpha$ must be a production of P.

Assume that, the relation is true for trees with up to $(k-1)$ interior vertices. Also, assume that α is the yield of an A-tree with k interior vertices for some $k > 1$. Let us consider the children $a_1, a_2, a_3, \cdots, a_n$ of the root in order from left. Let us assume their labels are $X_1, X_2, \cdots, X_n$. Therefore, by definition of derivation tree we get $A \to X_1 X_2 \cdots X_n$ is in the production P and so

$$A \Rightarrow X_1 X_2 \cdots X_n \tag{1}$$

Since $k > 1$, it is clear that at least one of the children is an interior vertex. Let us assume that the ith child X_i is an interior vertex. It implies that X_i is the root of a subtree, and X_i must be a variable. The subtree, X_i-tree must have some yield α_i that has obtained by the concatenation of the labels of the leaves which are descendents of the vertex a_i labelled X_i. Also, the number of interior vertices of the subtree is less than k. Since the subtree with root X_i is not the entire tree, so by method of induction $X_i \overset{*}{\Rightarrow} \alpha_i$. Also if the vertex a_i is not an interior vertex (a leaf), then we have $X_i \Rightarrow \alpha_i$. Hence by using equation (1) we have

$$A \Rightarrow X_1 X_2 X_3 \cdots X_n \Rightarrow \alpha_1 X_2 X_3 \cdots X_n$$
$$\Rightarrow \alpha_1 \alpha_2 X_3 \cdots X_n$$
$$\cdots \quad \cdots \quad \cdots$$
$$\Rightarrow \alpha_1 \alpha_2 \alpha_3 \cdots \alpha_n = \alpha$$

i.e., $A \overset{*}{\Rightarrow} \alpha$

Therefore, whenever α is the yield of an A-tree, by method of induction it can be shown that $A \overset{*}{\Rightarrow} \alpha$.

In order to show the only if part, let us assume that $A \overset{*}{\Rightarrow} \alpha$. Our aim is to construct a derivation tree that yield α. We prove this by method of induction on the number of steps in $A \overset{*}{\Rightarrow} \alpha$. When there is a derivation $A \Rightarrow \alpha$, then there exists a production $A \rightarrow \alpha$ in P. If $\alpha = X_1 X_2 \cdots X_n$ then the A-tree with yield α is given below.

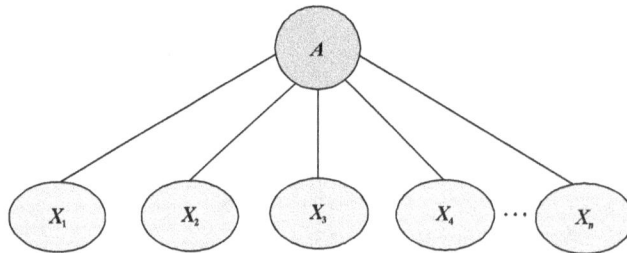

Now assume that, the result for derivations takes at most k-steps. Let $A \overset{k}{\Rightarrow} \alpha$. We can split this derivation as $A \Rightarrow X_1 X_2 \cdots X_n \overset{k-1}{\Rightarrow} \alpha$, where $A \Rightarrow X_1 X_2 \cdots X_n$ leads to the production $A \rightarrow X_1 X_2 \cdots X_n$ in P. Hence there arise two cases for the derivation of $X_1 X_2 \cdots X_n \overset{k-1}{\Rightarrow} \alpha$. Therefore, we get either of the following cases:

Case 1 X_i has not changed throughout the derivation and thus X_i is a terminal.

Case 2 X_i has changed in some subsequent step and so X_i is a variable.

Let α_i be the substring of α derived from X_i. It implies that $X_i = \alpha_i$ according to Case 1 whereas according to Case 2 $X_i \overset{*}{\Rightarrow} \alpha_i$. As G is a context-free grammar, therefore in every step of the derivation $X_1 X_2 \cdots X_n \overset{*}{\Rightarrow} \alpha$, we replace a single variable by a string. Since $\alpha_1, \alpha_2, \cdots, \alpha_n$ account for all the symbols in α we have $\alpha = \alpha_1 \alpha_2 \alpha_3 \cdots \alpha_n$.

Now, we construct the derivation tree with yield α as follows: As $A \to X_1 X_2 X_3 \cdots X_n$ is in P, we construct a derivation tree with n leaf nodes whose labels are $X_1, X_2, \cdots, X_n$ in order from the left. The tree is given in the following figure.

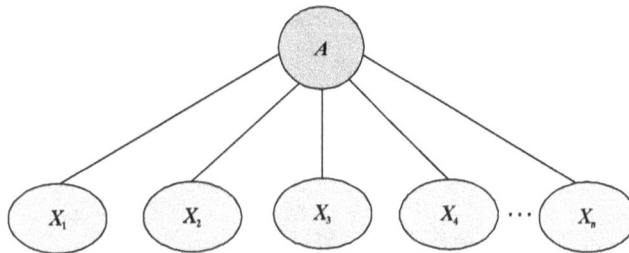

In Case 1, we leave the vertex as it is whereas in Case 2, for each X_i that is a variable, there is an X_i-tree with yield α_i. Let this tree be T_i. The resulting tree when some of X_i is a variable is given in the following figure. This completes the proof.

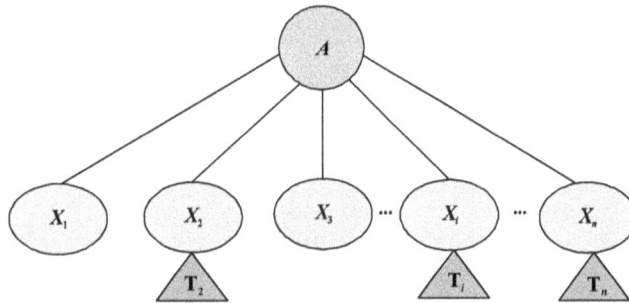

Consider a context-free grammar G having productions $Q \to bAQ | b$ and $A \to QbA | ab$. Our aim is to construct a derivation tree whose yield is "*bbbabb*". From the productions it is clear that

$$Q \Rightarrow bAQ \Rightarrow bQbAQ \Rightarrow bbbAQ \Rightarrow bbbabQ \Rightarrow bbbabb.$$

The derivation tree is as follows:

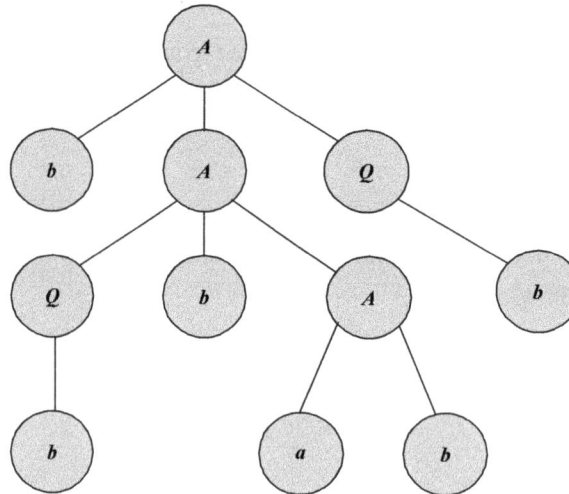

5.2.3   Leftmost and Rightmost Derivation

A derivation $Q_0 \overset{*}{\Rightarrow} \alpha$ is said to be leftmost if in each step the production is applied only to the leftmost variable. On the other hand a derivation $Q_0 \overset{*}{\Rightarrow} \alpha$ is said to be rightmost if the production is applied only to the rightmost variable at every step.

Consider the grammar with productions $Q \to 0AB$, $A \to 1B1$, $B \to A \,|\, \wedge$. Let us consider the string $w = 01111$. The leftmost derivation of the string 'w' is given as $Q_0 \Rightarrow 0AB \Rightarrow 01B1B \Rightarrow 01A1B \Rightarrow 011B11B \Rightarrow 01111B \Rightarrow 01111$.

The rightmost derivation of the string 'w' is given as $Q_0 \Rightarrow 0AB \Rightarrow 0A \Rightarrow 01B1 \Rightarrow 01A1 \Rightarrow 011B11 \Rightarrow 01111$.

5.2.4   Ambiguity in Context-free Grammars

In our usual language, sometimes we come across ambiguous sentences. For example, consider the sentence in English: In books selected examples are given. It indicates that the word selected can be parsed into two ways. Therefore, some sentences may be parsed into two different ways. This situation may arise in context-free languages also. In this case a given terminal string may be the yield of two derivation trees. Now, we give the definition of ambiguous sentences in context-free languages.

A terminal string $z \in L(G)$ is said to be ambiguous if there exists two or more derivation tree for z. In other words we say that, a terminal string $z \in L(G)$ is said to be ambiguous if there exists two or more distinct leftmost or rightmost derivations. A context-free grammar G is said to be ambiguous if there exists at least one string $z \in L(G)$, which has two or more distinct derivation trees.

Consider the context-free grammar G having productions $Q_0 \rightarrow aQ_0 | aA$ and $A \rightarrow b | Q_0$. Let us consider a string $z = aab$. Then, we get two distinct leftmost derivation trees for z as given below. Therefore, the context-free grammar G is ambiguous.

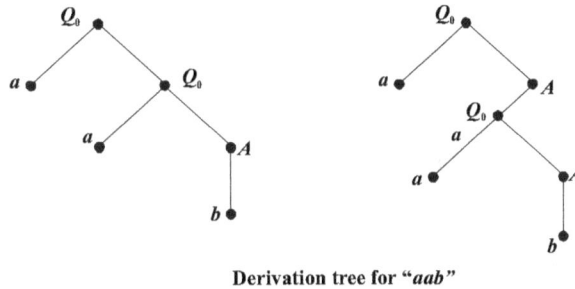

Derivation tree for "*aab*"

5.3 SIMPLIFICATION OF CONTEXT-FREE GRAMMARS

There are many ways in which we can restrict the format of productions in a context-free grammar without reducing the capabilities of context-free grammar. It indicates that, it may not be necessary to use all the variables V_N and terminals T or all the productions in P for generating the strings in a context-free grammar G. Hence, we try to eliminate symbols and productions in G that are not useful for derivation of strings. It can be better explained with the following example:

Consider the context-free grammar G as $G = (V_N, T, P, Q_0)$, where $V_N = \{Q_0, A, B, C, D, E\}$, $T = \{a, b, c\}$ and productions P as $Q_0 \rightarrow BD$, $B \rightarrow a$, $D \rightarrow c$, $A \rightarrow C$ and $E \rightarrow b | \wedge$. Therefore, the language generated by G is given as $L(G) = \{ac\}$.

Similarly, consider another context-free grammar $G_1 = (V_N, T, P, Q_0)$, where $V_N = \{Q_0, B, D\}$; $T = \{a, c\}$ with productions P as $Q_0 \rightarrow BD$, $B \rightarrow a$ and $D \rightarrow c$. Therefore, the language generated by G_1 is given as $L(G_1) = \{ac\}$ and hence $L(G) = L(G_1)$. In the later case G_1, we have eliminated the symbols A, C, b and the productions $A \rightarrow C$ and $E \rightarrow b | \wedge$. It leads to the following points regarding the productions and symbols that are eliminated.

(*a*) The variable C does not derive any terminal string.

(*b*) $E \rightarrow \wedge$ is a null production in G.

(*c*) E and b do not involve in any sentential form.

(*d*) $A \rightarrow C$ replaces A by C only.

Thus, it is necessary to eliminate the variables, terminals and productions having following characteristics for the simplification of context-free grammar. The different characteristic includes:

(*a*) Null productions of the form $A \rightarrow \wedge$.

(*b*) Unit productions of the form $A \rightarrow B$.

(*c*) Productions in which a variable does not derive a terminal string.

(*d*) The symbols $(V_N \cup T)$ not appearing in any sentential form.

5.3.1 Construction of Reduced Context-free Grammar

In this section we will discuss how to construct a reduced context-free grammar for a given context-free grammar. Let $G = (V_N, T, P, Q_0)$ be a context-free grammar. The grammar G is said to be reduced if every symbol in $(V_N \cup T)$ appears in the derivation of some terminal string w. It indicates that, for each $X \in (V_N \cup T)$, there exists a derivation $Q_0 \overset{*}{\Rightarrow} \alpha X \beta \overset{*}{\Rightarrow} w$; $w \in L(G)$.

Theorem *If $G = (V_N, T, P, Q_0)$ is a context-free grammar with $L(G) \neq \phi$, then we can find an equivalent context-free grammar G_1 such that each variable in G_1 derives some terminal string.*

Proof   Let $G = (V_N, T, P, Q_0)$ is a context-free grammar with $L(G) \neq \phi$. We define $G_1 = (V_N', T, P', Q_0)$ as follows:

(a) *Construction of V_N'*   We define $W_i \subseteq V_N$ by recursion as follows:
$$W_1 = \{Q \in V_N \mid Q \to w \text{ is in } P, \text{ where } w \in T^*\}$$

If $W_1 = \{\phi\}$ then some variables will remain after the application of any production. Therefore, we have $L(G) = \phi$. Now, we define W_{i+1} recursively as $W_{i+1} = W_i \cup \{Q \in V_N \mid Q \to \alpha \text{ with } \alpha \in (T \cup W_i)^*\}$

According to definition of W_i, we have $W_i \subseteq W_{i+1}$ for all i. It implies that all the variables in W_i are in W_{i+1} but all the variables in W_{i+1} may also be in W_i. As V_N contains only finite number of variables, then definitely we will arrive at $W_k = W_{k+1}$ for some $k \leq |V_N|$. Therefore, $W_k = W_{k+j}$ for $j \geq 1$. We define W_k as V_N', i.e., $V_N' = W_k$.

(b) *Construction of P'*   The set of productions P' is defined as: $P' = \{Q \to \alpha \mid Q, \alpha \in (V_N' \cup T)^*\}$

This implies that P' contains only those productions that involve the variables in V_N'. Therefore, we have $G_1 = (V_N', T, P', Q_0)$ where $Q_0 \in V_N'$. Now, we prove that every variable in V_N' derives some terminal string and $L(G) = L(G_1)$.

In order to prove the first part, we note that $W_k = W_1 \cup W_2 \cup \cdots \cup W_k$. We prove by induction on i that $Q \overset{*}{\underset{G_1}{\Rightarrow}} w$ for some $w \in T^*$ and $i = 1, 2, 3, \cdots, k$. If $Q \in W_1$, then $Q \underset{G}{\Rightarrow} w$. Thus the production $Q \to w$ is in P'. Therefore, $Q \overset{*}{\underset{G_1}{\Rightarrow}} w$ This is the basis for induction.

Assume that the result is true for i. Let $Q \in W_{i+1}$. Then either $Q \in W_i$ or $Q \to \alpha$ with $\alpha \in (T \cup W_i)^*$. If $Q \in W_i$, then $Q \overset{*}{\underset{G_1}{\Rightarrow}} w$ for some $w \in T^*$ by induction. If $Q \to \alpha$, then by definition of P', $Q \to \alpha$ is in P'. So, we can write $\alpha = X_1 X_2 \cdots X_m$ where $X_j \in (T \cup W_i)$. If $X_j \in W_i$, then by induction $X_j \overset{*}{\underset{G_1}{\Rightarrow}} w_j$ for some $w_j \in T^*$. Hence, $Q \overset{*}{\underset{G_1}{\Rightarrow}} w_1 w_2 \cdots w_m \in T^*$. Therefore, by method of induction the result is true for $i = 1, 2, 3, \cdots, k$. Thus, $L(G_1) \subseteq L(G)$ as $V_N' \subseteq V_N$ and $P' \subseteq P$.

Now our aim is to show $L(G) \subseteq L(G_1)$. In order to prove this we need an auxiliary result:

$$Q \underset{G_1}{\overset{*}{\Rightarrow}} w \text{ if } w \text{ for some} w \in T^* \tag{1}$$

We prove this by the method of induction on the number of steps in the derivation $Q \underset{G}{\overset{*}{\Rightarrow}} w.$

If $Q \underset{G}{\overset{*}{\Rightarrow}} w$, then $Q \to w$ is in P and $Q \in W_i \subseteq V_N'$. Again, as $Q \in V_N'$ and $w \in T^*$, we have $Q \to w$ is

in P'. Therefore, $Q \underset{G_1}{\overset{*}{\Rightarrow}} w$. Thus there exists a basis for induction. Assume that the result (1) is true for

derivations in at most k-steps. Let us take $Q \underset{G}{\overset{k+1}{\Rightarrow}} w.$ Thus we can split this as

$Q \underset{G}{\Rightarrow} X_1 X_2 X_3 \cdots X_m \underset{G}{\overset{k}{\Rightarrow}} w_1 w_2 w_3 \cdots w_m$ such that $X_j \underset{G}{\overset{*}{\Rightarrow}} w_j.$ If $X_j \in T$, then $w_j = X_j.$

If $X_j \in V_N$, then by discussion given above we have $X_j \in V_N'$. Since $X_j \underset{G}{\overset{*}{\Rightarrow}} w_j$, in at most k

number of steps, we have $X_j \underset{G_1}{\overset{*}{\Rightarrow}} w_j.$ Also, $X_1, X_2, \cdots, X_m \in (T \cup V_N')^*$ implies $Q \to X_1 X_2 \cdots X_m$ is

in P'. Therefore,

$$Q \underset{G_1}{\Rightarrow} X_1 X_2 \cdots X_m \underset{G_1}{\overset{*}{\Rightarrow}} w_1 w_2 \cdots w_m$$

Hence by induction, the result (1) is true for all derivations. In particular, $Q_0 \underset{G}{\overset{*}{\Rightarrow}} w$ implies that

$Q_0 \underset{G_1}{\overset{*}{\Rightarrow}} w$. This proves that $L(G) \subseteq L(G_1)$ and hence $L(G) = L(G_1).$

For example, consider the context-free grammar $G = (V_N, T, P, Q_0)$, where the productions are $Q_0 \to AB$, $A \to 0$, $B \to 1|C$ and $D \to 0$. Our aim is to construct a context-free grammar G_1 such that every variable in G_1 derives some terminal string.

(a) *Construction of* V_N' From the productions it is clear that $W_1 = \{A, B, D\}$ since $A \to 0, B \to 1$ and $D \to 0$ are productions in P that are having a terminal string on its right-hand side. Now,

$$\begin{aligned} W_2 &= W_1 \cup \{A_1 \in V_N \,|\, A_1 \to \alpha; \alpha \in (W_1 \cup T)^*\} \\ &= \{A, B, D\} \cup \{Q_0\} \\ &= \{A, B, D, Q_0\} \\ W_3 &= W_2 \cup \{A_1 \in V_N \,|\, A_1 \to \alpha; \alpha \in (W_2 \cup T)^*\} \\ &= \{A, B, D, Q_0\} \\ &= W_2 \end{aligned}$$

Therefore, $V_N' = W_2 = \{A, B, D, Q_0\}$

(*b*) *Construction of P'* The productions of P' are defined as:

$$P' = \{A_1 \to \alpha \mid A_1 ; \alpha \in (T \cup V_N')^*\}$$
$$= \{Q_0 \to AB, B \to 1, A \to 0, D \to 0\}$$

Therefore, the reduced grammar G_1 is given as $G_1 = (V_N', T, P, Q_0)$, where

$$V_N' = \{Q_0, A, B, D\},$$
$$T = \{0, 1\},$$
$$P' = \{Q_0 \to AB, A \to 0, B \to 1, D \to 0\}$$

Theorem *For every context-free grammar $G_1 = (V_N', T, P, Q_0)$ there exists an equivalent grammar $G_2 = (V_N'', T'', P'', Q_0)$ such that every symbol in $(V_N'' \cup T'')$ appears in some sentential form.*

Proof   Let $G_1 = (V_N', T, P, Q_0)$ be a context-free grammar, where every variable derives some terminal string. We construct $G_2 = (V_N'', T'', P'', Q_0)$ as follows:

(*a*) *Construction of W_i for $i \geq 1$*   We define W_i recursively as follows:

i. $W_1 = \{Q_0\}$

ii. $W_{i+1} = W_i \cup \{X \in (V_N' \cup T) \mid Q \to \alpha$ with $Q \in W_i$ and α containing $X\}$

It is clear that $W_i \subseteq (V_N' \cup T)$ and $W_i \subseteq W_{i+1}$. Since we have finite number of elements in $(V_N' \cup T)$, definitely we will arrive at a situation where $W_k = W_{k+1}$ for some positive integer k. It implies that $W_k = W_{k+j}$ for $j \geq 1$.

(*b*) *Construction of V_N'', T'' and P''*   We define V_N'', T'' and P'' as follows:

$$V_N'' = V_N' \cap W_k$$
$$T'' = T \cap W_k \qquad \text{and}$$
$$P'' = \{Q \to \alpha \mid Q \in W_k\}$$

Every symbol X in G_2 appears in some sentential form, say $\alpha X \beta$. Therefore,

$$Q_0 \overset{*}{\Rightarrow} \alpha X \beta \overset{*}{\Rightarrow} w \text{ for some } w \in T^*, \text{ i.e., } G_2 \text{ is reduced.}$$

For example consider the context-free grammar G, where every variable derives some terminal string as $G = (\{Q_0, A, B, D\}, \{0, 1\}, \{Q_0 \to AB, A \to 0, B \to 1, D \to 0\})$. Our aim is to construct an equivalent grammar $G_1 = (V_N'', T'', P'', Q_0)$, where every symbol in $(V_N'' \cup T'')$ appears in some sentential form.

(*a*) *Construction of W_i for $i \geq 1$*

$$W_1 = \{Q_0\}$$

$$W_2 = W_1 \cup \{X \in (V'_N \cup T) | Q \rightarrow \alpha \text{ with } Q \in W_1 \text{ and } \alpha \text{ containing } X\}$$
$$= \{Q_0\} \cup \{A, B\}$$
$$= \{Q_0, A, B\}$$

$$W_3 = W_2 \cup \{X \in (V'_N \cup T) | Q \rightarrow \alpha \text{ with } Q \in W_2 \text{ and } \alpha \text{ containing } X\}$$
$$= \{Q_0, A, B\} \cup \{0, 1\}$$
$$= \{Q_0, A, B, 0, 1\}$$
$$W_4 = W_3 = \{Q_0, A, B, 0, 1\}$$

(*b*) *Construction of V''_N, T'' and P''* We define V''_N, T'' and P'' as follows:

$$V''_N = V'_N \cap W_3 = \{Q_0, A, B, D\} \cap \{Q_0, A, B, 0, 1\} = \{Q_0, A, B\}$$
$$T'' = T \cap W_3 = \{0, 1\} \cap \{Q_0, A, B, 0, 1\} = \{0, 1\} \qquad \text{and}$$
$$P'' = \{Q \rightarrow \alpha | Q \in W_3\} \quad = \{Q_0 \rightarrow AB, A \rightarrow 0, B \rightarrow 1\}$$

Thus, the required grammar $G_1 = (V''_N, T'', P'', Q_0)$ is given as $V''_N = \{Q_0, A, B\}$; $T'' = \{0,1\}$ and $P'' = \{Q_0 \rightarrow AB, A \rightarrow 0, B \rightarrow 1\}$.

Theorem *For every context-free grammar G there is a reduced grammar G_1 which is equivalent to G.*

Proof The reduced grammar G_1 which is equivalent to the context-free grammar G is constructed in two steps. First we construct a grammar G' equivalent to the given grammar G so that each variable in G' derives some terminal string and then we construct a grammar $G_1 = (V''_N, T'', P'', Q_0)$ equivalent to G' so that every symbol in G_1 appears in some sentential form. Therefore, grammar G_1 is the required reduced grammar equivalent to the grammar G.

Note In order to get a reduced grammar for a given context-free grammar, we must apply first and second theorem in order discussed in this section. If we apply second theorem followed by first theorem, then we may not get a reduced context-free grammar.

5.3.2 Elimination of Null Productions

A context-free grammar may have productions of the form $Q \rightarrow \wedge, Q \in V_N$. Such productions are generally used to erase the variable Q. Therefore, a production of the form $Q \rightarrow \wedge, Q \in V_N$ is called a null production. Our aim is to give a construction to eliminate null productions in a context-free grammar. In this context, a variable X in a context-free grammar is said to be nullable if $X \overset{*}{\Rightarrow} \wedge$.

Theorem *If $G = (V_N, T, P, Q_0)$ is a context-free grammar, then we can find a context-free grammar G_1 having no null productions such that $L(G_1) = L(G) - \{\wedge\}$.*

Proof   Given $G = (V_N, T, P, Q_0)$ is a context-free grammar. Our aim is to construct a context-free grammar G_1 having no null productions such that $L(G_1) = L(G) - \{\wedge\}$. This implies that, the difference between context-free grammars G and G_1 is of only productions. Therefore, we construct $G_1 = (V_N, T, P', Q_0)$ as follows:

(*a*) *Construction of set of nullable variables W*   We determine the nullable variables recursively as follows.

$$W_1 = \{Q \in V_N \mid Q \to \wedge \text{ is in } P\}$$
$$W_{i+1} = W_i \cup \{Q \in V_N \mid Q \to \alpha \text{ is in } P; \alpha \in W_i^*\}$$

Therefore, it is clear that $W_i \subseteq W_{i+1}$. Since V_N has finite number of variables, so definitely we will arrive at $W_{k+1} = W_k$ for some $k \leq |V_N|$. Therefore, $W_{k+j} = W_k \; \forall \; j$. Therefore, we define the set of nullable variables as $W = W_k$.

(*b*) *Construction of P'*   The productions P' of the context-free grammar G_1 can be derived by using the following steps.

   i. If $Q \to w, w \in T^* - \{\wedge\}$ is a production in P, then the production is also included in P'.

   ii. Any production of P which does not have any nullable variable in the right hand side is included in P'.

   iii. If $Q \to X_1 X_2 X_3 \cdots X_m$ is a production in P, then the productions in P' are of the form $Q \to \alpha_1 \alpha_2 \alpha_3 \cdots \alpha_m$, where

      1. $\alpha_i = X_i$ if $X_i \notin W$

      2. $\alpha_i = X_i$ or $\wedge$ if $X_i \in W$

      3. Not all α_i's are null ($\wedge$)

(*c*) Now we have to show that $L(G_1) = L(G) - \{\wedge\}$. In order to prove this, we prove an auxiliary result given by the following relation.

$$Q \underset{G_1}{\overset{*}{\Rightarrow}} w \text{ if and only if } Q \underset{G}{\overset{*}{\Rightarrow}} w \text{ and } w \neq \wedge; \; \forall \; Q \in V_N \text{ and } w \in T^*.$$

(If Part) Let us assume that $Q \underset{G}{\overset{*}{\Rightarrow}} w$ and $w \neq \wedge$. Our aim is to show that $Q \underset{G_1}{\overset{*}{\Rightarrow}} w$. We show this by the method of induction on G_1 the number of steps present in the derivation $Q \underset{G}{\overset{*}{\Rightarrow}} w$.

If $Q \underset{G}{\overset{*}{\Rightarrow}} w$ and $w \neq \wedge$, then $Q \to w$ is a production in P'. Therefore, $Q \underset{G_1}{\Rightarrow} w$. Thus there is a basis for induction. Suppose that the result is true for derivations in at most k steps. Let $Q \underset{G}{\overset{k+1}{\Rightarrow}} w$ and $w \neq \wedge$. So the derivation can be splitted as

$$Q \underset{G}{\Rightarrow} X_1 X_2 \cdots X_n \overset{k}{\underset{G}{\Rightarrow}} w_1 w_2 \cdots w_n,$$

where $w = w_1 w_2 \cdots w_n$ and $Q_j \overset{*}{\underset{G}{\Rightarrow}} w_j$. Since $w \neq \wedge$; so not all w_j's are $\wedge$. If $w_j \neq \wedge$, then by induction $X_j \overset{*}{\underset{G_1}{\Rightarrow}} w_j$. If $w_j = \wedge$, then $X_j \in W$. Therefore, using the production $Q \to X_1 X_2 X_3 \cdots X_n$ in P, we construct $Q \to \alpha_1 \alpha_2 \alpha_3 \cdots \alpha_n$ in P', where $\alpha_j = X_j$ if $w_j \neq \wedge$ and $\alpha_j = \wedge$ if $w_j = \wedge$. Hence, we get

$$Q \underset{G_1}{\Rightarrow} \alpha_1 \alpha_2 \cdots \alpha_n \overset{*}{\underset{G_1}{\Rightarrow}} w_1 \alpha_2 \cdots \alpha_n \overset{*}{\underset{G_1}{\Rightarrow}} w_1 w_2 \cdots \alpha_n \overset{*}{\underset{G_1}{\Rightarrow}} w_1 w_2 \cdots w_n = w$$

i.e., $Q \overset{*}{\underset{G_1}{\Rightarrow}} w$

(Only if Part) Assume that $Q \overset{*}{\underset{G_1}{\Rightarrow}} w$. Our aim is to show that $Q \overset{*}{\underset{G}{\Rightarrow}} w$. and $w \neq \wedge$. We prove this by induction on the number of steps in the derivation $Q \overset{*}{\underset{G_1}{\Rightarrow}} w$.

If $Q \underset{G_1}{\Rightarrow} w$, then there exists a production $Q \to w$ in P' and by construction of P', $Q \to w$ is obtained from some production $Q \to X_1 X_2 X_3 \cdots X_n$ in P by erasing some or none of nullable variables. Therefore, $Q \underset{G}{\Rightarrow} X_1 X_2 X_3 \cdots X_n \overset{*}{\underset{G}{\Rightarrow}} w$. So, there is a basis for induction. Assume that the result is true for derivation in at most k steps. Let $Q \overset{k+1}{\underset{G_1}{\Rightarrow}} w$. We can split the derivation as

$$Q \underset{G_1}{\Rightarrow} X_1 X_2 \cdots X_n \overset{k}{\underset{G_1}{\Rightarrow}} w_1 w_2 \cdots w_n, \text{ where } X_i \overset{*}{\underset{G_1}{\Rightarrow}} w_i.$$ The first production $Q \to X_1 X_2 X_3 \cdots X_n$ in P' is obtained from some production $Q \to \alpha$ in P by erasing some or none of nullable variables in α. Therefore,

$$Q \underset{G}{\Rightarrow} \alpha \overset{*}{\underset{G}{\Rightarrow}} X_1 X_2 \cdots X_n$$

If $X_i \in T$, then $X_i \overset{0}{\underset{G}{\Rightarrow}} X_i = w_i$. If $X_i \in V_N$, then $X_i \overset{*}{\underset{G}{\Rightarrow}} w_i$ by induction. Thus we get $Q \overset{*}{\underset{G}{\Rightarrow}} X_1 X_2 \cdots X_n \overset{*}{\underset{G}{\Rightarrow}} w_1 w_2 \cdots w_n$. Therefore, by induction whenever $Q \overset{*}{\underset{G_1}{\Rightarrow}} w$, we have $Q \overset{*}{\underset{G}{\Rightarrow}} w$ and $w \neq \wedge$.

On applying the above result to Q_0, we have $w \in L(G_1)$ if and only if $w \in L(G)$ and $w \neq \wedge$. Therefore, we have $L(G_1) = L(G) - \{\wedge\}$.

For example consider a context-free grammar $G = (V_N, T, P, Q_0)$ given by the productions $Q_0 \to ABD | BC$, $B \to b | \wedge$, $C \to a | \wedge$, $A \to a$ and $D \to b$.

Our aim is to construct a context-free grammar G_1 equivalent to G by eliminating null productions. Therefore, we have to construct a set of nullable variables followed by the productions P' of G_1.

(*a*) *Construction of the set W of nullable variables*

$$W_1 = \{Q \in V_N | Q \to \wedge \text{ is a production in } P\}$$
$$= \{B, C\}$$
$$W_2 = W_1 \cup \{Q_0\} \qquad [\because Q_0 \to BC \text{ is in } P \text{ and } BC \in W_1^*]$$
$$= \{Q_0, B, C\}$$

$$W_3 = W_2 = \{Q_0, B, C\}$$

Thus, $W = W_2 = \{Q_0, B, C\}$

(*b*) *Construction of P'*   We define the productions of G_1 as follows:

i.   Productions $B \to b, C \to a, A \to a$ and $D \to b$ of P are included in P'.

ii. Production $Q_0 \to BC$ of P gives rise to $Q_0 \to BC, Q_0 \to B$ and $Q_0 \to C$ in P'.

iii. Production $Q_0 \to ABD$ of P gives rise to $Q_0 \to ABD$ and $Q_0 \to AD$ in P'.

So, the reduced context-free grammar G_1 is defined as $G_1 = (V_N, T, P', Q_0)$, where

$$V_N = \{Q_0, A, B, C, D\},$$
$$T = \{a, b\} \quad \text{and}$$
$$P' = \{Q_0 \to ABD|AD|BC|B|C, A \to a, D \to b, B \to b, C \to a\}$$

5.3.3   Elimination of Unit Productions

A context-free grammar may have productions of the form $Q_1 \to Q_2$, where $Q_1, Q_2 \in V_N$. Such type of productions is known as unit productions. In this section we give a construction to eliminate unit productions in a context-free grammar.

Theorem *If G be a context-free grammar with no null productions, then we can find a context-free grammar G_1 that has no unit productions such that $L(G_1) = L(G)$.*

Proof   Let us consider G be a context-free grammar with no null productions. Our aim is to construct a context-free grammar G_1 that has no unit productions. Let Q be any variable in V_N, i.e., $Q \in V_N$. We construct G_1 as follows:

(*a*)   *Construction of the set of variables derivable W(Q), $Q \in V_N$.* We define $W_i(Q)$ recursively as follows.

$$W_0(Q) = \{Q\}$$
$$W_{i+1}(Q) = W_i(Q) \cup \{A \in V_N | B \to A \text{ is in } P \text{ with } B \in W_i(Q)\}$$

This indicates that $W_i(Q) \subseteq W_{i+1}(Q)$. Since V_N is finite, so definitely we will arrive at situation $W_{k+1}(Q) = W_k(Q)$ for some $k \leq |V_N|$. Therefore, $W_{k+j}(Q) = W_k(Q)$ for all $j \geq 0$. Let us define $W(Q) = W_k(Q)$, where $W(Q)$ the set of all variables derivable from Q.

(*b*) *Construction of Q productions in G_1* The Q productions in G_1 are of the form $Q \to \alpha$ whenever $A \to \alpha$ is in G with $A \in W(Q)$ and $\alpha \notin V_N$.

Now, we define $G_1 = (V_N, T, P', Q_0)$, where P' is constructed using step (*b*) for every $Q \in V_N$.

(*c*) Finally, we have to show that $L(G_1) = L(G)$. If $Q \to \alpha$ is in $P' - P$, then it is induced by $A \to \alpha$ in P with $A \in W(Q)$ and $\alpha \notin V_N$. $A \in W(Q)$ implies $Q \overset{*}{\underset{G}{\Rightarrow}} A$ and so, $Q \overset{*}{\underset{G}{\Rightarrow}} A \underset{G}{\Rightarrow} a$. Therefore, if $Q \underset{G_1}{\Rightarrow} \alpha$, then $Q \overset{*}{\underset{G}{\Rightarrow}} \alpha$. Hence, $L(G_1) \subseteq L(G)$. Similarly, it can be shown that $L(G) \subseteq L(G_1)$ and hence $L(G_1) = L(G)$. For example, consider a context-free grammar $G = (V_N, T, P, Q_0)$ given by the productions $Q_0 \to ABF$, $A \to a, B \to b|C, C \to D, D \to E, E \to a$ and $F \to b$.

Our aim is to construct a context-free grammar G_1 equivalent to G by eliminating unit productions. Therefore, we have $W(Q_0) = \{Q_0\}$.

Similarly, we get $W(A) = \{A\}; W(E) = \{E\}; W(F) = \{F\}$. Also we have

$$W_0(B) = \{B\}$$
$$W_1(B) = W_0(B) \cup \{C\} = \{B, C\}$$
$$W_2(B) = W_1(B) \cup \{D\} = \{B, C, D\}$$
$$W_3(B) = W_3(B) \cup \{E\} = \{B, C, D, E\}$$
$$W_4(B) = W_3(B) = \{B, C, D, E\}$$

Hence, we get

$$W(B) = W_3(B) = \{B, C, D, E\}.$$

Similarly, we get

$$W(C) = \{C, D, E\}$$
$$W(D) = \{D, E\}$$

Finally, the productions P' of G_1 are given as $Q_0 \to ABF$, $A \to a$, $B \to b|a$, $C \to a$, $D \to a$, $E \to a$ and $F \to b$. Thus, the reduced context-free grammar G_1 is defined as $G_1 = (V_N, T, P', Q_0)$, where

$$V_N = \{Q_0, A, B, C, D, E, F\},\ T = \{a, b\} \quad \text{and}$$
$$P' = \{Q_0 \to ABF, A \to a, D \to a, B \to b|a, C \to a, E \to a, F \to b\}$$

Theorem *If G is a context-free grammar, then there exists an equivalent context-free grammar G_1, which is reduced and has no null, unit productions.*

Proof Let G be a context-free grammar. Our aim is to construct G_1 that is reduced and has no null, unit productions. We construct G_1 as follows:

(a) First eliminate null productions from G so as to get the context-free grammar G' equivalent to G.

(b) Secondly, eliminate unit productions from G' so as to get the context-free grammar G'' equivalent to G'.

(c) Finally, construct a reduced grammar G_1 equivalent to the context-free grammar G''.

Therefore, G_1 is the required grammar equivalent to the G.

Note In order to get a simplified grammar, we have to apply the constructions only in the order given in the above theorem. If we change the order, we may not get the most simplified form of the grammar.

5.4 NORMAL FORMS

The right-hand side of a production in a context-free grammar can be of any string of terminals and non-terminals. If we impose certain restrictions in the productions of G, then G is said to be in a "normal form". There are many kinds of normal forms we can establish for context-free grammars. Some of these have been studied extensively because of their wide usefulness. In this section we discuss two of them. These are the Chomsky Normal Form and Greibach Normal Form.

5.4.1 Chomsky Normal Form

In the Chomsky normal form (CNF), we impose restrictions on the length and the nature of symbols in the right-hand side of productions. A context-free grammar $G = (V_N, T, P, Q_0)$ is said to be in Chomsky normal form if every productions are of the form $Q \rightarrow AB$ or $Q \rightarrow a$, where $A, B \in V_N$ and $a \in T$.

For example, the grammar G having productions $Q_0 \rightarrow AQ_0 | b, A \rightarrow AA | a$ is in Chomsky normal form whereas the grammar G_1 having productions $Q_0 \rightarrow AQ_0 | ABQ_0, A \rightarrow a$ and $B \rightarrow BQ_0 | bb$ is not in Chomsky normal form. This is because of the productions $Q_0 \rightarrow ABQ_0$ and $B \rightarrow bb$.

Theorem *For every context-free grammar, there is an equivalent grammar in Chomsky normal form.*

Proof Let $G = (V_N, T, P, Q_0)$ be a context-free grammar. A context-free grammar is converted into Chomsky normal form by using following steps:

1. Eliminate null and unit productions as discussed earlier. Let the modified reduced grammar thus obtained be

$$G_1 = (V_N', T', P', Q_0)$$

2. *Elimination of terminals from right-hand side*   We define a modified grammar $G_2 = (V_N'', T', P', Q_0)$ equivalent to G_1, where V_N'' and P'' are constructed as follows:

(a) All the productions of the form $Q \to AB$ or $Q \to a$ in P' are included in P''. All the variables in V_N' are included in V_N''.

(b) If a production in P' is of the form $Q \to x_1 x_2 \cdots x_i \cdots x_n; n \geq 2,$ where each x_i is a symbol either in V_N' or T', then introduce new variables C_{x_i} for each $x_i \in T'$. Introduce these new variables C_{x_i} to V_N'' and productions $C_{x_i} \to x_i$ to P''.

This part of the algorithm removes all terminals from productions whose right-hand side has length more than one by replacing newly introduced variables. At the end of this step we have the grammar $G_2 = (V_N'', T', P'', Q_0)$ in which each production has the form

$$Q \to a \quad \text{or}$$
$$Q \to C_1 C_2 C_3 \cdots C_n \quad \text{where } C_1, C_2, C_3, \cdots, C_n \in V_N''$$

3. *Restricting the number of variables on right-hand side*   In this step we reduce the length of the right sides of the productions where necessary by introducing additional variables. We define grammar $G' = (V_N''', T', P''', Q_0)$ in Chomsky normal form as follows:

(a) If the production in P'' is in the required form, then include these productions in P'''. Also include the variables of V_N'' to V_N'''.

(b) If the production is in the form of $Q \to C_1 C_2 C_3 \cdots C_n; n > 2,$ then introduce new variables $D_1, D_2, D_3, \cdots, D_{n-2}$ to V_N''' and put into P''' the productions

$$Q \to C_1 D_1$$
$$D_1 \to C_2 D_2$$
$$D_2 \to C_3 D_3$$
$$\cdots\cdots\cdots\cdots$$
$$D_{n-3} \to C_{n-2} D_{n-2}$$
$$D_{n-2} \to C_{n-1} C_n$$

Thus the grammar $G' = (V_N''', T', P''', Q_0)$ is in Chomsky normal form. Also it can be shown that $L(G) = L(G')$.

Consider a context-free grammar $G = (V_N, T, P, Q_0)$ given by the productions $Q_0 \to a|b|cQ_0Q_0$. Our aim is to construct a context-free grammar G_1 equivalent to G in Chomsky normal form.

It is clear from the productions that there is no null or unit productions. Therefore, it has to go through the following steps.

(a) *Elimination of terminals from RHS*   We define a modified grammar $G' = (V_N', T, P', Q_0)$ equivalent to G as follows:

i.   Productions $Q_0 \rightarrow a$ and $Q_0 \rightarrow b$ of P are included in P'.

ii. Production $Q_0 \rightarrow cQ_0Q_0$ of P reduces to $Q_0 \rightarrow AQ_0Q_0$ and $A \rightarrow c$.

Therefore, the modified grammar G' is defined as $G' = (V'_N, T, P', Q_0)$, where $V'_N = \{Q_0, A\}$, $T = \{a, b, c\}$ and $P' = \{Q_0 \rightarrow a|b|AQ_0Q_0, A \rightarrow c\}$

(b) Restricting the number of variables on RHS We define the grammar $G_1 = (V''_N, T, P'', Q_0)$ in Chomsky normal form as follows:

i.   Productions $Q_0 \rightarrow a|b$ and $A \rightarrow c$ of P' are included in P''.

ii. The production $Q_0 \rightarrow AQ_0Q_0$ of P' reduces to $Q_0 \rightarrow AB$ and $B \rightarrow Q_0Q_0$.

Therefore, the context-free grammar G_1 in Chomsky normal form is defined as $G_1 = (V''_N, T, P'', Q_0)$ where $V''_N = \{Q_0, A, B\}$, $T = \{a, b, c\}$ and $P'' = \{Q_0 \rightarrow a|b|AB, A \rightarrow c, B \rightarrow Q_0Q_0\}$

5.4.2   Greibach Normal Form

Unlike Chomsky normal form Greibach normal form (GNF) has also many theoretical and practical applications. In Chomsky normal form, we put restrictions on the length of right sides of a production whereas in Greibach normal form, we put restrictions on the position in which terminals and nonterminals can appear. A context-free grammar G is said to be in Greibach normal form if every production is of the form $Q \rightarrow aX$, where $a \in T$ and $X \in V_N^*$.

For example, the grammar G having productions $Q_0 \rightarrow aBQ_0A|aB$, $A \rightarrow b$ and $B \rightarrow a$ is in Greibach normal form whereas the grammar having productions $Q_0 \rightarrow aA$, $A \rightarrow aB|BB$ and $B \rightarrow b$ is not in Greibach normal form. This is because of the production $A \rightarrow BB$.

Greibach normal form basically depends on two important lemmas. In this section we discuss these lemmas before construction of Greibach normal form.

Lemma 1   Let $G = (V_N, T, P, Q_0)$ be a context-free grammar. If $Q \rightarrow B\gamma$ be a Q-production in P and $B \rightarrow \alpha_1|\alpha_2|\alpha_3|\cdots|\alpha_k$ be B productions in P, then define $P_1 = (P - \{Q \rightarrow B\gamma\}) \cup \{Q \rightarrow \alpha_i\gamma|1 \leq i \leq k\}$ such that $G_1 = (V_N, T, P_1, Q_0)$ is a context-free grammar equivalent to G.

Proof   Consider the production $Q \rightarrow B\gamma$ be used in some derivation for $w \in L(G)$. So, definitely we have to apply $B \rightarrow \alpha_i$ for some i at a later step. Therefore, we get $Q \overset{*}{\underset{G}{\Rightarrow}} \alpha_i\gamma$. It indicates that applying $Q \rightarrow B\gamma$ and eliminating B in grammar G is same as applying $Q \rightarrow \alpha_i$ for some i in grammar G_1. Therefore, $w \in L(G_1)$ and hence $L(G) \subseteq L(G_1)$. Similarly, applying $Q \rightarrow \alpha_i$ in G_1 is same as applying $Q \rightarrow B\gamma$ and $B \rightarrow \alpha_i$ for some i to get $Q \overset{*}{\underset{G}{\Rightarrow}} \alpha_i\gamma$. Thus, $L(G_1) \subseteq L(G)$ and hence $L(G) = L(G_1)$.

Therefore by using this Lemma, we eliminate a variable B that appears as the first symbol on the right-hand side of some Q-production provided no B production has B as the first symbol on right-hand side.

For example, consider the context-free grammar G having productions $Q_0 \to Ba$ and $B \to aA|a|AA$. On applying the above Lemma 1 the production $Q_0 \to Ba$ of G will be replaced by $Q_0 \to aAa, Q_0 \to aa$ and $Q_0 \to AAa$.

Lemma 2 Let $G = (V_N, T, P, Q_0)$ be a context-free grammar. Let $Q \to Q\alpha_1|Q\alpha_2|\cdots|Q\alpha_r$ be the set of Q-productions of P for which Q is the leftmost symbol in the right-hand side. Further let $Q \to \beta_1|\beta_2|\cdots|\beta_s$ be the remaining Q-productions. Let Z be a new variable. Let $G_1 = (V_N \cup \{Z\}, T, P', Q_0)$ be the context-free grammar, where P' is defined as follows:

 i. The set of Q-productions in P' are
$$Q \to \beta_i \text{ and } Q \to \beta_i Z \quad \text{for } 1 \le i \le s$$
i.e., $Q \to \beta_1|\beta_2|\cdots|\beta_s$ and $Q \to \beta_1 Z|\beta_2 Z|\cdots|\beta_s Z$

 ii. The set of Z-productions in P' are
$$Z \to \alpha_i \text{ and } Z \to \alpha_i Z \quad \text{for } 1 \le i \le r$$
i.e., $Z \to \alpha_1|\alpha_2|\cdots|\alpha_r$ and $Z \to \alpha_1 Z|\alpha_2 Z|\cdots|\alpha_r Z$

Therefore, G_1 is a context-free grammar equivalent G, i.e., $L(G) = L(G_1)$.

Proof Let $G = (V_N, T, P, Q_0)$ be a context-free grammar. Let $Q \to Q\alpha_i$, $1 \le i \le r$ be the set of Q-productions of P for which Q is the leftmost symbol in the right-hand side. Therefore it is clear that, in a leftmost derivation a sequence of productions $Q \to Q\alpha_i$ will terminate with a production $Q \to \beta_j; 1 \le j \le s$.
The sequence of replacement in G is given below.

$$Q \Rightarrow Q\alpha_{i_1} \Rightarrow Q\alpha_{i_2}\alpha_{i_1} \Rightarrow \cdots \Rightarrow Q\alpha_{i_r}\alpha_{i_{r-1}}\cdots\alpha_{i_2}\alpha_{i_1} \Rightarrow \beta_j\alpha_{i_r}\alpha_{i_{r-1}}\cdots\alpha_{i_2}\alpha_{i_1}$$

The sequence of replacements in G can be obtained by G_1 as below.

$$Q \Rightarrow \beta_j Z \Rightarrow \beta_j\alpha_{i_r}Z \Rightarrow \cdots \Rightarrow \beta_j\alpha_{i_r}\alpha_{i_{r-1}}\cdots\alpha_{i_2}Z \Rightarrow \beta_j\alpha_{i_r}\alpha_{i_{r-1}}\cdots\alpha_{i_2}\alpha_{i_1}$$

Similarly, the reverse transformation can be made and hence $L(G) = L(G_1)$.

For example, consider the context-free grammar G having Q-productions $Q \to aB|bE|c$ and $Q \to QB|QE$. From the above productions, it is clear that $\alpha_1 = B, \alpha_2 = E, \beta_1 = aB, \beta_2 = bE$ and $\beta_3 = c$. On applying above Lemma 2 the new set of productions in the modified grammar are and $Z \to BZ|EZ$.

Theorem (*Reduction to GNF*) *Every context-free language L can be generated by a context-free grammar G in Greibach normal form.*

Proof We prove this when $\wedge \notin L$ and then extend the idea of construction when $\wedge \in L$. So, there arises two cases, i.e., $\wedge \notin L$ and $\wedge \in L$.

Case 1   (Construction of G when $\wedge \notin L$) The following steps are used for construction of G in Greibach normal form.

1.   Eliminate the null productions and then construct a grammar G in Chomsky normal form.

2.   Rename the variables of V_N as $Q_1, Q_2, Q_3, \cdots, Q_n$ with $Q_0 = Q_1$. Therefore, grammar $G = (V_N, T, P, Q_0)$ is defined as $V_N = \{Q_1, Q_2, Q_3, \cdots, Q_n\}$ and $Q_0 = Q_1$. Rewrite the productions in terms of $Q_1, Q_2, Q_3, \cdots, Q_n$.

3.   Obtain the productions in the form $Q_i \to a\gamma$ or $Q_i \to Q_j\gamma$, where $i < j$. We convert the Q_i-production in the form of $Q_i \to Q_j\gamma$ such that $i < j$ for $i = 1, 2, 3, \cdots, (n-1)$. We prove this by induction on i.

Consider Q_1. Productions. If some of the Q_1 productions are in the form of $Q_1 \to Q_1\gamma$, then we can apply Lemma 2 to eliminate such productions. In such case, we introduce a new variable, say Z_1 and we get Q_1 productions of the form $Q_1 \to a$ or $Q_1 \to Q_j\gamma'$; $j > 1$. Thus there exists a basis for induction.

Assume that we have modified Q_k productions for $k = 1, 2, \cdots, i$. Let us consider the $(i+1)^{\text{th}}$ production Q_{i+1}. If the production is in the form of $Q_{i+1} \to a\gamma$, then there is no need of any modification. Let us consider the first symbol on the right-hand side of the remaining Q_{i+1} productions is a variable. Let the smallest index among the indices of such variables is 't'. If $t > (i+1)$, then there is no need of any proof. Otherwise, apply method of induction to Q_t productions for $t \le i$. Therefore, we get any Q_t production is of the form $Q_t \to Q_j\gamma$, where $j > t$ or $Q_t \to a\gamma'$. Now apply Lemma 1 to the production Q_{i+1} whose right-hand side starts with Q_t. Thus the resulting Q_{i+1} productions are of the form $Q_{i+1} \to Q_j\gamma$, where $j > t$.

Repeat the above process of construction by finding t for the new set of Q_{i+1} productions. Finally, the Q_{i+1} productions are reduced to the form $Q_{i+1} \to Q_j\gamma$, where $j > (i+1)$ or $Q_{i+1} \to a\gamma'$. If the production is in the form $Q_{i+1} \to Q_{i+1}\gamma$, then it can be modified by using Lemma 2. The construction is carried out for $i = 1, 2, 3, \cdots, n$ by induction. Therefore, any Q_i production is of the form $Q_i \to Q_j\gamma$, where $i < j$ or $Q_i \to a\gamma'$. With Q_n production $Q_n \to Q_n\gamma$ or $Q_n \to a\gamma'$.

4.   Convert Q_n productions to the form $Q_n \to a\gamma$. The productions of the form $Q_n \to Q_n\gamma$ are removed by using Lemma 2. It leads to the fact that Q_n productions are of the form $Q_n \to a\gamma$.

5.   Now we modify Q_i productions to the form of $Q_i \to a\gamma$ for $i = 1, 2, \cdots, (n-1)$. At the end of step 4, the Q_n productions to the form $Q_n \to a\gamma$. Again, the Q_{n-1} productions are of the form $Q_{n-1} \to a\gamma'$ or $Q_{n-1} \to Q_n\gamma$. We remove the productions of the form $Q_{n-1} \to Q_n\gamma$ by applying Lemma 1. This implies that, the resulting Q_{n-1} productions are in the required form. Repeat the construction by considering $Q_{n-2}, Q_{n-3}, \cdots, Q_1$.

6.   In this step we modify Z_i productions. Every time we apply Lemma 2, we get a new variable. As a result all Z_i productions are of the form $Z_i \to \alpha$ or $Z_i \to \alpha Z_i$, where α is obtained from $Q_i \to Q_i\alpha$. Therefore, Z_i productions are of the form $Z_i \to a\gamma$ or $Z_i \to Q_k\gamma$ for some k. At the end of step 5, the right-hand side of any Q_k production starts with a terminal. Hence, we can apply Lemma 1 to remove $Z_i \to Q_k\gamma$.

This leads to an equivalent grammar G_1 in Greibach normal form.

Case 2 (Construction of G when $\wedge \in L$)   The following steps are used for construction of G in Greibach normal form.

1. Apply all the steps of case 1. So we get a grammar $G_1 = (V'_N, T, P', Q_0)$ in Greibach normal form such that $L(G_1) = L(G) - \{\wedge\}$.

2. Define a new grammar $G_2 = (V''_N, T, P'', Q'_0)$ such that $V''_N = (V'_N \cup Q'_0)$ and $P'' = P' \cup \{Q'_0 \to Q_0, Q'_0 \to \wedge\}$.

3. Eliminate the unit production $Q'_0 \to Q_0$ as discussed earlier. As Q_0 productions are in the required form, so Q'_0 productions are also in the required form. Therefore, $L(G_2) = L(G)$ and the grammar G_1 is in Greibach normal form.

For example, consider the grammar $G = (V_N, T, P, Q_0)$ having productions $Q_0 \to AA|b$, $A \to BB|a$ and $B \to AQ_0$. Our aim is to construct a grammar in Greibach normal form equivalent to the grammar G.

1. It is clear that, the above given grammar is in Chomsky normal form. So, we omit first step of modification to Chomsky normal form. Now rename the variables Q_0, A, B as Q_1, Q_2 and Q_3 respectively. So, the productions are given as:

    $$Q_1 \to Q_2 Q_2 |b, Q_2 \to Q_3 Q_3 |a \text{ and } Q_3 \to Q_2 Q_1$$

2. The Q_1 production $Q_1 \to Q_2 Q_2 |b$ is in the form of $Q_i \to a\gamma$ or $Q_i \to Q_j\gamma$, where $i < j$. Similarly the Q_2 production, i.e., $Q_2 \to Q_3 Q_3 |a$ is also in required form. The Q_3 production $Q_3 \to Q_2 Q_1$ is not in the required form. Therefore, apply Lemma 1 to the production $Q_3 \to Q_2 Q_1$. So, the resulting Q_3 productions are $Q_3 \to Q_3 Q_3 Q_1 |aQ_1$.

3. We have to apply Lemma 2 to Q_3 productions as we have $Q_3 \to Q_3 Q_3 Q_1 |aQ_1$. Let us introduce a variable Z_3 such that $Q_3 \to aQ_1, Q_3 \to aQ_1 Z_3, Z_3 \to Q_3 Q_1$ and $Z_3 \to Q_3 Q_1 Z_3$.

    Therefore Q_3 productions are $Q_3 \to aQ_1 |aQ_1 Z_3$.

4. Now, consider Q_2 productions. The production $Q_2 \to a$ is in required form whereas the production $Q_2 \to Q_3 Q_3$ is not in the required form. Therefore, on applying Lemma 1, the modified Q_2 productions are $Q_2 \to a|aQ_1 Q_3 |aQ_1 Z_3 Q_3$.

    Similarly, consider Q_1 productions. The production $Q_1 \to b$ is in required form whereas the production $Q_1 \to Q_2 Q_2$ is not in the required form. Thus, on applying Lemma 1, the production $Q_1 \to Q_2 Q_2$ reduces to $Q_1 \to aQ_2 |aQ_1 Q_3 Q_2 |aQ_1 Z_3 Q_3 Q_2$. Therefore, the modified Q_1 productions are $Q_1 \to b|aQ_2 |aQ_1 Q_3 Q_2 |aQ_1 Z_3 Q_3 Q_2$.

5. The Z_3 productions to be modified are $Z_3 \to Q_3 Q_1 |Q_3 Q_1 Z_3$. On applying Lemma 1 the productions $Z_3 \to Q_3 Q_1 |Q_3 Q_1 Z_3$ reduces to $Z_3 \to aQ_1 Q_1 |aQ_1 Z_3 Q_1 |aQ_1 Q_1 Z_3 |aQ_1 Z_3 Q_1 Z_3$.

Therefore, the equivalent grammar G_1 is given as $G_1 = (V'_N, T, P', Q_0)$, where

$$V'_N = \{Q_1, Q_2, Q_3, Z_3\}; \mathrm{T} = \{a, b\} \text{ and}$$
$$P' = \{Q_1 \rightarrow b|aQ_1|aQ_1Q_3Q_2|aQ_1Z_3Q_3Q_2,\ Q_2 \rightarrow a|aQ_1Q_3|aQ_1Z_3Q_3,$$
$$Q_3 \rightarrow aQ_1|aQ_1Z_3,\ Z_3 \rightarrow aQ_1Q_1|aQ_1Z_3Q_1|aQ_1Q_1Z_3|aQ_1Z_3Q_1Z_3\}$$

5.5 PUMPING LEMMA

The pumping lemma for regular sets is a tool for proving certain languages are not regular. It states that every sufficiently long string contains a short substring that can be pumped as many times in a regular set so as to yield a string in the regular set. The pumping lemma for context-free languages states that there always exist two substrings close together that can be pumped, both same numbers of times as we like. This indicates that, it gives a method for generating infinite number of strings from a sufficiently large string in a context-free language L. Before coming to the formal statement, we prove the following lemma.

Lemma Let G be a grammar in Chomsky normal form and T be a derivation tree in G. If the length of the longest path in T is less than or equal to i, then the yield of T is of length less than or equal to 2^{i-1}.

Proof Let T be a derivation tree of grammar G in Chomsky normal form. We prove the lemma by induction on i, the length of the longest path for all Q_0 trees. When the longest path in a Q_0 tree is 1, then definitely the root has only one child labelled with a terminal. This is as shown in the following Figure 5.1a. So, there is basis for induction. Assume that the result is true for $(i-1)$ with $i > 1$.

Let us consider a Q_0 tree with a longest path of length less than or equal to i. As $i > 1$, the Q_0 tree has exactly two sons labelled with Q_1 and Q_2. This indicates that Q_1 and Q_2 have longest paths of length less than or equal to $(i-1)$. This is as shown in the following Figure 5.1b.

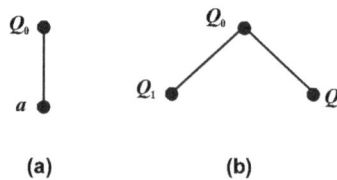

(a) (b)

Figure 5.1 Derivation tree

Let the yields of trees Q_1 and Q_2 be w_1 and w_2 respectively. Therefore, we have $|w_1| \leq 2^{i-2}$ and $|w_2| \leq 2^{i-2}$. So, the yield of Q_0 tree is given as w_1w_2. Thus, we get

$$|w_1w_2| = |w_1| + |w_2|$$
$$\leq 2^{i-2} + 2^{i-2} = 2^{i-1}$$

Therefore, by induction the result is true for all Q_0 trees and hence true for all derivation trees.

5.5.1 Pumping Lemma For Context-free Languages

Let L be any context-free language. Then there is a natural number n, depending on L, such that if z is in L and $|z| \geq n$, then we may write $z = uvwxy$ such that

   i. $|vx| \geq 1$

   ii. $|vwx| \leq n$

   iii.   for all $i \geq 0, uv^i wx^i y \in L$

Proof   Let G is a grammar in Chomsky normal form generating $L - \{\wedge\}$. Let $|V_N| = k$ and $n = 2^k$. In order to show the required number is n we start with $z \in L, |z| \geq 2^k$ and then construct a derivation tree T of z. Let the length of a longest path in T is at most k. Therefore, by above Lemma $|z| \leq 2^{k-1}$. But k is assumed such that $|z| \geq 2^k > 2^{n-1}$.

Therefore, the derivation tree T has a path P of length greater than or equal to $(k + 1)$. It implies that the path P has at least $(k + 2)$ vertices and the vertex is a leaf only. So, in path P all the labels except the last node are variables. But we assume that $|V_N| = k$, so some label is repeated.

A repeated label may be chosen as follows: We start with the leaf node of path P and traverse upwards along P. We stop the traversing, when some label is repeated say S. Let us consider v_1 and v_2 be the vertices labelled with S and v_1 is nearer to the root. In P, the portion of the path from v_1 to the leaf node has only one label, S, that is repeated and therefore its length is at most $(k + 1)$.

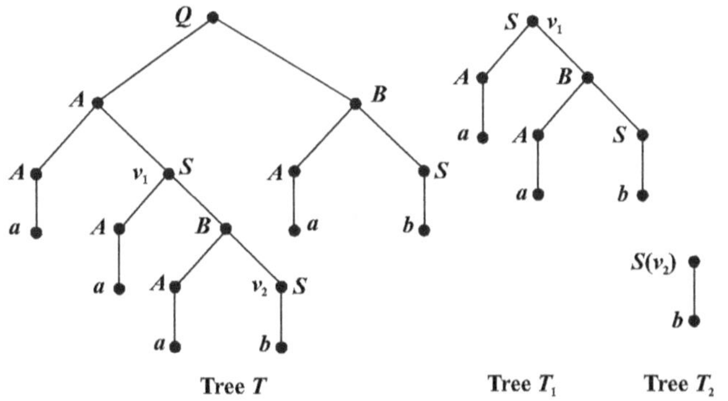

Figure 5.2 An illustration of derivation tree

Let us consider T_1 and T_2 be subtree with v_1 and v_2 as roots. Let the yields be z_1, w respectively. Since P is a longest path in T, the portion of P from v_1 to the leaf node is the longest path in T_1 having length at most $(k + 1)$. Therefore, $|z_1| \leq 2^k$, since z_1 is the yield of T_1.

For example, we illustrate the construction for the grammar whose productions are $Q_0 \rightarrow AB, A \rightarrow AS, S \rightarrow AB, B \rightarrow AS, S \rightarrow b$ and $A \rightarrow a$ as shown in the above Figure 5.2. From the Figure 5.2 it is clear that

$$P = Q \rightarrow A \rightarrow S \rightarrow B \rightarrow S \rightarrow b$$

$$z = aaabab, z_1 = aab, w = b, u = a, v = aa, x = \wedge, y = ab$$

It is clear that, z and z_1 is the yields of T and T_1 respectively and T_1 is a proper subtree of T. Therefore, we can write $z = uz_1y$. Similarly, z_1 and w are the yields of T_1 and T_2 respectively and T_2 is a proper subtree of T_1. Therefore, we can write $z_1 = vwx$. Also, $|vwx| > |w|$. This leads to $|vx| \geq 1$. Thus we have $z = uvwxy$ with $|vx| \geq 1$ and $|vwx| \leq n$.

Now our aim is to show that $uv^i wx^i y \in L$ for all $i \geq 0$. From the figure, it is clear that T is a Q-tree and T_1, T_2 are S-trees. Therefore, we get $Q \overset{*}{\Rightarrow} uSy, S \overset{*}{\Rightarrow} vSx$ and $S \overset{*}{\Rightarrow} w$. Since $Q \overset{*}{\Rightarrow} uSy \Rightarrow uwy$, we have $uv^0 wx^0 y \in L$. For $i \geq 1$, we have $Q \overset{*}{\Rightarrow} uSy \overset{*}{\Rightarrow} uv^i Sx^i y \overset{*}{\Rightarrow} uv^i wx^i y \in L$.

5.5.2   Applications of the Pumping Lemma

We use pumping lemma to show a variety of languages L is not a context-free, using the same argument as for the regular set pumping lemma. The following steps are used for proving that a given language is not context-free.

1. Assume that the given language L is context-free. Let n be a natural number obtained by using the pumping lemma.

2. Choose $z \in L$ such that $|z| \geq n$. Use pumping lemma to write $z = uvwxy$, with $|vx| \geq 1$ and $|vwx| \leq n$.

3. Find a suitable integer i so that $uv^i wx^i y \notin L$. So, we get a contradiction. Therefore, the language L is not context-free.

For example, consider the language $L = \{1^p \,|\, p$ is a prime number$\}$. Let us assume that L is context-free. Let n be the natural number obtained by using pumping lemma. Suppose p is a prime number greater than n, i.e., $p > n$.

Let us take $z = 1^p$. Therefore, $|z| = p > n$. So, z can be expressed as $uvwxy$ i.e., $z = uvwxy$. Since z is a context-free language, we have $uv^i wx^i y \in L$ for all values of i. On taking $i = 0$, we get $uv^0 wx^0 y = uwy \in L$. Therefore, $|uwy|$ is a prime number say m. Suppose that $|vx| = r$. Thus, we have

$$|uv^m wx^m y| = |uwy| + m|vx|$$

$$= m + mr = m(1 + r)$$

As $m(1 + r)$ is not a prime number, so $uv^m wx^m y \notin L$. Hence, it is a contradiction. Thus, the given language $L = \{1^p \,|\, p$ is a prime number$\}$ is not a context-free language.

5.5.3   Pumping Lemma For Linear Language

Let L be an infinite linear language. Then there is a natural number n, such that if z is in L and $|z| \geq n$, then we can decompose z as $z = uvwxy$ such that

i. $|vx| \geq 1$

ii. $|uvxy| \leq n$

iii. for all $i \geq 0, uv^i wx^i y \in L$

The conclusions of this pumping lemma differ from that of pumping lemma for context-free languages, since $|vwx| \leq n$ is replaced by $|uvxy| \leq n$. It implies that the strings v and x to be pumped must now be located within n symbols of the left and right ends of z, respectively. The middle string w can be of arbitrary length. The proof of the above pumping lemma is beyond the scope of this book.

5.6 CLOSURE PROPERTIES OF CONTEXT-FREE LANGUAGES

Earlier we have discussed closure properties for regular sets under certain operations. In this section we will discuss closure properties of context-free languages. These properties are useful not only in constructing or proving that certain languages are context-free but also in proving certain languages that are not context-free.

Theorem *The family of context-free languages is closed under union, i.e., if L_1 and L_2 are context-free languages, then $(L_1 \cup L_2)$ is also a context-free language.*

Proof Let L_1 and L_2 be two context-free languages generated by the context-free grammars $G_1 = (V_{N_1}, T_1, P_1, Q_0')$ and $G_2 = (V_{N_2}, T_2, P_2, Q_0'')$ respectively. Without loss of generality, we can assume that $(V_{N_1} \cap V_{N_2}) = \phi$.

Now, we define a grammar G_3 generating the context-free language L such that $G_3 = (V_N, T, P, Q_0)$, where

$$V_N = V_{N_1} \cup V_{N_2} \cup \{Q_0\}$$
$$T = T_1 \cup T_2$$
$$P = P_1 \cup P_2 \cup \{Q_0 \to Q_0' | Q_0''\} \quad \text{and}$$

Q_0 is a new start symbol such that $Q_0 \notin (V_{N_1} \cup V_{N_2})$.

Now, our aim is to show that $L = L_1 \cup L_2$. In order to show this, let us consider $z \in L_1$. Then we have, $Q_0 \Rightarrow Q_0' \overset{*}{\Rightarrow} z$ is a possible derivation in grammar G_3. This is because we can start with the production rule $Q_0 \to Q_0'$ and continue with the derivation of z in G_1. A similar argument can be made for $z \in L_2$. Therefore, $(L_1 \cup L_2) \subseteq L$.

Also, if $z \in L$, then either $Q_0 \Rightarrow Q_0'$ or $Q_0 \Rightarrow Q_0''$ must be the initial step of derivation. If the initial production is $Q_0 \Rightarrow Q_0'$, then all subsequent productions used must be productions in G_1 as no variable in G_2 are involved. Therefore, $z \in L_1$. Similarly, if we use the initial production $Q_0 \Rightarrow Q_0''$, then z must be in L_2. Therefore, $L \subseteq (L_1 \cup L_2)$. Hence, we have $L = L_1 \cup L_2$.

Theorem *The family of context-free languages is closed under concatenation, i.e., if L_1 and L_2 are context-free languages, then $L_1 L_2$ is also a context-free language.*

Proof   Let L_1 and L_2 be two context-free languages generated by the context-free grammars $G_1 = (V_{N_1}, T_1, P_1, Q_0')$ and $G_2 = (V_{N_2}, T_2, P_2, Q_0'')$ respectively. Without loss of generality, we can assume that $(V_{N_1} \cap V_{N_2}) = \phi$.

Now, we define a grammar G_3 generating the context-free language L such that $G_3 = (V_N, T, P, Q_0)$, where

$$V_N = V_{N_1} \cup V_{N_2} \cup \{Q_0\}$$
$$T = T_1 \cup T_2$$
$$P = P_1 \cup P_2 \cup \{Q_0 \to Q_0' Q_0''\} \quad \text{and}$$

Q_0 is a new start symbol such that $Q_0 \notin (V_{N_1} \cup V_{N_2})$.

Suppose for instance $z \in L$. Then definitely, $Q_0 \to Q_0' Q_0''$ must be the initial step of derivation. This implies that, z must be derived form As $(V_{N_1} \cap V_{N_2}) = \phi$, so definitely z can be expressed as $z_1 z_2$, where z_1 is derived from Q_0' and z_2 is derived from Q_0''. It indicates that $z_1 \in L_1$ and $z_2 \in L_2$. Therefore, $L \subseteq L_1 L_2$.

Conversely, $z \in L_1 L_2$. Then definitely z can be as $z = z_1 z_2$, where z_1 is in L_1 and z_2 is in L_2. We may then derive z in grammar G_3 as follows:

$$Q_0 \Rightarrow Q_0' Q_0'' \stackrel{*}{\Rightarrow} z_1 Q_0'' \stackrel{*}{\Rightarrow} z_1 z_2 = z$$

So, we get $L_1 L_2 \subseteq L$ and hence $L = L_1 L_2$.

Theorem   *The family of context-free languages is closed under star closure, i.e., if L is a context-free language, then L* is also a context-free language.*

Proof   Let L_1 be a context-free language generated by the context-free grammars $G_1 = (V_{N_1}, T_1, P_1, Q_0')$.

Now, we define a grammar G generating the context-free language L such that $G = (V_N, T, P, Q_0)$, where

$$V_N = V_{N_1} \cup \{Q_0\}; T = T_1$$
$$P = P_1 \cup \{Q_0 \to Q_0' Q_0 | \wedge\} \quad \text{and}$$

Q_0 is a new start symbol such that $Q_0 \notin V_{N_1}$.

Suppose for instance that $z \in L$. Then definitely $z = \wedge$ or can be derived from the production $Q_0 \to Q_0' Q_0$ in G. If the initial production is $Q_0 \to Q_0' Q_0$, then we get $Q_0 \to Q_0' Q_0' Q_0' \cdots Q_0'$ (*m-times*) in G. Therefore, $z \in L_1^*$ and hence $L \subseteq L_1^*$. Similarly, it can be shown that $L_1^* \subseteq L$. Therefore, $L = L_1^*$.

Theorem   *The context-free languages are closed under substitution.*

Proof   Suppose that L be a context-free language generated by the context-free grammar G such that $L \subseteq T^*$. Let for each $a \in T$, L_a be a context-free language generated by the context-free grammar G_a. Therefore, $L = L(G)$ and $L_a = L(G_a)$ for each $a \in T$. We can assume without loss of generality that the variables of G and G_a are disjoint. Now we construct a grammar G' as follows:

i. The variables of G' are all the variables of G and the variables of G_a.

ii. The terminals of G' are the terminal of G_a.

iii. The start symbol of G' is the start symbol of G.

iv. The productions of G' are defined as the productions of the grammar G_a with those productions formed by taking a production $A \to \alpha$ of G and substituting Q_a, the start symbol of G_a for each $a \in T$ appearing in α.

Therefore, the language generated by G' is closed under substitution.

Theorem *The family of context-free languages is not closed under intersection.*

Proof Consider two languages L_1 and L_2 as

$$L_1 = \{a^n b^n a^m \mid n, m \geq 1, n \text{ is not necessarily be same as } m\}$$

$$L_2 = \{a^n b^m a^m \mid n, m \geq 1, n \text{ is not necessarily be same as } m\}$$

There are several ways one can show that L_1 and L_2 are context-free. For instance, a grammar for L_1 is $Q_0 \to AB, A \to aAb \mid ab$ and $B \to aB \mid a$. Similarly, a grammar for L_2 is given as $Q_0 \to AB, A \to aA \mid a$ and $B \to bBa \mid ba$. Therefore, the intersection $(L_1 \cap L_2)$ is given as

$$(L_1 \cap L_2) = \{a^n b^n a^n \mid n \geq 1\}$$

This is the case obtained when $n = m$ from L_1 and L_2. But, it can be proved by using pumping lemma that $(L_1 \cap L_2)$ is not a context-free language. Therefore, the family of context-free languages is not closed under intersection.

Theorem *The family of context-free languages is not closed under complementation.*

Proof Suppose that the complement of every context-free language is context-free. Let us consider L_1 and L_2 are two context-free languages. Therefore, by assumption $\overline{L_1}$ and $\overline{L_2}$ are also context-free. But we know that context-free languages are closed under union. Therefore, their union $(\overline{L_1} \cup \overline{L_2})$ is context-free. Again by assumption we get $\overline{(\overline{L_1} \cup \overline{L_2})}$ is context-free. But by De Morgan's law we have

$$\overline{(\overline{L_1} \cup \overline{L_2})} = (L_1 \cap L_2)$$

Therefore, $(L_1 \cap L_2)$ is context-free and it is a contradiction.

Thus, our assumption is wrong. Therefore, context-free languages are not closed under complementation.

5.7 DECISION ALGORITHM FOR CONTEXT-FREE LANGUAGES

There are many questions in context-free languages which are unanswerable. This is due to lack of algorithms to solve some questions. Few of them are given below.

    (*a*) Whether a given context-free grammar is ambiguous or not?

    (*b*) Whether the intersection of two context-free languages is context-free or not?

    (*c*) Whether two context-free grammars are equivalent or not?

    (*d*) Whether the complement of a context-free language is context-free or not?

    However there are some questions that are answerable. These include whether a given context-free language is empty, finite or infinite and whether a given string is in a given context-free language.

Theorem *Given a context-free grammar* $G = (V_N, T, P, Q_0)$, *there is an algorithm for deciding whether or not L(G) is (a) empty, (b) finite or (c) infinite.*

Proof   Assume that for simplicity $\wedge \notin L(G)$. If $\wedge \in L(G)$, then slight changes have to be made in the argument. Actually, we have already discussed an algorithm for removing useless symbols and productions. If the start symbol Q_0 is found to be useless, then definitely $L(G)$ is empty. On the other side, $L(G)$ contains at least one element.

    Assume that $G = (V_N, T, P, Q_0)$ contains no null productions, no unit productions and no useless symbols. Suppose that, the grammar has a repeating variable $A \in V_N$ such that $A \overset{*}{\Rightarrow} vAx$. Since G has no null and unit productions, v and x can not be simultaneously empty. As A is neither nullable nor a useless symbol, we have

$$Q_0 \overset{*}{\Rightarrow} uAy \overset{*}{\Rightarrow} z \text{ and } A \overset{*}{\Rightarrow} w,$$

where $u, y, w \in T^*$. Therefore, we get

$$Q_0 \overset{*}{\Rightarrow} uAy \overset{*}{\Rightarrow} uvAxy \overset{*}{\Rightarrow} uv^2 Ax^2 y \overset{*}{\Rightarrow} \cdots \overset{*}{\Rightarrow} uv^i Ax^i y \overset{*}{\Rightarrow} uv^i wx^i y = z$$

    It indicates that $Q_0 \overset{*}{\Rightarrow} uv^i wx^i y = z$ is possible for all values of i. Thus $L(G)$ is infinite. If no variable can ever repeat, then the length of any derivation is bounded by $|V_N|$ and hence $L(G)$ is finite. To test whether $L(G)$ is finite, find a context-free grammar $G' = (V_N', T, P', Q_0)$ in Chomsky normal form generating $L(G) - \{\wedge\}$ with no useless symbols. Now $L(G')$ is finite if and only if $L(G)$ is finite. It can be done by drawing a dependency graph for the variables in such a way that there is an edge from A to B if there is a production of the form $A \to BC$ or $A \to CB$ for any C. Then the language generated is finite if and only if the dependency graph has no cycles.

    If the dependency graph has at least one cycle, then the generated language is infinite, which is already discussed. If the dependency graph has no cycles, define the rank of a variable A as the length of the longest path in the dependency graph beginning at A. Since there is no cycle, so the rank of A is finite. Again as $A \to BC$ is a production, so the rank of B and C is strictly less than the rank of A. We show by induction on 'm' that if A has rank 'm', then no terminal string has length greater than $2m$ if the string is derived from A. It indicates that the language generated is finite.

5.8   MEMBERSHIP ALGORITHM

We have concentrated on the generative aspects of grammars. It indicates that given a grammar G, we studied the set of strings that can be generated using G. On the other end, given a string z of terminals, we are interested to know whether z is in $L(G)$ or not. The algorithm that finds z is in $L(G)$ or not is known as membership algorithm. In this section we will discuss a membership algorithm for context-free grammars. Here we will describe an important and fundamental algorithm called the CYK algorithm. This is named after its originators J. Cocke, D. H. Younger and T. Kasami. It works only when the grammar is in Chomsky normal form. Assume that $G = (V_N, T, P, Q_0)$ be a grammar in Chomsky normal form and $z = a_1 a_2 a_3 \cdots a_n$ be a string. The algorithm succeeds by breaking the problem into a sequence of smaller ones in the following way.

(a) Define substrings z_{ij} and subsets of V_N, i.e., V_{ij} as $z_{ij} = a_i \cdots a_j$ and respectively

$$V_{ij} = \left\{ A \in V_N \,\middle|\, A \overset{*}{\Rightarrow} z_{ij} \right\}.$$

(b) If G contains a production $A \to a_i$, then $A \in V_{ii}$. Therefore, compute V_{ii} for all $1 \leq i \leq n$ by inspection of z and the productions of the grammar.

(c) Compute V_{ij} for $j > i$ by using the relation

$$V_{ij} = \bigcup_{k \in \{i, i+1, \cdots, j-1\}} \left\{ A \,\middle|\, A \to BC; B \in V_{ik}, C \in V_{k+1,j} \right\}$$

In other words, A derives Z_{ij} for $j > i$ if and only if there is a production $A \to BC$ with $B \overset{*}{\Rightarrow} Z_{ik}$ and $C \overset{*}{\Rightarrow} Z_{k+1,j}$ for some k with $i \leq k < j$. An inspection of the indices shows that, we can compute all these V_{ij}'s if we proceed in the following sequence:

1.   Compute $V_{12}, V_{23}, V_{34}, \cdots, V_{n-1,n}$
2.   Compute $V_{13}, V_{24}, V_{35}, \cdots, V_{n-2,n}$
3.   Compute $V_{14}, V_{25}, V_{36}, \cdots, V_{n-3,n}$
     And so on …

(d) The given string $z \in L(G)$ if and only if $Q_0 \in V_{1n}$.

For example consider the string $z = aabba$ and the language generated by the grammar $Q_0 \to AB$, $A \to BB \,|\, a$ and $B \to AB \,|\, b$.

Assume that string $z = aabba = a_1 a_2 a_3 a_4 a_5$. Therefore, $n = 5$. We define substrings z_{ij} as $z_{ij} = a_i \cdots a_j$. First note that $z_{11} = a_1 = a$, so V_{11} is the set of all variables that derive 'a' immediately. From the productions it is clear that $A \to a$ and so $V_{11} = \{A\}$. Also, from the productions it is clear that $A \to a_1 = a$ and so $V_{11} = \{A\}$. Since G contains a production $A \to a_2 = a$, so $V_{22} = \{A\}$. Similarly, $V_{33} = \{B\}, V_{44} = \{B\}$ and $V_{55} = \{A\}$.

Now, we have to compute $V_{12}, V_{23}, V_{34}, V_{45}$ by using the relation

$$V_{ij} = \bigcup_{k \in \{i, i+1, \cdots, j-1\}} \left\{ A \middle| A \to BC; B \in V_{ik}, C \in V_{k+1,j} \right\}, \, j > i.$$

Therefore, we get $V_{12} = \{A | A \to BC, B \in V_{11} = \{A\}, C \in V_{22} = \{A\}\} = \phi$, since there is no production whose right side is AA. Similarly, we have

$$V_{23} = \{A | A \to BC, B \in V_{22} = \{A\}, C \in V_{33} = \{B\}\} = \{B\}$$
$$V_{34} = \{A | A \to BC, B \in V_{33} = \{B\}, C \in V_{44} = \{B\}\} = \{A\}$$
$$V_{45} = \{A | A \to BC, B \in V_{44} = \{B\}, C \in V_{55} = \{A\}\} = \phi$$

Now, compute $V_{13}, V_{24}, V_{35}, V_{14}, V_{25}$ and V_{15} by using the above relation. So, we get

$$V_{13} = \{A | A \to BC, B \in V_{11} = \{A\}, C \in V_{23} = \{B\}\} \cup$$
$$\{A | A \to BC, B \in V_{12} = \phi, C \in V_{33} = \{B\}\}$$
$$= \{Q_0, B\} \cup \phi = \{Q_0, B\}$$
$$V_{24} = \{A | A \to BC, B \in V_{22} = \{A\}, C \in V_{34} = \{A\}\} \cup$$
$$\{A | A \to BC, B \in V_{23} = \{B\}, C \in V_{44} = \{B\}\} = \phi \cup \{A\} = \{A\}$$
$$V_{35} = \{A | A \to BC, B \in V_{33} = \{B\}, C \in V_{45} = \phi\} \cup$$
$$\{A | A \to BC, B \in V_{34} = \{A\}, C \in V_{55} = \{A\}\} = \phi$$
$$V_{14} = \{A | A \to BC, B \in V_{11} = \{A\}, C \in V_{24} = \{A\}\} \cup$$
$$\{A | A \to BC, B \in V_{12} = \phi, C \in V_{34} = \{A\}\} \cup$$
$$\{A | A \to BC, B \in V_{13} = \{Q_0, B\}, C \in V_{44} = \{B\}\} = \phi \cup \phi \cup \{A\} = \{A\}$$
$$V_{25} = \{A | A \to BC, B \in V_{22} = \{A\}, C \in V_{35} = \phi\} \cup$$
$$\{A | A \to BC, B \in V_{23} = \{B\}, C \in V_{45} = \phi\} \cup$$
$$\{A | A \to BC, B \in V_{24} = \{A\}, C \in V_{55} = \{A\}\} = \phi$$
$$V_{15} = \{A | A \to BC, B \in V_{11} = \{A\}, C \in V_{25} = \phi\} \cup$$
$$\{A | A \to BC, B \in V_{12} = \phi, C \in V_{35} = \phi\} \cup$$
$$\{A | A \to BC, B \in V_{13} = \{Q_0, B\}, C \in V_{45} = \phi\}$$
$$\{A | A \to BC, B \in V_{14} = \{A\}, C \in V_{55} = \{A\}\} = \phi$$

Hence, it is clear that $Q_0 \notin V_{15}$. Therefore, the string $z = aabba$ can not be generated by the given grammar.

SOLVED EXAMPLES

Example 1

Determine whether the string $z = 11000$ is in the language generated by the grammar $Q_0 \to BC, B \to CC|1$ and $C \to BC|0$.

Solution   Assume that string $z = 11000 = a_1 a_2 a_3 a_4 a_5$. Therefore, $n = 5$. We define substrings z_{ij} as $z_{ij} = a_i \cdots a_j$. From the productions it is clear that $B \to a_1 = 1$ and so $V_{11} = \{B\}$. Similarly, $V_{22} = \{B\}, V_{33} = \{C\}, V_{44} = \{C\}$ and $V_{55} = \{C\}$.

Now, we have to compute $V_{12}, V_{23}, V_{34}, V_{45}$ by using the relation

$$V_{ij} = \bigcup_{k \in \{i, i+1, \cdots, j-1\}} \left\{ A \Big| A \to BC; B \in V_{ik}, C \in V_{k+1, j} \right\}, \; j > i.$$

Therefore, we get the following:

$V_{12} = \{A | A \to BC, B \in V_{11} = \{B\}, C \in V_{22} = \{B\}\} = \phi$

$V_{23} = \{A | A \to BC, B \in V_{22} = \{B\}, C \in V_{33} = \{C\}\} = \{Q_0, C\}$

$V_{34} = \{A | A \to BC, B \in V_{33} = \{C\}, C \in V_{44} = \{C\}\} = \{B\}$

$V_{45} = \{A | A \to BC, B \in V_{44} = \{C\}, C \in V_{55} = \{C\}\} = \{B\}$

Now, compute $V_{13}, V_{24}, V_{35}, V_{14}, V_{25}$ and V_{15} by using the given relation. So, we get

$V_{13} = \{A | A \to BC, B \in V_{11} = \{B\}, C \in V_{23} = \{Q_0, C\}\} \cup$
$\quad \{A | A \to BC, B \in V_{12} = \phi, C \in V_{33} = \{C\}\} = \{Q_0, C\}$

$V_{24} = \{A | A \to BC, B \in V_{22} = \{B\}, C \in V_{34} = \{B\}\} \cup$
$\quad \{A | A \to BC, B \in V_{23} = \{Q_0, C\}, C \in V_{44} = \{C\}\} = \{B\}$

$V_{35} = \{A | A \to BC, B \in V_{33} = \{C\}, C \in V_{45} = \{B\}\} \cup$
$\quad \{A | A \to BC, B \in V_{34} = \{B\}, C \in V_{55} = \{C\}\} = \{Q_0, C\}$

$V_{14} = \{A | A \to BC, B \in V_{11} = \{B\}, C \in V_{24} = \{B\}\} \cup$
$\quad \{A | A \to BC, B \in V_{12} = \phi, C \in V_{34} = \{B\}\} \cup$
$\quad \{A | A \to BC, B \in V_{13} = \{Q_0, C\}, C \in V_{44} = \{C\}\} = \{B\}$

$V_{25} = \{A | A \to BC, B \in V_{22} = \{B\}, C \in V_{35} = \{Q_0, C\}\} \cup$
$\quad \{A | A \to BC, B \in V_{23} = \{Q_0, C\}, C \in V_{45} = \{B\}\} \cup$
$\quad \{A | A \to BC, B \in V_{24} = \{B\}, C \in V_{55} = \{C\}\} = \{Q_0, C\}$

$$V_{15} = \{A \mid A \rightarrow BC, B \in V_{11} = \{B\}, C \in V_{25} = \{Q_0, C\}\} \cup$$
$$\{A \mid A \rightarrow BC, B \in V_{12} = \phi, C \in V_{35} = \{Q_0, C\}\} \cup$$
$$\{A \mid A \rightarrow BC, B \in V_{13} = \{Q_0, C\}, C \in V_{45} = \{B\}\} \cup$$
$$\{A \mid A \rightarrow BC, B \in V_{14} = \{B\}, C \in V_{55} = \{C\}\} = \{Q_0, C\}$$

Hence, it is clear that $Q_0 \in V_{15}$ and so, $z = 11000 \in L(G)$.

Example 2

If $G = (\{Q_0\}, \{a, b\}, \{Q_0 \rightarrow a Q_0 b, Q_0 \rightarrow \wedge\})$, then derive the language $L(G)$ representing the grammar G.

Solution   Given that the context-free grammar G as

$$G = (\{Q_0\}, \{a, b\}, \{Q_0 \rightarrow a Q_0 b, Q_0 \rightarrow \wedge\})$$

The productions are $Q_0 \rightarrow a Q_0 b$ and $Q_0 \rightarrow \wedge$. From the production $Q_0 \rightarrow \wedge$, it is clear that the smallest string generated by the grammar G is empty string. The next string generated is 'ab' and it is obtained by substituting $\wedge$ in the right-hand side of the production $Q_0 \rightarrow a Q_0 b$ on place of non-terminal Q_0. Similarly, for $n \geq 1$ we have $Q_0 \underset{G}{\Rightarrow} a Q_0 b \underset{G}{\Rightarrow} aa Q_0 bb \underset{G}{\Rightarrow} \cdots \underset{G}{\Rightarrow} a^n Q_0 b^n = a^n b^n$. Therefore, the language $L(G)$ representing the grammar G is given as below:

$$L(G) = \{a^n b^n \mid n \geq 0\}$$

Example 3

Let $G = (V_N, T, P, Q_0)$ such that $V_N = \{Q_0, A\}, T = \{0, 1\}$ and P consists of $Q_0 \rightarrow 0A0$ and $A \rightarrow 0A0 \mid 1$. Find the language generated by the grammar G.

Solution   Given that $G = (V_N, T, P, Q_0)$ is the context-free grammar. Here, $V_N = \{Q_0, A\}, T = \{0, 1\}$ and P consists of $Q_0 \rightarrow 0A0$ and $A \rightarrow 0A0 \mid 1$. Now, $Q_0 \Rightarrow 0A0 \Rightarrow 010$. Thus, $010 \in L(G)$. Again,

$$Q_0 \Rightarrow 0A0$$
$$\Rightarrow 0^n A 0^n \quad \text{[By application of } A \rightarrow 0A0, (n-1) \text{ times]}$$
$$\Rightarrow 0^n 1 0^n \quad \text{[By application of } A \rightarrow 1]$$

Hence, $0^n 1 0^n \in L(G)$, where $n \geq 1$. Therefore, we get

$$L(G) = \{0^n 1 0^n \mid n \geq 1\}$$

Example 4

Let $G = (\{Q_0\}, \{a,b\}, P, Q_0)$, where P is defined as $Q_0 \rightarrow a, Q_0 \rightarrow b, Q_0 \rightarrow aQ_0, Q_0 \rightarrow bQ_0$. Find the language generated by the grammar G.

Solution Let $G = (\{Q_0\}, \{a,b\}, P, Q_0)$, where P is defined as $Q_0 \rightarrow a, Q_0 \rightarrow b, Q_0 \rightarrow aQ_0,$ $Q_0 \rightarrow bQ_0$. From the production it is clear that the smallest string generated by the grammar G is either 'a' or 'b'. Also in the productions $Q_0 \rightarrow aQ_0, Q_0 \rightarrow bQ_0$, the right hand side Q_0 can be replaced by a, b, aQ_0 or bQ_0 in any order.

This implies that we will get all strings having any combination of a's and b's except the empty string. Therefore,

$$L(G) = (a+b)^* - \wedge = (a+b)^+$$

In order to show $(a+b)^+ \subseteq G$, let us consider any string $x_1 x_2 \cdots x_n$, where each x_i is either 'a' or 'b'. The first production in the derivation of $x_1 x_2 \cdots x_n$ is either $Q_0 \rightarrow aQ_0$ or $Q_0 \rightarrow bQ_0$ according to $x_1 = a$ or $x_1 = b$ respectively. Similarly, other productions can be obtained. The last production is $Q_0 \rightarrow a$ or $Q_0 \rightarrow b$ according to $x_n = a$ or $x_n = b$ respectively.

Example 5

If $G = (\{Q_0\}, \{1\}, P, Q_0)$, where P is the production $Q_0 \rightarrow Q_0 Q_0$, then find the language $L(G)$ generated by G.

Solution Here, we have the production $Q_0 \rightarrow Q_0 Q_0$, where there is no terminal in the right-hand side. Therefore, $L(G) = \phi$.

Example 6

Determine the yield for the following derivation tree.

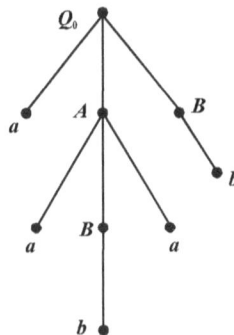

Solution From the above derivation tree, we get a sequence of labels of leaf nodes from left to right as *a, a, b, a* and *b*. On concatenation of these symbols we get '*aabab*'. So, the yield for the derivation tree is '*aabab*'.

Example 7

Consider the context-free grammar G with productions $Q_0 \to aB|bA$, $A \to bAA|aQ_0|a$, $B \to aBB|bQ_0|b$. Obtain derivation trees for the string '*aabbabab*' by using leftmost and rightmost derivations.

Solution   Given productions are

$$Q_0 \to aB|bA, A \to bAA|aQ_0|a, B \to aBB|bQ_0|b.$$

Leftmost derivation   We start with the production $Q_0 \to aB$ and in each step we apply production to the leftmost variable as shown below.

$$
\begin{aligned}
Q_0 &\Rightarrow aB \\
&\Rightarrow aaBB && [\because \ B \to aBB] \\
&\Rightarrow aabQ_0B && [\because \ B \to bQ_0] \\
&\Rightarrow aabbAB && [\because \ Q_0 \to bA] \\
&\Rightarrow aabbaB && [\because \ A \to a] \\
&\Rightarrow aabbabQ_0 && [\because \ B \to bQ_0] \\
&\Rightarrow aabbabaB && [\because \ Q_0 \to aB] \\
&\Rightarrow aabbabab && [\because \ B \to b]
\end{aligned}
$$

The derivation tree can be obtained very easily by considering the leftmost derivation steps. First, we will draw derivation tree for $Q_0 \to aB$ by assigning Q_0 as root with left child 'a' and right child 'B'. The leftmost derivation tree is given below.

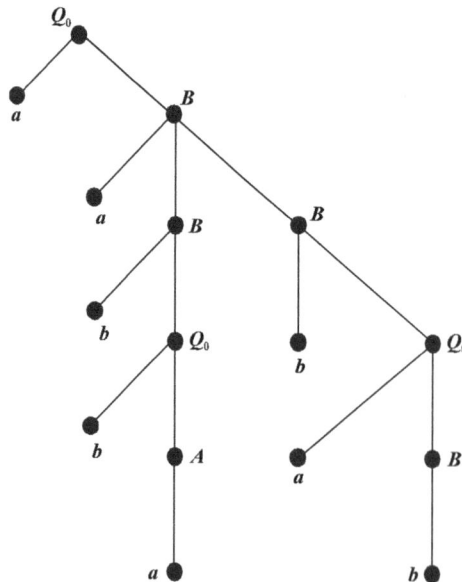

Rightmost derivation We start with the production $Q_0 \rightarrow aB$ and in each step we apply production to the rightmost variable as shown below.

$$Q_0 \Rightarrow aB$$
$$\Rightarrow aaBB \qquad [\because \ B \rightarrow aBB]$$
$$\Rightarrow aaBb \qquad [\because \ B \rightarrow b]$$
$$\Rightarrow aabQ_0b \qquad [\because \ B \rightarrow bQ_0]$$
$$\Rightarrow aabbAb \qquad [\because \ Q_0 \rightarrow bA]$$
$$\Rightarrow aabbaQ_0b \qquad [\because \ A \rightarrow aQ_0]$$
$$\Rightarrow aabbabAb \qquad [\because \ Q_0 \rightarrow bA]$$
$$\Rightarrow aabbabab \qquad [\because \ A \rightarrow a]$$

The derivation tree can be obtained very easily by considering the rightmost derivation steps. The derivation tree is given below.

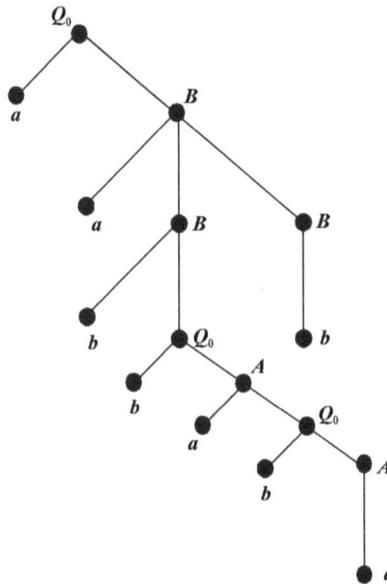

Example 8

Show the derivation steps and construct derivation tree for the string '*ababbb*' by using leftmost derivation with the grammar $Q_0 \rightarrow AB|\wedge$, $A \rightarrow aB$ and $B \rightarrow Q_0b$.

Solution Given productions are $Q_0 \rightarrow AB|\wedge$, $A \rightarrow aB$ and $B \rightarrow Q_0b$. We start with the production $Q_0 \rightarrow AB$. In each step we apply production on the leftmost variable. The derivation steps are given below.

$$Q_0 \Rightarrow AB$$
$$\Rightarrow aBB \qquad [\because \quad A \rightarrow aB]$$
$$\Rightarrow aQ_0bB \qquad [\because \quad B \rightarrow Q_0b]$$
$$\Rightarrow a \wedge bB \qquad [\because \quad Q_0 \rightarrow \wedge]$$
$$\Rightarrow abQ_0b \qquad [\because \quad B \rightarrow Q_0b]$$
$$\Rightarrow abABb \qquad [\because \quad Q_0 \rightarrow AB]$$
$$\Rightarrow abaBBb \qquad [\because \quad A \rightarrow aB]$$
$$\Rightarrow abaQ_0bBb \qquad [\because \quad B \rightarrow Q_0b]$$
$$\Rightarrow aba \wedge bBb \qquad [\because \quad Q_0 \rightarrow \wedge]$$
$$\Rightarrow ababQ_0bb \qquad [\because \quad B \rightarrow Q_0b]$$
$$\Rightarrow ababbb \qquad [\because \quad Q_0 \rightarrow \wedge]$$

The derivation tree can be constructed easily by considering the leftmost derivation steps. It is given below.

Example 9

Construct the derivation tree for the string '*babba*' by using rightmost derivation with the grammar $Q_0 \rightarrow Aa|bB|\wedge$, $A \rightarrow bQ_0|\wedge$ and $B \rightarrow aA$.

Solution   Here the productions are $Q_0 \rightarrow Aa|bB|\wedge$, $A \rightarrow bQ_0|\wedge$ and $B \rightarrow aA$. Our aim is to generate the string '*babba*' by using rightmost derivation. We start with the production $Q_0 \rightarrow bB$. In each step we apply production on rightmost variable. The derivation steps and the derivation tree are given below.

$Q_0 \Rightarrow bB$

$\quad \Rightarrow baA \qquad [\because B \rightarrow aA]$

$\quad \Rightarrow babQ_0 \qquad [\because A \rightarrow bQ_0]$

$\quad \Rightarrow babbB \qquad [\because Q_0 \rightarrow bB]$

$\quad \Rightarrow babbaA \qquad [\because B \rightarrow aA]$

$\quad \Rightarrow babba \qquad [\because A \rightarrow \wedge]$

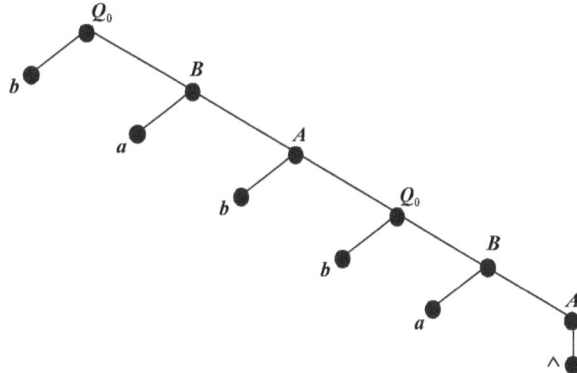

Example 10

Show that the language defined by a context-free grammar $G = (\{Q_0, A, B\}, \{a, b\}, P, Q_0)$, where P is the set of productions $Q_0 \rightarrow Aa|bB|\wedge, A \rightarrow bQ_0|\wedge$ and $B \rightarrow aA$ is ambiguous.

Solution Here the productions of context-free grammar G are $Q_0 \rightarrow Aa|bB|\wedge$, $A \rightarrow bQ_0|\wedge$ and $B \rightarrow aA$. Our aim is to show that the grammar G is ambiguous. In order to show this, we have to find at least one string that is having either two distinct leftmost derivations or rightmost derivations. Now consider the string '*babba*'. The two different rightmost derivation for the string '*babba*' are given below.

Rightmost Derivation-1

$Q_0 \Rightarrow bB$

$\quad \Rightarrow baA \qquad [\because B \rightarrow aA]$

$\quad \Rightarrow babQ_0 \qquad [\because A \rightarrow bQ_0]$

$\quad \Rightarrow babbB \qquad [\because Q_0 \rightarrow bB]$

$\quad \Rightarrow babbaA \qquad [\because B \rightarrow aA]$

$\quad \Rightarrow babba \qquad [\because A \rightarrow \wedge]$

Rightmost Derivation-2

$Q_0 \Rightarrow bB$

$\quad \Rightarrow baA \qquad [\because B \rightarrow aA]$

$\quad \Rightarrow babQ_0 \qquad [\because A \rightarrow bQ_0]$

$\quad \Rightarrow babAa \qquad [\because Q_0 \rightarrow Aa]$

$\quad \Rightarrow babbQ_0a \qquad [\because A \rightarrow bQ_0]$

$\quad \Rightarrow babba \qquad [\because Q_0 \rightarrow \wedge]$

The derivation trees are as follows:

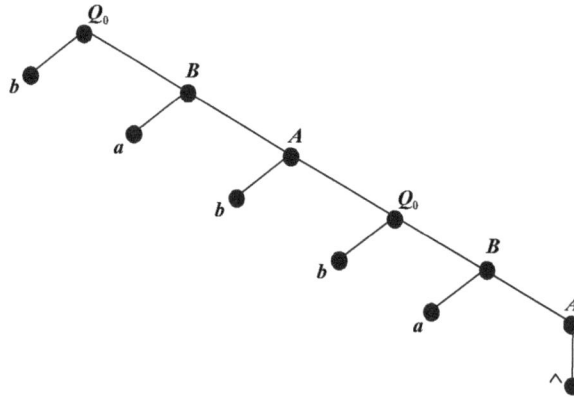

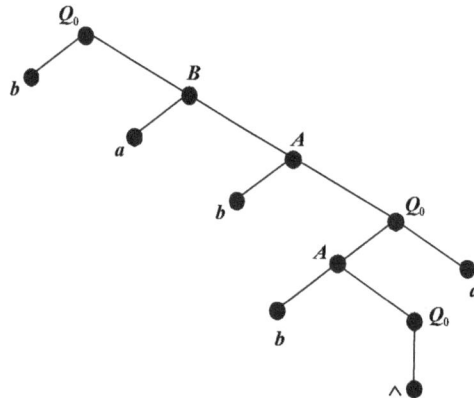

A set of two derivation trees for the string '*babba*' shows that the grammar G is ambiguous.

Example 11

Show that the language of all non null-strings of 1's defined by a context-free grammar $G = (\{Q_0\}, \{1\}, \{Q_0 \rightarrow 1Q_0 | Q_0 1 | 1\}, Q_0)$ is ambiguous.

Solution   Given that the context-free grammar G is

$$G = (\{Q_0\}, \{1\}, \{Q_0 \rightarrow 1Q_0 | Q_0 1 | 1\}, Q_0).$$

The productions of G are $Q_0 \rightarrow 1Q_0 | Q_0 1 | 1$. By using the above productions of the grammar G, the string 1^3 can be derived by the following ways whose derivation trees are as follows:

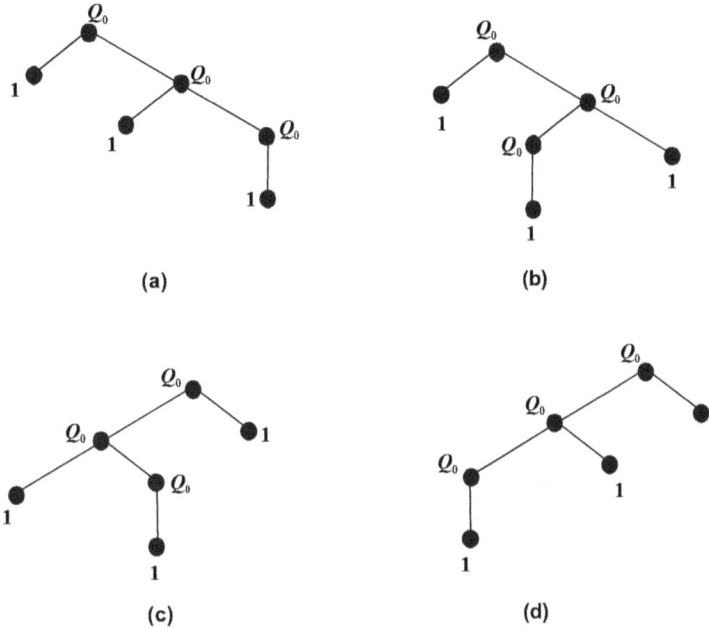

(a)

(b)

(c)

(d)

From the above derivation trees, it is clear that the string '111' can be parsed in many ways. Therefore, the given context-free grammar G is ambiguous.

Example 12

Consider the grammar $G = (V_N, T, P, Q_0)$ with productions $Q_0 \rightarrow bQ_0 | AB$, $A \rightarrow b | \wedge$, $B \rightarrow a | \wedge$ and $D \rightarrow c$. Construct an equivalent grammar G_1 without null productions generating $L(G) - \{\wedge\}$.

Solution Given that $G = (V_N, T, P, Q_0)$ be a context-free grammar, where

$V_N = \{Q_0, A, B, D\}$; $T = \{a, b, c\}$ and $P = \{Q_0 \rightarrow bQ_0 | AB, A \rightarrow b | \wedge, B \rightarrow a | \wedge, D \rightarrow c\}$. We construct an equivalent grammar $G_1 = (V_N, T, P', Q_0)$ that has no null productions as follows:

1. *Construction of the set of nullable variables W:*

$W_1 = \{Q \in V_N | Q \rightarrow \wedge \text{ is a production in } P\}$

$= \{A, B\}$

$W_2 = W_1 \cup \{Q \in V_N | Q \rightarrow \alpha \text{ with } \alpha \in W_1^*\}$

$= \{A, B\} \cup \{Q_0\}$ $[\because Q_0 \rightarrow AB \text{ and } AB \in W_1^*]$

$= \{A, B, Q_0\}$

$W_3 = W_2 \cup \{Q \in V_N | Q \rightarrow \alpha \text{ with } \alpha \in W_2^*\}$

$= W_2 \cup \phi = W_2$

Therefore, $W = W_2 = \{Q_0, A, B\}$

2. *Construction of* P'

(a) The productions $A \to b$, $B \to a$ and $D \to c$ are included in P'.

(b) The production $Q_0 \to bQ_0$ leads to the productions $Q_0 \to bQ_0$ and $Q_0 \to b$.

(c) The production $Q_0 \to AB$ leads to the productions $Q_0 \to AB$, $Q_0 \to A$ and $Q_0 \to B$.

Therefore, the required grammar without null productions is given as $G_1 = (V_N, T, P', Q_0)$, where

$$V_N = \{Q_0, A, B, D\}, \ T = \{a, b, c\} \text{ and}$$
$$P' = \{Q_0 \to bQ_0 | AB | b | A | B, \ A \to b, B \to a, D \to c\}$$

Example 13

Consider the context-free grammar G with productions $Q_0 \to a | AA$, $A \to ab | aB | \wedge$, $B \to C | \wedge$ and $C \to bb$. Find a context-free grammar without null productions generating $L(G) - \{\wedge\}$.

Solution   Consider the context-free grammar $G = (V_N, T, P, Q_0)$ with productions $Q_0 \to a | AA$, $A \to ab | aB | \wedge$, $B \to C | \wedge$ and $C \to bb$.

Here, $V_N = \{Q_0, A, B, C\}$ and $T = \{a, b\}$. Our aim is to construct an equivalent grammar $G_1 = (V_N, T, P', Q_0)$ that has no null productions as follows:

1. *Construction of the set of nullable variables* W

$$W_1 = \{Q \in V_N | Q \to \wedge \text{ is a production in } P\}$$
$$= \{A, B\}$$
$$W_2 = W_1 \cup \{Q \in V_N | Q \to \alpha \text{ with } \alpha \in W_1^*\}$$
$$= \{A, B\} \cup \{Q_0\} \qquad\qquad [\because Q_0 \to AB \text{ and } AB \in W_1^*]$$
$$= \{A, B, Q_0\}$$
$$W_3 = W_2 \cup \{Q \in V_N | Q \to \alpha \text{ with } \alpha \in W_2^*\}$$
$$= W_2 \cup \phi = W_2$$

Therefore, $W = W_2 = \{Q_0, A, B\}$

2. *Construction of* P'

(a) The productions $Q_0 \to a$, $A \to ab$, $B \to C$ and $C \to bb$ are included in P'.

(b) The production $Q_0 \to AA$ leads to the productions $Q_0 \to AA$ and $Q_0 \to A$.

(c) The production $A \to aB$ leads to the productions $A \to aB$ and $A \to a$.

Therefore, the required grammar G_1 without null productions is given as $G_1 = (V_N, T, P', Q_0)$, where $V_N = \{Q_0, A, B, C\}$, $T = \{a, b\}$ and $P' = \{Q_0 \to a | AA | A, \ A \to ab | aB | a, \ B \to C, C \to bb\}$

Example 14

Construct a reduced context-free grammar equivalent to the context-free grammar having productions $Q_0 \to CA|AB$, $A \to a$, $B \to AB|BC$ and $C \to aB|b$.

Solution   Let us consider the context-free grammar G having productions $Q_0 \to CA|AB$, $A \to a$, $B \to AB|BC$ and $C \to aB|b$. In order to get the reduced context-free grammar G_1, we have to reduce the context-free grammar G to $G' = (V'_N, T, P', Q_0)$ such that every variable in G' derives some terminal string and then G' to $G_1 = (V''_N, T'', P'', Q_0)$ such that every symbol in $(V''_N \cup T'')$ appears in some sentential form.

1. Reduction of G to $G' = (V'_N, T, P', Q_0)$

(a) *Construction of V'_N*   From the productions it is clear that $A \to a$ and $C \to b$. It implies that the variables A and C derive a terminal string. Therefore, we have

$$W_1 = \{A, C\}$$
$$W_2 = W_1 \cup \{Q \in V_N \,|\, Q \to \alpha; \alpha \in (W_1 \cup T)^*\}$$
$$= \{A, C\} \cup \{Q_0\} = \{Q_0, A, C\}$$
$$W_3 = W_2 \cup \{Q \in V_N \,|\, Q \to \alpha; \alpha \in (W_2 \cup T)^*\}$$
$$= W_2 \cup \phi = W_2$$

Therefore, $V'_N = W_2 = \{Q_0, A, C\}$

(b) *Construction of P'*   The productions P' are defined as

$$P' = \{Q \to \alpha \,|\, Q; Q \in V'_N, \alpha \in (T \cup V'_N)^*\}$$

i.e., $P' = \{Q_0 \to AC, A \to a, C \to b\}$

Therefore, G' is defined as $G' = (V'_N, T, P', Q_0)$ such that $V'_N = \{Q_0, A, C\}$; $T = \{a, b\}$ and $P' = \{Q_0 \to AC, A \to a, C \to b\}$

2. Reduction of G' to $G_1 = (V''_N, T'', P'', Q_0)$

(a) *Construction of W_i for $i \geq 1$*   We compute W_i recursively as follows.

$$W_1 = \{Q_0\}$$
$$W_2 = W_1 \cup \{X \in (V'_N \cup T) \,|\, A \to \alpha, A \in W_1 \text{ and } \alpha \text{ containing } X\}$$
$$= \{Q_0\} \cup \{A, C\}$$
$$= \{Q_0, A, C\}$$
$$W_3 = W_2 \cup \{X \in (V'_N \cup T) \,|\, A \to \alpha, A \in W_2 \text{ and } \alpha \text{ containing } X\}.$$
$$= \{Q_0, A, C\} \cup \{a, b\} = \{Q_0, A, C, a, b\}$$
$$W_4 = W_3 = \{Q_0, A, C, a, b\}$$

(b) *Construction of V''_N, T'' and P''*   We construct V''_N, T'' and P'' as follows.

$$V''_N = V'_N \cap W_3 = \{Q_0, A, C\}$$
$$T'' = T \cap W_3 = \{a, b\}$$
$$P'' = \{Q \to \alpha | Q \in W_3\} = \{Q_0 \to AC, A \to a, C \to b\}$$

Thus the reduced grammar $G_1 = (V''_N, T'', P'', Q_0)$ is given as

$$V''_N = \{Q_0, A, C\}; \; T'' = \{a, b\} \text{ and } P'' = \{Q_0 \to AC, A \to a, C \to b\}$$

Example 15

Consider the context-free grammar $G = (V_N, T, P, Q_0)$ with the productions $Q_0 \to BC, A \to E, B \to a | D, E \to a$ and $C \to b$. Find the grammar G_1 such that every symbol in G_1 derives some terminal string.

Solution   Given that the productions of context-free grammar $G = (V_N, T, P, Q_0)$ are $Q_0 \to BC, A \to E, B \to a | D, E \to a$ and $C \to b$. We construct $G_1 = (V'_N, T, P', Q_0)$ as follows:

(a) *Construction of V'_N*   From the productions it is clear that $B \to a, E \to a$ and $C \to b$. It indicates that the variables B, E and C derive a terminal string. Therefore, we get $W_1 = \{B, C, E\}$. Similarly, we get

$$\begin{aligned} W_2 &= W_1 \cup \{Q \in V_N | Q \to \alpha, \alpha \in (W_1 \cup T)^*\} \\ &= \{B, C, E\} \cup \{Q_0, A\} \\ &= \{Q_0, A, B, C, E\} \\ W_3 &= W_2 \cup \{Q \in V_N | Q \to \alpha, \alpha \in (W_2 \cup T)^*\} \\ &= W_2 = \{Q_0, A, B, C, E\} \end{aligned}$$

Therefore, $V'_N = W_2 = \{Q_0, A, B, C, E\}$

(b) *Construction of P'*   The productions P' are defined as

$$\begin{aligned} P' &= \{Q \to \alpha | Q; Q \in V'_N \text{ and } \alpha \in (V'_N \cup T)^*\} \\ &= \{Q_0 \to BC, A \to E, B \to a, E \to a, C \to b\} \end{aligned}$$

So, the reduced grammar $G_1 = (V'_N, T, P', Q_0)$ is defined as $V'_N = \{Q_0, A, B, C, E\}$, $T = \{a, b\}$ and $P' = \{Q_0 \to BC, A \to E, B \to a, E \to a, C \to b\}$.

Example 16

Consider the context-free grammar $G = (V_N, T, P, Q_0)$ with productions $Q_0 \to BC$, $A \to E$, $B \to a, E \to a$ and $C \to b$. Construct the grammar G_1 such that every symbol appears in some sentential form.

Solution   The productions of $G = (V_N, T, P, Q_0)$ are $Q_0 \to BC$, $A \to E, B \to a$, $E \to a$ and $C \to b$. Our aim is to construct the grammar $G_1 = (V_N', T, P', Q_0)$ such that every symbol appears in some sentential form.

(a) *Construction of* W_i *for* $i \geq 1$ We compute W_i recursively as follows.

$$W_1 = \{Q_0\}$$
$$W_2 = W_1 \cup \{X \in (V_N \cup T) | A \to \alpha, A \in W_1 \text{ and } \alpha \text{ containing } X\}$$
$$= \{Q_0, B, C\}$$
$$W_3 = W_2 \cup \{X \in (V_N \cup T) | A \to \alpha, A \in W_2 \text{ and } \alpha \text{ containing } X\}$$
$$= \{Q_0, B, C\} \cup \{a, b\} = \{Q_0, B, C, a, b\}$$
$$W_4 = W_3 = \{Q_0, B, C, a, b\}$$

(b) *Construction of* V_N', T' *and* P' We construct V_N', T' and P' as follows.

$$V_N' = V_N \cap W_3 = \{Q_0, B, C\}$$
$$T' = T \cap W_3 = \{a, b\}$$
$$P' = \{Q \to \alpha | Q \in W_3\} = \{Q_0 \to BC, B \to a, C \to b\}$$

Thus the required reduced grammar G_1 is given as $G_1 = (V_N', T', P', Q_0)$, where $V_N' = \{Q_0, B, C\}$; $T' = \{a, b\}$ and $P' = \{Q_0 \to BC, B \to a, C \to b\}$.

Example 17

Eliminate unit productions and get an equivalent grammar for the grammar G having productions $Q_0 \to AB, A \to a, B \to D, D \to E, E \to b$ and $F \to c$.

Solution   The productions of given context-free grammar are $Q_0 \to AB$, $A \to a$, $B \to D$, $D \to E$, $E \to b$ and $F \to c$.

(a) *Construction of* $W(Q)$ *for each* $Q \in V_N$ We construct $W(Q)$ for each $Q \in V_N$ recursively as follows:

$$W_0(Q_0) = \{Q_0\}$$
$$W_1(Q_0) = W_0(Q_0) \cup \phi = W_0(Q_0) = \{Q_0\}$$

Therefore, we get $W(Q_0) = \{Q_0\}$. Similarly, we have $W(A) = \{A\}$, $W(E) = \{E\}$ and $W(F) = \{F\}$. Again, we have

$$W_0(B) = \{B\}$$
$$W_1(B) = W_0(B) \cup \{D\} = \{B, D\}$$
$$W_2(B) = W_1(B) \cup \{E\} = \{B, D, E\}$$
$$W_3(B) = W_2(B) = \{B, D, E\}$$

Therefore, we get $W(B) = \{B, D, E\}$. Similarly, we have $W(D) = \{D, E\}$.

(b) The productions in the equivalent grammar G_1 are $Q_0 \to AB, A \to a, B \to b, D \to b, E \to b$ and $F \to c$.

Hence, the equivalent grammar without unit productions is given as $G_1 = (V_N, T, P', Q_0)$, where $V_N = \{Q_0, A, B, D, E, \text{F}\}, T = \{a, b, c\}$ and $P' = \{Q_0 \to AB, \quad A \to a, \quad B \to b, \quad D \to b, \quad E \to b, F \to c\}$.

Example 18

Eliminate the unit productions from the context-free grammar G having productions $Q_0 \to Aa|B, B \to A|cc, A \to D$ and $D \to a|bc|B$.

Solution   The productions of the given context-free grammar G are $Q_0 \to Aa|B$, $B \to A|cc, A \to D$ and $D \to a|bc|B$. Our aim is to construct an equivalent grammar $G_1 = (V_N, T, P', Q_0)$ that has no unit productions.

(a) Construction of $W(Q)$ for each $Q \in V_N$: We construct $W(Q)$ for each $Q \in V_N$ recursively as follows:

$$W_0(Q_0) = \{Q_0\}$$
$$W_1(Q_0) = W_0(Q_0) \cup \{B\} = \{Q_0, B\}$$
$$W_2(Q_0) = W_1(Q_0) \cup \{A\} = \{Q_0, B, A\}$$
$$W_3(Q_0) = W_2(Q_0) \cup \{D\} = \{Q_0, B, A, D\}$$
$$W_4(Q_0) = W_3(Q_0) = \{Q_0, B, A, D\}$$

Therefore, we get $W(Q_0) = \{Q_0, B, A, D\}$. Similarly, it can be shown that $W(B) = \{B, A, D\}, W(A) = \{A, D, B\}$ and $W(D) = \{D, B, A\}$.

(b) The productions in the equivalent grammar G_1 are $Q_0 \to Aa|cc|a|bc, \quad A \to a|bc|cc$ and $B \to cc|a|bc$.

Hence, the equivalent grammar without unit productions is given as $G_1 = (V_N, T, P', Q_0)$, where $V_N = \{Q_0, A, B, D\}, T = \{a, b, c\}$ and $P' = \{A \to a|bc|cc, \quad B \to cc|a|bc, Q_0 \to Aa|cc|a|bc\}$.

Example 19

Reduce the following grammar G to Chomsky normal form. It is given that the productions of G are $Q_0 \rightarrow bAD$, $A \rightarrow aB|bAB$, $B \rightarrow b$ and $D \rightarrow c$.

Solution The productions of the context-free grammar $G = (V_N, T, P, Q_0)$ are $Q_0 \rightarrow bAD$, $A \rightarrow aB|bAB$, $B \rightarrow b$ and $D \rightarrow c$. In the above productions, there is no null and unit production. Our aim is to obtain the grammar in Chomsky normal form.

    (a) *Elimination of terminals from right-hand side* We define a modified grammar $G_1 = (V'_N, T, P', Q_0)$ equivalent to G as follows.

    i.   The productions $B \rightarrow b$ and $D \rightarrow c$ are included in P'.

    ii.   The production $Q_0 \rightarrow bAD$ reduces to $Q_0 \rightarrow C_b AD$ and $C_b \rightarrow b$.

    iii.   The production $A \rightarrow aB$ reduces to $A \rightarrow C_a B$ and $C_a \rightarrow a$.

    iv.   The production $A \rightarrow bAB$ reduces to $A \rightarrow C_b AB$ and $C_b \rightarrow b$.

Therefore, $G_1 = (V'_N, T, P', Q_0)$ is defined as $V'_N = \{Q_0, A, B, D, C_a, C_b\}$; $T = \{a, b, c\}$ and

$P' = \{Q_0 \rightarrow C_b AD, C_b \rightarrow b, A \rightarrow C_a B, C_a \rightarrow a, A \rightarrow C_b AB, B \rightarrow b, D \rightarrow c\}$.

(b) *Restricting the number of variables on right-hand side* Here, we reduce the length of the right sides of the productions where necessary by introducing additional variables. We define the grammar $G_2 = (V''_N, T, P'', Q_0)$ in Chomsky normal form as follows.

    i.   The productions $B \rightarrow b, D \rightarrow c, C_a \rightarrow a, C_b \rightarrow b, A \rightarrow C_a B$ are included in P''.

    ii.   The production $Q_0 \rightarrow C_b AD$ is replaced by $Q_0 \rightarrow C_b Q_1$ and $Q_1 \rightarrow AD$.

    iii.   The production $A \rightarrow C_b AB$ is replaced by $A \rightarrow C_b Q_2$ and $Q_2 \rightarrow AB$.

Therefore, the Chomsky normal form $G_2 = (V''_N, T, P'', Q_0)$ is given as $V''_N = \{Q_0, A, B, D, C_a, C_b, Q_1, Q_2\}$; $T = \{a, b, c\}$ and $P'' = \{Q_0 \rightarrow C_b Q_1, \quad Q_1 \rightarrow AD, A \rightarrow C_b Q_2 | C_a B, Q_2 \rightarrow AB, C_b \rightarrow b, B \rightarrow b, D \rightarrow c\}$.

Example 20

Reduce the following grammar G to Chomsky normal form. It is given that the productions of G are $Q_0 \rightarrow Aa|B$, $B \rightarrow aa|C$, $C \rightarrow a|bd|c$ and $A \rightarrow b$.

Solution The productions of the context-free grammar $G = (V_N, T, P, Q_0)$ are $Q_0 \rightarrow Aa|B$, $B \rightarrow aa|C$, $C \rightarrow a|bd|c$ and $A \rightarrow b$. In the above productions, there is no null production whereas there are two unit productions. Our aim is to obtain the grammar in Chomsky normal form.

(a) *Elimination of unit productions*   In this step we eliminate the unit productions $Q_0 \rightarrow B$ and $B \rightarrow C$.

1. *Construction of $W(Q)$ for each $Q \in V_N$*   We construct $W(Q)$ for each $Q \in V_N$ recursively as follows:

$$W_0(Q_0) = \{Q_0\}$$
$$W_1(Q_0) = W_0(Q_0) \cup \{B\} = \{Q_0, B\}$$
$$W_2(Q_0) = W_1(Q_0) \cup \{C\} = \{Q_0, B, C\}$$
$$W_3(Q_0) = W_2(Q_0) = \{Q_0, B, C\}$$

Therefore, we get $W(Q_0) = \{Q_0, B, C\}$. Similarly, it can be shown that $W(B) = \{B, C\}$, $W(C) = \{C\}$ and $W(A) = \{A\}$.

2. The productions in the equivalent grammar $G_1 = (V_N, T, P', Q_0)$ are $Q_0 \rightarrow Aa|aa|a|bd|c$, $B \rightarrow aa|a|bd|c$, $C \rightarrow a|bd|c$ and $A \rightarrow b$.

(b) *Elimination of terminals from right-hand side*   We define a modified grammar $G_2 = (V_N'', T, P'', Q_0)$ equivalent to G_1 as follows.

i. The productions $Q_0 \rightarrow a|c, B \rightarrow a|c, C \rightarrow a|c$ and $A \rightarrow b$ are included in P''.

ii. The production $Q_0 \rightarrow Aa$ reduces to $Q_0 \rightarrow AC_a$ and $C_a \rightarrow a$.

iii. The production $Q_0 \rightarrow aa$ reduces to $Q_0 \rightarrow C_a C_a$ and $C_a \rightarrow a$.

iv. The production $Q_0 \rightarrow bd$ reduces to $Q_0 \rightarrow C_b C_d, C_b \rightarrow b$ and $C_d \rightarrow d$.

v. The production $B \rightarrow aa$ reduces to $B \rightarrow C_a C_a$ and $C_a \rightarrow a$.

vi. The production $B \rightarrow bd$ reduces to $B \rightarrow C_b C_d, C_b \rightarrow b$ and $C_d \rightarrow d$.

vii. The production $C \rightarrow bd$ reduces to $C \rightarrow C_b C_d, C_b \rightarrow b$ and $C_d \rightarrow d$.

Therefore, $G_2 = (V_N'', T, P'', Q_0)$ is defined as $V_N'' = \{Q_0, A, B, C, C_a, C_b, C_d\}$; $T = \{a, b, c, d\}$ and $P'' = \{Q_0 \rightarrow AC_a|C_a C_a|a|C_b C_d|c, B \rightarrow C_a C_a|a|C_b C_d|c, A \rightarrow b, C \rightarrow a|C_b C_d|c, C_a \rightarrow a, C_b \rightarrow b, C_d \rightarrow d\}$.

(c) *Restricting the number of variables on right-hand side*   Here, we reduce the length of the right sides of the productions by introducing additional variables. It is clear that all productions in P'' are in Chomsky normal form. Therefore, the grammar in Chomsky normal form is given as

$G_2 = (V_N'', T, P'', Q_0)$, where $V_N'' = \{Q_0, A, B, C, C_a, C_b, C_d\}$; $T = \{a, b, c, d\}$ and $P'' = \{Q_0 \rightarrow AC_a | C_a C_a | a | C_b C_d | c,\ B \rightarrow C_a C_a | a | C_b C_d | c,\ A \rightarrow b,\ C \rightarrow a | C_b C_d | c,\ C_a \rightarrow a,$ $C_b \rightarrow b,\ C_d \rightarrow d\}$.

Example 21

Obtain an equivalent grammar in Chomsky normal form for the given context-free grammar having productions $Q_0 \rightarrow a | AAB$, $A \rightarrow ab | aB | \wedge$ and $B \rightarrow aba | \wedge$.

Solution   Consider the context-free grammar $G = (V_N, T, P, Q_0)$ with productions $Q_0 \rightarrow a | AAB$, $A \rightarrow ab | aB | \wedge$ and $B \rightarrow aba | \wedge$. Here, $V_N = \{Q_0, A, B\}$ and $T = \{a, b\}$. Our aim is to obtain the grammar in Chomsky normal form.

(a) *Elimination of null productions*   First, we eliminate the null productions from the grammar G so as to get the reduced grammar $G_1 = (V_N, T, P', Q_0)$ as below.

1. *Construction of the set of nullable variables W*   We construct the set of nullable variables recursively as follows:

$$W_1 = \{Q \in V_N | Q \rightarrow \wedge \text{ is a production in } P\} = \{A, B\}$$
$$W_2 = W_1 \cup \{Q \in V_N | Q \rightarrow \alpha; \alpha \in W_1^*\}$$
$$= \{Q_0, A, B\}$$
$$W_3 = W_2 \cup \{Q \in V_N | Q \rightarrow \alpha; \alpha \in W_2^*\}$$
$$= W_2 \cup \phi = W_2$$

Therefore, $W = W_2 = \{Q_0, A, B\}$.

2. *Construction of reduced productions P'*   The reduced productions are

   i. The productions $Q_0 \rightarrow a$, $A \rightarrow ab$ and $B \rightarrow aba$ are included in P'.

   ii. The production $Q_0 \rightarrow AAB$ leads to the productions $Q_0 \rightarrow AAB, Q_0 \rightarrow AA$, $Q_0 \rightarrow AB, Q_0 \rightarrow A$ and $Q_0 \rightarrow B$.

   iii. The production $A \rightarrow aB$ leads to the productions $A \rightarrow aB$ and $A \rightarrow a$.

Therefore, the reduced grammar $G_1 = (V_N, T, P', Q_0)$ without null productions is given as $V_N = \{Q_0, A, B\}$, $T = \{a, b\}$ and $P' = \{Q_0 \rightarrow AAB | AB | AA | B | A | a, B \rightarrow aba, A \rightarrow ab | aB | a\}$.

(b) *Elimination of unit productions*   Secondly, we eliminate the unit productions from the reduced grammar G_1 so as to get the reduced grammar $G_2 = (V_N, T, P'', Q_0)$ as below.

1. *Construction of $W(Q)$ for each $Q \in V_N$*   We construct $W(Q)$ for each $Q \in V_N$ recursively as follows:

$$W_0(Q_0) = \{Q_0\}$$
$$W_1(Q_0) = W_0(Q_0) \cup \{A, B\} = \{Q_0, A, B\}$$
$$W_2(Q_0) = W_1(Q_0) \cup \phi = W_1(Q_0)$$

Therefore, we get $W(Q_0) = \{Q_0, A, B\}$. Similarly, it can be shown that $W(A) = A$ and $W(B) = B$.

2. The productions P'' in the equivalent grammar $G_2 = (V_N, T, P'', Q_0)$ are given as $Q_0 \to AAB|AB|AA|aba|ab|aB|a$, $A \to ab|aB|a$ and $B \to aba$.

(c) *Elimination of terminals from right-hand side*   We define a modified grammar $G_3 = (V_N', T, P''', Q_0)$ equivalent to G_2 as follows:

   i. The productions $Q_0 \to AB|AA|a$ and $A \to a$ are included in P'''.

   ii. The production $Q_0 \to AAB$ is included in P'''.

   iii. The production $Q_0 \to aba$ reduces to $Q_0 \to C_a C_b C_a$, $C_a \to a$ and $C_b \to b$.

   iv. The production $Q_0 \to ab$ reduces to $Q_0 \to C_a C_b$, $C_b \to b$ and $C_a \to a$.

   v. The production $Q_0 \to aB$ reduces to $Q_0 \to C_a B$ and $C_a \to a$.

   vi. The production $A \to ab$ reduces to $A \to C_a C_b$, $C_a \to a$ and $C_b \to b$.

   vii. The production $A \to aB$ reduces to $A \to C_a B$ and $C_a \to a$.

   viii. The production $B \to aba$ reduces to $B \to C_a C_b C_a$, $C_a \to a$ and $C_b \to b$.

Therefore, the modified grammar $G_3 = (V_N', T, P''', Q_0)$ equivalent to G_2 is given as $V_N' = \{Q_0, A, B, C_a, C_b\}$, $T = \{a, b\}$ and $P''' = \{Q_0 \to a|AA|AB|C_a C_b|C_a B|C_a C_b C_a$, $Q_0 \to AAB$, $A \to C_a C_b|C_a B|a$, $B \to C_a C_b C_a$, $C_a \to a, C_b \to b\}$

(d) *Restricting the number of variables on right-hand side*   We define the modified grammar $G_4 = (V_N'', T, P'''', Q_0)$ equivalent to G_3 as follows:

   i. Productions $Q_0 \to a|AA|AB|C_a C_b|C_a B$, $A \to C_a C_b|C_a B|a$, $C_a \to a$, and $C_b \to b$ are included in P''''.

   ii. Production $Q_0 \to C_a C_b C_a$ reduces to $Q_0 \to C_a D_1$ and $D_1 = C_b C_a$.

   iii. Production $Q_0 \to AAB$ reduces to $Q_0 \to AD_2$ and $D_2 = AB$.

   iv. Production $B \to C_a C_b C_a$ reduces to $B \to C_a D_3$ and $D_3 = C_b C_a$.

Therefore, the Chomsky normal form of the grammar G is given as $G_4 = (V_N'', T, P'''', Q_0)$, where $V_N'' = \{Q_0, A, B, C_a, C_b, D_1, D_2, D_3\}, T = \{a, b\}$ and $P'''' = \{Q_0 \to a|AA|AB|C_a C_b|C_a B|C_a D_1|A D_2$, $D_1 \to C_b C_a$, $D_2 \to AB$, $C_a \to a, C_b \to b$, $A \to C_a C_b|C_a B | a$, $B \to C_a D_3$, $D_3 \to C_b C_a\}$.

Example 22

Obtain an equivalent grammar in Greibach normal form for the given context-free grammar having productions $Q_0 \to a|AB$, $A \to a|BC$, $B \to b$ and $C \to b$.

Solution Consider the context-free grammar $G = (V_N, T, P, Q_0)$ with productions $Q_0 \to a|AB$, $A \to a|BC$, $B \to b$ and $C \to b$.

It is clear that, the given productions are in Chomsky normal form and hence we can omit first step of reduction to Chomsky normal form. Rename the variables Q_0, A, B, C as Q_1, Q_2, Q_3 and Q_4 respectively. So, the renamed productions are given as $Q_1 \to a|Q_2Q_3$, $Q_2 \to a|Q_3Q_4$, $Q_3 \to b$ and $Q_4 \to b$. The above productions are in the form of $Q_i \to a\gamma$ or $Q_i \to Q_j\gamma$, where $i < j$ and $a \in T$. The productions $Q_1 \to a$, $Q_2 \to a$, $Q_3 \to b$ and $Q_4 \to b$ are in Greibach normal form. Now, consider the production $Q_2 \to Q_3Q_4$. On applying Lemma 1 to this production, the modified Q_2 production is given as $Q_2 \to bQ_4$. So, we get Q_2 productions as $Q_2 \to a|bQ_4$.

Similarly, consider the production $Q_1 \to Q_2Q_3$. On applying Lemma 1 to this production, we get the modified Q_1 production as $Q_1 \to aQ_3|bQ_4Q_3$. Therefore, the Q_1 productions are $Q_1 \to a|aQ_3|bQ_4Q_3$.

Thus, the reduced equivalent grammar $G_1 = (V_N, T, P', Q_1)$ in Greibach normal form is given as $V_N = \{Q_1, Q_2, Q_3, Q_4\}$, $T = \{a, b\}$ and $P' = \{Q_1 \to a|aQ_3|bQ_4Q_3, Q_2 \to a|bQ_4, Q_3 \to b, Q_4 \to b\}$.

Example 23

Reduce the following grammar G to Greibach normal form. It is given that the productions of G are $Q_0 \to b|ABE$, $A \to a|BE$, $B \to b$ and $E \to d$.

Solution Given that the productions of $G = (V_N, T, P, Q_0)$ are $Q_0 \to b|ABE$, $A \to a|BE$, $B \to b$ and $E \to d$. From the given productions it is clear that $V_N = \{Q_0, A, B, E\}$ and $T = \{a, b, d\}$. Our aim is to get the grammar in Greibach normal form. We obtain this by using the following steps.

(a) *Reduction to CNF* It is clear that all the productions except $Q_0 \to ABE$ are in Chomsky normal form. Also, there is no terminal in the right-hand side of the production $Q_0 \to ABE$. Hence, we have to reduce the number of variables on right-hand side of the production $Q_0 \to ABE$. In order to reduce the number of variables in right-hand side of the production, introduce a variable D to V_N such that $Q_0 \to AD$ and $D \to BE$.

Therefore, the grammar $G_1 = (V_N', T, P', Q_0)$ equivalent to the grammar G in Chomsky normal form is given as $V_N' = \{Q_0, A, B, E, D\}$, $T = \{a, b, d\}$ and $P' = \{Q_0 \to b|AD, A \to a|BE, B \to b, E \to d, D \to BE\}$.

(b) *Renaming of variables*   Rename the variables Q_0, A, B, E, D as Q_1, Q_2, Q_3, Q_4 and Q_5 respectively. So, the reduced productions are $P' = \{Q_1 \rightarrow b | Q_2 Q_5, \ Q_2 \rightarrow a | Q_3 Q_4, \ Q_3 \rightarrow b, \ Q_4 \rightarrow d, Q_5 \rightarrow Q_3 Q_4\}$

(c) Obtain productions in the form $Q_i \rightarrow a\gamma$ or $Q_i \rightarrow Q_j \gamma$ with $i < j$ From the productions it is clear that

    i. The Q_1 productions $Q_1 \rightarrow b | Q_2 Q_5$ are in required form.

    ii. The Q_2 productions $Q_2 \rightarrow a | Q_3 Q_4$ are in required form.

    iii. The Q_3 production $Q_3 \rightarrow b$ is in required form.

    iv. The Q_4 production $Q_4 \rightarrow d$ is in required form.

    v. The Q_5 production $Q_5 \rightarrow Q_3 Q_4$ is not according to required form.

Therefore, on applying Lemma 1 to this production, we get $Q_5 \rightarrow b Q_4$.

(d) Obtain productions in the form $Q \rightarrow aX; a \in T$ and $X \in (V_N')^*$.

    i. The reduced Q_5 production is given as $Q_5 \rightarrow b Q_4$.

    ii. The Q_3, Q_4 production is given as $Q_3 \rightarrow b$ and $Q_4 \rightarrow d$ respectively.

    iii. Consider the Q_2 production $Q_2 \rightarrow Q_3 Q_4$. On applying Lemma 1, the above Q_2 production reduces to $Q_2 \rightarrow b Q_4$. So, the modified Q_2 productions are given as $Q_2 \rightarrow a | b Q_4$.

    iv. Consider the Q_1 production $Q_1 \rightarrow b | Q_2 Q_5$. On applying Lemma 1, the above Q_1 production reduces to $Q_1 \rightarrow b | a Q_5 | b Q_4 Q_5$.

Therefore, the grammar $G_2 = (V_N'', T, P'', Q_1)$ equivalent to the grammar G in Greibach normal form is given as $V_N'' = \{Q_1, Q_2, Q_3, Q_4, Q_5\}, T = \{a, b, d\}$ and $P'' = \{Q_1 \rightarrow b, \ Q_1 \rightarrow a Q_5 | b Q_4 Q_5, \ Q_2 \rightarrow a | b Q_4, \ Q_3 \rightarrow b, Q_4 \rightarrow d, \ Q_5 \rightarrow b Q_4\}$.

Example 24

Obtain an equivalent grammar in Greibach normal form for the context-free grammar having productions $Q_0 \rightarrow Aa | B, B \rightarrow aa | C, C \rightarrow a | bd | c$ and $A \rightarrow b$.

Solution   The productions of the context-free grammar $G = (V_N, T, P, Q_0)$ are $Q_0 \rightarrow Aa | B, B \rightarrow aa | C, C \rightarrow a | bd | c$ and $A \rightarrow b$. In the above productions, there is no null production whereas there are two unit productions $Q_0 \rightarrow B$ and $B \rightarrow C$.

(a) *Elimination of unit productions*   In this step we eliminate the unit productions $Q_0 \rightarrow B$ and $B \rightarrow C$.

1. *Construction of $W(Q)$ for each $Q \in V_N$*   We construct $W(Q)$ for each $Q \in V_N$ recursively as follows:

$$W_0(Q_0) = \{Q_0\}$$
$$W_1(Q_0) = W_0(Q_0) \cup \{B\} = \{Q_0, B\}$$
$$W_2(Q_0) = W_1(Q_0) \cup \{C\} = \{Q_0, B, C\}$$
$$W_3(Q_0) = W_2(Q_0) = \{Q_0, B, C\}$$

Therefore, we get $W(Q_0) = \{Q_0, B, C\}$. Similarly, it can be shown that $W(B) = \{B, C\}$, $W(C) = \{C\}$ and $W(A) = \{A\}$.

2. The productions in the equivalent grammar $G_1 = (V_N, T, P', Q_0)$ are $Q_0 \rightarrow Aa|aa|a|bd|c$, $B \rightarrow aa|a|bd|c$, $C \rightarrow a|bd|c$ and $A \rightarrow b$.

(b) *Reduction to Chomsky normal form*   In this step we have to reduce the grammar G to Chomsky normal form.

1. *Elimination of terminals from right-hand side*   We define a modified grammar $G_2 = (V_N', T, P'', Q_0)$ equivalent to G_1 as follows.
   i. The productions $Q_0 \rightarrow a|c$, $B \rightarrow a|c$, $C \rightarrow a|c$ and $A \rightarrow b$ are included in P''.
   ii. The production $Q_0 \rightarrow Aa$ reduces to $Q_0 \rightarrow AC_a$ and $C_a \rightarrow a$.
   iii. The production $Q_0 \rightarrow aa$ reduces to $Q_0 \rightarrow C_aC_a$ and $C_a \rightarrow a$.
   iv. The production $Q_0 \rightarrow bd$ reduces to $Q_0 \rightarrow C_bC_d$, $C_b \rightarrow b$ and $C_d \rightarrow d$.
   v. The production $B \rightarrow aa$ reduces to $B \rightarrow C_aC_a$ and $C_a \rightarrow a$.
   vi. The production $B \rightarrow bd$ reduces to $B \rightarrow C_bC_d$, $C_b \rightarrow b$ and $C_d \rightarrow d$.
   vii. The production $C \rightarrow bd$ reduces to $C \rightarrow C_bC_d$, $C_b \rightarrow b$ and $C_d \rightarrow d$.

Therefore, $G_2 = (V_N', T, P'', Q_0)$ is defined as $V_N' = \{Q_0, A, B, C, C_a, C_b, C_d\}$; $T = \{a, b, c, d\}$ and $P'' = \{Q_0 \rightarrow AC_a|C_aC_a|a|C_bC_d|c, \ B \rightarrow C_aC_a|a|C_bC_d|c, \ A \rightarrow b, \ C \rightarrow a|C_bC_d|c, \ C_a \rightarrow a, \ C_a \rightarrow a, \ C_b \rightarrow b, \ C_d \rightarrow d\}$.

2. *Restricting the number of variables on right-hand side*   It is clear that all productions in P'' are in required form as per the requirement of Chomsky normal form. Therefore, the Chomsky normal form of the given grammar G is given as $G_2 = (V_N', T, P'', Q_0)$, where $V_N'' = \{Q_0, A, B, C, C_a, C_b, C_d\}$; $T = \{a, b, c, d\}$ and $P'' = \{Q_0 \rightarrow a|c|AC_a|C_aC_a|C_bC_d, \ B \rightarrow a|c|C_aC_a|C_bC_d, \ C \rightarrow a|c|C_bC_d, \ A \rightarrow b, C_a \rightarrow a, \text{ and } C_b \rightarrow b, C_d \rightarrow d\}$.

(c) *Renaming of variables* Rename the variables Q_0, A, B, C, C_a, C_b, C_d as $Q_1, Q_2, Q_3, Q_4, Q_5, Q_6$ and Q_7 respectively. So, the renamed productions are given as $Q_1 \to a|c|Q_2 Q_5|Q_5 Q_5|Q_6 Q_7$, $Q_2 \to b$, $Q_3 \to a|c|Q_5 Q_5|Q_6 Q_7$, $Q_4 \to a|c|Q_6 Q_7, Q_5 \to a$, $Q_6 \to b$ and $Q_7 \to d$ respectively.

(d) Obtain productions in the form $Q_i \to a\gamma$ or $Q_i \to Q_j \gamma$ with $i < j$ All the productions given above are in required form and hence no need to change them by using Lemma 2.

(e) Obtain productions in the form $Q \to aX; a \in T$ and $X \in (V'_N)^*$

    i. The productions $Q_5 \to a, Q_6 \to b, Q_7 \to d$ and $Q_2 \to b$ are in required form.

    ii. Consider the Q_4 production $Q_4 \to Q_6 Q_7$. On applying Lemma 1, the above Q_4 production reduces to $Q_4 \to bQ_7$. So, the modified Q_4 productions are given as $Q_4 \to a|c|bQ_7$.

    iii. Consider the Q_3 production $Q_3 \to Q_5 Q_5$. On applying Lemma 1, the above Q_3 production reduces to $Q_3 \to aQ_5$. Similarly, the Q_3 production $Q_3 \to Q_6 Q_7$ reduces to $Q_3 \to bQ_7$. Therefore, the modified Q_3 productions are given as $Q_3 \to a|c|aQ_5|bQ_7$.

    iv. Consider the Q_1 productions $Q_1 \to Q_2 Q_5|Q_5 Q_5|Q_6 Q_7$. On applying Lemma 1, the above Q_1 productions reduces to $Q_1 \to bQ_5|aQ_5|bQ_7$. Therefore, the modified Q_1 productions are given as $Q_1 \to a|c|bQ_5|aQ_5|bQ_7$.

Therefore, the equivalent grammar in Greibach normal form is given as $G_3 = (V''_N, T, P''', Q_1)$, where $V''_N = \{Q_1, Q_2, Q_3, Q_4, Q_5, Q_6, Q_7\}$, $T = \{a, b, c, d\}$ and $P''' = \{Q_1 \to a|c|bQ_5|aQ_5|bQ_7,$ $Q_2 \to b$, $Q_3 \to a|c|aQ_5|bQ_7$, $Q_4 \to a|c|bQ_7$, $Q_5 \to a \; Q_6 \to b, Q_7 \to d\}$.

Example 25

Show that the language $L = \{a^n b^n c^n \mid n \geq 1\}$ is not context-free by using pumping lemma.

Solution   Consider the language $L = \{a^n b^n c^n \mid n \geq 1\}$. Assume that the language L is context-free. Let n be the natural number obtained by using pumping lemma. Let us take $z = a^n b^n c^n$. Therefore, $|z| = 3n > n$.

Therefore, z can be expressed as $uvwxy$, i.e., $z = uvwxy$ with $|vx| \geq 1$ and $|vwx| \leq n$. As $|vx| \geq 1$, so both v and x can not be $\wedge$. Now, $z = uvwxy = a^n b^n c^n$. As $1 \leq |vx| \leq n$, so it is clear that v or x can not contain all the three symbols a, b and c. Therefore, we have two cases.

Case 1 v or x is of the form $a^k b^l$ or $b^k c^l$ for some k, l such that $(k + l) \leq n$.

Case 2 v or x is a string formed by the repetition of only one symbol among a, b and c.

Let us assume that $v = a^k b^l$. On taking $i = 2$ we can not express $uv^i wx^i y = uv^2 wx^2 y$ in the form of $a^n b^n c^n$. Therefore, $uv^i wx^i y \notin L$. On the other hand consider $v = a^k$. Therefore, we get $u = a^{n-k}$. On taking $i = 0$ we have $uv^i wx^i y = uwy \notin L$. This is because of the number of occurrences of a, b and c are unequal.

Therefore, our assumption is wrong and hence the language $L = \{a^n b^n c^n \mid n \geq 1\}$ is not context-free.

Example 26

Show that the language $L = \{a^n b^n a^n \mid n \geq 1\}$ is not context-free by using pumping lemma.

Solution Assume that the language $L = \{a^n b^n a^n \mid n \geq 1\}$ is context-free. Let n be the natural number obtained by using pumping lemma. Let us take $z = a^n b^n a^n$. Therefore, we get $|z| = 3n > n$.

Therefore, z can be expressed as $uvwxy$, i.e., $z = uvwxy$ with $|vx| \geq 1$ and $|vwx| \leq n$. As $|vx| \geq 1$, so both v and x can not be $\wedge$. Also, it is clear that $z = a^n b^n a^n$ contains exactly one occurance of the substring '*ab*' and '*ba*' for all values of n. As z is decomposed into '$uvwxy$', so we must have either of the following cases.

Case 1 v or x contains a substring '*ab*'.

Case 2 v or x contains a substring '*ba*'.

Case 3 v or x is a string formed by the repetition of only one symbol a and b.

If v or x contains a substring '*ab*', then on taking $i = 2$ we have $uv^i wx^i y \notin L$. This is because the string $uv^2 wx^2 y$ contains more than one substring of '*ab*'. The same argument holds for the case when v or x contains a substring '*ba*'. If v or x contains only one symbol a or b, then on taking $i = 2$ we have $uv^i wx^i y = uv^2 wx^2 y \notin L$. This is because the string contains more number of a's than b's or vice versa.

Therefore, our assumption is wrong and hence the language $L = \{a^n b^n a^n \mid n \geq 1\}$ is not context-free.

Example 27

Show that the language $L = \{a^{n^2} \mid n \geq 1\}$ is not context-free by using pumping lemma.

Solution sume that the language $L = \{a^{n^2} \mid n \geq 1\}$ is context-free. Let n be the natural number obtained by using pumping lemma. Let us take $z = a^{n^2}$. Therefore, we get $|z| = n^2 > n$.

Therefore, z can be expressed as $uvwxy$, i.e., $z = uvwxy$ with $1 \leq |vx| \leq n$ and $|vwx| \leq n$. Let us choose $|vx| = m$ with $m \leq n$. On taking $i = 2$ we have

$$\left|uv^2wx^2y\right| = |u| + 2|vx| + |w| + |y|$$
$$> |u| + |vx| + |w| + |y|$$
$$= |uvwxy| = n^2$$

i.e., $\left|uv^2wx^2y\right| > n^2$ (1)

Again, we have

$$\left|uv^2wx^2y\right| = |uvwxy| + |vx|$$
$$= n^2 + m < n^2 + 2n + 1 \quad [\because m < n]$$

i.e., $\left|uv^2wx^2y\right| < (n+1)^2$ (2)

From equations (1) and (2) we get $n^2 < \left|uv^2wx^2y\right| < (n+1)^2$.

Therefore, we get $uv^2wx^2y \notin L$. This is because there is no square number which lies between n^2 and $(n+1)^2$. So, our supposition is wrong and hence the given language $L = \{a^{n^2} \mid n \geq 1\}$ is not context-free.

Example 28

Determine whether the string $z = 1100110$ is in the language generated by the context-free grammar $Q_0 \rightarrow AB$, $A \rightarrow BB|0$ and $B \rightarrow BA|1$.

Solution　Assume that string $z = 1100110 = a_1a_2a_3a_4a_5a_6a_7$. Therefore, $n = 7$. We define substrings z_{ij} as $z_{ij} = a_i \cdots a_j$. From the productions it is clear that $B \rightarrow a_1 = 1$ and so $V_{11} = \{B\}$. Similarly, $V_{22} = \{B\}, V_{33} = \{A\}, V_{44} = \{A\}, V_{55} = \{B\}, V_{66} = \{B\}$ and $V_{77} = \{A\}$.

Now, we have to compute $V_{12}, V_{23}, V_{34}, V_{45}, V_{56}$ and V_{67} by using the relation

$$V_{ij} = \bigcup_{k \in \{i, i+1, \cdots, j-1\}} \left\{A \mid A \rightarrow BC; B \in V_{ik}, C \in V_{k+1, j}\right\}, \ j > i$$ (1)

Therefore, we get the following:

$$V_{12} = \{A | A \to BC, B \in V_{11} = \{B\}, C \in V_{22} = \{B\}\} = \{A\}$$

$$V_{23} = \{A | A \to BC, B \in V_{22} = \{B\}, C \in V_{33} = \{A\}\} = \{B\}$$

$$V_{34} = \{A | A \to BC, B \in V_{33} = \{A\}, C \in V_{44} = \{A\}\} = \phi$$

$$V_{45} = \{A | A \to BC, B \in V_{44} = \{A\}, C \in V_{55} = \{B\}\} = \{Q_0\}$$

$$V_{56} = \{A | A \to BC, B \in V_{55} = \{B\}, C \in V_{66} = \{B\}\} = \{A\}$$

$$V_{67} = \{A | A \to BC, B \in V_{66} = \{B\}, C \in V_{77} = \{A\}\} = \{B\}$$

Now, compute $V_{13}, V_{24}, V_{35}, V_{46}$ and V_{57} by using the above relation (1). So, we get

$$V_{13} = \{A | A \to BC, B \in V_{11} = \{B\}, C \in V_{23} = \{B\}\} \cup$$
$$\{A | A \to BC, B \in V_{12} = \{A\}, C \in V_{33} = \{A\}\} = \{A\}$$

$$V_{24} = \{A | A \to BC, B \in V_{22} = \{B\}, C \in V_{34} = \phi\} \cup$$
$$\{A | A \to BC, B \in V_{23} = \{B\}, C \in V_{44} = \{A\}\}$$
$$= \phi \cup \{B\} = \{B\}$$

$$V_{35} = \{A | A \to BC, B \in V_{33} = \{A\}, C \in V_{45} = \{Q_0\}\} \cup$$
$$\{A | A \to BC, B \in V_{34} = \phi, C \in V_{55} = \{B\}\} = \phi$$

$$V_{46} = \{A | A \to BC, B \in V_{44} = \{A\}, C \in V_{56} = \{A\}\} \cup$$
$$\{A | A \to BC, B \in V_{45} = \{Q_0\}, C \in V_{66} = \{B\}\} = \phi$$

$$V_{57} = \{A | A \to BC, B \in V_{55} = \{B\}, C \in V_{67} = \{B\}\} \cup$$
$$\{A | A \to BC, B \in V_{56} = \{A\}, C \in V_{77} = \{A\}\} = \{A\}$$

Now, compute V_{14}, V_{25}, V_{36} and V_{47} by using the given relation (1). So, we get

$$V_{14} = \{A | A \to BC, B \in V_{11} = \{B\}, C \in V_{24} = \{B\}\} \cup$$
$$\{A | A \to BC, B \in V_{12} = \{A\}, C \in V_{34} = \phi\} \cup$$
$$\{A | A \to BC, B \in V_{13} = \{A\}, C \in V_{44} = \{A\}\} = \{A\}$$

$$V_{25} = \{A | A \to BC, B \in V_{22} = \{B\}, C \in V_{35} = \phi\} \cup$$
$$\{A | A \to BC, B \in V_{23} = \{B\}, C \in V_{45} = \{Q_0\}\} \cup$$
$$\{A | A \to BC, B \in V_{24} = \{B\}, C \in V_{55} = \{B\}\} = \{A\}$$

$$V_{36} = \{A | A \to BC, B \in V_{33} = \{A\}, C \in V_{46} = \phi\} \cup$$
$$\{A | A \to BC, B \in V_{34} = \phi, C \in V_{56} = \{A\}\} \cup$$
$$\{A | A \to BC, B \in V_{35} = \phi, C \in V_{66} = \{B\}\} = \phi$$

$$V_{47} = \{A \mid A \to BC, B \in V_{44} = \{A\}, C \in V_{57} = \{A\}\} \cup$$
$$\{A \mid A \to BC, B \in V_{45} = \{Q_0\}, C \in V_{67} = \{B\}\} \cup$$
$$\{A \mid A \to BC, B \in V_{46} = \phi, C \in V_{77} = \{A\}\} = \phi$$

Now, compute V_{15}, V_{26} and V_{37} by using the given relation (1). So, we get

$$V_{15} = \{A \mid A \to BC, B \in V_{11} = \{B\}, C \in V_{25} = \{A\}\} \cup$$
$$\{A \mid A \to BC, B \in V_{12} = \{A\}, C \in V_{35} = \phi\} \cup$$
$$\{A \mid A \to BC, B \in V_{13} = \{A\}, C \in V_{45} = \{Q_0\}\} \cup$$
$$\{A \mid A \to BC, B \in V_{14} = \{A\}, C \in V_{55} = \{B\}\}$$
$$= \{Q_0, B\}$$

$$V_{26} = \{A \mid A \to BC, B \in V_{22} = \{B\}, C \in V_{36} = \phi\} \cup$$
$$\{A \mid A \to BC, B \in V_{23} = \{B\}, C \in V_{46} = \phi\} \cup$$
$$\{A \mid A \to BC, B \in V_{24} = \{B\}, C \in V_{56} = \{A\}\} \cup$$
$$\{A \mid A \to BC, B \in V_{25} = \{A\}, C \in V_{66} = \{B\}\}$$
$$= \{Q_0, B\}$$

$$V_{37} = \{A \mid A \to BC, B \in V_{33} = \{A\}, C \in V_{47} = \phi\} \cup$$
$$\{A \mid A \to BC, B \in V_{34} = \phi, C \in V_{57} = \{A\}\} \cup$$
$$\{A \mid A \to BC, B \in V_{35} = \phi, C \in V_{67} = \{B\}\} \cup$$
$$\{A \mid A \to BC, B \in V_{36} = \phi, C \in V_{77} = \{A\}\} = \phi$$

Now, compute V_{16}, V_{27} and V_{17} by using the above relation (1). So, we get

$$V_{16} = \{A \mid A \to BC, B \in V_{11} = \{B\}, C \in V_{26} = \{Q_0, B\}\} \cup$$
$$\{A \mid A \to BC, B \in V_{12} = \{A\}, C \in V_{36} = \phi\} \cup$$
$$\{A \mid A \to BC, B \in V_{13} = \{A\}, C \in V_{46} = \phi\} \cup$$
$$\{A \mid A \to BC, B \in V_{14} = \{A\}, C \in V_{56} = \{A\}\} \cup$$
$$\{A \mid A \to BC, B \in V_{15} = \{Q_0, B\}, C \in V_{66} = \{B\}\} = \{A\}$$

$$V_{27} = \{A|A \rightarrow BC, B \in V_{22} = \{B\}, C \in V_{37} = \phi\} \cup$$
$$\{A|A \rightarrow BC, B \in V_{23} = \{B\}, C \in V_{47} = \phi\} \cup$$
$$\{A|A \rightarrow BC, B \in V_{24} = \{B\}, C \in V_{57} = \{A\}\} \cup$$
$$\{A|A \rightarrow BC, B \in V_{25} = \{A\}, C \in V_{67} = \{B\}\} \cup$$
$$\{A|A \rightarrow BC, B \in V_{26} = \{Q_0, B\}, C \in V_{77} = \{A\}\} = \{Q_0, B\}$$
$$V_{17} = \{A|A \rightarrow BC, B \in V_{11} = \{B\}, C \in V_{27} = \{Q_0, B\}\} \cup$$
$$\{A|A \rightarrow BC, B \in V_{12} = \{A\}, C \in V_{37} = \phi\} \cup$$
$$\{A|A \rightarrow BC, B \in V_{13} = \{A\}, C \in V_{47} = \phi\} \cup$$
$$\{A|A \rightarrow BC, B \in V_{14} = \{A\}, C \in V_{57} = \{A\}\} \cup$$
$$\{A|A \rightarrow BC, B \in V_{15} = \{Q_0, B\}, C \in V_{67} = \{B\}\} \cup$$
$$\{A|A \rightarrow BC, B \in V_{16} = \{A\}, C \in V_{77} = \{A\}\} = \{A\}$$

Therefore, it is clear that $Q_0 \notin V_{17}$ and hence, $z = 1100110 \notin L(G)$.

Example 29

Determine whether the string $z = 1101$ is in the language generated by the context-free grammar $Q_0 \rightarrow AB|0, A \rightarrow BBB|1$ and $B \rightarrow AB|BA|0$.

Solution Consider the context-free grammar $G = (V_N, T, P, Q_0)$ such that $P = \{Q_0 \rightarrow AB|0, A \rightarrow BBB|1, B \rightarrow AB|BA|0\}$. In order to apply CYK algorithm, we have to reduce the given context-free grammar to Chomsky normal form.

(a) *Reduction to CNF* In this step we have to reduce the grammar to Chomsky normal form.

1. *Elimination of terminals from right-hand side of productions* From the productions it is clear that there is no terminal in the right-hand side of the productions when the length of right-hand side of the productions exceeds one.

2. *Restricting the number of variables on right-hand side* We define a modified grammar $G_1 = (V_N', T, P', Q_0)$ as follows.
   i. Productions $Q_0 \rightarrow AB|0, B \rightarrow AB|BA|0$ and $A \rightarrow 1$ are included in P'.
   ii. Production $A \rightarrow BBB$ reduces to $A \rightarrow BD$ and $D \rightarrow BB$, where D is a newly introduced variable.

Therefore, the modified grammar $G_1 = (V_N', T, P', Q_0)$ equivalent to the grammar G in Chomsky normal form is given as $V_N' = \{Q_0, A, B, D\}$; $T = \{0, 1\}$ and $P' = \{Q_0 \rightarrow AB|0, A \rightarrow BD|1, B \rightarrow AB|BA|0, D \rightarrow BB\}$.

(b) *CYK Algorithm*　Assume that the string $z = 1101 = a_1a_2a_3a_4$. Thus, $n = 4$. We define substrings z_{ij} as $z_{ij} = a_i \cdots a_j$. First we note that $z_{11} = a_1 = 1$. Therefore, the set of variables that derive 1 immediately are included in V_{11}. Thus, $V_{11} = \{A\}$. Similarly, $V_{22} = \{A\}, V_{33} = \{B\}$ and $V_{44} = \{A\}$.

Now, we have to compute $V_{12}, V_{23}, V_{34}, V_{13}, V_{24}$ and V_{14} in order by using the relation

$$V_{ij} = \bigcup_{k \in \{i, i+1, \cdots, j-1\}} \left\{ A \middle| A \rightarrow BC; B \in V_{ik}, C \in V_{k+1,j} \right\}, \; j > i \tag{1}$$

Therefore, we get the following

$$V_{12} = \{A | A \rightarrow BC, B \in V_{11} = \{A\}, C \in V_{22} = \{A\}\} = \phi$$
$$V_{23} = \{A | A \rightarrow BC, B \in V_{22} = \{A\}, C \in V_{33} = \{B\}\} = \{Q_0, B\}$$
$$V_{34} = \{A | A \rightarrow BC, B \in V_{33} = \{B\}, C \in V_{44} = \{A\}\} = \{B\}$$
$$V_{13} = \{A | A \rightarrow BC, B \in V_{11} = \{A\}, C \in V_{23} = \{Q_0, B\}\} \cup$$
$$\{A | A \rightarrow BC, B \in V_{12} = \phi, C \in V_{33} = \{B\}\} = \{Q_0, B\}$$
$$V_{24} = \{A | A \rightarrow BC, B \in V_{22} = \{A\}, C \in V_{34} = \{B\}\} \cup$$
$$\{A | A \rightarrow BC, B \in V_{23} = \{Q_0, B\}, C \in V_{44} = \{A\}\} = \{Q_0, B\}$$
$$V_{14} = \{A | A \rightarrow BC, B \in V_{11} = \{A\}, C \in V_{24} = \{Q_0, B\}\} \cup$$
$$\{A | A \rightarrow BC, B \in V_{12} = \phi, C \in V_{34} = \{B\}\} \cup$$
$$\{A | A \rightarrow BC, B \in V_{13} = \{Q_0, B\}, C \in V_{44} = \{A\}\} = \{Q_0, B\}$$

Therefore, it is clear that $Q_0 \in V_{14}$ and hence the string $z = 1101$ is in the language generated by the given context-free grammar.

Example 30

Show that the context-free grammar G having productions $Q_0 \rightarrow AB$, $A \rightarrow DD | DA$, $B \rightarrow CC | CB$, $C \rightarrow 0$ and $D \rightarrow 1$ produces at least one word.

Solution　We show this with the help of following three steps.

1. For each variable A that has some production of the form $A \rightarrow t, t \in \Sigma^*$, we choose one of these productions and drop the productions for which A is on the left side. We then replace A by t in all productions in which A is on the right side.

2. Repeat step 1 until either it eliminates Q_0 or it eliminates no new variables.

3. If Q_0 is eliminated, then the context-free grammar produces some words otherwise it does not produce any word at all.

Consider the context-free grammar G having productions $Q_0 \rightarrow AB$, $A \rightarrow DD|DA$, $B \rightarrow CC|CB$, $C \rightarrow 0$ and $D \rightarrow 1$. From the productions it is clear that, there exists two productions $C \rightarrow 0$ and $D \rightarrow 1$ of the form $A \rightarrow t, t \in \Sigma^*$. Now consider the production $D \rightarrow 1$. Therefore, the productions obtained by using step 1 is given as $Q_0 \rightarrow AB, A \rightarrow 11|1A, B \rightarrow CC|CB, C \rightarrow 0$.

Again on considering the production $C \rightarrow 0$ and using step 1, we get the new set of equivalent productions as

$$Q_0 \rightarrow AB, \ A \rightarrow 11|1A, B \rightarrow 00|0B$$

Similarly, consider the productions $A \rightarrow 11$ and $B \rightarrow 00$. On using step 1 the reduced production is given as $Q_0 \rightarrow 1100$. Now, we can replace Q_0's by 1100. As a result, we eliminate the start symbol Q_0. Therefore, it is clear that the context-free grammar produces at least one word.

Example 31

Show that the context-free grammar G having productions $Q_0 \rightarrow AB$, $A \rightarrow a, B \rightarrow DB$ and $D \rightarrow b$ is empty.

Solution Consider the context-free grammar G having productions $Q_0 \rightarrow AB$, $A \rightarrow a, B \rightarrow DB$ and $D \rightarrow b$. From the productions it is clear that, there exists two productions $A \rightarrow a$ and $D \rightarrow b$ of the form $A \rightarrow t, t \in \Sigma^*$. Therefore, the reduced productions are is given as $Q_0 \rightarrow aB, B \rightarrow bB$.

Since, there is no production of the form $A \rightarrow t, t \in \Sigma^*$, we terminate the process. It indicates that the start symbol Q_0 cannot be eliminated. Therefore, it is clear that the context-free grammar is empty or does not produce any word at all.

Example 32

Construct the context-free grammar G generating the language defined by regular expression bba^*.

Solution Assume that the context-free grammar is $G = (V_N, T, P, Q_0)$, where $T = \{a, b\}$ and Q_0 is the assumed start symbol.

Our aim is to construct a context-free grammar that generates bba^*. Let us devide the regular expression bba^* into two parts such that bb is followed by a^*. For the string a^*, we can design the production as $Q_0 \rightarrow Q_0 a|\wedge$. But the smallest string belonging to bba^* is bb. Therefore, we must replace $Q_0 \rightarrow \wedge$ by $Q_0 \rightarrow bb$. Therefore, the constructed context-free grammar is given as $P = \{Q_0 \rightarrow Q_0 a, Q_0 \rightarrow bb\}$.

Example 33

Construct a context-free grammar for the language $L = \{0^n 1^n \mid n \geq 1\}$.

Solution Assume that the context-free grammar is $G = (V_N, T, P, Q_0)$, where $T = \{0,1\}$ and Q_0 is the assumed start symbol.

Our aim is to construct a context-free grammar that generates $L = \{0^n 1^n \mid n \geq 1\}$. Therefore, the smallest string belonging to L is '01'. So, we can design the production as $Q_0 \rightarrow 0Q_0 1 \mid 01$. Therefore, the constructed context-free grammar is given as $G = (V_N, T, P, Q_0)$, where $V_N = \{Q_0\}$, $T = \{0,1\}$ and $P = \{Q_0 \rightarrow 0Q_0 1 \mid 01\}$.

Example 34

Construct the context-free grammar G generating the language $L = \{0^n 1^m 0^m 1^n \mid m, n \geq 1\}$.

Solution Assume that the context-free grammar is $G = (V_N, T, P, Q_0)$, where $T = \{0,1\}$ and Q_0 is the assumed start symbol.

Our aim is to construct a context-free grammar that generates $L = \{0^n 1^m 0^m 1^n \mid m, n \geq 1\}$. Let us devide the language into three parts, such that 0^n is followed by $1^m 0^m$ which further followed by 1^n. Therefore, we can design the production as $Q_0 \rightarrow 0Q_0 1 \mid 0A1$, $A \rightarrow 1A0 \mid 10$. Therefore, the constructed context-free grammar is given as $G = (V_N, T, P, Q_0)$, where $V_N = \{Q_0, A\}$, $T = \{0,1\}$ and $P = \{Q_0 \rightarrow 0Q_0 1 \mid 0A1, A \rightarrow 1A0 \mid 10\}$.

Consider the string $z = 0^n 101^n$. Therefore, the derivation steps is given as

$$Q_0 \rightarrow 0Q_0 1$$
$$\xrightarrow{*} 0^{n-1} Q_0 1^{n-1} \qquad [Q_0 \rightarrow 0Q_0 1]$$
$$\xrightarrow{*} 0^{n-1} 0A1 1^{n-1} \qquad [Q_0 \rightarrow 0A1]$$
$$\xrightarrow{*} 0^n 101^n \qquad [A \rightarrow 10]$$

Similarly, the strings $01^m 0^m 1$ and 0101 can also be generated.

Example 35

Construct a context-free grammar G generating the language $(L_1 \cup L_2)$, where $L_1 = \{a^n b^m a^m b^n \mid m, n \geq 1\}$ and $L_2 = \{bba^n \mid n \geq 0\}$.

Solution Our aim is to construct a context-free grammar for $(L_1 \cup L_2)$. In order to get a context-free grammar for $(L_1 \cup L_2)$, first we have to design the context-free grammar for L_1 and L_2 independently.

Suppose the context-free grammar generating L_1 is $G_1 = (V_N', T_1, P_1, Q_0')$, where $T_1 = \{a, b\}$ and Q_0' is the assumed start symbol. Let us devide the language into three parts, such that a^n is followed by $b^m a^m$ which further followed by b^n.

Therefore, we can design the production as $Q_0' \to a Q_0' b | a A b$, $A \to b A a | ba$. Therefore, the constructed context-free grammar is given as $G_1 = (V_N', T_1, P_1, Q_0')$, where $V_N' = \{Q_0', A\}$, $T_1 = \{a, b\}$ and $P_1 = \{Q_0' \to a Q_0' b | a A b, A \to b A a | ba\}$.

Similarly, assume that the context-free grammar generating L_2 is $G_2 = (V_N'', T_2, P_2, Q_0'')$, where $T_2 = \{a, b\}$ and Q_0'' is the assumed start symbol. Let us devide the language into two parts, such that 'bb' is followed by a^n.

Therefore, we can design the production as $Q_0'' \to Q_0'' a | bb$. Therefore, the constructed context-free grammar is given as $G_2 = (V_N'', T_2, P_2, Q_0'')$, where $V_N'' = \{Q_0''\}$, $T_2 = \{a, b\}$ and $P_2 = \{Q_0'' \to Q_0'' a | bb\}$.

Let the context-free grammar generating $(L_1 \cup L_2)$ is $G = (V_N, T, P, Q_0)$. Therefore, we have

$$V_N = V_N' \cup V_N'' \cup \{Q_0\} = \{Q_0, Q_0', Q_0'', A\}$$
$$T = T_1 \cup T_2 = \{a, b\} \quad \text{and}$$
$$P = P_1 \cup P_2 \cup \{Q_0 \to Q_0' | Q_0''\}$$
$$= \{Q_0 \to Q_0' | Q_0'', \quad Q_0' \to a Q_0' b | a A b, \quad A \to b A a | ba, \quad Q_0'' \to Q_0'' a | bb\}$$

REVIEW QUESTIONS

1. Define context-free grammar. Give an example that explains context-free grammar.
2. What do you mean by context-free languages? Write different closure properties of context-free language.
3. Define ambiguity in context-free grammar. Explain with the help of an example.
4. Write a note on simplification of context-free grammar. Explain why simplification is essential in context-free grammar. Give an example.
5. What do you mean by null production and unit production? Give an example.
6. Define Chomsky normal form. Prove that for every context-free grammar; there exists an equivalent grammar in Chomsky normal form.

7. Define Griebach normal form. Show that for every context-free grammar; there is an equivalent grammar in Greibach normal form.

8. State and prove pumping lemma for context-free languages.

9. State the different closure properties of context-free languages.

10. Show that the family of context-free languages is not closed under intersection.

PROBLEMS

1. Consider the context-free grammar G having productions, $Q_0 \rightarrow AB|b$, $A \rightarrow aQ_0|b$ and $B \rightarrow BQ_0|a$. Determine whether the string $z = aabbab$ is in the language $L(G)$.

2. Determine whether the string $z = 1101001$ is in the language generated by the grammar $G = (V_N, T, P, Q_0)$, where $V_N = \{Q_0, A, B, C\}$, $T = \{0, 1, 2\}$ and $P = \{Q_0 \rightarrow 1Q_0|ABC|0$, $A \rightarrow 10|BB$, $B \rightarrow 0|1$, $C \rightarrow AB|1\}$. Use CYK algorithm to justify your answer.

3. Determine the yield for the following derivation trees.

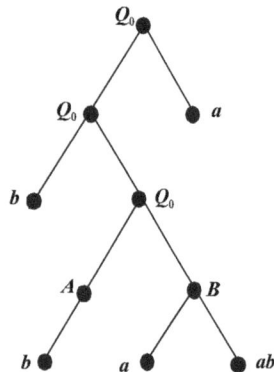

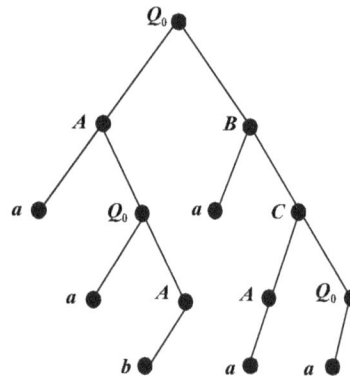

4. Consider the context-free grammar G with productions $Q_0 \rightarrow aB|bA$, $A \rightarrow bAA|aQ_0|0$ and $B \rightarrow aBB|bQ_0|b$. Construct derivation trees for string $z = aabbabab$ by using both leftmost and rightmost derivations.

5. Show that the context-free grammar G with productions $Q_0 \rightarrow Q_0aQ_0|b$ is ambiguous.

6. Consider the productions $Q_0 \rightarrow 0Q_0|1A|2BB$, $A \rightarrow 0AA|1B|2$ and $B \rightarrow BQ_0|2A|1$. Find the leftmost, rightmost derivation and parse tree for the string $z = 02211122022$.

7. Prove that the grammar $Q_0 \rightarrow 0|01Q_01|0A1$, $A \rightarrow 1Q_0|0AA1$ is ambiguous.

8. Construct a context-free grammar G that generates all integers.

9. Let G be the grammar with productions $Q_0 \to aA|bB, B \to a|aQ_0|bBB$ and $A \to b|bQ_0|aAA$. For the string $z = aabbabab$, find

   i.   rightmost derivation

   ii.   leftmost derivation and

   iii.   parse tree

10. Show that the language generated by the context-free grammar $G = (V_N, T, P, Q_0)$, where $V_N = \{Q_0\}, T = \{a, b\}$ and $P = \{Q_0 \to Q_0 Q_0\}$ is empty, i.e., $L(G) = \phi$.

11. Show that the language of all non-null strings of 0's defined by a context-free grammar $G = (V_N, T, P, Q_0)$ having productions $Q_0 \to 0|0Q_0|Q_0 0$ is ambiguous.

12. Show that a context-free grammar $G = (V_N, T, P, Q_0)$ having productions $Q_0 \to 1Q_0|AB, A \to 1$ and $B \to 0$ is a grammar having no null productions.

13. Consider a context-free grammar $G = (V_N, T, P, Q_0)$ having productions $Q_0 \to aQ_0|AB|aB|C, A \to b|\wedge, B \to a$ and $C \to \wedge$. Construct a grammar G_1 equivalent to G having no null productions.

14. A context-free grammar $G = (V_N, T, P, Q_0)$ is given by following productions: $Q_0 \to AB|aA|bA|C, A \to a, C \to B$ and $B \to \wedge|a$. Construct a grammar G_1 equivalent to G having no null productions.

15. Consider a context-free grammar $G = (V_N, T, P, Q_0)$ having productions $Q_0 \to AQ_0|BQ_0|a|b$, $A \to B, B \to C$ and $C \to a|b$. Construct a grammar G_1 equivalent to G having no unit productions.

16. Consider the context-free grammar $G = (\{Q_0, A, B, C, D\}, \{a, b, c, d\}, P, Q_0)$, $P = \{Q_0 \to BC|ad|bc|aA, B \to D, D \to C|b, A \to aB|c, C \to d\}$. Construct a grammar equivalent to G having no unit productions.

17. Let us consider a context-free grammar $G = (V_N, T, P, Q_0)$ given by productions $Q_0 \to aQ_0|bA|cB|BC, A \to a|\wedge, B \to a, C \to a, D \to d, E \to F|a$ and $F \to b$. Construct a grammar $G_1 = (V_N', T', P', Q_0)$ equivalent to G such that every symbol in $(V_N' \cup T')$ appears in some sentential form.

18. A context-free grammar $G = (V_N, T, P, Q_0)$ is given by following productions: $Q_0 \to AB|BQ_0|aQ_0|bE|dH, A \to B, B \to c, E \to H|\wedge, H \to a$. Construct an equivalent grammar G_1 having no, null and unit productions.

19. Consider a context-free grammar $G = (V_N, T, P, Q_0)$ with productions $Q_0 \to aQ_0$, $Q_0 \to bA \mid cB \mid BC$, $A \to a \mid \wedge$, $B \to a$, $C \to a, D \to d$, $E \to F \mid a$ and $F \to b$. Find an equivalent grammar G_1 that derives some terminal string.

20. Reduce the context-free grammar G given by productions $Q_0 \to aAB \mid b$, $B \to b$, $A \to aB \mid aABC$, $C \to d$ into Chomsky normal form.

21. Reduce the grammar $G = (\{A, B, C, D, E\}, \{0, 1, 2\}, P, A)$, $P = \{A \to 0ABC$, $A \to BCD \mid 2 \mid 1B \mid 2DE$, $B \to CE \mid 1$, $C \to 0 \mid 2$, $D \to 2$, $E \to 1\}$ into Chomsky normal form.

22. A context-free grammar $G = (V_N, T, P, Q_0)$ is given by following productions: $Q_0 \to ABCDQ_0 \mid 1Q_0A \mid BQ_0 \mid 2$, $A \to 0BC \mid 1$, $B \to C$, $C \to D$, $D \to 0 \mid 2$. Reduce the grammar G into Chomsky normal form.

23. Consider a context-free grammar $G = (V_N, T, P, Q_0)$ having productions $Q_0 \to aA \mid bBC \mid \wedge$, $A \to Aa \mid b \mid \wedge$, $B \to BBC \mid a \mid \wedge$ and $C \to b$. Construct a grammar equivalent to G in Chomsky normal form.

24. Consider a context-free grammar $G = (V_N, T, P, Q_0)$ having productions $Q_0 \to Aa \mid Bbab \mid CD \mid d \mid AQ_0 \mid \wedge$, $A \to D \mid d$, $B \to b$, $C \to D \mid a$ and $D \to b$. Reduce the grammar G into Chomsky normal form.

25. Reduce the context-free grammar G having productions $Q_0 \to AAB \mid b$, $A \to a$ and $B \to Q_0Q_0 \mid b$ into Greibach normal form.

26. Let us consider the context-free grammar $G = (V_N, T, P, Q_0)$ having productions $Q_0 \to aAB$, $A \to aC \mid bAC$, $C \to b$ and $B \to c$. Reduce the above context-free grammar G into Greibach normal form.

27. Consider a context-free grammar $G = (V_N, T, P, Q_0)$ with productions $Q_0 \to ABCQ_0b \mid aABb \mid bD$, $A \to BC \mid d$, $B \to C$, $C \to D$ and $D \to a$. Obtain a grammar G_1 equivalent to G in Greibach normal form.

28. Obtain a grammar in Greibach normal form for the context-free grammar $G = (V_N, T, P, Q_0)$ having productions $Q_0 \to AB$, $A \to BE \mid a$, $B \to a \mid b \mid d$ and $D \to a$.

29. Consider the context-free grammar $G = (V_N, T, P, Q_0)$ given by productions $Q_0 \to AB \mid \wedge$, $A \to DE \mid a$, $B \to b \mid \wedge$, $D \to b$ and $E \to a$. Reduce the grammar G into Greibach normal form.

30. Show by using pumping lemma that the language $L = \{a^n b^n c^m \mid m \neq n\}$ is not context-free.

31. Prove that the following languages are not context-free. Use pumping lemma to support your answer.

    i. $L = \{a^k b^l c^k a^l \,|\, k \geq 1, l \geq 1\}$

    ii. $L = \{0^i 1^j 2^k \,|\, i < j < k\}$

32. Show that the set of all strings over $\{a, b, c\}$ in which the number of occurrences of a, b and c is the same is not context-free.

33. Prove that the language $L = \{a^i b^j c^k \,|\, 0 \leq i \leq j \leq k\}$ is not context-free.

34. Show that the following languages are not context-free.

    (a) $L = \{a^p \,|\, p$ is a prime number$\}$

    (b) $L = \{a^{k^2} \,|\, k$ is an integer$\}$

    (c) $L = \{a^k b^l \,|\, l = k^2\}$

35. Determine whether the string $z = bbaaaa$ is in the language generated by the grammar $Q_0 \rightarrow AB, A \rightarrow BB|a, B \rightarrow AB|b$. Use CYK algorithm to support your answer.

36. Show that the string $z = 22211$ can be generated by the grammar $Q_0 \rightarrow AB$, $A \rightarrow BC|0|1, B \rightarrow BB|2|1$ and $C \rightarrow 2$. Verify the solution by using CYK algorithm.

37. Obtain a context-free grammar generating the language $L = \{0^n 1^n 2^m \,|\, n \geq 1, m \geq 0\}$.

38. Construct a grammar G such that $L(G) = \{a^k ba^l \,|\, k, l \geq 1\}$.

39. Construct the context-free grammar G that accepts the set $L = \{a^n b^{2n} \,|\, n \geq 1\}$.

40. Let L be the language corresponding to the regular expression $(abb + b)^* (ba)^*$. Construct a context-free grammar G generating L.

41. Let G be a context-free grammar having productions $Q_0 \rightarrow Q_0 * Q_0 | Q_0 + Q_0 | 0 | 1$. Show that $0 * 1 + 1 * 0$ is in $L(G)$.

42. Construct a context-free grammar generating the language $L = a^m b^n c^k$, where $n = m$ or $n \leq k$.

43. Obtain a context-free grammar generating the language $L = \{a^m b^n \,|\, m \leq 2n\}$.

44. Prove that the language generated by the grammar G having productions $Q_0 \rightarrow AB, A \rightarrow BC|a, B \rightarrow CC|b, C \rightarrow a$ is infinite.

45. Let G be a context-free grammar having productions $Q_0 \to XY$, $X \to XX|a$ and $Y \to YY|b$. Find the language in terms of regular expressions.

46. Consider the context-free grammar given by productions $Q_0 \to 0XQ_0|X\gamma$, $X \to YZ|1, Y \to XX|0$. State whether the grammar produces any word. If not, justify your answer.

47. Construct context-free grammars for the following regular expressions.

   (a) a^*b^*

   (b) $(abb + baa)^*$

   (c) $a^n b^n; n \geq 0$

48. Find the language generated by the context-free grammar $G = (\{Q_0, A\}, \{0, 1\}, \{Q_0 \to 0A0AA, A \to 1A|0A|A\}, Q_0)$.

49. Let $G = (V_N, T, P, Q_0)$ be a context-free grammar, where $P = \{Q_0 \to AB, \ A \to BE|BC|BQ_0|a, \ B \to b|BQ_0, \ E \to a|b, C \to AQ_0\}$. Reduce the context-free grammar G into Greibach normal form.

50. Show by using pumping lemma that the following languages are not context-free.

   (a) $L = \{0^n 1^n 2^m \mid n \leq m \leq 2n\}$

   (b) $L = \{a^p b^q a^r \mid q = Max\{p, r\}\}$

6

PUSHDOWN AUTOMATA

6.0 INTRODUCTION

We have studied that the finite automaton accepts the regular languages but not every context-free languages. This is because some context-free languages are not regular. Intuitively, we understand that this is because finite automata have finite memories, whereas a context-free language may require storing an unbounded amount of information for its recognition. Therefore, it is important to recognize a context-free language for both practical and theoretical reasons. This suggests a new type of computation model with an extra component called stack. On introduction of stack, it provides additional memory beyond finite amount available in the control. This leads to a class of machines called pushdown automata.

In this chapter we discuss pushdown automata, acceptance of sets by pushdown automata, auxiliary and two-stack pushdown automata. We also explore the relation between pushdown automata and context-free languages.

6.1 PUSHDOWN AUTOMATA

In theory of computation, pushdown automata are a new type of computation model and are like nondeterministic finite automaton but have an extra component called stack. On introducing a stack, we get additional memory for processing context-free languages that require unbounded amount of information for its recognition. The main advantage of introducing a stack into a pushdown automaton is that it allows recognizing some non regular languages.

Analytically, a pushdown automaton is a seven-tuple $(S, \Sigma, \Gamma, \delta, s_0, z_0, A)$ consisting of

S = Finite non-empty set of states of the control unit

Σ = Finite non-empty set of input symbols called the alphabet.

Γ = Finite non-empty set of pushdown symbols called stack alphabet.

s_0 = Initial state of control unit, i.e., $s_0 \in S$.

z_0 = Initial symbol in the stack alphabet, i.e., $z_0 \in \Gamma$

A = Set of final or accepting states, i.e., $A \subseteq S$.

δ = Transition function from $S \times (\Sigma \cup \{\wedge\}) \times \Gamma \rightarrow$ finite subsets of $S \times \Gamma^*$

The transition function δ, explains that a pushdown automaton is in a state of S, takes an input symbol from $(\Sigma \cup \{\wedge\})$ when top of pushdown symbol is Γ, to make a transition to next state and writes a string denoted by Γ^* in top of stack alphabet.

6.1.1 Model of Pushdown Automata

Here we will study the components of a pushdown automaton and its operations. The different components are read-only input tape, input alphabet, finite state control, a set of final states, an initial state and a stack or pushdown store.

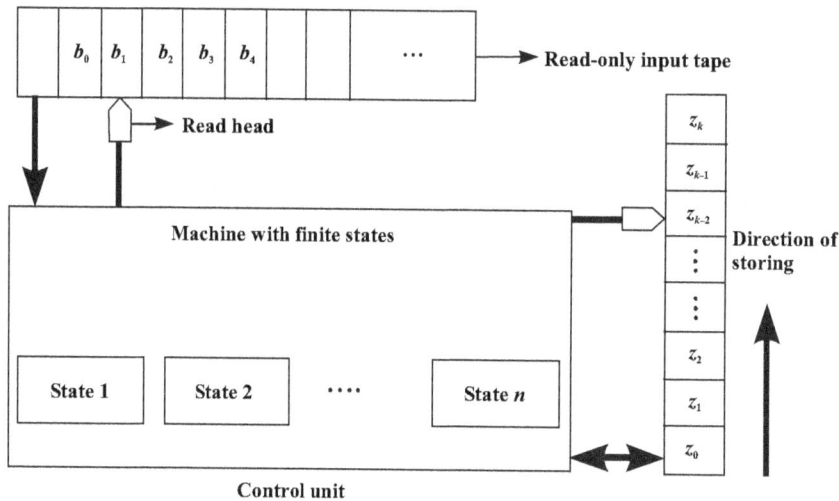

Figure 6.1 Model of pushdown automaton

In pushdown automata both pop (read) and push (write) operations are performed in pushdown store (stack). Inserting an input to the pushdown store is known as write operation whereas deleting an input from the pushdown store is known as read operation. The transition from one state to another in pushdown automaton is almost same as in case of finite automaton. The model of a pushdown automaton is illustrated in the Figure 6.1.

6.1.2 Deterministic Pushdown Automata

A pushdown automaton is said to be deterministic if for every input string there exists a unique path through the machine. It indicates that there exists at most one move for any instantaneous description (s, w, γ), where s is a state, w is a string of input symbols and γ is a string of stack symbols.

Analytically, a pushdown automaton $(S, \Sigma, \Gamma, \delta, s_0, z_0, A)$ is said to be deterministic if it satisfies the following conditions.

i. $\delta(s,a,z)$ is either empty or single move, and

ii. whenever $\delta(s,\wedge,z) \neq \phi$, then $\delta(s,a,z) = \phi$ for all $a \in \Sigma$.

For example, consider a pushdown automaton $P = (S,\Sigma,\Gamma,\delta,s_0,z_0,A)$ where $S = \{s_0,s_1,s_2\}, \Sigma = \{a,b,c\}, \Gamma = \{0,1,z_0\}, A = \{s_2\}$ and the transition δ is defined as:

$\delta(s_0,a,z_0) = \{(s_0,0z_0)\};$ $\delta(s_0,a,0) = \{(s_1,00)\};$ $\delta(s_0,a,1) = \{(s_1,10)\};$

$\delta(s_0,b,z_0) = \{(s_0,11)\};$ $\delta(s_0,b,0) = \{(s_0,01)\};$ $\delta(s_0,b,1) = \{(s_1,1)\};$

$\delta(s_1,c,z_0) = \{(s_1,0)\};$ $\delta(s_1,c,0) = \{(s_1,\wedge)\};$ $\delta(s_1,\wedge,1) = \{(s_2,z_0)\}$

It is clear that, all the transitions given above have single moves. Also, $\delta(s_1,a,1) = \phi$, $\delta(s_1,b,1) = \phi$, and $\delta(s_1,c,1) = \phi$. So, the given pushdown automaton is deterministic.

6.1.3   Nondeterministic Pushdown Automata

A pushdown automata is said to be nondeterministic if at certain time we may have to select a particular path among all possible paths. It indicates that for particular input we have at least two transitions to go to next state. Therefore, it is clear that a nondeterministic pushdown automaton accepts an input string if any sequence of choices leads to an accept state. If all possible sequence of choices for an input string leads to a reject state, then the string will be rejected. Here, we observed that nondeterministic pushdown automata have greater modelling power than deterministic pushdown automata. Again, deterministic pushdown automata have greater modelling power than finite automaton. The following Figure 6.2 represents the relative strength of nondeterministic pushdown automata, deterministic pushdown automata and finite automaton.

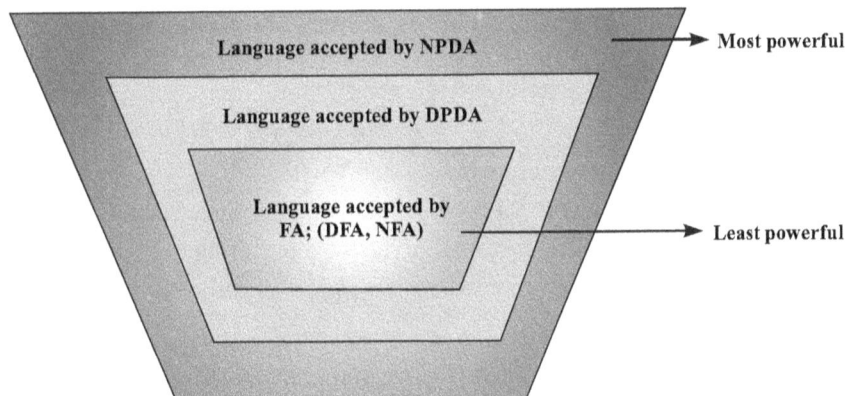

Figure 6.2   Relative strength of FA, DPDA, and NPDA

6.2   WORKING PRINCIPLE OF PUSHDOWN AUTOMATA

In this section we will discuss the working principle of a pushdown automaton that includes instantaneous descriptions and move relation.

6.2.1   Interpretation of Moves

The meaning of $\delta(s_0, a, z) = \{(s_1, \gamma_1), (s_2, \gamma_2), \cdots, (s_n, \gamma_n)\}$ where $s_0, s_1, s_2, \cdots, s_n \in S$, a is in Σ, z is a stack symbol and $\gamma_1, \gamma_2, \cdots, \gamma_n \in \Gamma^*$, is that the pushdown automata is in state s_0, with input symbol a and z the top symbol on the pushdown store (stack) can enter state s_i for any i and replace the symbol z by γ_i. It also advances the input head one symbol.

The meaning of $\delta(s_0, \wedge, z) = \{(s_1, \gamma_1), (s_2, \gamma_2), \cdots, (s_n, \gamma_n)\}$ where $s_i \in S, 0 \le i \le n; z$ is a stack symbol and $\gamma_1, \gamma_2, \cdots, \gamma_n \in \Gamma^*$ is that the pushdown automata is in state s_0, independent of the input symbol and z the top symbol on the stack can enter state s_i for any $1 \le i \le n$ and replace the symbol z by $\gamma_i, 1 \le i \le n$. In such case the input head is not advanced. In the above interpretations if $\gamma_i = \wedge$, then the topmost symbol z is erased.

6.2.2   Instantaneous Descriptions

Let $M = (S, \Sigma, \Gamma, \delta, s_0, z_0, A)$ be a pushdown automata. Thus we define an instantaneous description (ID) to be a triple (s, x, γ), where s is a state, x is a string of input symbols and γ is a string of stack symbols, i.e., $s \in S, x \in \Sigma^*$ and $\gamma \in \Gamma^*$. For example, consider an instantaneous description $(s_1, b_1 b_2 ... b_n, z_1 z_2 ... z_m)$. It describes that the pushdown automata is in state s_1 and the string to be processed in order is $b_1 b_2 \cdots b_n$. The stack has symbols $z_1, z_2, \cdots, z_m$ with z_1 at the top, z_2 is the second element from the top and so on.

6.2.3   Move Relation

Let $M = (S, \Sigma, \Gamma, \delta, s_0, z_0, A)$ be a pushdown automata. If the transition $\delta(s, a, z)$ contains (s', β), i.e., $(s', \beta) \in \delta(s, a, z)$, then the move relation between instantaneous descriptions is defined as

$$(s, ax, z\alpha) \vdash (s', x, \beta\alpha)$$

i.e., $(s, a_1 a_2 \cdots a_n, \alpha_1 \alpha_2 \cdots \alpha_n) \vdash (s', a_2 \cdots a_n, \beta\alpha_2 \cdots \alpha_n)$ if $\delta(s, a_1, \alpha_1)$ contains (s', β). Therefore it is clear that, the pushdown automata is in state s with $\alpha_1 \alpha_2 \alpha_3 \cdots \alpha_n$ in stack reads the input symbol a_1. If $\delta(s, a_1, \alpha_1)$ contains (s', β), then the pushdown moves to a state s' and write (push) β on the top of $\alpha_2 \alpha_3 \cdots \alpha_n$. As soon as the transition is over, the input string to be processed is $a_2 a_3 \cdots a_n$. A pictorial representation of the move relation when $\beta = \beta_1 \beta_2 \cdots \beta_m$ is given in Figure 6.3.

The symbol $\vdash$ represents a move from one instantaneous description to another. We use $\vdash^*$ for the reflexive and transitive closure of $\vdash$. The reflexive and transitive closure $\vdash^*$ represents a definite sequence of m-moves, where $m \ge 0$ is an integer.

If $(s, x, \alpha) \vdash^* (s', y, \beta)$ represents m-moves, then we write $(s, x, \alpha) \vdash^m (s', y, \beta)$. Therefore, it is clear that $(s, x, \alpha) \vdash^0 (s, x, \alpha)$. Also, we can write $(s, x, \alpha) \vdash^m (s', y, \beta)$ as $(s, x, \alpha) \vdash (s_1, x_1, \alpha_1) \vdash (s_2, x_2, \alpha_2) \vdash \cdots \vdash (s', y, \beta)$ for some $x_1, x_2, x_3, x_4, \cdots \in \Sigma^*$ and $\alpha_1, \alpha_2, \cdots \in \Gamma^*$.

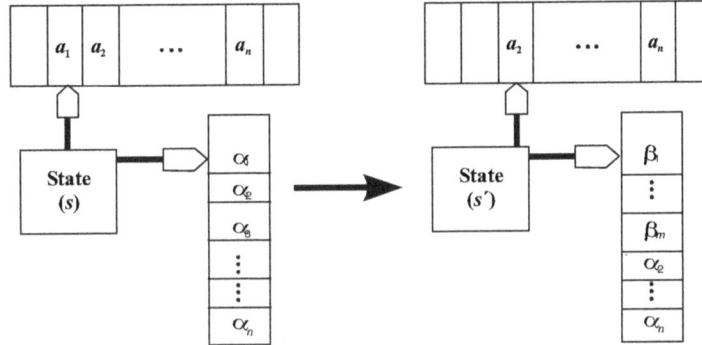

Figure 6.3 Move relation

6.2.4   Properties of Move Relation

In this section we prove two important properties of move relation that are frequently used in construction of pushdown automata.

Property 1 If $(s, x, \alpha) \vdash^* (s', \wedge, \beta)$, then for every $y \in \Sigma^*$, we have $(s, xy, \alpha) \vdash^* (s', y, \beta)$. Conversely, if $(s, xy, \alpha) \vdash^* (s', y, \beta)$ for some $y \in \Sigma^*$, then $(s, x, \alpha) \vdash^* (s', \wedge, \beta)$.

Proof Assume that $(s, x, \alpha) \vdash^* (s', \wedge, \beta)$ (i)

It indicates that, the pushdown automaton is in state s with α in stack moves to state s' with β in stack on processing the string 'x'. The same transition is effected by starting with the input string 'xy' and processing only 'x'. Therefore, the string to be processed is 'y' and hence we get $(s, xy, \alpha) \vdash^* (s', y, \beta)$.

Similarly, the converse part can be proved.

Property 2   If $(s, x, \alpha) \vdash^* (s', \wedge, \beta)$, then for every $\gamma \in \Gamma^*$, $(s, x, \alpha\gamma) \vdash^* (s', \wedge, \beta\gamma)$.

Proof Assume that $(s, x, \alpha) \vdash^* (s', \wedge, \beta)$. The sequence of move relation given above can be split as

$$(s, x, \alpha) \vdash (s_1, x_1, \alpha_1) \vdash (s_2, x_2, \alpha_2) \vdash (s_3, x_3, \alpha_3) \vdash \cdots \vdash (s', \wedge, \beta)$$

Let us assume, $(s_i, x_i, \alpha_i) \vdash (s_{i+1}, x_{i+1}, \alpha_{i+1})$ and $\alpha_i = z_1 z_2 z_3 \cdots z_n$. Therefore, according to the above move z_1 will be replaced by some string. It indicates that $z_2 z_3 \cdots z_n$ will not be affected. It implies that, if we place γ below $z_2 z_3 \cdots z_n$, then $z_2 z_3 \cdots z_n \gamma$ will also not be affected. Thus we get, $(s_i, x_i, \alpha_i \gamma) \vdash (s_{i+1}, x_{i+1}, \alpha_{i+1} \gamma)$. Therefore, we get the following sequence

$$(s, x, \alpha\gamma) \vdash (s_1, x_1, \alpha_1\gamma) \vdash (s_2, x_2, \alpha_2\gamma) \vdash (s_3, x_3, \alpha_3\gamma) \vdash \cdots \vdash (s', \wedge, \beta\gamma)$$

i.e., $(s, x, \alpha\gamma) \vdash^* (s', \wedge, \beta\gamma)$

Note It is to be noted that the converse is not true. In order to show this consider the pushdown automata $M = (\{s_0\}, \{0,1\}, \{z_0\}, \delta, s_0, z_0, \phi)$ such that $\delta(s_0, 1, z_0) = \{(s_0, \wedge)\}$ and $\delta(s_0, 0, z_0) = \{(s_0, z_0 z_0)\}$. Therefore, we have

$$(s_0, 110, z_0 z_0 z_0 z_0) \vdash (s_0, 10, z_0 z_0 z_0)$$
$$\vdash (s_0, 0, z_0 z_0)$$
$$\vdash (s_0, \wedge, z_0 z_0 z_0)$$
$$\text{i.e.,} \quad (s_0, 110, z_0 z_0 z_0 z_0) \vdash (s_0, \wedge, z_0 z_0 z_0)$$

However, $(s_0, 110, z_0) \overset{*}{\vdash} (s_0, 10, \wedge)$. It indicates that the pushdown automata cannot make any more transition as the stack is empty. It shows that the converse of the above property is not true if we assume $\alpha = z_0 z_0 z_0, \beta = z_0 z_0$ and $\gamma = z_0$.

6.3 ACCEPTANCE OF STRINGS BY PUSHDOWN AUTOMATA

The acceptance of input strings by pushdown automata can be defined in two ways. This is because, a pushdown automata has final states and has also the additional structure stack. The two different ways by which a string can be accepted by pushdown automata are in terms of final states or in terms of empty stack. In the following subsection we discuss in detail the acceptance of strings by pushdown automata.

6.3.1 Acceptance by Final State

Let $M = (S, \Sigma, \Gamma, \delta, s_0, z_0, A)$ be a pushdown automaton. The language $L(M)$ accepted by final state is defined as $L(M) = \{w \in \Sigma^* : (s_0, w, z_0) \overset{*}{\vdash} (p, \wedge, \alpha)\}$ for some $p \in A$ and $\alpha \in \Gamma^*$.

For example, consider the pushdown automata $M = (S, \Sigma, \Gamma, \delta, s_0, z_0, A)$ such that $S = \{s_0, s_1\}$, $\Sigma = \{a, b\}$, $\Gamma = \{z_0, z_1, z_2\}$, $A = \{s_1\}$ and the transition function δ is given as $\delta(s_0, a, z_0) = \{(s_0, z_1 z_0)\}, \delta(s_0, b, z_0) = \{(s_0, z_2 z_0)\}, \delta(s_0, a, z_1) = \{(s_0, z_1 z_1)\}, \delta(s_0, b, z_1) = \{(s_0, z_2 z_1)\}$, $\delta(s_0, a, z_2) = \{(s_0, z_1 z_2)\} \delta(s_0, b, z_2) = \{(s_1, z_2 z_2)\} \delta(s_1, a, z_2) = \{(s_1, \wedge)\}$ and $\delta(s_1, b, z_0) = \{(s_1, z_0 z_2)\}$

Let us consider a string $w = bbaabb$. Therefore, we have

$$(s_0, bbaabb, z_0) \vdash (s_0, baabb, z_2 z_0)$$
$$\vdash (s_1, aabb, z_2 z_2 z_0)$$
$$\vdash (s_1, abb, z_2 z_0)$$
$$\vdash (s_1, bb, z_0)$$
$$\vdash (s_1, b, z_0 z_2)$$
$$\vdash (s_1, \wedge, z_0 z_2 z_2)$$

It implies that $(s_0, bbaabb, z_0) \overset{*}{\vdash} (s_1, \wedge, z_0 z_2 z_2)$ with s_1 as final state and $z_0 z_2 z_2 \in \Gamma^*$. Therefore, it is clear that the string $w = bbaabb$ is accepted by pushdown automata by final states.

6.3.2   Acceptance by Null Stack

Let $M = (S, \Sigma, \Gamma, \delta, s_0, z_0, A)$ be a pushdown automaton. The language $N(M)$ accepted by null stack or empty stack is defined as

$$N(M) = \{w \in \Sigma^* : (s_0, w, z_0) \overset{*}{\vdash} (p, \wedge, \wedge)\} \text{ for some } p \in S.$$

Therefore, it is clear that the set of final states is irrelevant when we consider the acceptance by null stack. Hence, we usually let the set of final states be the empty set, i.e., $A = \phi$.

For example, consider the pushdown automata $M = (S, \Sigma, \Gamma, \delta, s_0, z_0, \phi)$ such that $S = \{s_0, s_1, s_2\}$, $\Sigma = \{a, b\}$, $\Gamma = \{z_0, z_1, z_2\}$ and the transition function δ is defined as $\delta(s_0, a, z_0) = \{(s_1, z_0 z_1)\}$, $\delta(s_1, a, z_1) = \{(s_2, z_2 z_1)\}$, $\delta(s_1, a, z_0) = \{(s_1, z_1)\}$, $\delta(s_2, b, z_2) = \{(s_2, \wedge)\}$ and $\delta(s_2, b, z_1) = \{(s_2, \wedge)\}$.

Let us consider a string $w = a^3 b^3$. Therefore, we have

$$(s_0, a^3 b^3, z_0) \vdash (s_1, a^2 b^3, z_0 z_1)$$
$$\vdash (s_1, ab^3, z_1 z_1)$$
$$\vdash (s_2, b^3, z_2 z_1 z_1)$$
$$\vdash (s_2, b^2, z_1 z_1)$$
$$\vdash (s_2, b, z_1)$$
$$\vdash (s_2, \wedge, \wedge)$$

So, $(s_0, a^3 b^3, z_0) \overset{*}{\vdash} (s_2, \wedge, \wedge)$. Therefore, it is clear that the string $w = a^3 b^3$ is accepted by pushdown automata by null stack.

6.3.3   Equivalence of Acceptance by Final State and Null Stack

In this section, we show that the languages accepted by pushdown automata's by final state and by empty stack are equivalent. In the following theorem we prove that the set accepted by a pushdown automaton M_2 by final state is accepted by some pushdown automata M_1 by null stack.

Theorem *If $M_2 = (S, \Sigma, \Gamma, \delta, s_0, z_0, A)$ is a pushdown automaton accepting L by final state, then we can find a pushdown automaton M_1 accepting L by null stack, i.e., $L = L(M_2) = N(M_1)$.*

Proof Let $M_2 = (S, \Sigma, \Gamma, \delta, s_0, z_0, A)$ be a pushdown automaton accepting L by final state, i.e., $L = L(M_2)$.

Our aim is to construct a pushdown automaton M_1 to simulate M_2, with the option for M_1 to erase its stack whenever M_2 reaches a final state. Here we use a dead state s_d of M_1 to erase the stack whereas we use a bottom of stack z_0' for M_1. Therefore, M_1 does not accidentally accept if M_2 empties its stack without entering a final state.

Let us define M_1 as $M_1 = (S', \Sigma, \Gamma', \delta', q_0', z_0', \phi)$, where $S' = S \cup \{s_0', s_d\}$, $\Gamma' = \Gamma \cup \{z_0'\}$. The transition function δ' is defined by the rules R_1, R_2, R_3 and R_4.

$R_1 :\quad \delta'(s_0', \wedge, z_0') = \{(s_0, z_0 z_0')\}$

$R_2 :\quad \delta'(s, a, z) = \delta(s, a, z) \quad \forall \quad s \in S, a \in \Sigma \cup \{\wedge\}$ and $z \in \Gamma$

$R_3 :\quad \delta'(s, \wedge, z) = \delta(s, \wedge, z) \cup \{(s_d, \wedge)\} \quad \forall \quad z \in \Gamma' = \Gamma \cup \{z_0'\}$ and $s \in A$

$R_4 :\quad \delta'(s_d, \wedge, z) = \{(s_d, \wedge)\} \; \forall \; z \in \Gamma' = \Gamma \cup \{z_0'\}$

Rule 1 (R_1) causes M_1 to enter an instantaneous description of M_2 and the start symbol z_0 is placed on top of the stack or pushdown store. Rule 2 (R_2) allows M_1 to simulate M_2 until it reaches a final state of M_2. Once M_2 enters a final state, M_1 makes a guess whether the input string is exhausted or not. If the string is exhausted, then M_1 enters the dead state s_d by erasing its stack. Otherwise M_1 once again simulate M_2. Rule 4 (R_4) causes a $\wedge$ move. On using these $\wedge$ moves, M_1 erases all symbols on the stack.

Let $w \in L(M_2)$. Therefore, we have

$$(s_0, w, z_0) \overset{*}{\underset{M_2}{\vdash}} (s, \wedge, \alpha) \text{ for some } s \in A \text{ and } \alpha \in \Gamma^*$$

Now consider the pushdown automata M_1 with input w. Therefore, we get

$$(s_0', w, z_0') \overset{*}{\underset{M_1}{\vdash}} (s_0, w, z_0 z_0') \qquad [\text{By using } R_1]$$

Again by R_2, it is clear that every move of M_2 is a legal move for M_1. So, we get

$$(s_0, w, z_0) \overset{*}{\underset{M_1}{\vdash}} (s, \wedge, \alpha) \text{ for some } s \in A \text{ and } \alpha \in \Gamma^*$$

Also, if a pushdown automaton can make a sequence of moves from a given instantaneous description, then it can make the same sequence of moves from any instantaneous description obtained from the first by inserting a fixed string of stack symbols below the original stack symbols. Therefore, we get

$$(s_0', w, z_0') \overset{*}{\underset{M_1}{\vdash}} (s_0, w, z_0 z_0') \overset{*}{\underset{M_1}{\vdash}} (s, \wedge, \alpha z_0')$$

$$\overset{*}{\underset{M_1}{\vdash}} (s_d, \wedge, z_0') \qquad [\text{by } R_3]$$

$$\overset{*}{\underset{M_1}{\vdash}} (s_d, \wedge, \wedge) \qquad [\text{by } R_4]$$

$$\text{i.e.,} \quad (s_0', w, z_0') \overset{*}{\underset{M_1}{\vdash}} (s_d, \wedge, \wedge)$$

Therefore, we get $w \in N(M_1)$ and hence we have

$$L(M_2) \subseteq N(M_1) \tag{1}$$

Now, our aim is to show that $N(M_1) \subseteq L(M_2)$. Let $w \in N(M_1)$. Therefore, we have

$$(s_0', w, z_0') \overset{*}{\underset{M_1}{\vdash}} (s, \wedge, \wedge) \text{ for some } s \in S' \tag{2}$$

Also, the initial move of M_1 can be made only by using R_1. It indicates that the first move of equation (2) is $(s_0', w, z_0') \underset{M_1}{\vdash} (s_0, w, z_0 z_0')$. It is clear that z_0' in the stack can be erased only when M_1 enters s_d. Again M_1 enters s_d only when it reaches a final state s of A in an earlier step. Therefore, equation (2) can be written as

$$(s_0', w, z_0') = (s_0', \wedge w, z_0') \underset{M_1}{\vdash} (s_0, w, z_0 z_0') \overset{*}{\underset{M_1}{\vdash}} (s, \wedge, \alpha z_0') \overset{*}{\underset{M_1}{\vdash}} (s, \wedge, \wedge)$$

for some $s \in A$ and $\alpha \in \Gamma^*$.

Thus, it is clear that moves involved in the above instantaneous description are those induced by the moves of M_2. Since, z_0' is not a pushdown symbol in M_2, so z_0' lying at the bottom is not affected by these moves. Thus we have

$$(s_0', \wedge w, z_0') \overset{*}{\underset{M_2}{\vdash}} (s, \wedge, a) \text{ for some } s \in A \text{ and } \alpha \in \Gamma^*.$$

Therefore, $w \in L(M_2)$ and hence we have $N(M_1) \subseteq L(M_2)$ $\tag{3}$

So, from equations (1) and (3) we get $L(M_2) = N(M_1)$.

Theorem *If $M_1 = (S, \Sigma, \Gamma, \delta, s_0, z_0, \phi)$ is a pushdown automata accepting L by null stack, then we can find a pushdown automata M_2 accepting L by final state i.e., $L = N(M_1) = L(M_2)$. In other words, if L is $N(M_1)$ for some pushdown automata M_1, then there exists a pushdown automata M_2 such that L is $L(M_2)$.*

Proof Let $M_1 = (S, \Sigma, \Gamma, \delta, s_0, z_0, \phi)$ be a pushdown automata accepting L by null stack, i.e., $L = N(M_1)$.

Our aim is to construct a pushdown automaton M_2 to simulate M_1 and identify when M_1 empties its stack. The pushdown automaton M_2 enters an accepting state if and only if this occurs.

Let us define pushdown automaton M_2 as $M_2 = (S', \Sigma, \Gamma', \delta', s_0', z_0', A)$, where $S' = S \cup \{s_0', s_f\}, \Gamma' = \Gamma \cup \{z_0'\}$ and $A = \{s_f\}$. It is to be noted that s_0' and s_f are newly introduced states in S whereas z_0' is a new stack symbol of M_2. The transition function δ' is defined by the rules R_1, R_2 and R_3.

$R_1:$ $\delta'(s_0', \wedge, z_0') = \{(s_0, z_0 z_0')\}$
$R_2:$ $\delta'(s, a, z) = \delta(s, a, z)$ $\forall$ $s \in S, a \in \Sigma \cup \{\wedge\}$ and $z \in \Gamma$
$R_3:$ $\delta'(s, \wedge, z_0') = \{(s_f, \wedge)\}$ $\forall$ $s \in S$

By rule 1 (R_1), the pushdown automata M_2 enter the initial instantaneous description of M_1. Rule 1 (R_1) gives a null move, except that M_2 will have its own bottom of stack marker z'_0, which is below the symbols of pushdown automata M_1's stack. Once pushdown automata M_2 reaches an initial instantaneous description of M_1, rule 2 (R_2) allows to simulate moves of M_1. The rule 2 (R_2) is applied repeatedly until z'_0 is pushed to the top of stack. Once z'_0 is top of the stack, we use rule 3 (R_3). Rule 3 (R_3) causes M_2 to move a final state s_f by erasing z'_0 in stack, thereby accepting the input string w.

Let $w \in N(M_1)$. Therefore, by definition of M_1 we have

$$(s_0, w, z_0) \overset{*}{\underset{M_1}{\vdash}} (s, \wedge, \wedge) \text{ for some } s \in S.$$

Now consider the pushdown automata M_2 with input w. Therefore, by R_1 we have

$$(s'_0, w, z'_0) = (s'_0, \wedge w, z'_0) \overset{*}{\underset{M_2}{\vdash}} (s_0, w, z_0 z'_0) \tag{1}$$

Again by rule 2 (R_2), every move of M_1 is a legal move for M_2. Therefore, we get

$$(s_0, w, z_0 z'_0) \overset{*}{\underset{M_2}{\vdash}} (s, \wedge, z'_0) \qquad [\text{by } R_2]$$

$$\overset{*}{\underset{M_2}{\vdash}} (s_f, \wedge, \wedge) \qquad [\text{by } R_3]$$

Hence, on combining the above results we get from equation (1)

$$(s'_0, w, z'_0) \overset{*}{\underset{M_2}{\vdash}} (s_0, w, z_0 z'_0) \overset{*}{\underset{M_2}{\vdash}} (s_f, \wedge, \wedge).$$

Therefore, we have $w \in L(M_2)$ and hence $N(M_1) \subseteq L(M_2)$.

Now our aim is to show that $L(M_2) \subseteq N(M_1)$. In order to prove this, consider a string $w \in L(M_2)$. Therefore, by definition of M_2 we get

$$(s'_0, w, z'_0) \overset{*}{\underset{M_2}{\vdash}} (s_f, \wedge, \alpha) \text{ for } \alpha \in (\Gamma')^* \tag{2}$$

But M_2 can reach s_f only by application of rule 3. In order to apply rule 3, z'_0 must be the topmost symbol on the stack. As z'_0 is placed initially and thus there is no other element in stack when z'_0 is on the top. Therefore, $\alpha = \wedge$. Thus, equation 2 reduces to

$$(s'_0, w, z'_0) \overset{*}{\underset{M_2}{\vdash}} (s_f, \wedge, \wedge) \tag{3}$$

The intermediate steps of equation 3 are given as below.

$$(s_0', w, z_0') = (s_0', \wedge w, z_0') \underset{M_2}{\overset{*}{\vdash}} (s_0, w, z_0 z_0')$$

$$\underset{M_2}{\overset{*}{\vdash}} (s, \wedge, z_0') \quad \text{for some } s \in S$$

$$\underset{M_2}{\overset{*}{\vdash}} (s_f, \wedge, \wedge)$$

As we get $(s_0, w, z_0 z_0') \underset{M_2}{\overset{*}{\vdash}} (s, \wedge, z_0')$, by applying rule 2 repeatedly z_0' does not get affect at the bottom. Therefore, we get

$$(s_0, w, z_0) \underset{M_2}{\overset{*}{\vdash}} (s, \wedge, \wedge) \text{ for some } s \in S.$$

Again by construction of rule 2, we have $(s_0, w, z_0) \underset{M_1}{\overset{*}{\vdash}} (s, \wedge, \wedge)$.

Therefore, we get $w \in N(M_1)$ and hence $L(M_2) \subseteq N(M_1)$. Thus we have $N(M_1) \subseteq L(M_2)$ and $L(M_2) \subseteq N(M_1)$. Hence, $N(M_1) = L(M_2) = L$.

6.4   PUSHDOWN AUTOMATA AND CONTEXT-FREE LANGUAGES

There is a general relation between pushdown automata and context-free languages. In this section we show that for every context-free language there exists a nondeterministic pushdown automaton that accepts it. Also we prove that, the language accepted by any nondeterministic pushdown automata is context free.

6.4.1   Pushdown Automata for Context-free Language

We first prove that for every context-free language there is a nondeterministic pushdown automaton that accepts it. In this section we construct a nondeterministic pushdown automaton that can, carry out a leftmost derivation of any string in the language. For simplicity, we assume that the language is in Greibach normal form. The following theorem provides a construction of nondeterministic pushdown automata for context-free language.

Theorem *If L is a context-free language, then there exists a pushdown automata M accepting L by empty stack, i.e., $L = N(M)$.*

Proof   We assume that $\wedge$ is not in $L(G)$. Let $G = (V_N, T, P, Q_0)$ be a context-free grammar in Greibach normal form such that $L = L(G)$. We construct a pushdown automaton M as $M = (S, \Sigma, \Gamma, \delta, s_0, z_0, \phi)$ such that $S = \{s_0\}$, $\Sigma = T$, $\Gamma = V_N \cup T$ and $z_0 = Q_0$. The transition function δ is defined as

1. $\delta(s_0, \wedge, Q) = \{(s_0, \alpha) \mid Q \to \alpha \text{ is in } P\}$ $\forall Q \in V_N$
2. $\delta(s_0, a, a) = \{(s_0, \wedge)\}$ $\forall \ a \in T$

From the transition function it is clear that the pushdown symbols Γ are variables and terminals. If the pushdown automata reads a variable Q on the top of the stack, then it makes a null ($\wedge$) move by placing α, where α is the right-hand side of the Q-production after erasing Q. If the pushdown automata reads a terminal 'a' on stack and if it matches with the current input symbol, then the pushdown automata erases the terminal 'a' on the stack. Apart from these cases the pushdown automata halts.

Let $w \in L(G)$. Therefore, w is obtained by the following leftmost derivation.

$$Q_0 \Rightarrow u_1 Q_1 \alpha_1 \Rightarrow u_1 u_2 Q_2 \alpha_2 \alpha_1 \Rightarrow \quad \cdots \quad \Rightarrow w$$

It indicates that, M can empty the stack on application of input string w. The first move of M is by a $\wedge$ move corresponding to $Q_0 \Rightarrow u_1 Q_1 \alpha_1$. As a result the pushdown automata erases Q_0 and stores $u_1 Q_1 \alpha_1$. Then, we use rule 2 to erase the symbols of pushdown automata in u_1 by processing a prefix of w. After this, we will get Q_1 as the topmost symbol in the stack. Again we apply rule 1 by making a $\wedge$ move corresponding to the production $Q_1 \to u_2 Q_2 \alpha_2$. This erases Q_1 and stores $u_2 Q_2 \alpha_2$ above α_1. Proceeding in this manner, the pushdown automata empties the stack by processing the string w.

Now, our claim is $L(G) = N(M)$. Therefore, we have to show that $L(G) \subseteq N(M)$ and $N(M) \subseteq L(G)$. First we show that $L(G) \subseteq N(M)$.

Let $w \in L(G)$. Therefore, w can be derived by a leftmost derivation. We know that, any sentential form in a leftmost derivation is of the form $uQ\alpha$, where $u \in T^*, Q \in V_N$ and $\alpha \in (V_N \cup T)^*$. Now we prove the following result:

If $Q_0 \overset{*}{\Rightarrow} uQ\alpha$ by a leftmost derivation, then

$$(s_0, uv, Q_0) \overset{*}{\underset{M_1}{\vdash}} (s_0, v, Q\alpha) \ \forall \ v \in T^* \tag{1}$$

We prove this result by method of induction on the number of steps in the derivation of $uQ\alpha$. If $Q_0 \overset{0}{\Rightarrow} uQ$, then $u = \wedge, \alpha = \wedge$ and $Q_0 = Q$. As $(s_0, v, Q_0) \vdash (s_0, v, Q_0)$, there is a basis for induction. Assume that $Q_0 \overset{i+1}{\Rightarrow} uQ\alpha$ by a leftmost derivation. Therefore, it can be split as $Q_0 \overset{i}{\Rightarrow} u_1 Q_1 \alpha_1 \Rightarrow uQ\alpha$. If the Q_1-production is $Q_1 \to u_2 Q \alpha_2$, then we have $u = u_1 u_2$ and $\alpha = \alpha_2 \alpha_1$. As $Q_0 \overset{i}{\Rightarrow} u_1 Q_1 \alpha_1$, by induction hypothesis we have

$$(s_0, u_1 u_2 v, Q_0) \overset{*}{\vdash} (s_0, u_1 u_2 v, u_1 Q_1 \alpha_1) \overset{*}{\vdash} (s_0, u_2 v, Q_1 \alpha_1) \tag{2}$$

Since $Q_1 \rightarrow u_2 Q \alpha_2$ is a production in P, by using rule 1 we get $(s_0, \wedge, Q_1) \vdash (s_0, \wedge, u_2 Q \alpha_2)$. Therefore, we have

$$(s_0, u_2 v, Q_1 \alpha_1) \vdash (s_0, u_2 v, u_2 Q \alpha_2 \alpha_1) \vdash (s_0, v, Q \alpha_2 \alpha_1) \tag{3}$$

Thus, on combining equations (2) and (3) we get

$$(s_0, u_1 u_2 v, Q_0) \overset{*}{\vdash} (s_0, u_2 v, Q_1 \alpha_1) \overset{*}{\vdash} (s_0, v, Q \alpha_2 \alpha_1)$$

$$\text{i.e., } (s_0, uv, Q_0) \overset{*}{\vdash} (s_0, v, Q \alpha) \qquad [\because \ u = u_1 u_2 \text{ and } \alpha = \alpha_2 \alpha_1]$$

Hence, equation (1) is true for $Q_0 \overset{i+1}{\Rightarrow} uQ\alpha$. Now, we have to show that $L(G) \subseteq N(M)$.

Let $w \in L(G)$. So, w can be derived by a leftmost derivation. Therefore, we have

$$Q_0 \overset{*}{\vdash} uQv \Rightarrow uu'v = w$$

This derivation includes the Q-production $Q \rightarrow u'$. From equation (1) we have

$$(s_0, uu'v, Q_0) \vdash (s_0, uu'v, uQv) \qquad \text{[By rule 1]}$$
$$\vdash (s_0, u'v, Qv) \qquad \text{[By rule 2]}$$
$$\vdash (s_0, u'v, u'v) \qquad [\because Q \rightarrow u']$$
$$\vdash (s_0, v, v) \qquad \text{[By rule 2]}$$
$$\vdash (s_0, \wedge, \wedge) \qquad \text{[By rule 2]}$$

$$\text{i.e, } (s_0, w, Q_0) \overset{*}{\underset{M}{\vdash}} (s_0, \wedge, \wedge) \qquad [\because uu'v = w]$$

It implies that $w \in N(M)$ and hence $L(G) \subseteq N(M)$. Now, our aim is to show that $N(M) \subseteq L(G)$. In order to prove this, first we prove an auxiliary result:

$$Q_0 \overset{*}{\vdash} u\alpha \text{ if } (s_0, uv, Q_0) \overset{*}{\underset{M}{\vdash}} (s_0, v, \alpha) \tag{4}$$

We prove this by method of induction on the number of moves in $(s_0, uv, Q_0) \overset{*}{\underset{M}{\vdash}} (s_0, v, \alpha)$.

If $(s_0, uv, Q_0) \overset{0}{\underset{M}{\vdash}} (s_0, v, \alpha)$, then $u = \wedge$ and $Q_0 = \alpha$. Thus, obviously, $Q_0 \overset{0}{\Rightarrow} \wedge \alpha$. Therefore, there is a basis for induction. Assume that equation (4) holds for i number of moves. Assume that

$$(s_0, uv, Q_0) \overset{i+1}{\underset{M}{\vdash}} (s_0, v, \alpha) \tag{5}$$

Hence it is clear that, the last move is obtained either from $(s_0, \wedge, Q) \vdash (s_0, \wedge, \alpha')$ or $(s_0, a, a) \vdash (s_0, \wedge, \wedge)$. Consider the first case $(s_0, \wedge, Q) \vdash (s_0, \wedge, \alpha')$. Therefore, equation (5) can be split as

$$(s_0, uv, Q_0) \overset{i}{\underset{M}{\vdash}} (s_0, v, Q\alpha_2) \vdash (s_0, v, \alpha_1\alpha_2) = (s_0, v, \alpha)$$

This includes the productions $Q_0 \overset{*}{\Rightarrow} uQ\alpha_2$ and $Q_0 \rightarrow \alpha_1$. Therefore, $Q_0 \overset{*}{\Rightarrow} uQ\alpha_2$ implies $\alpha_1\alpha_2 = \alpha$. Therefore, we have $Q_0 \overset{*}{\Rightarrow} uQ\alpha_2 \Rightarrow u\alpha_1\alpha_2 = u\alpha$.

Consider the second case $(s_0, a, a) \vdash (s_0, \wedge, \wedge)$. Therefore, equation (5) can be split as $(s_0, uv, Q_0) \overset{i}{\underset{M}{\vdash}} (s_0, av, a\alpha) \vdash (s_0, v, \alpha)$.

Also, $u = u'a$ for some $u' \in \Sigma$. So, we have

$$(s_0, u'av, Q_0) \overset{i}{\underset{M}{\vdash}} (s_0, av, a\alpha) \Rightarrow Q_0 \overset{*}{\Rightarrow} u'a\alpha = u\alpha.$$

It indicates that equation (4) holds true for any number of moves in $(s_0, uv, Q_0) \overset{*}{\underset{M}{\vdash}} (s_0, v, \alpha)$.

Now, we show that $N(M) \subseteq L(G)$.

Let $w \in N(M)$. Therefore, we have $(s_0, w, Q_0) \overset{*}{\vdash} (s_0, \wedge, \wedge)$. On taking $u = w, v = \wedge$, $\alpha = \wedge$ and on applying result (4) we get $Q_0 \overset{*}{\Rightarrow} u\alpha = w \wedge = w$. So, $w \in L(G)$ and hence $N(M) \subseteq L(G)$. Therefore, it is clear that $L = N(M) = L(G)$.

6.4.2   Context-free Grammar for Pushdown Automata

In the previous section, we have shown that for every context-free grammar there exists an equivalent pushdown automaton. But the converse is also true. It indicates that the grammar simulates the moves of the pushdown automata. In this section, we show that for every pushdown automata there exists an equivalent context-free grammar. The following theorem establishes the above relation.

Theorem   *If $L = N(M)$ for some pushdown automata $M = (S, \Sigma, \Gamma, \delta, s_0, z_0, \phi)$, then L is a context-free language. In other words, if $M = (S, \Sigma, \Gamma, \delta, s_0, z_0, \phi)$ is a pushdown automata, then there exists a context-free grammar G such that $L(G) = N(M)$.*

Proof   Let $M = (S, \Sigma, \Gamma, \delta, s_0, z_0, \phi)$ be a pushdown automata accepting L by null stack. We define the grammar G equivalent to the pushdown automata M as $G = (V_N, T, P, Q_0)$, where

$$V_N = \{Q_0\} \cup \{[s, z, s'] \mid s, s' \in S, z \in \Gamma\}$$
$$T = \Sigma$$

The set of productions P are defined by the following rules R_1, R_2 and R_3.

R_1: The Q_0 productions are given by $Q_0 \rightarrow [s_0, z_0, s]$ $\forall$ $s \in S$.

R_2: Each move erasing a pushdown symbol given by $\delta(s, a, z) = (s', \wedge)$ induces the production $[s, z, s'] \rightarrow a$.

R_3: Each move not erasing a pushdown symbol of the form $\delta(s, a, z) = (s_1, z_1 z_2 z_3 \cdots z_n)$ induces many productions given by

$$[s, z, s'] = a[s_1, z_1, s_2][s_2, z_2, s_3][s_3, z_3, s_4] \cdots [s_n, z_n, s'],$$

where $s', s_2, s_3, s_4, \cdots, s_n$ can be any state in S. It implies that, each move yields many productions. Now our claim is $N(M) = L(G)$.

Before proving $N(M) = L(G)$, it is to be noted that a variable $[s, z, s']$ of pushdown automata is at current state s with top of the stack z. In the course of derivation, a state s' is chosen in such a manner that the stack is emptied ultimately. This corresponds to rule 2.

Now we prove an auxiliary result: $(s, z, s') \overset{*}{\underset{G}{\Rightarrow}} w$ if and only if $(s, w, z) \overset{*}{\vdash} (s', \wedge, \wedge)$.

We prove the if part by method of induction on the number of steps in $(s, w, z) \overset{*}{\vdash} (s', \wedge, \wedge)$. If $(s, w, z) \overset{*}{\vdash} (s', \wedge, \wedge)$, then w is either $a \in \Sigma$ or $\wedge$. Therefore, we have $(s', \wedge) \in \delta(s, w, z)$. Thus, by rule 2 ($R_2$) we get a production $[s, z, s'] \rightarrow w$. So, $[s, z, s'] \underset{G}{\Rightarrow} w$. Therefore, there exists a basis for induction. Let us assume that $(s, w, z) \overset{i}{\vdash} (s', \wedge, \wedge)$. This can be split as:

$$(s, aw', z) \vdash (s_1, w', z_1 z_2 \cdots z_n) \overset{i+1}{\vdash} (s', \wedge, \wedge) \tag{1}$$

where $w = aw'$ and $a \in \Sigma$ or $a = \wedge$ depending on the first move.

Now, consider the second part of equation (1). It indicates that the stack has $z_1 z_2 \cdots z_n$ initially and on application of w', the stack is emptied. Therefore, each move of pushdown automata either replaces the topmost symbol by some nonempty string or erases the topmost symbol on the stack. Thus, several moves are required for getting z_2 on the top of stack.

Let w_1 be the prefix of w' such that the stack has $z_2 z_3 \cdots z_n$ after application of w_1. We assume that $z_2 z_3 \cdots z_n$ are not distributed while applying w_1. Similarly, let w_k be the substring of w' such that the stack has $z_{k+1} z_{k+2} \cdots z_n$ on application of w_k. We assume that $z_{k+1} z_{k+2} \cdots z_n$ are not distributed while applying $w_1, w_2, \cdots, w_k$. A pictorial approach is as follows.

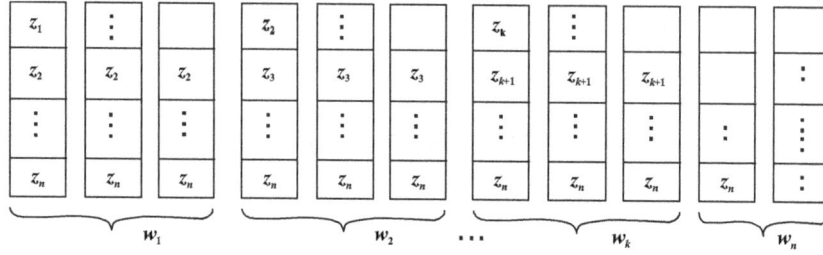

In terms of instantaneous descriptions, we get

$$(s_k, w_k, z_k) \overset{*}{\vdash} (s_{k+1}, \wedge, \wedge) \text{ for } k = 1, 2, \cdots, n \text{ and } s_{n+1} = s'.$$

Therefore, we get $[s_k, z_k, s_{k+1}] \overset{*}{\underset{G}{\vdash}} w_k$ for $k = 1, 2, \cdots, n$ \hfill (2)

From the first part of equation (1) we have $\delta(s, a, z) = (s_1, z_1 z_2 z_3 \cdots z_n)$. By using rule 3 ($R_3$) we get a production

$$[s, z, s'] \vdash a[s_1, z_1, s_2][s_2, z_2, s_3][s_3, z_3, s_4] \cdots [s_n, z_n, s']$$
$$\vdash a w_1 w_2 w_3 \cdots w_n$$
$$= a w' = w$$
$$\Rightarrow [s, z, s'] \underset{G}{\Rightarrow} w$$

Therefore, by method of induction this completes the proof of if part. Now we prove the only if part, i.e., if $(s, z, s') \overset{*}{\underset{G}{\Rightarrow}} w$, then $(s, w, z) \overset{*}{\vdash} (s', \wedge, \wedge)$. We prove this result by method of induction on the number of steps involved in the derivation of $(s, z, s') \overset{*}{\underset{G}{\Rightarrow}} w$. If $(s, z, s') \Rightarrow w$, then $(s, z, s') \to w$ is a production of P. This is obtained by rule 2 (R_2). So, $w = \wedge$ or $w \in \Sigma$ and $\delta(s, w, z) = (s', \wedge)$. This gives $(s, w, z) \vdash (s', \wedge, \wedge)$. Hence there exists a basis for induction.

Assume that the result for derivations is true where the number of steps involved is less than i. Now, consider $(s, z, s') \overset{i}{\Rightarrow} w$. It can be derived as

$$(s, z, s') \Rightarrow a[s_1, z_1, s_2][s_2, z_2, s_3][s_3, z_3, s_4] \cdots [s_n, z_n, s'] \overset{i-1}{\Rightarrow} w \quad (3)$$

As G is context free, we can write $w = a w_1 w_2 w_3 \cdots w_n$, where $[s_k, z_k, s_{k+1}] \overset{*}{\underset{G}{\Rightarrow}} w_k$ for $k = 1, 2, 3, \cdots, n$ and $s_{n+1} = s'$. Therefore, by induction hypothesis we get

$$(s_k, w_k, z_k) \overset{*}{\vdash} (s_{k+1}, \wedge, \wedge) \text{ for } k = 1, 2, 3, \cdots, n.$$

Thus we get $(s_k, w_k, z_k z_{k+1} z_{k+2} \cdots z_n) \overset{*}{\vdash} (s_{k+1}, \wedge, z_{k+1} z_{k+2} \cdots z_n)$ (4)

and $(s_k, w_k w_{k+1} \cdots w_n, z_k z_{k+1} z_{k+2} \cdots z_n) \overset{*}{\vdash} (s_{k+1}, w_{k+1} \cdots w_n, z_{k+1} z_{k+2} \cdots z_n)$ (5)

On combining moves (4) and (5) we get

$$(s_1, w_1 w_2 \cdots w_n, z_1 z_2 \cdots z_n) \overset{*}{\vdash} (s', \wedge, \wedge)$$ (6)

The first step of equation (3), i.e., $(s, z, s') \Rightarrow a[s_1, z_1, s_2][s_2, z_2, s_3] \cdots [s_n, z_n, s']$ is induced by the transition $\delta(s, a, z) = (s_1, z_1 z_2 z_3 \cdots z_n)$. The corresponding move is given as $(s, a, z) \vdash (s_1, \wedge, z_1 z_2 z_3 \cdots z_n)$. It implies that

$$(s, a w_1 w_2 \cdots w_n, z) \vdash (s_1, w_1 w_2 \cdots w_n, z_1 z_2 z_3 \cdots z_n)$$ (7)

Therefore, from equations (7) and (6) we get

$$(s, a w_1 w_2 \cdots w_n, z) \vdash (s_1, w_1 w_2 \cdots w_n, z_1 z_2 \cdots z_n)$$
$$\vdash (s', \wedge, \wedge)$$
$$\text{i.e.,} \quad (s, w, z) \overset{*}{\vdash} (s', \wedge, \wedge)$$

By method of induction, this completes the proof of only if part, i.e., if $(s, z, s') \Rightarrow w$, then $(s, w, z) \overset{*}{\vdash} (s', \wedge, \wedge)$.

Finally, we proved the auxiliary result, i.e., $(s_0, z_0, s') \overset{*}{\Rightarrow} w$ if and only if $(s_0, w, z_0) \overset{*}{\vdash} (s', \wedge, \wedge)$. Now we show that $N(M) = L(G)$.

Let $w \in L(G)$. Therefore, we have $Q_0 \overset{*}{\Rightarrow} w$

i.e., $Q_0 \rightarrow [s_0, z_0, s'] \overset{*}{\Rightarrow} w$ [by rule 1 for some s']

i.e., $(s_0, w, z_0) \overset{*}{\vdash} (s', \wedge, \wedge)$ [by auxiliary result]

i.e., $w \in N(M)$

Thus, $N(M) = L(G)$. This completes the proof.

Corollary If M be a pushdown automata, then there exists a context-free grammar G such that $L(M) = L(G)$.

Proof Let M be a pushdown automaton. Then we can find an equivalent automaton M_1 such that $L(M) = N(M_1)$. By previous theorem, we can construct a grammar G such that $N(M_1) = L(G)$. Thus, we have $L(M) = L(G)$.

6.5   AUXILIARY PUSHDOWN AUTOMATA

An auxiliary pushdown automaton is a pushdown automaton that has two-way input with additional general purpose storage. The general purpose storage is in the form of a space bounded Turing tape. The interesting and important characteristic of auxiliary pushdown automata is that for a fixed amount of extra storage, the deterministic and nondeterministic versions are equivalent in language recognizing power, and the class of languages accepted by auxiliary pushdown automata with a given space bound is equivalent to the class of languages accepted by Turing machines of time complexity exponential in that space bound.

The block diagram of an auxiliary pushdown automaton is given below in Figure 6.4. It consists of a read-only input tape with end markers $\mathcal{c}$ and $\mathcal{s}$, a finite state control, a read-write storage tape of length $S(n)$, where n is the length of the input string w and a stack or pushdown store.

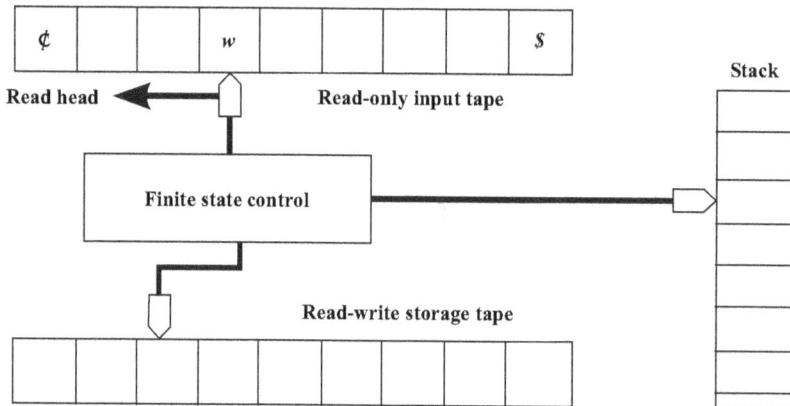

Figure 6.4 Auxiliary pushdown automata

A transition in an auxiliary pushdown automaton is decided by the state of the finite control, along with the symbols scanned by the read head of input tape, storage tape and stack. In one transition, an auxiliary pushdown automaton may do any or all of the following:

(*a*) Change state

(*b*) Move its input read head one position left or right, but not off the input.

(*c*) Write a symbol on the cell scanned by the storage head and move that head one position either left or right.

(*d*) Push a symbol on to the stack or pop the symbol off the stack.

In case of nondeterministic auxiliary pushdown automaton, it has a finite number of alternatives of the above type. Initially the tape heads of read-only input tape and read-write storage tape are at the left end, with the finite state control in a designated initial state and the stack consisting of an appropriate start symbol.

It is discovered, that the deterministic and nondeterministic auxiliary pushdown automata's with the same space bound are equivalent. Also $S(n)$, space on an auxiliary pushdown automata is equivalent to $kS(n)$ time on a Turing machine for some constant k. It leads to following three equivalent statements and its proof of equivalence is beyond the scope of this book.

1. L is accepted by a deterministic $S(n)$ auxiliary pushdown automaton.

2. L is accepted by a nondeterministic $S(n)$ auxiliary pushdown automaton.

3. L is in DTIME ($kS(n)$) for some constant k, where DTIME is the time complexity.

6.6   TWO-STACK PUSHDOWN AUTOMATA

If we allow a finite automaton access to two stacks instead of just one, we obtain a more powerful device equivalent in power to a Turing machine, which we will study in the next chapter.

A two-stack pushdown automaton is just like a pushdown automaton except that it has two stack, $STACK_1$ and $STACK_2$. Since the pushdown automaton contains two stacks, we have to specify the stack while pushing a symbol 'a' onto a stack. Generally, we use $PUSH_1$ a or $PUSH_2$ a to push a symbol 'a' onto $STACK_1$ or $STACK_2$ respectively. Similarly, if we wish to pop a symbol 'a' from a stack for the purpose of branching, we have to specify the stack. Generally, we use POP_1 a and POP_2 a to pop a symbol from $STACK_1$ and $STACK_2$ respectively. The input string is placed on the read-only input tape. The main and important difference is that we shall insist that a two-stack pushdown automata be deterministic, i.e., branching will only occur at the READ and POP states and there will be at most one edge from any state for any given symbol. We define the transition function δ of two-stack pushdown automaton as:

$$\delta : \quad S \times (\Sigma \cup \{\wedge\}) \times \Gamma \times \Gamma \rightarrow \text{finite subsets of } (S \times \Gamma^* \times \Gamma^*)$$

The transition function states that the move in a two-stack pushdown automata depends on the tops of the two stacks and pushes new values to two-stacks. This two-stack pushdown automaton is otherwise known as two-pushdown stack machine or two-pushdown store machine. More details on two stack pushdown automaton are beyond the scope of this book.

$$\equiv \text{SOLVED EXAMPLES} \equiv$$

Example 1

Consider the pushdown automaton $M = (\{s_0, s_1\}, \{0,1\}, \{0, z_0\}, \ \delta, s_0, z_0, \phi)$ where δ is defined as:

1. $\delta(s_0, 0, z_0) = \{(s_0, 0z_0)\}$ 2. $\delta(s_0, 0, 0) = \{(s_0, 00)\}$ 3. $\delta(s_0, 1, 0) = \{(s_1, \wedge)\}$
4. $\delta(s_1, 1, 0) = \{(s_1, \wedge)\}$ and   5. $\delta(s_1, \wedge, z_0) = \{(s_1, \wedge)\}$

Show that the string $0^n 1^n, n > 0$ is accepted by null stack.

Solution   Here, s_0 is the initial state with z_0 as the top of the stack. Therefore, we start with instantaneous description representing s_0 as initial state and z_0 as top of the stack to process the string $0^n1^n, n > 0$. Thus, we get

$$
\begin{aligned}
(s_0, 0^n1^n, z_0) &\vdash (s_0, 0^{n-1}1^n, 0z_0) && \text{[by 1]} \\
&\vdash (s_0, 0^{n-2}1^n, 0^2z_0) && \text{[by 2]} \\
&\overset{n-2}{\vdash} (s_0, 1^n, 0^nz_0) && \text{[by 2]} \\
&\vdash (s_1, 1^{n-1}, 0^{n-1}z_0) && \text{[by 3]} \\
&\overset{n-1}{\vdash} (s_1, \wedge, z_0) && \text{[by 4]} \\
&\vdash (s_1, \wedge, \wedge) && \text{[by 5]}
\end{aligned}
$$

Therefore, we conclude that the string $w = 0^n1^n, n > 0$ is accepted by pushdown automaton by null stack.

Example 2

Show that the string $0^21^20^2$ is accepted by the following pushdown automaton by null stack. Given pushdown automaton is $M = (\{s_1, s_2\}, \{0,1\}, \{z_0, z_1, z_2\}, \delta, s_1, z_0, \phi)$ where δ is defined as:

1. $\delta(s_1, 0, z_0) = \{(s_1, z_1z_0)\}$,   2. $\delta(s_1, 0, z_1) = \{(s_1, z_1z_1)\}$,

3. $\delta(s_1, 1, z_1) = \{(s_1, z_2z_1)\}$,   4. $\delta(s_1, 1, z_2) = \{(s_2, \wedge)\}$,

5. $\delta(s_2, 0, z_1) = \{(s_2, \wedge)\}$,   6. $\delta(s_2, \wedge, z_0) = \{(s_2, \wedge)\}$

Solution   Here, s_1 is the initial state with z_0 as the top of the stack. Therefore, we start with instantaneous description representing s_1 as initial state and z_0 as top of the stack to process the string $0^21^20^2$. Therefore, we have

$$
\begin{aligned}
(s_1, 0^21^20^2, z_0) &\vdash (s_1, 01^20^2, z_1z_0) && \text{[By 1]} \\
&\vdash (s_1, 1^20^2, z_1z_1z_0) && \text{[By 2]} \\
&\vdash (s_1, 10^2, z_2z_1z_1z_0) && \text{[By 3]} \\
&\vdash (s_2, 0^2, z_1z_1z_0) && \text{[By 4]} \\
&\vdash (s_2, 0, z_1z_0) && \text{[By 5]} \\
&\vdash (s_2, \wedge, z_0) && \text{[By 5]} \\
&\vdash (s_2, \wedge, \wedge) && \text{[By 6]}
\end{aligned}
$$

Therefore, we conclude that the string $w = 0^21^20^2$ is accepted by pushdown automaton by null stack.

Example 3

Show that the string 0112110 is accepted by final state. Consider the pushdown automaton $M = (\{s_0, s_1, s_2\}, \{0,1,2\}, \{0,1,z_0\}, \delta, s_0, z_0, \{s_2\})$ where the transition δ is defined as:

1. $\delta(s_0, 0, z_0) = \{(s_0, 0z_0)\};$ 2. $\delta(s_0, 0, 0) = \{(s_0, 00)\};$

3. $\delta(s_0, 0, 1) = \{(s_0, 01)\};$ 4. $\delta(s_0, 2, 0) = \{(s_1, 0)\};$

5. $\delta(s_0, 2, z_0) = \{(s_1, z_0)\};$ 6. $\delta(s_1, 1, 1) = \{(s_1, \wedge)\};$

7. $\delta(s_0, 1, z_0) = \{(s_0, 1z_0)\};$ 8. $\delta(s_0, 1, 0) = \{(s_0, 10)\};$

9. $\delta(s_0, 1, 1) = \{(s_0, 11)\}$ 10. $\delta(s_0, 2, 1) = \{(s_1, 1)\}$

11. $\delta(s_1, 0, 0) = \{(s_1, \wedge)\}$ 12. $\delta(s_1, \wedge, z_0) = \{(s_2, z_0)\}$

Solution   In this case, s_0 is the initial state whereas z_0 is the top symbol of the stack. The final state of the machine is given as s_2. Therefore, we have

$$(s_1, 0112110, z_0) \vdash (s_0, 112110, 0z_0)$$
$$\vdash (s_0, 12110, 10z_0)$$
$$\vdash (s_0, 2110, 110z_0)$$
$$\vdash (s_1, 110, 110z_0)$$
$$\vdash (s_1, 10, 10z_0)$$
$$\vdash (s_1, 0, 0z_0)$$
$$\vdash (s_1, \wedge, z_0)$$
$$\vdash (s_2, \wedge, z_0)$$

i.e., $(s_0, 0112110, z_0) \overset{*}{\vdash} (s_2, \wedge, z_0)$

It implies that the string $w = 0112110$ is accepted by pushdown automata by final state.

Example 4

Construct a pushdown automaton that accepts the following language L on $\Sigma = \{a, b\}$ by empty

$$L = \left\{ ww^R \,\middle|\, w \in \Sigma^+ \right\}$$

Solution   Here we use the fact that the symbols are retrieved from a stack in reverse order of their insertion. While reading the first part w of the string, we push consecutive symbols on the stack. For the second part w^R of the string, we compare the current input symbol with the top of the stack and then retrieve the top symbol from the stack. This process is repeated as long as the two matches occur. As the symbols are retrieved from the stack in the reverse order in which they were inserted, a complete match will be obtained if and only if the input string is of the form ww^R.

Now we formally define the pushdown automaton $M = (S, \Sigma, \Gamma, \delta, s_0, z_0, \phi)$ given by $S = \{s_0, s_1, s_2\}$, $\Sigma = \{a, b\}$ and $\Gamma = \{a, b, z_0\}$. The transition function δ can be visualized into several parts as below.

1. A set of transitions to push '*w*' into the stack:

    $\delta(s_0, a, z_0) = \{(s_0, az_0)\}; \quad \delta(s_0, a, a) = \{(s_0, aa)\}; \quad \delta(s_0, a, b) = \{(s_0, ab)\};$
    $\delta(s_0, b, z_0) = \{(s_0, bz_0)\}; \quad \delta(s_0, b, a) = \{(s_0, ba)\}; \quad \delta(s_0, b, b) = \{(s_0, bb)\}.$

2. A set of transitions to guess the middle of the string, where the pushdown automaton changes its state from s_0 to s_1:

    $\delta(s_0, \wedge, a) = \{(s_1, a)\}; \quad \delta(s_0, \wedge, b) = \{(s_1, b)\}$

3. A set of transitions to match w^R against the contents of the stack:

    $\delta(s_1, a, a) = \{(s_1, \wedge)\}; \quad \delta(s_1, b, b) = \{(s_1, \wedge)\}$

4. Finally, the transition to make a recognize match:

    $\delta(s_1, \wedge, z_0) = \{(s_2, \wedge)\}$

For example, consider the string '$w = bbaabb$'. The sequence of moves in accepting the string $w = bbaabb$ is given below.

$$(s_0, bbaabb, z_0) \vdash (s_0, baabb, bz_0) \vdash (s_0, aabb, bbz_0)$$
$$\vdash (s_0, abb, abbz_0) \vdash (s_1, abb, abbz_0)$$
$$\vdash (s_1, bb, bbz_0) \vdash (s_1, b, bz_0)$$
$$\vdash (s_1, \wedge, z_0) \vdash (s_2, \wedge, \wedge)$$

Therefore, it is clear that the string $w = bbaabb$ is accepted by the pushdown automaton by null stack.

Example 5

Construct a pushdown automaton that accepts the language $L = \left\{0^n 1^{2n} \mid n \geq 0\right\}$ on $\Sigma = \{0, 1\}$ by final state.

Solution   Given that the language $L = \left\{0^n 1^{2n} \mid n \geq 0\right\}$. From the language it is clear that the number of 1's is twice in number of 0's. Also, all 0's are followed by all 1's. When reading 0's of the string, we push two symbols each time into the stack whereas we pop one symbol each time while reading 1's.

Now we formally define the pushdown automaton $M = (S, \Sigma, \Gamma, \delta, s_0, z_0, A)$ given by $S = \{s_0, s_1, s_2\}$, $\Sigma = \{0, 1\}$, $A = \{s_2\}$ and $\Gamma = \{a, z_0\}$. The transition function δ can be visualized into several parts as below.

1. The transition to accept a null string:

    $\delta(s_0, \wedge, z_0) = \{(s_2, z_0)\}$

2. A set of transitions to push a's into the stack while reading 0's, where the pushdown automaton changes its state from s_0 to s_1:

    $\delta(s_0, 0, z_0) = \{(s_1, aaz_0)\}; \; \delta(s_1, 0, a) = \{(s_1, aaa)\};$

3. A set of transitions to pop a's from the stack while reading 1's:

    $\delta(s_1, 1, a) = \{(s_1, \wedge)\}$

4. Finally, the transition to make a recognize match:

    $\delta(s_1, \wedge, z_0) = \{(s_2, z_0)\}$

For example, consider the string '$w = 011$'. The sequence of moves in accepting the string $w = 011$ is given below.

$$(s_0, 011, z_0) \vdash (s_1, 11, aaz_0)$$
$$\vdash (s_1, 1, az_0) \;\; \vdash (s_1, \wedge, z_0) \;\; \vdash (s_2, \wedge, z_0)$$

Therefore, it is clear that the string $w = 011$ is accepted by the pushdown automaton by final state.

Example 6

Construct a pushdown automaton to accept a given language L by empty store, where $L = \{0^n 1^{2n} \mid n \geq 1\}$.

Solution　　Given that the language $L = \{0^n 1^{2n} \mid n \geq 1\}$. From the language it is clear that the number of 1's is twice in number of 0's. Also, all 0's are followed by all 1's. When reading 0's of the string, we push two symbols each time into the stack whereas we pop one symbol each time while reading 1's.

Now we formally define the pushdown automaton $M = (S, \Sigma, \Gamma, \delta, s_0, z_0, \phi)$ given by $S = \{s_0, s_1\}, \Sigma = \{0, 1\}$ and $\Gamma = \{a, z_0\}$. The transition function δ can be visualized into several parts as below.

1. A set of transitions to push a's into the stack when reading 0's, where the pushdown automaton changes its state from s_0 to s_1 :

    $\delta(s_0, 0, z_0) = \{(s_1, aaz_0)\}; \; \delta(s_1, 0, a) = \{(s_1, aaa)\}$

2. A set of transitions to pop a's from the stack when reading 1's:

    $\delta(s_1, 1, a) = \{(s_1, \wedge)\}$

3. Finally, the transition to recognize the string by empty stack:

    $\delta(s_1, \wedge, z_0) = \{(s_1, \wedge)\}$

For example, consider the string '$w = 011$'. The sequence of moves in accepting the string $w = 011$ is given below.

$$(s_0, 011, z_0) \vdash (s_1, 11, aaz_0) \vdash (s_1, 1, az_0)$$
$$\vdash (s_1, \wedge, z_0) \vdash (s_1, \wedge, \wedge)$$

Therefore, it is clear that the string $w = 011$ is accepted by the pushdown automaton by null stack.

Example 7

Construct a pushdown automaton to accept a given language L by final state, where $L = \left\{ waw^R \,\middle|\, w \in \{b, c\}^+ \right\}$.

Solution   Given that $L = \left\{ waw^R \,\middle|\, w \in \{b, c\}^+ \right\}$. Here we use the fact that the symbols are retrieved from a stack in reverse order of their insertion after consuming the symbol 'a'. When reading the first part 'w' of the string, we push consecutive symbols into the stack. When reading the symbol 'a' of the string, the pushdown automaton changes its state without pushing any symbol into the stack. For the final part w^R of the string, we retrieve the top symbol from the stack.

Now we formally define the pushdown automaton $M = (S, \Sigma, \Gamma, \delta, s_0, z_0, A)$ given by $S = \{s_0, s_1, s_2\}$, $\Sigma = \{b, c\}$, $A = \{s_2\}$ and $\Gamma = \{b, c, z_0\}$. The transition function can be visualized into several parts as below.

1. A set of transitions to push 'w' on the stack:

    $\delta(s_0, b, z_0) = \{(s_0, bz_0)\}$; $\delta(s_0, b, b) = \{(s_0, bb)\}$; $\delta(s_0, b, c) = \{(s_0, bc)\}$;

    $\delta(s_0, c, z_0) = \{(s_0, cz_0)\}$; $\delta(s_0, c, b) = \{(s_0, cb)\}$; $\delta(s_0, c, c) = \{(s_0, cc)\}$.

2. A set of transitions when reading symbol 'a' of the string, where the pushdown automaton changes its state from s_0 to s_1:

    $\delta(s_0, a, b) = \{(s_1, b)\}$; $\delta(s_0, a, c) = \{(s_1, c)\}$

3. A set of transitions to match w^R against the contents of the stack:

    $\delta(s_1, b, b) = \{(s_1, \wedge)\}$; $\delta(s_1, c, c) = \{(s_1, \wedge)\}$

4. Finally, the transition to make a recognize match: $\delta(s_1, \wedge, z_0) = \{(s_2, z_0)\}$

For example, consider the string '$w = bcacb$'. The sequence of moves in accepting the string $w = bcacb$ is given below.

$$(s_0, bcacb, z_0) \vdash (s_0, cacb, bz_0) \vdash (s_0, acb, cbz_0)$$
$$\vdash (s_1, cb, cbz_0) \vdash (s_1, b, bz_0)$$
$$\vdash (s_1, \wedge, z_0) \vdash (s_2, \wedge, z_0)$$

Therefore, it is clear that the string $w = bcacb$ is accepted by the pushdown automaton by final state.

Example 8

Construct a pushdown automaton to accept a given language L by final state, where $L = \left\{ a^n b^m c^{n+m} \mid n \geq 0, m \geq 0 \right\}$.

Solution   Given that $L = \left\{ a^n b^m c^{n+m} \mid n \geq 0, m \geq 0 \right\}$. From the language it is clear that the cardinality of the symbol 'c' is same as the sum of cardinality of the symbols 'a' and 'b'. Therefore, when reading a's as well as b's of the string, we push a symbol each time into the stack whereas we pop one symbol each time when reading the symbol 'c'.

Now we formally define the pushdown automaton $M = (S, \Sigma, \Gamma, \delta, s_0, z_0, A)$ given by $S = \{s_0, s_1, s_2\}$, $\Sigma = \{a, b, c\}$, $A = \{s_2\}$ and $\Gamma = \{1, z_0\}$. The transition function δ can be visualized into several parts as below.

1. The transition to accept a null string: $\delta(s_0, \wedge, z_0) = \{(s_2, z_0)\}$

2. A set of transitions to push a symbol 1 into the stack, when reading a and b:

   $\delta(s_0, a, z_0) = \{(s_0, 1z_0)\};$ $\delta(s_0, a, 1) = \{(s_0, 11)\};$
   $\delta(s_0, b, z_0) = \{(s_0, 1z_0)\};$ $\delta(s_0, b, 1) = \{(s_0, 11)\}.$

3. A set of transitions to pop a symbol 1 from the stack while reading the symbol 'c', where the pushdown automaton changes its state from s_0 to s_1:

   $\delta(s_0, c, 1) = \{(s_1, \wedge)\}; \; \delta(s_1, c, 1) = \{(s_1, \wedge)\}.$

4. Finally the transition to make a recognize match:

   $\delta(s_1, \wedge, z_0) = \{(s_2, z_0)\}$

For example, consider the string $w = abc^2$. The sequence of moves in accepting $w = abc^2$ is given below.

$$(s_0, abc^2, z_0) \vdash (s_0, bc^2, 1z_0) \vdash (s_0, c^2, 11z_0)$$
$$\vdash (s_1, c, 1z_0) \vdash (s_1, \wedge, z_0)$$
$$\vdash (s_2, \wedge, z_0)$$

Therefore, it is clear that the string $w = abc^2$ is accepted by the pushdown automaton by final state.

Example 9

Consider the pushdown automaton M such that $M = (\{s_0, s_1, s_2\}, \{0, 1\}, \{a, z_0\}, \delta, s_0, z_0, \{s_2\})$, where δ is defined as

1. $\delta(s_0, 0, z_0) = \{(s_0, az_0)\};$
2. $\delta(s_0, 0, a) = \{(s_0, aa)\};$
3. $\delta(s_0, 1, a) = \{(s_1, \wedge)\};$
4. $\delta(s_1, 1, a) = \{(s_1, \wedge)\};$
5. $\delta(s_1, \wedge, z_0) = \{(s_2, z_0)\};$
6. $\delta(s_2, \wedge, z_0) = \{(s_2, \wedge)\}.$

Show that the string $w = 0^n 1^n, n > 0$ is accepted by pushdown automaton by both final state and empty stack.

Solution   The sequence of moves in accepting $w = 0^n 1^n, n > 0$ is given below.

$$(s_0, 0^n 1^n, z_0) \vdash (s_0, 0^{n-1} 1^n, az_0) \quad \text{[by 1]}$$
$$\vdash (s_0, 0^{n-2} 1^n, a^2 z_0) \quad \text{[by 2]}$$
$$\vdash^* (s_0, 1^n, a^n z_0) \quad \text{[by 2]}$$
$$\vdash (s_1, 1^{n-1}, a^{n-1} z_0) \quad \text{[by 3]}$$
$$\vdash (s_1, 1^{n-2}, a^{n-2} z_0) \quad \text{[by 4]}$$
$$\vdash^* (s_1, \wedge, z_0) \quad \text{[by 4]}$$
$$\vdash (s_2, \wedge, z_0) \quad \text{[by 5]}$$
$$\vdash (s_2, \wedge, \wedge) \quad \text{[by 6]}$$

Therefore, it is clear that the string is accepted by both final state and empty stack.

Example 10

Construct a pushdown automaton to accept the language L by final state. It is given that $L = \{a^n b^m c^m d^n \mid m, n \geq 1\}$.

Solution   Given that $L = \{a^n b^m c^m d^n \mid m, n \geq 1\}$. From the language it is clear that the number of symbol a's is equal to the number of symbols d's. Similarly, the number of b's is equal to the number of c's. Therefore, when reading symbols a and b of the string, we push a symbol into the stack each time whereas we pop one symbol each time when reading symbols c and d.

Now, we define the pushdown automaton $M = (S, \Sigma, \Gamma, \delta, s_0, z_0, A)$ given by $S = \{s_0, s_1, s_2, s_3\}, \Sigma = \{a, b, c, d\}, A = \{s_3\}$ and $\Gamma = \{a, b, z_0\}$. The transition function δ can be visualized into several parts as follows:

1. A set of transitions to push a symbol into the stack when reading symbols a and b:

$$\delta(s_0, a, z_0) = \{(s_0, az_0)\}; \quad \delta(s_0, b, a) = \{(s_0, ba)\};$$
$$\delta(s_0, a, a) = \{(s_0, aa)\}; \quad \delta(s_0, b, b) = \{(s_0, bb)\}.$$

2. A set of transitions to pop a symbol from the stack when reading symbol c, where the pushdown automaton changes its state from s_0 to s_1:

$$\delta(s_0, c, b) = \{(s_1, \wedge)\}; \delta(s_1, c, b) = \{(s_1, \wedge)\};$$

3. A set of transitions to pop a symbol from the stack when reading symbol d, where the pushdown automaton changes its state from s_1 to s_2:

$$\delta(s_1, d, a) = \{(s_2, \wedge)\}; \delta(s_2, d, a) = \{(s_2, \wedge)\}.$$

4. Finally a transition to recognize the string by final state:

$$\delta(s_2, \wedge, z_0) = \{(s_3, z_0)\}.$$

For example consider the string $w = ab^2c^2d$. The sequence of moves in accepting $w = ab^2c^2d$ is given below.

$$(s_0, ab^2c^2d, z_0) \vdash (s_0, b^2c^2d, az_0) \vdash (s_0, bc^2d, baz_0)$$
$$\vdash (s_0, c^2d, bbaz_0) \vdash (s_1, cd, baz_0)$$
$$\vdash (s_1, d, az_0) \vdash (s_2, \wedge, z_0)$$
$$\vdash (s_3, \wedge, z_0)$$

Therefore, it is clear that the string $w = ab^2c^2d$ is accepted by the pushdown automaton by final state.

Example 11

Construct a pushdown automaton M to accept the language L by null stack, where L is given as $L = \{b^n c^m b^n \mid n, m \geq 1\}$.

Solution Given that $L = \{b^n c^m b^n \mid n, m \geq 1\}$. Here we construct a pushdown automaton M to accept the language L by null store such that it stores b's until a symbol 'c' occurs. After storing b's, we read 'c' without changing the status of stack and then we pop all b's stored earlier from the stack.

Now, we define the pushdown automaton $M = (S, \Sigma, \Gamma, \delta, s_0, z_0, \phi)$ given by $S = \{s_0, s_1\}, \Sigma = \{b, c\}$ and $\Gamma = \{b, z_0\}$. The transition function δ can be visualized into several parts as below.

1. A set of transitions to push a symbol 'b' into the stack when reading symbol b:

$$\delta(s_0, b, z_0) = \{(s_0, bz_0)\}; \delta(s_0, b, b) = \{(s_0, bb)\}$$

2. A set of transitions when the status of stack is unchanged while reading the symbol 'c'. In this case the pushdown automaton changes its state from s_0 to s_1:

$$\delta(s_0, c, b) = \{(s_1, b)\}; \delta(s_1, c, b) = \{(s_1, b)\}$$

3. A set of transitions to pop a symbol from the stack when reading symbol '*b*' after reading the symbol '*c*':

$$\delta(s_1, b, b) = \{(s_1, \wedge)\}$$

4. Finally a transition to recognize the string by null stack:

$$\delta(s_1, \wedge, z_0) = \{(s_1, \wedge)\}$$

The sequence of moves in accepting the string $b^n c^m b^n$ is given as below.

$$(s_0, b^n c^m b^n, z_0) \vdash (s_0, b^{n-1} c^m b^n, bz_0)$$
$$\overset{n-1}{\vdash} (s_0, c^m b^n, b^n z_0) \vdash (s_1, c^{m-1} b^n, b^n z_0)$$
$$\overset{m-1}{\vdash} (s_1, b^n, b^n z_0) \vdash (s_1, b^{n-1}, b^{n-1} z_0)$$
$$\overset{n-1}{\vdash} (s_1, \wedge, z_0) \vdash (s_1, \wedge, \wedge)$$

Hence, it is clear that the language $L = \{b^n c^m b^n \mid n, m \geq 1\}$ is accepted by the pushdown automaton *M* by null stack.

Example 12

Construct a pushdown automaton to accept the language *L* by final state. It is given that $L = \{w \in \{a, b\}^* \mid n_a(w) = n_b(w)\}$, where $n_a(w)$ and $n_b(w)$ represents the number of *a*'s and *b*'s in the string *w* respectively.

Solution   Given that the language $L = \{w \in \{a, b\}^* \mid n_a(w) = n_b(w)\}$.

We formally define the pushdown automaton *M* as $M = (S, \Sigma, \Gamma, \delta, s_0, z_0, A)$ given by $S = \{s_0, s_1\}, \Sigma = \{a, b\}, A = \{s_1\}$ and $\Gamma = \{0, 1, z_0\}$. The transition function δ can be visualized into several parts as below.

1. A transition to accept a null string:

$$\delta(s_0, \wedge, z_0) = \{(s_1, z_0)\}$$

2. A set of transitions to push a symbol 0 into the stack when reading the symbol '*a*':
$$\delta(s_0, a, z_0) = \{(s_0, 0z_0)\}; \delta(s_0, a, 0) = \{(s_0, 00)\}$$

3. A set of transitions to push a symbol 1 into the stack when reading the symbol '*b*':
$$\delta(s_0, b, z_0) = \{(s_0, 1z_0)\}; \delta(s_0, b, 1) = \{(s_0, 11)\}$$

4. A transition to pop a symbol, when reading symbol '*b*' with top of the stack 0: $\delta(s_0, b, 0) = \{(s_0, \wedge)\}$. Similarly, a transition to pop a symbol, when reading '*a*' with top the stack 1 is given as $\delta(s_0, a, 1) = \{(s_0, \wedge)\}$

For example, consider a string $w = abba$. The sequence of moves to accept the string is given as below.

$$\delta(s_0, abba, z_0) \vdash (s_0, bba, 0z_0)$$
$$\vdash (s_0, ba, z_0) \vdash (s_0, a, 1z_0)$$
$$\vdash (s_0, \wedge, z_0) \vdash (s_1, \wedge, z_0)$$

Hence, it is clear that, the language $L = \{w \in \{a,b\}^* | n_a(w) = n_b(w)\}$ is accepted by pushdown automaton by final state.

Example 13

Show that a pushdown automaton M given below is deterministic, where $M = (\{s_0, s_1\}, \{0, 1\}, \{0, 1, z_0\}, \delta, s_0, z_0, \{s_1\})$. The transition function δ is defined as below.

$$\delta(s_0, 0, z_0) = \{(s_1, z_0)\}; \qquad \delta(s_0, 1, z_0) = \{(s_0, 1z_0)\};$$
$$\delta(s_0, 0, 1) = \{(s_0, \wedge)\}; \qquad \delta(s_0, 1, 1) = \{(s_0, 11)\};$$
$$\delta(s_1, 0, z_0) = \{(s_1, 0z_0)\}; \qquad \delta(s_1, 1, z_0) = \{(s_0, z_0)\};$$
$$\delta(s_1, 0, 0) = \{(s_1, 00)\}; \qquad \delta(s_1, 1, 0) = \{(s_1, \wedge)\}.$$

Solution   It is clear that, all the transitions given above have single moves. Also there is no such transition of the form $\delta(s, \wedge, z) \neq \phi$ for $s \in \{s_0, s_1\}$ and $z \in \{0, 1, z_0\}$. Therefore, the given pushdown automaton is deterministic.

Example 14

Let $M = (\{s_0, s_1\}, \{a, b\}, \{a, z_0\}, \delta, s_0, z_0, \phi)$ be a pushdown automaton, where δ is given by the transitions $\delta(s_0, a, z_0) = \{(s_0, az_0)\}$, $\delta(s_0, a, a) = \{(s_0, aa)\}$, $\delta(s_0, b, a) = \{(s_1, \wedge)\}$, $\delta(s_1, b, a) = \{(s_1, \wedge)\}$ and $\delta(s_1, \wedge, z_0) = \{(s_1, \wedge)\}$. What is the instantaneous description that the pushdown automaton M reaches after processing the strings (i) $w = a^2 b^3$ and (ii) $w = ababa$ if the pushdown automaton M starts with the initial instantaneous description?

Solution   (i) The instantaneous description that the pushdown automaton M reaches after processing $w = a^2 b^3$ is $(s_1, b, \wedge)$. The sequence of steps is given below.

$$\delta(s_0, a^2 b^3, z_0) \vdash (s_0, ab^3, az_0)$$
$$\vdash (s_0, b^3, aaz_0) \vdash (s_1, b^2, az_0)$$
$$\vdash (s_1, b, z_0) \vdash (s_1, b, \wedge)$$

(ii) The instantaneous description that the pushdown automaton M reaches after processing $w = ababa$ is $(s_1, aba, \wedge)$. The sequence of steps is given below.

$$\delta(s_0, ababa, z_0) \vdash (s_0, baba, az_0)$$
$$\vdash (s_1, aba, z_0)$$
$$\vdash (s_1, aba, \wedge)$$

Example 15

Consider the pushdown automaton M accepting $0^n 1^n, n > 0$ by final state given by $M = (\{s_0, s_1, s_2\}, \{0, 1\}, \{a, z_0\}, \delta, s_0, z_0, \{s_2\})$ where the transition function δ is defined as $\delta(s_0, 0, z_0) = \{(s_0, az_0)\}$, $\delta(s_0, 0, a) = \{(s_0, aa)\}$, $\delta(s_0, 1, a) = \{(s_1, \wedge)\}$, $\delta(s_1, 1, a) = \{(s_1, \wedge)\}$ and $\delta(s_1, \wedge, z_0) = \{(s_2, z_0)\}$. Construct a pushdown automaton M_1 equivalent to M by null stack.

Solution Let the pushdown automaton M_1 given by $M_1 = (S', \Sigma, \Gamma', \delta', s_0', z_0', \phi)$, where $S' = \{s_0, s_1, s_2, s_0', s_d\}$, $\Sigma = \{0, 1\}$ and $\Gamma' = \{a, z_0, z_0'\}$. The transition function δ' is defined as below.

$$\delta'(s_0', \wedge, z_0') = \{(s_0, z_0 z_0')\}$$
$$\delta'(s_0, 0, z_0) = \delta(s_0, 0, z_0) = \{(s_0, az_0)\}$$
$$\delta'(s_0, 0, a) = \delta(s_0, 0, a) = \{(s_0, aa)\}$$
$$\delta'(s_0, 1, a) = \delta(s_0, 1, a) = \{(s_1, \wedge)\}$$
$$\delta'(s_1, 1, a) = \delta(s_1, 1, a) = \{(s_1, \wedge)\}$$
$$\delta'(s_1, \wedge, z_0) = \delta(s_1, \wedge, z_0) = \{(s_2, z_0)\}$$
$$\delta'(s_2, \wedge, a) = \{(s_d, \wedge)\}; \delta'(s_2, \wedge, z_0) = \{(s_d, \wedge)\};$$
$$\delta'(s_2, \wedge, z_0') = \{(s_d, \wedge)\}; \delta'(s_d, \wedge, a) = \{(s_d, \wedge)\};$$
$$\delta'(s_d, \wedge, z_0) = \{(s_d, \wedge)\}; \delta'(s_d, \wedge, z_0') = \{(s_d, \wedge)\}.$$

Example 16

Construct a pushdown automaton M equivalent to the context-free grammar $Q_0 \to aAA$, $A \to aQ_0 | bQ_0 | a$. Show that the string aba^4 is accepted by pushdown automaton by null stack.

Solution The productions of context-free grammar are $A \to aQ_0 | bQ_0 | a$, $Q_0 \to aAA$. We define pushdown automaton M as $M = (S, \Sigma, \Gamma, \delta, s_0, z_0, \phi)$, where $S = \{s_0\}$, $\Sigma = \{a, b\}$, $\Gamma = \{Q_0, A, a, b\}$ and $z_0 = Q_0$. The transition function δ is defined as below.

1. $\delta(s_0, \wedge, Q_0) = \{(s_0, aAA)\}$;
2. $\delta(s_0, \wedge, A) = \{(s_0, aQ_0), (s_0, bQ_0), (s_0, a)\}$;
3. $\delta(s_0, a, a) = \{(s_0, \wedge)\}$;
4. $\delta(s_0, b, b) = \{(s_0, \wedge)\}$

Now we have to show that aba^4 is accepted by pushdown automaton by null stack. The sequence of steps is given below.

$$(s_0, aba^4, Q_0) \vdash (s_0, aba^4, aAA)$$
$$\vdash (s_0, ba^4, AA) \vdash (s_0, ba^4, bQ_0A)$$
$$\vdash (s_0, a^4, Q_0A) \vdash (s_0, a^4, aAAA)$$
$$\vdash (s_0, a^3, AAA) \vdash (s_0, a^3, aAA)$$
$$\vdash (s_0, a^2, AA) \vdash (s_0, a^2, aA)$$
$$\vdash (s_0, a, A) \vdash (s_0, a, a)$$
$$\vdash (s_0, \wedge, \wedge)$$

Therefore, the string aba^4 is accepted by pushdown automaton by null stack.

Example 17

Construct a pushdown automaton M equivalent to the context-free grammar $Q_0 \rightarrow 0A|1BB|0Q_0$, $A \rightarrow 1|0B$ and $B \rightarrow 1Q_0|1$.

Solution   The productions of context-free grammar are $Q_0 \rightarrow 0A|1BB|0Q_0$, $A \rightarrow 1|0B$ and $B \rightarrow 1Q_0|1$. Here, $V_N = \{Q_0, A, B\}$ and $T = \{0,1\}$. We define pushdown automaton M as $M = (S, \Sigma, \Gamma, \delta, s_0, z_0, \phi)$, where $S = \{s_0\}, \Sigma = T = \{0, 1\}$, $\Gamma = \{Q_0, A, B, 0, 1\}$ and $z_0 = Q_0$. The transition function δ is defined as below.

1. $\delta(s_0, \wedge, Q_0) = \{(s_0, 0A), (s_0, 1BB), (s_0, 0Q_0)\}$;

2. $\delta(s_0, \wedge, A) = \{(s_0, 1), (s_0, 0B)\}$;

3. $\delta(s_0, \wedge, B) = \{(s_0, 1Q_0), (s_0, 1)\}$;

4. $\delta(s_0, 0, 0) = \{(s_0, \wedge)\}$;

5. $\delta(s_0, 1, 1) = \{(s_0, \wedge)\}$.

Example 18

Construct a pushdown automaton to accept the language $L = \{a^n b^n | n \geq 1\}$ by null stack.

Solution   Given that $L = \{a^n b^n | n \geq 1\}$. We construct a context-free grammar $G = (V_N, T, P, Q_0)$ such that $V_N = \{Q_0\}$, $T = \{a, b\}$ and $P = \{Q_0 \rightarrow aQ_0b|ab\}$. Therefore, it is clear that $L(G) = \{a^n b^n | n \geq 1\}$.

Now, we define pushdown automaton M as $M = (S, \Sigma, \Gamma, \delta, s_0, z_0, \phi)$, where $S = \{s_0\}$, $\Sigma = T = \{a, b\}$, $\Gamma = (V_N \cup T) = \{Q_0, a, b\}$ and $z_0 = Q_0$. The transition function δ is defined as below.

1. $\delta(s_0, \wedge, Q_0) = \{(s_0, ab), (s_0, aQ_0b)\}$;
2. $\delta(s_0, a, a) = \{(s_0, \wedge)\}$;
3. $\delta(s_0, b, b) = \{(s_0, \wedge)\}$.

Example 19

Construct a pushdown automaton to accept the language L, by null stack, where $L = \{a^n b^n \mid n \geq 1\} \cup \{a^m b^{2m} \mid m \geq 1\}$.

Solution   Given that $L = \{a^n b^n \mid n \geq 1\} \cup \{a^m b^{2m} \mid m \geq 1\}$. In order to construct a pushdown automata M, we construct a context-free grammar $G = (V_N, T, P, Q_0)$ such that $V_N = \{Q_0, A, B\}$, $T = \{a, b\}$ and $P = \{Q_0 \rightarrow A \mid B, \quad A \rightarrow aAb \mid ab, \quad B \rightarrow aBbb \mid abb\}$. Therefore, it is easily seen that $L = \{a^n b^n \mid n \geq 1\} \cup \{a^m b^{2m} \mid m \geq 1\}$.

Now, we define a pushdown automaton M as $M = (S, \Sigma, \Gamma, \delta, s_0, z_0, \phi)$, where $S = \{s_0\}$, $\Sigma = T = \{a, b\}$, $\Gamma = (V_N \cup T) = \{Q_0, A, B, a, b\}$ and $z_0 = Q_0$. The transition function δ is defined as below.

1. $\delta(s_0, \wedge, Q_0) = \{(s_0, A), (s_0, B)\}$;
2. $\delta(s_0, \wedge, A) = \{(s_0, aAb), (s_0, ab)\}$;
3. $\delta(s_0, \wedge, B) = \{(s_0, aBbb), (s_0, abb)\}$;
4. $\delta(s_0, a, a) = \{(s_0, \wedge)\}$;
5. $\delta(s_0, b, b) = \{(s_0, \wedge)\}$.

Example 20

Construct a pushdown automaton to accept the language L by final state, where $L = \{a^n b^{2n} \mid n \geq 1\}$.

Solution   Given that $L = \{a^n b^{2n} \mid n \geq 1\}$. First we construct a context-free grammar G that accepts L. Let us define $G = (V_N, T, P, Q_0)$ such that $V_N = \{Q_0\}$, $T = \{a, b\}$ and $P = \{Q_0 \rightarrow aQ_0bb \mid abb\}$. Therefore, it is clear that $L = \{a^n b^{2n} \mid n \geq 1\}$.

Now, we define pushdown automaton M as $M = (S, \Sigma, \Gamma, \delta, s_0, z_0, \phi)$, where $S = \{s_0\}$, $\Sigma = T = \{a, b\}$, $\Gamma = (V_N \cup T) = \{Q_0, a, b\}$ and $z_0 = Q_0$. The transition function δ is defined as follows:

1. $\delta(s_0, \wedge, Q_0) = \{(s_0, abb), (s_0, aQ_0bb)\};$
2. $\delta(s_0, a, a) = \{(s_0, \wedge)\};$
3. $\delta(s_0, b, b) = \{(s_0, \wedge)\}.$

It is clear that the above pushdown automaton M accepts the language L by null stack. Therefore, we have to construct an equivalent pushdown automaton M_1 that accepts L by final state. Let us define the pushdown automaton M_1 given by $M_1 = (S', \Sigma, \Gamma', \delta', s_0', z_0', A)$, where $S' = S \cup \{s_0', s_f\} = \{s_0, s_0', s_f\}$, $\Gamma' = \Gamma \cup \{z_0'\} = \{Q_0, a, b, z_0'\}$ and $A = s_f$. The transition function δ' is defined as follows:

(a) $\delta'(s_0', \wedge, z_0') = \{(s_0, Q_0 z_0')\}$

(b) $\delta'(s_0, \wedge, Q_0) = \{(s_0, abb), (s_0, aQ_0bb)\}$

(c) $\delta'(s_0, a, a) = \delta(s_0, a, a) = \{(s_0, \wedge)\}$

(d) $\delta'(s_0, b, b) = \delta(s_0, b, b) = \{(s_0, \wedge)\}$

(e) $\delta'(s_0, \wedge, z_0') = \{(s_f, \wedge)\}$

Example 21

Construct a context-free grammar G that accepts pushdown automaton M, where $M = (\{s_0, s_1\}, \{a, b\}, \{z_0, a\}, \delta, s_0, z_0, \phi)$ and the transition function δ is given by $\delta(s_0, a, z_0) = \{(s_0, az_0)\}$, $\delta(s_0, a, a) = \{(s_0, aa)\}$, $\delta(s_0, b, a) = \{(s_1, \wedge)\}$, $\delta(s_1, b, a) = \{(s_1, \wedge)\}$ and $\delta(s_1, \wedge, z_0) = \{(s_1, \wedge)\}$.

Solution   Let the context-free grammar G is defined as $G = (V_N, T, P, Q_0)$, where $T = \{a, b\}$ and $V_N = \{Q_0, [s_0, z_0, s_0], [s_0, z_0, s_1], [s_1, z_0, s_0], [s_1, z_0, s_1], [s_0, a, s_0], [s_0, a, s_1], [s_1, a, s_0], [s_1, a, s_1]\}$.

Note that the elements of V_N except Q_0 are triple constructed by all combinations of s_0 and s_1 (i.e., $(s_0, s_0), (s_0, s_1), (s_1, s_0), (s_1, s_1)$) with a or z_0 in middle. The set of productions P are defined as follows:

1. The Q_0 productions are given as below.

    $P_1 : \quad Q_0 \rightarrow [s_0, z_0, s_0]$

    $P_2 : \quad Q_0 \rightarrow [s_0, z_0, s_1]$

2. The transition $\delta(s_0, b, a) = \{(s_1, \wedge)\}$ leads to the production P_3, where $P_3 : \quad [s_0, a, s_1] \rightarrow b$

3. The transition $\delta(s_1, b, a) = \{(s_1, \wedge)\}$ leads to the production P_4, where $P_4 : \quad [s_1, a, s_1] \rightarrow b$

4. The transition $\delta(s_1, \wedge, z_0) = \{(s_1, \wedge)\}$ leads to the production P_5, where $P_5 : \quad [s_1, z_0, s_1] \rightarrow \wedge$

5.   The transition $\delta(s_0, a, z_0) = \{(s_0, az_0)\}$ leads to the following productions.

P_6: $[s_0, z_0, s_0] \rightarrow a[s_0, a, s_0][s_0, z_0, s_0]$

P_7: $[s_0, z_0, s_0] \rightarrow a[s_0, a, s_1][s_1, z_0, s_0]$

P_8: $[s_0, z_0, s_1] \rightarrow a[s_0, a, s_0][s_0, z_0, s_1]$

P_9: $[s_0, z_0, s_1] \rightarrow a[s_0, a, s_1][s_1, z_0, s_1]$

6.   The transition $\delta(s_0, a, a) = \{(s_0, aa)\}$ leads to the following productions.

P_{10}: $[s_0, a, s_0] \rightarrow a[s_0, a, s_0][s_0, a, s_0]$

P_{11}: $[s_0, a, s_0] \rightarrow a[s_0, a, s_1][s_1, a, s_0]$

P_{12}: $[s_0, a, s_1] \rightarrow a[s_0, a, s_0][s_0, a, s_1]$

P_{13}: $[s_0, a, s_1] \rightarrow a[s_0, a, s_1][s_1, a, s_1]$

Therefore, the productions in P are P_1 to P_{13}. These productions can be reduced by using the techniques discussed earlier in chapter 5.

Example 22

Construct a context-free grammar G that accepts pushdown automaton M, where $M = (\{s_0, s_1\}, \{a, b\}, \{0, 1, z_0\}, \delta, s_0, z_0, \phi)$ and the transition function δ is given by $\delta(s_0, \wedge, z_0) = \{(s_1, \wedge)\}$, $\delta(s_0, a, 0) = \{(s_0, 00)\}$ $\delta(s_0, a, z_0) = \{(s_0, 0z_0)\}$, $\delta(s_0, b, 0) = \{(s_0, \wedge)\}$, $\delta(s_0, b, z_0) = \{(s_0, 1z_0)\}$, $\delta(s_0, a, 1) = \{(s_0, \wedge)\}$, and $\delta(s_0, b, 1) = \{(s_0, 11)\}$.

Solution   Let the context-free grammar G is defined as $G = (V_N, T, P, Q_0)$, where $T = \{a, b\}$ and $V_N = \{Q_0$, $[s_0, 0, s_0]$, $[s_0, 1, s_0], [s_0, z_0, s_0]$, $[s_0, 0, s_1]$, $[s_0, 1, s_1], [s_0, z_0, s_1]$, $[s_1, 0, s_0]$, $[s_1, 1, s_0]$, $[s_1, z_0, s_0]$, $[s_1, 0, s_1]$, $[s_1, 1, s_1]$, $[s_1, z_0, s_1]\}$.

The elements of V_N except Q_0 are triple constructed by all combinations of s_0 and s_1 with 0, 1 or z_0 in middle. The set of productions P are defined as follows:

1.   The Q_0 productions are given as below.

P_1: $Q_0 \rightarrow [s_0, z_0, s_0]$

P_2: $Q_0 \rightarrow [s_0, z_0, s_1]$

2.   The transition $\delta(s_0, \wedge, z_0) = \{(s_1, \wedge)\}$ leads to the production P_3, where P_3: $[s_0, z_0, s_1] \rightarrow \wedge$.

3.   The transition $\delta(s_0, a, z_0) = \{(s_0, 0z_0)\}$ leads to the following productions.

P_4: $[s_0, z_0, s_0] \rightarrow a[s_0, 0, s_0][s_0, z_0, s_0]$

P_5: $[s_0, z_0, s_0] \rightarrow a[s_0, 0, s_1][s_1, z_0, s_0]$

P_6: $[s_0, z_0, s_1] \rightarrow a[s_0, 0, s_0][s_0, z_0, s_1]$

P_7: $[s_0, z_0, s_1] \rightarrow a[s_0, 0, s_1][s_1, z_0, s_1]$

4. The transition $\delta(s_0, b, z_0) = \{(s_0, 1z_0)\}$ leads to the following productions.

P_8: $[s_0, z_0, s_0] \to b[s_0, 1, s_0][s_0, z_0, s_0]$
P_9: $[s_0, z_0, s_0] \to b[s_0, 1, s_1][s_1, z_0, s_0]$
P_{10}: $[s_0, z_0, s_1] \to b[s_0, 1, s_0][s_0, z_0, s_1]$
P_{11}: $[s_0, z_0, s_1] \to b[s_0, 1, s_1][s_1, z_0, s_1]$

5. The transition $\delta(s_0, a, 0) = \{(s_0, 00)\}$ leads to the following productions.

P_{12}: $[s_0, 0, s_0] \to a[s_0, 0, s_0][s_0, 0, s_0]$
P_{13}: $[s_0, 0, s_0] \to a[s_0, 0, s_1][s_1, 0, s_0]$
P_{14}: $[s_0, 0, s_1] \to a[s_0, 0, s_0][s_0, 0, s_1]$
P_{15}: $[s_0, 0, s_1] \to a[s_0, 0, s_1][s_1, 0, s_1]$

6. The transition $\delta(s_0, b, 0) = \{(s_0, \wedge)\}$ leads to the production P_{16}, where P_{16}: $[s_0, 0, s_0] \to b$

7. The transition $\delta(s_0, a, 1) = \{(s_0, \wedge)\}$ leads to the production P_{17}, where P_{17}: $[s_0, 1, s_0] \to a$

8. The transition $\delta(s_0, b, 1) = \{(s_0, 11)\}$ leads to the following productions.

P_{18}: $[s_0, 1, s_0] \to b[s_0, 1, s_0][s_0, 1, s_0]$
P_{19}: $[s_0, 1, s_0] \to b[s_0, 1, s_1][s_1, 1, s_0]$

P_{20}: $[s_0, 1, s_1] \to b[s_0, 1, s_0][s_0, 1, s_1]$
P_{21}: $[s_0, 1, s_1] \to b[s_0, 1, s_1][s_1, 1, s_1]$

Therefore, the productions in P are P_1 to P_{21}. These productions can be reduced by using the techniques discussed earlier in chapter 5.

Example 23

Construct a context-free grammar G that accepts pushdown automaton M, where $M = (\{s_0, s_1\}, \{a, b\}, \{a, z_0\}, \delta, s_0, z_0, \phi)$ and the transition function δ is given by $\delta(s_0, a, z_0) = \{(s_0, az_0)\}$, $\delta(s_0, a, a) = \{(s_0, aa)\}$, $\delta(s_0, b, a) = \{(s_1, a)\}$, $\delta(s_1, b, a) = \{(s_1, a)\}$, $\delta(s_1, a, a) = \{(s_1, \wedge)\}$ and $\delta(s_1, \wedge, z_0) = \{(s_1, \wedge)\}$.

Solution   Let the context-free grammar G is defined as $G = (V_N, T, P, Q_0)$, where $T = \{a, b\}$ and $V_N = \{Q_0, [s_0, a, s_0], [s_0, z_0, s_0], [s_0, a, s_1], [s_0, z_0, s_1], [s_1, a, s_0], [s_1, z_0, s_0], [s_1, a, s_1], [s_1, z_0, s_1]\}$.

The elements of V_N except Q_0 are triple constructed by all combinations of s_0 and s_1 with a or z_0 in middle. The set of productions P are defined as follows:

1. The Q_0 productions are given as below.

   P_1 : $Q_0 \to [s_0, z_0, s_0]$

   P_2 : $Q_0 \to [s_0, z_0, s_1]$

2. The transition $\delta(s_0, a, z_0) = \{(s_0, az_0)\}$ leads to the following productions.

   P_3 : $[s_0, z_0, s_0] \to a[s_0, a, s_0][s_0, z_0, s_0]$

   P_4 : $[s_0, z_0, s_0] \to a[s_0, a, s_1][s_1, z_0, s_0]$

   P_5 : $[s_0, z_0, s_1] \to a[s_0, a, s_0][s_0, z_0, s_1]$

   P_6 : $[s_0, z_0, s_1] \to a[s_0, a, s_1][s_1, z_0, s_1]$

3. The transition $\delta(s_0, a, a) = \{(s_0, aa)\}$ leads to the following productions.

   P_7 : $[s_0, a, s_0] \to a[s_0, a, s_0][s_0, a, s_0]$

   P_8 : $[s_0, a, s_0] \to a[s_0, a, s_1][s_1, a, s_0]$

   P_9 : $[s_0, a, s_1] \to a[s_0, a, s_0][s_0, a, s_1]$

   P_{10} : $[s_0, a, s_1] \to a[s_0, a, s_1][s_1, a, s_1]$

4. The transition $\delta(s_0, b, a) = \{(s_1, a)\}$ yields production P_{11} and P_{12}, where

   P_{11} : $[s_0, a, s_0] \to b[s_1, a, s_0]$

   P_{12} : $[s_0, a, s_1] \to b[s_1, a, s_1]$

5. The transition $\delta(s_1, b, a) = \{(s_1, a)\}$ yields production P_{13} and P_{14}, where

   P_{13} : $[s_1, a, s_0] \to b[s_1, a, s_0]$

   P_{14} : $[s_1, a, s_1] \to b[s_1, a, s_1]$

6. The transition $\delta(s_1, a, a) = \{(s_1, \wedge)\}$ leads to the production P_{15}, where P_{15} : $[s_1, a, s_1] \to a$

7. The transition $\delta(s_1, \wedge, z_0) = \{(s_1, \wedge)\}$ leads to the production P_{16}, where

   P_{16} : $[s_1, z_0, s_1] \to \wedge$

Therefore, the productions in P are P_1 to P_{16}. These productions can be reduced by using the techniques discussed earlier in chapter 5.

REVIEW QUESTIONS

1. Define pushdown automata and explain briefly its different components.
2. State the difference between deterministic and nondeterministic pushdown automata.
3. Explain in detail the working principle of pushdown automata.

4. State the properties of move relation in pushdown automata.

5. Explain with suitable examples the acceptance of strings by pushdown automata.

6. Show that the acceptance by final state is equivalent to the acceptance by null stack.

7. Prove that for every context-free language L, there is a pushdown automaton accepting L by empty stack.

8. Prove that for every pushdown automaton M, there is a context-free grammar G such that $L(G) = N(M)$.

9. Write a note on auxiliary pushdown automata.

10. Write a note on two-stack pushdown automata.

PROBLEMS

1. Construct a pushdown automaton accepting L by null stack each of the languages.

   (a) $L = \{a^n b^{2n} \mid n \geq 0\}$

   (b) $L = \{waw^R \mid w \in \{b, c\}^*\}$

   (c) $L = \{a^n b^m c^{n+m} \mid n \geq 0, m \geq 0\}$

   (d) $L = \{0^n 1^m 2^m 3^n \mid m, n \geq 1\}$

2. Construct a pushdown automaton accepting L by final state each of the following languages.

   (a) $L = \{b^n c^m b^n \mid n, m \geq 1\}$

   (b) $L = \{ww^R \mid w \in \{a, b\}^+\}$

   (c) $L = \{a^m b^n \mid m > n \geq 1\}$

   (d) $L = \{a^n b^{n+m} c^m \mid m \geq 1, n \geq 0\}$

3. Construct a pushdown automaton accepting L by null stack each of the following languages.

   (a) $L = \{w \in \{a, b\}^+ \mid n_a(w) = n_b(w)\}$

   (b) $L = \{w \in \{a, b\}^+ \mid n_a(w) = n_b(w) + 1\}$

   (c) $L = \{w \in \{a, b\}^+ \mid n_a(w) = 2n_b(w)\}$

   (d) $L = \{w \in \{a, b\}^+ \mid n_a(w) < n_b(w)\}$

(e) $L = \{w \in \{a, b, c\}^{+} \mid n_a(w) + n_b(w) = n_c(w)\}$

·Here, $n_a(w), n_b(w)$ and $n_c(w)$ are number of a's, b's and c's in string w respectively.

4. Construct a pushdown automaton accepting by final state each of the languages given in solved example 3.

5. Construct a pushdown automaton accepting L by null stack each of the following languages.

(a) $L = \{0^n 1^m \mid n \leq m \leq 3n\}$

(b) $L = \{0^3 1^n 2^n \mid n \geq 0\}$

6. Construct a nondeterministic pushdown automaton that accepts the language $L = \{0^n 1^m \mid n \neq m, n \geq 0\}$ by final state.

7. Construct a pushdown automaton to accept the language of palindromes by final state.

8. Construct a pushdown automaton accepting L by null stack each of the following languages on $\Sigma = \{0, 1, 2\}$.

(a) $L = \{0^m 1^m 2^n \mid n, m \geq 1\}$

(b) $L = \{0^m 1^n \mid n > m\}$

9. Construct a pushdown automaton to accept the language $L = \{0^{3n} 1 2 \mid n \geq 0\}$ on $\Sigma = \{0, 1, 2\}$ by final state.

10. Construct a pushdown automaton accepting L by final state each of the languages given in solved example 8.

11. Construct a context-free grammar G that accepts pushdown automaton M, where $M = (\{s_0, s_1\}, \{0, 1\}, \{a, z_0\}, \delta, s_0, z_0, \{s_1\})$ with   transitions $\delta(s_0, 0, z_0) = \{(s_0, a z_0)\}$, $\delta(s_0, 1, a) = \{(s_0, aa)\}$ and $\delta(s_0, 0, a) = \{(s_1, \wedge)\}$.

12. Construct a pushdown automaton M equivalent to the context-free grammar $Q_0 \rightarrow 0AB \mid 1A$, $A \rightarrow 1Q_0 \mid 0B \mid 1B$ and $B \rightarrow 0 \mid 1$. Show that the string '011000' is accepted by pushdown automaton by null stack.

13. Construct a pushdown automaton M equivalent to the context-free grammar having productions $Q_0 \rightarrow a Q_0 bb \mid abb$.

14. Construct a context-free grammar G that accepts pushdown automaton M, where $M = (\{s_0, s_1\}, \{a, b, c\}, \{z_0, z_1\}, \delta, s_0, z_0, \phi)$ with   transitions $\delta(s_0, b, z_1) = \{(s_1, \wedge)\}$, $\delta(s_0, a, z_0) = \{(s_0, z_1 z_0)\}$, $\delta(s_0, a, z_1) = \{(s_0, z_1 z_1)\}$, $\delta(s_1, b, z_1) = \{(s_1, \wedge)\}$, $\delta(s_1, c, z_0) = \{(s_1, z_0)\}$ and $\delta(s_1, \wedge, z_0) = \{(s_1, \wedge)\}$.

15. Construct a context-free grammar generating the following language $L = \{a^n b^n \mid n \geq 1\} \cup \{a^m b^n \mid n < m\}$ and hence a pushdown automaton accepting L by null stack.

16. Construct a pushdown automaton corresponding to the grammar having productions $Q_0 \rightarrow 0ABB \mid 0AA$, $A \rightarrow 0BB \mid 0$ and $B \rightarrow 1BB \mid A$.

17. Construct a pushdown automaton that accepts the language generated by the grammar $G = (\{Q_0, A\}, \{0, 1\}, P, Q_0)$ with productions $Q_0 \rightarrow AA \mid 0$ and $A \rightarrow Q_0 A \mid 1$.

18. Construct a pushdown automaton that accept the following language L, where $L = \{a^n b^{2n} \mid n \geq 1\} \cup \{a^n b^{n+1} \mid n \geq 0\}$.

19. Construct a nondeterministic pushdown automaton that accepts the language generated by the context-free grammar $Q_0 \rightarrow aQ_0 A \mid abbb$ and $A \rightarrow bbb$.

20. Construct a pushdown automaton that accept the following language L, where $L = \{a^n b^{3n} \mid n \geq 1\} \cup \{(ab)^n \mid n \geq 1\}$.

21. Construct a context-free grammar G that accepts the pushdown automaton M, where $M = (\{s_0, s_1\}, \{a, b\}, \{A, z_0\}, \delta, s_0, z_0, \{s_1\})$ with transitions $\delta(s_0, a, z_0) = \{(s_0, Az_0)\}$, $\delta(s_0, b, A) = \{(s_0, AA)\}$ and $\delta(s_0, a, A) = \{(s_1, \wedge)\}$.

22. Construct a context-free grammar G that accepts the pushdown automaton M, where $M = (\{s_0\}, \{0,1\}, \{z_0, A, 0, 1\}, \delta, s_0, z_0, \phi)$ with transitions $\delta(s_0, \wedge, z_0) = \{(s_0, AA0)\}$, $\delta(s_0, 0, 0) = \{(s_0, \wedge)\}, \delta(s_0, 1, 1) = \{(s_0, \wedge)\}$ and $\delta(s_0, \wedge, A) = \{(s_0, z_0 0), (s_0, z_0 1), (s_0, 0)\}$.

23. Show that the string $a^2 bc$ is accepted by pushdown automata by null stack, where the pushdown automaton M is given as $M = (\{s_0, s_1, s_2\}, \{a, b, c\}, \{0, z_0\}, \delta$, $s_0, z_0, \phi)$ with transitions $\delta(s_0, a, z_0) = \{(s_0, 0z_0)\}$, $\delta(s_0, a, 0) = \{(s_0, 00)\}$, $\delta(s_0, b, 0) = \{(s_1, \wedge)\}$, $\delta(s_1, c, 0) = \{(s_1, \wedge)\}$ and $\delta(s_1, \wedge, z_0) = \{(s_2, \wedge)\}$.

24. Show that the string $0^2 1^3 0^2$ is accepted by pushdown automata by null stack, where the pushdown automaton M is given as $M = (\{s_0, s_1\}, \{0,1\}, \{z_0, z_1\}, \delta, s_0, z_0, \phi)$ with transitions $\delta(s_0, 0, z_0) = \{(s_0, z_1 z_0)\}$, $\delta(s_0, 0, z_1) = \{(s_0, z_1 z_1)\}$, $\delta(s_0, 1, z_1) = \{(s_1, z_1)\}$, $\delta(s_1, 1, z_1) = \{(s_1, z_1)\}$, $\delta(s_1, 0, z_1) = \{(s_1, \wedge)\}$ and $\delta(s_1, \wedge, z_0) = \{(s_1, \wedge)\}$.

25. Consider the following pushdown automaton M, where $M = (\{s_0, s_1\}, \{a, b\}, \{a, z_0\}, \delta, s_0, z_0, \phi)$ with transitions $\delta(s_0, a, z_0) = \{(s_0, az_0)\}$, $\delta(s_0, a, a) = \{(s_0, aa)\}$, $\delta(s_0, b, a) = \{(s_1, \wedge)\}$ and $\delta(s_1, b, a) = \{(s_1, \wedge)\}$. What is the instantaneous description that the pushdown automata M reach after processing the following strings?

   (i) $a^2 b^2$ (ii) $a^3 b^3$ (iii) $a^2 b^3 a$ (iv) a^4 (v) b^4

   (vi) *ababab*

26. Show that the set of all strings over $\{0, 1\}$ consisting of equal number of 0's and 1's is accepted by a deterministic pushdown automata.

27. Show that every regular set accepted by a finite automaton with m-states is accepted by a deterministic pushdown automaton with single state and m-pushdown symbols.

28. If a language L is accepted by a deterministic pushdown automaton M, then prove that L is accepted by deterministic pushdown automata M that always removes the topmost symbol.

29. Show that the string '$abaabba$' is accepted by pushdown automata by final state, where the pushdown automaton $M = (\{s_0, s_1\}, \{a, b\}, \{z_0, z_1, z_2\}, \delta, s_0, z_0, \{s_1\})$ with transitions $\delta(s_0, a, z_0) = \{(s_1, z_0)\}, \delta(s_0, b, z_0) = \{(s_0, z_2 z_0)\}, \delta(s_0, a, z_2) = \{(s_0, \wedge)\}, \delta(s_0, b, z_2) = \{(s_0, z_2 z_2)\}, \delta(s_1, a, z_0) = \{(s_1, z_1 z_0)\}, \delta(s_1, b, z_0) = \{(s_0, z_0)\}, \delta(s_1, a, z_1) = \{(s_1, z_1 z_1)\}$ and $\delta(s_1, b, z_1) = \{(s_1, \wedge)\}$.

30. Construct a context-free grammar G accepting $L = \{a^m b^n \mid n < m\}$ and construct a pushdown automaton accepting L by final state.

7

TURING MACHINE

7.0   INTRODUCTION

In the previous two models of computation we have seen that neither finite automata nor pushdown automata can be regarded as general models for computers. This is because they are not capable of recognizing even simple languages such as $\{a^n b^n c^n \mid n \geq 0\}$ and $\{ww \mid w \in \Sigma^*\}$. This leads to look beyond regular and context-free languages. As a result the study of devices that can recognize this and more complicated languages came into picture. Such devices are called Turing machines. It was named after the scientist Alan Turing in 1936. Historically, the order of invention of these ideas is as follows:

1. Finite automata and regular languages were developed by Kleene, Moore, Mealy and Scott in the 1950's.

2. Pushdown automata and context-free languages were developed later, by Chomsky, Greibach, Oettinger, Evey and Schutzenberger in the 1960's.

3. Turing machines and their theory were developed by Alan Turing and Emil Post in the 1930's.

It is surprising that these dates are out of order than that Turing's work predated the computer itself. Turing was not analysing a computer; he was engaged in inventing the beast. It was from the ideas in his work on mathematical models that the first computers were built. Therefore, the Turing machine is a simple mathematical model of a general computer. It is an extremely basic abstract symbol-manipulating device which, despite their simplicity, can be adapted to simulate the logic of any computer that could possibly be constructed. Though they were intended to be technically feasible, Turing machines were not meant to be a practical computing technology, but a thought experiment about the limits of mechanical computation; thus they were not actually constructed. Studying their abstract properties yields many insights into computer science and complexity theory.

The languages accepted by Turing machine are said to be recursively enumerable. It is precisely those languages whose strings can be listed by a Turing machine. The Turing machine is studied both for the class of languages it defines and the class of integer functions it computes called the partial recursive functions. In particular as today, the Turing machine is equivalent to the computational power of a digital computer.

In this chapter we describe the basic Turing machine model and illustrate some of the problems that can be executed by Turing machine. Later in the chapter we discuss certain variations on the basic model of Turing machine and will examine ways in which one variation of Turing machine can simulate the behaviour of another variation.

7.1   TURING MACHINE MODEL

The basic Turing machine model can be visualized as an indefinitely long tape coupled to a finite control unit. The tape, which acts as the machine's memory is divided into cells or squares. Each square may be inscribed with one symbol from a designated finite alphabet or it may be blank. The different tape symbols will usually be represented by numerals or capital letters, with the restriction that the letter B will be reserved to denote a blank cell. A read-write head is associated with this tape that can travel right or left on the tape and that can read and write a single symbol on each move.

The control unit of a Turing machine is capable of assuming any one of a fixed, finite number of states. At a given time, the state of the control unit together with the tape symbol scanned by the read-write head, uniquely determines how the machine will behave at that time. The actions available to a Turing machine are quite limited; either it may halt, thereby terminating its operation, or a basic move may be carried out. Each move consists in writing a symbol in the currently scanned cell, shifting the tape one cell to the back or forth, and causing the control unit to enter a new state. The symbol that the Turing machine writes on its tape need not be different then the symbol that is already there and the new state need not be different from the current state. A block diagram of the Turing machine model is given in Figure 7.1.

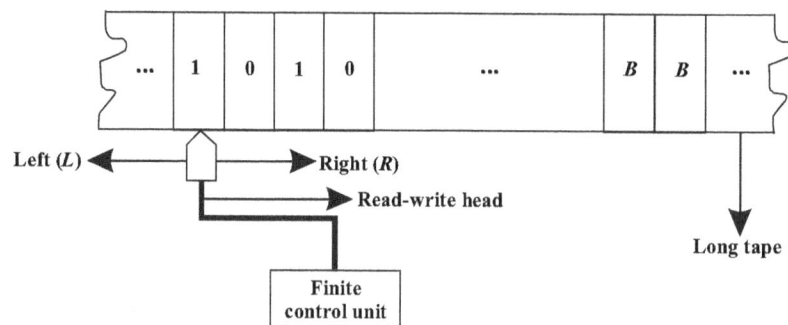

Figure 7.1   A basic Turing machine model

7.2   TURING MACHINE

A Turing machine M is defined as a seven-tuple $(S, \Sigma, \Gamma, \delta, s_0, B, A)$, consisting of

S = Non-empty finite set of internal states

Γ = Non-empty finite set of tape symbols

Σ = Non-empty finite set of input symbols, which is a subset of Γ but not including B, i.e., $\Sigma \subseteq \Gamma - \{B\}$

B = Symbol of Γ, known as blank

δ = Transition function from $S \times \Gamma$ to $S \times \Gamma \times \{L, R\}$, i.e., $S \times \Gamma \to S \times \Gamma \times \{L, R\}$, however δ may be undefined for some arguments.

s_0 = Initial or start state, i.e., $s_0 \in S$

A = Set of final or accepting states, i.e., $A \subseteq S$

The transition function $\delta : S \times \Gamma \to S \times \Gamma \times \{L, R\}$ states that, if the Turing machine is in some state (from set S), by taking a tap symbol (from set Γ), moves to some next state by replacing the current symbol by another or same symbol and moves the read-write head one cell either left (L) or right (R) along the tape. The new state need not be different from the current state.

For example: if $\delta(s, x) = (s', y, R)$, then the machine gets into the next state s', writes the symbol 'y' on the tape by replacing the symbol 'x' and moves one cell right. Similarly, the transition $\delta(s, 0) = (s', B, L)$ describes that the Turing machine changes its state from s to s' on replacing the current input symbol '0' by blank 'B'. The read-write head moves one cell left.

7.2.1   Instantaneous Description

An instantaneous description (ID) of a Turing machine is defined in terms of the entire input string and the current state. We denote an instantaneous description of a Turing machine M by $\alpha s \beta$, where 's' is the present state of M is in S; α and β are substrings of string '$\alpha \beta$'. The substring 'α' is the processed string whereas the substring 'β' is the remaining string to be processed. The substring 'α' is also known as left sequence whereas the substring 'β' is known as right sequence. Observe that the blank 'B' may occur in '$\alpha \beta$'.

For example, consider the Turing machine is in state s_3 with the input string $a_1 a_2 a_3 a_4 a_5 a_6$. Let the symbol under read-write head be a_4. Therefore, the instantaneous description is given as '$a_1 a_2 a_3 s_3 a_4 a_5 a_6$'. Here, we have $\alpha = a_1 a_2 a_3$ and $\beta = a_4 a_5 a_6$. A diagrammatic representation is given in Figure 7.2.

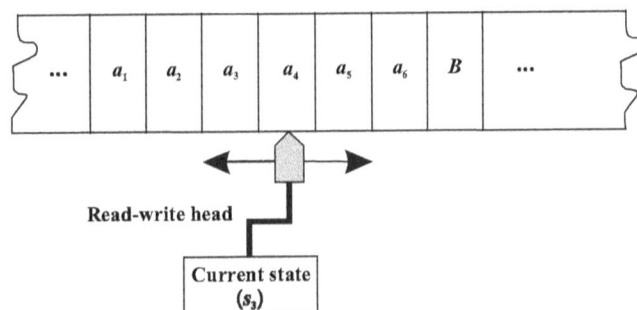

An Instantaneous Description

Figure 7.2   An instaneous description of Turing machine

7.3   REPRESENTATION OF TURING MACHINES

A Turing machine is generally represented by the following methods:

    i.   Representation by transition table

   ii.   Representation by transition graph

  iii.   Representation by ID using move relation

7.3.1   Representation by Transition Table

In this subsection we define the transition function δ in terms of a table. For the transition $\delta(s_i, a_i) = (s_j, a_j, D)$, we write $a_j D s_j$ under a_i-column and s_i-row in the transition table. Therefore, if we get $a_j D s_j$ in the transition table, then a_j is written in the current cell in place of a_i, D gives the direction of the read-write head either left or right whereas the Turing machine enters the next state 's_j'.

For example, consider the following Turing machine represented by a transition table. This Turing machine has five states with s_0 as initial state and s_4 as final state. The tape symbols are a, b, X, Y and B. From the table it is clear that, the Turing machine is in state s_0 on input 'a' moves to next state s_1 and replace the current symbol 'a' by X. The read-write head moves one cell in right direction. Similarly, if the Turing machine is in state s_2, then on input X, the machine moves to the state s_0 and replace X by X. The read-write head moves one cell in right direction.

State	Tape symbols				
	a	b	X	Y	B
s_0	XRs_1	$\bullet\bullet\bullet$	$\bullet\bullet\bullet$	YRs_3	$\bullet\bullet\bullet$
s_1	aRs_1	YLs_2	$\bullet\bullet\bullet$	XRs_1	$\bullet\bullet\bullet$
s_2	aLs_2	$\bullet\bullet\bullet$	XRs_1	YLs_2	$\bullet\bullet\bullet$
s_3	$\bullet\bullet\bullet$	$\bullet\bullet\bullet$	$\bullet\bullet\bullet$	YRs_3	BRs_4
s_4	$\bullet\bullet\bullet$	$\bullet\bullet\bullet$	$\bullet\bullet\bullet$	$\bullet\bullet\bullet$	$\bullet\bullet\bullet$

7.3.2   Representation by Transition Diagram

We have discussed transition graph for finite automaton in Chapter 2. Here, in this subsection we define transition function δ in terms of a transition graph to represent the Turing machine. In this transition graph, the edges are labelled with triples of the form (a_i, a_j, D), where $a_i, a_j \in \Gamma$ and $D \in \{L, R\}$. For the transition $\delta(s_i, a_i) = (s_j, a_j, D)$, we label the directed edge from the state s_i to s_j with the triple (a_i, a_j, D), where D is either right (R) or left (L).

The triple (a_i, a_j, D) between two states s_i to s_j indicate that, the symbol a_i is under read-write head of the Turing machine. The symbol a_i is replaced by the symbol a_j whereas the movement of read-write head is decided by the value of D. Therefore, the transition system of a Turing machine can be defined as quintuple $(s_p, \alpha, \beta, D, s_n)$, consisting of

s_p = Present state

α = Current symbol under read-write head

β = Replaced symbol

D = Direction of read-write head movement

s_n = Next state

Therefore, it is clear that each Turing machine can be represented by a collection of five-tuples representing all the directed edges. For example, the Turing machine discussed in the previous subsection can be represented as below.

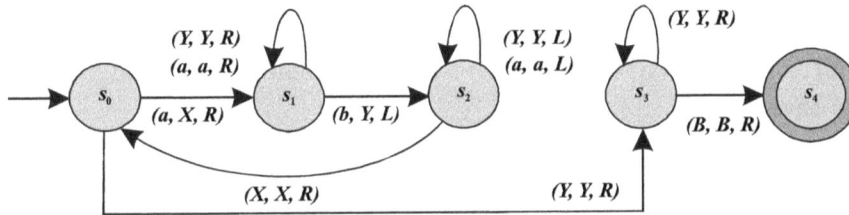

7.3.3 Representation by Instantaneous Description using Move Relation

In this subsection we define the transitions in a Turing machine by using move relation. Let us consider an instantaneous description $X_1 X_2 \cdots X_{i-1} s X_i \cdots X_n$. Suppose that $\delta(s, X_i) = (q, Y, L)$. If $(i-1) = n$, then X_i is taken to be B. If $i = 1$, then the tape head is to fall off the left end of the tape. This is not allowed and there is no next instantaneous description. If $i > 1$, then we have

$$X_1 X_2 \cdots X_{i-1} s X_i \cdots X_n \vdash X_1 X_2 \cdots X_{i-2} q X_{i-1} Y X_{i+1} \cdots X_n .$$

In this case, the symbol X_i is replaced by Y and there is a left move. Therefore, the current symbol under read-write head is X_{i-1} and the state s is changed to q.

Alternatively, suppose that $\delta(s, X_i) = (q, Y, R)$. If $i > 1$, then we have

$$X_1 X_2 \cdots X_{i-1} s X_i \cdots X_n \vdash X_1 X_2 \cdots X_{i-2} X_{i-1} Y q X_{i+1} \cdots X_n .$$

In this case, the symbol X_i is replaced by Y and there is a right move. Therefore, the current symbol under read-write head is X_{i+1} and the state s is changed to q.

7.4 ACCEPTANCE OF LANGUAGE BY TURING MACHINE

Turing machines can be viewed as language acceptors in the following sense. A string 'w' is written on the tape with blanks on either side. The Turing machine is started in the initial state s_0 with the read-write head at the leftmost symbol of 'w'. If the Turing machine enters a final state and halts after a sequence of moves, then the string 'w' is considered to be accepted. Mathematically,

the language accepted by Turing machine $M = (S, \Sigma, \Gamma, \delta, s_0, B, A)$ is given as $L(M)$. It is defined as
$$L(M) = \left\{ w \mid w \in \Sigma^* \text{ and } s_0 w \vdash^* x_1 s_a x_2 \text{ for some } s_a \in A;\ x_1, x_2 \in \Gamma^* \right\}$$

From the definition it is clear that, when $w \notin L(M)$, then either the machine can halt in a non-final state or it can enter an infinite loop and never halt. Therefore, any string for which M does not halt is not in $L(M)$. It should be noted that there are many other equivalent definitions of acceptance by the Turing machine. It is beyond the scope of this text.

For example, consider the Turing machine $M = (S, \Sigma, \Gamma, \delta, s_0, B, A)$ consisting of $S = \{s_0, s_1, s_2, s_3, s_4\}$, $\Sigma = \{a, b\}$, $\Gamma = \{a, b, X, Y, B\}$ and $A = \{s_4\}$. The transitions δ are defined as $\delta(s_0, a) = (s_1, X, R)$, $\delta(s_0, Y) = (s_3, Y, R)$, $\delta(s_1, a) = (s_1, a, R)$, $\delta(s_1, b) = (s_2, Y, L)$, $\delta(s_1, Y) = (s_1, Y, R)$, $\delta(s_2, a) = (s_2, a, L)$, $\delta(s_2, X) = (s_0, X, R)$, $\delta(s_2, Y) = (s_2, Y, L)$, $\delta(s_3, Y) = (s_3, Y, R)$ and $\delta(s_3, B) = (s_4, B, R)$. Let us consider the string $w = a^2 b^2 = aabb$. Therefore, we have

$$
\begin{aligned}
s_0 w = s_0 aabb &\vdash X s_1 abb \\
&\vdash X a s_1 bb \vdash X s_2 aYb \vdash s_2 X aYb \vdash X s_0 aYb \\
&\vdash XX s_1 Yb \vdash XXY s_1 b \vdash XX s_2 YY \vdash X s_2 XYY \\
&\vdash XX s_0 YY \vdash XXY s_3 Y \vdash XXYY s_3 B \vdash XXYYB s_4
\end{aligned}
$$

Therefore we get, $s_0 aabb \vdash XXYYBs_4$ with $s_4 \in A$. Thus, the string $w = aabb$ is accepted by the Turing machine M.

7.5 DESIGN OF TURING MACHINES

The basic guidelines for designing a Turing machine are given below.

(*a*) The basic objective in scanning a symbol from input tape by read-write head is to know what to do in the future. Also, the machine must remember the scanned past symbol. This can be done by moving to the next unique state on each input.

(*b*) The total number of states must be minimized. This can be done in a Turing machine by changing the states only when there is a change in the written symbol or when there is a change in the movement of read-write head.

However, these guidelines are not sufficient for complicated Turing machine constructions. In this section, we explain the basic guidelines by considering a simple example. Let us design a Turing machine to recognize all strings consisting of odd number of 0's. We explain the construction by defining moves in the following manner.

   i. Let the initial state be s_0. The Turing machine M enters state s_1 on scanning 0 and writes B.

   ii. If Turing machine M is in state s_1 and scans symbol 0, it enters the state s_0 and writes B.

   iii. The state s_1 is the only final state.

Therefore, the Turing machine M accepts a string if it exhausts all input symbols and finally in state s_1. Thus M is defined as:

$$M = (S, \Sigma, \Gamma, \delta, s_0, B, A)$$

where $S = \{s_0, s_1\}$, $\Sigma = \{0\}$, $\Gamma = \{0, B\}$, $A = \{s_1\}$ and δ is defined by following table.

State	Input symbol (0)
$\rightarrow s_0$	$B R s_1$
s_1	$B R s_0$

Let us consider a string $w = 00000$. Therefore, we have

$$s_0 00000 \vdash B s_1 0000 \vdash B B s_0 000 \vdash B B B s_1 00$$
$$\vdash B B B B s_0 0 \vdash B B B B B s_1 B$$

As we reached an accepting state s_1, the string $w = 00000$ is accepted. Similarly, it can be shown that the strings having even number of 0's is not accepted by M as s_0 is not a final state. It is to be noted that in both cases the Turing machine M halts, but in different states.

7.6   TWO-STACK PUSHDOWN AUTOMATA AND TURING MACHINE

In the introduction it has mentioned that, there is a natural extension from finite automata to pushdown automata. It made easy to prove that all regular languages could also be accepted by pushdown automata. At the same time, there is no such natural connection between pushdown automata and Turing machines.

In the previous Chapter, we discussed that the power of a pushdown automaton can be increased by introducing extra stacks. It is observed that, a two-stack pushdown automata can accept several languages that a pushdown automaton can not. This leads to many general questions. Is there any language that is not acceptable by two-stack pushdown automata? Is three-stack pushdown automata stronger than two-stack pushdown automata? However many of these questions are answered by a theorem of Marvin Minsky (1961). Here we quote the statement of the theorem however the proof of this theorem is beyond the scope of this book.

Minsky's Theorem   Any language accepted by a two-stack pushdown automaton can also be accepted by some Turing machine and any language accepted by Turing machine (TM) can also be accepted by some two-stack pushdown automaton (2 PDA). In other words, 2PDA = TM.

7.7   TURING MACHINE AND TYPE-0 GRAMMAR

A type-0 grammar is a four-tuple $G = (V_N, T, P, Q_0)$, where V_N and T are disjoint sets of variables and terminals, respectively; Q_0 is an element of V_N called the start symbol; and P is the set of production rules of the form $\alpha \rightarrow \beta$, where $\alpha, \beta \in (V_N \cup T)^*$ and α contains at least one variable. A type-0 grammar is otherwise known as unrestricted grammar or phrase structure grammar. In this section we construct an unrestricted grammar (type-0) grammar generating the set accepted by a Turing machine M. This construction of productions can be carried out in the following two steps.

Step 1   We construct production rules that transform the string $[s_0 \not c w \not s]$ to $[s_f B]$, where s_0 is the initial state, $\not c$ is the left-end marker, $\not s$ is the right-end marker, and s_f is an accepting state, i.e., $s_f \in A$. The reduced grammar obtained by applying step 1 is called as transformational grammar.

Step 2   Obtain the required unrestricted (type-0) grammar G by reversing the productions of the transformational grammar. The construction is in such a way that w is accepted by M if and only if $w \in L(G)$.

7.7.1   Construction of Type-0 Grammar

It is clear that in the construction of a type-0 grammar a transition of instantaneous description (ID) corresponds to a production. So, the acceptance of a string w by Turing machine M corresponds to the transformation of initial instantaneous description $[s_0 \not c w \not s]$ into final instantaneous description $[s_f B]$. The length of instantaneous description may change if the read-write head reaches the left end or right end. So, the productions are related to transitions of instantaneous description with change in length and no change in length. The construction of type-0 grammar corresponding to standard Turing machine given by a transition table is given below. It includes the following two steps.

Step 1

i.   *No change in length of instantaneous description*

(a) If there is a right move $a_k R s_l$ corresponding to s_i- row and a_j- column in the transition table, then the induced production is given as

$$s_i a_j \rightarrow a_k s_l.$$

(b) If there is a left move $a_k L s_l$ corresponding to s_i- row and a_j- column in the transition table, then the induced productions are given as

$$a_m s_i a_j \rightarrow s_l a_m a_k \quad \forall \quad a_m \in \Gamma.$$

ii.   *Change in length of instantaneous description*

(a) In case of right end, if B occurs to the left of right bracket $(])$, then it can be removed by introducing the production

$$a_j B] \rightarrow a_j] \quad \forall \, a_j \in \Gamma.$$

If the read-write head moves to the right of right bracket $(])$, then the length increases. The corresponding productions are given as

$$s_i] \rightarrow s_i B] \quad \forall \quad s_i \in S.$$

(*b*) In case of left end, if there is a move $a_k L s_l$ corresponding to s_i- row and a_j- column in the transition table, then the induced production is given as

$$[s_i a_j \rightarrow [s_l B a_k \, .$$

If B occurs next to the left bracket $($[$)$, then it can be removed by introducing the following production

$$[B \rightarrow [\, .$$

iii. *Introduction of end markers* We include the following productions to introduce end markers.

$$a_i \rightarrow [s_0 \text{¢} a_i \quad \forall \quad a_i \in \Gamma, a_i \neq B$$
$$a_i \rightarrow a_i \text{\$}] \quad \forall \quad a_i \in \Gamma, a_i \neq B$$

The brackets from $[s_f B]$ can be removed by introducing the production $[s_f B] \rightarrow Q_0$, where Q_0 is the newly introduced symbol called the start symbol of the grammar constructed and $s_f \in A$ is an accepting state.

Step 2 Reverse all the arrows of the productions obtained in Step 1 to get the required grammar. The productions that are generated are called inverse productions. The grammar thus obtained is also called generative grammar.

For illustration, consider the example of a Turing machine described by the transition table given below. Our aim is to obtain the inverse production rules. In the following given transition table, s_0 is considered as the initial state whereas s_4 is considered as the final state.

Present state	Tape symbols		
	¢	1	B
→ s_0	¢ $R s_0$	$1 R s_0$	$1 R s_1$
s_1	...	$1 R s_1$	$B L s_2$
s_2	...	$B L s_3$	...
s_3	...	$1 L s_3$	$B R s_4$
s_4	...	...	...

Step 1

i. *No change in length of instantaneous description*

(*a*) Productions corresponding to right moves are given as below.

$$s_0 \text{¢} \rightarrow \text{¢} s_0, \quad s_0 1 \rightarrow 1 s_0, \quad s_0 B \rightarrow 1 s_1, \quad s_1 1 \rightarrow 1 s_1, \quad s_3 B \rightarrow B s_4$$

(*b*) Productions corresponding to left moves are given as below.

$$Bs_1 B \rightarrow s_2 BB, \qquad Bs_2 1 \rightarrow s_3 BB, \qquad Bs_3 1 \rightarrow s_3 B1,$$
$$1s_1 B \rightarrow s_2 1B, \qquad 1s_2 1 \rightarrow s_3 1B, \qquad 1s_3 1 \rightarrow s_3 11$$

ii.   *Change in length of instantaneous description*

(a) Productions corresponding to right end are given below.

$BB] \to B]$, $s_0] \to s_0 B]$, $s_2] \to s_2 B]$, $s_4] \to s_4 B]$,

$1B] \to 1]$, $s_1] \to s_1 B]$, $s_3] \to s_3 B]$.

(b) Productions corresponding to left end are given below.

$[s_1 B \to [s_2 BB$, $[s_2 1 \to [s_3 BB$, $[s_3 1 \to [s_3 B1$, $[B \to [$

iii.   Productions corresponding to end markers are $1 \to [s_0 \phi 1, 1 \to 1 \text{\$}]$ and $[s_4 B] \to Q_0$, where Q_0 is the newly introduced symbol called the start symbol.

Step 2   The inverse productions are obtained by reversing the arrows of the above obtained productions.

$\phi s_0 \to s_0 \phi$, $1 s_0 \to s_0 1$, $1 s_1 \to s_0 B$, $1 s_1 \to s_1 1$, $B s_4 \to s_3 B$,

$s_2 BB \to B s_1 B$, $s_3 B1 \to B s_3 1$, $s_2 1B \to 1 s_1 B$, $s_3 1B \to 1 s_2 1$, $s_3 11 \to 1 s_3 1$,

$B] \to BB]$, $1] \to 1B]$, $s_0 B] \to s_0]$, $s_1 B] \to s_1]$, $s_2 B] \to s_2]$,

$s_3 B] \to s_3]$, $s_4 B] \to s_4]$, $[s_2 BB \to [s_1 B$, $[s_3 BB \to [s_2 1$, $[s_3 B1 \to [s_3 1$,

$[\to [B$, $[s_0 \phi 1 \to 1$, $1 \text{\$}] \to 1$, $Q_0 \to [s_4 B]$

Thus the grammar obtained is the type-0 grammar corresponding to the given Turing machine. The converse is also true, i.e., given a type-0 grammar *G*, there exists a Turing machine accepting *L* (*G*). This is beyond the scope of this book.

7.8   VARIATIONS OF TURING MACHINE

The standard Turing machine is not the only possible one; there are many alternative definitions that could serve equally well. In this section we will study several ways in which the standard Turing machine model might be modified. The modifications to be considered include both generalizations and restrictions. Variations on the basic Turing machine model providing with more than one tape or with tapes that extend in several dimensions is generalizations whereas limiting the number of tape symbols that a machine may use is known as restrictions. The basic reason for studying such variations is to determine how they affect the computing capabilities of the standard Turing machine model ignoring the question of efficiency. We will see that none of the generalizations increases the computing power of the standard Turing machine model. So, any computation that can be performed on such a variation will still fall under the category of a mechanical computation and hence can be done by a standard Turing machine model. We are therefore free to employ whichever variant of the standard model is most convenient for a given problem.

7.8.1   Nondeterministic Turing Machine

Nondeterminism plays vital roles in the two simpler models of computation that we discussed earlier. Standard Turing machines have enough computing power that the nondeterminism fail to reduce the computing power of the basic model. But we study this as a variation of standard Turing machine. A nondeterministic Turing machine is a parallel Turing machine that can take many computational paths simultaneously, with the restriction that the parallel Turing machines cannot communicate.

A nondeterministic Turing machine M is defined as seven-tuple, $M = (S, \Sigma, \Gamma, \delta, s_0, B, A)$, consisting of

S = Non-empty finite set of internal states

Γ = Non-empty finite set of tape symbols

B = Special symbol in Γ, called blank

Σ = Non-empty finite set of input symbols, which is a subset of Γ but not including B, i.e., $\Sigma \subseteq \Gamma - \{B\}$

s_0 = Initial or start state, i.e., $s_0 \in S$

A = Set of accepting states, i.e., $A \subseteq S$

δ = Transition function from $S \times \Gamma \to 2^{S \times \Gamma \times \{L, R\}}$

The transition function δ makes the Turing machine nondeterministic. This is because, the transition function δ contains a set of possible transitions and any one of which can be chosen by the machine. For example, consider the transition:

$$\delta(s, x) = \{(s_1, y, R), (s_2, z, L)\}$$

It is clear from the transition that, the Turing machine is nondeterministic. Let us consider a string $w = xyyx$. Therefore, by applying this transition we have

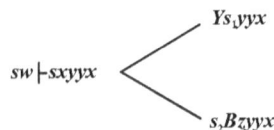

A nondeterministic Turing machine accepts a string w if there exists a possible sequence such that $s_0 w \vdash \alpha s_a \beta$; $\alpha, \beta \in \Gamma^*$ and $s_a \in A$. A nondeterministic Turing machine may have moves available that lead to a non-final state or to an infinite loop. But these alternatives are not relevant as with nondeterminism always.

7.8.2   Multitape Turing Machine

Another way to extend the standard Turing machine model is to provide it with several tapes. Although the resulting multitape Turing machines are most convenient to work, the availability of extra tapes does not provide in an increase in computing power. In this multitape Turing machine a finite state control unit is coupled to two or more different tapes by means of two or more different read-write heads.

Typically, a n-tape Turing machine M is defined as seven-tuple, $M = (S, \Sigma, \Gamma, \delta, s_0, B, A)$ consisting of

S = Non-empty finite set of internal states

Γ = Non-empty finite set of tape symbols

B = Special symbol in Γ, called blank

Σ = Non-empty finite set of input symbols, which is a subset of Γ but not including B, i.e.,
$\Sigma \subseteq \Gamma - \{B\}$

s_0 = Initial or start state, i.e., $s_0 \in S$

A = Set of accepting states, i.e., $A \subseteq S$

δ = Transition function from $S \times \Gamma^n \rightarrow S \times \Gamma^n \times \{L, R\}^n$

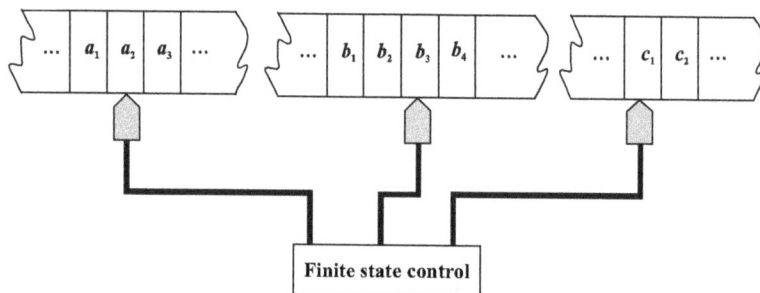

Figure 7.3   Three-tape Turing machine

For $n = 3$, a three-tape Turing machine model is given above in Figure 7.3. A basic move of such a multitape machine consists in writing a new symbol in the currently scanned cell of each tape, shifting all tapes one cell to the left or right, and entering a new internal state. It is to be noted that all tapes need not be shifted in the same direction. The particular action that the machine takes at any given step is determined by the machine's current state and the symbols appearing in the currently scanned cells of the tapes. However this Turing machine can be simulated to a standard Turing machine.

7.8.3   Multidimensional Turing Machine

Another way to extend the standard Turing machine that adds no additional power—the multidimensional Turing machine. A Turing machine is said to be multidimensional if its tape can be viewed as extending infinitely in more than one dimension. This machine has the usual finite control, but the tape consists of a k-dimensional array of cells infinite in all $2k$ directions, for some finite value of k. We establish this fact for the case of two-dimensional machines, the generalization to multidimensional machines being straightforward.

The formal definition of a two-dimensional Turing machine is seven-tuple M such that $M = (S, \Sigma, \Gamma, \delta, s_0, B, A)$ consisting of

S = Non-empty finite set of internal states

Γ = Non-empty finite set of tape symbols

B = Special symbol in Γ, called blank

Σ = Non-empty finite set of input symbols, which is a subset of Γ but not including B, i.e.,
$\Sigma \subseteq \Gamma - \{B\}$

s_0 = Initial or start state, i.e., $s_0 \in S$

A = Set of accepting states, i.e., $A \subseteq S$

δ = Transition function from $S \times \Gamma \to S \times \Gamma \times \{L, R, U, D\}$

Here, U and D specify movement of read-write head in up and down directions respectively. A basic move of such a machine consists in writing a new symbol in the currently scanned cell, shifting its read-write head in one of 4 directions (in general $2k$), along one of the 2 axes (in general k). The particular action that the machine takes at any step is determined by the current state and the symbol scanned. Initially, the input is along one axis, and the read-write head is at the left end of the input. However a standard Turing machine can simulate a multidimensional Turing machine. A model of two-dimensional Turing machine is given below in Figure 7.4.

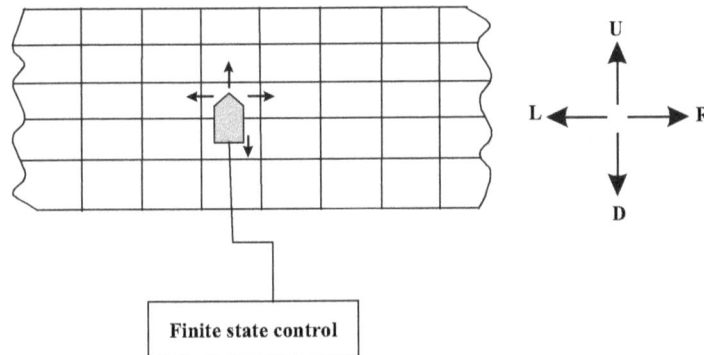

Figure 7.4 Two-dimensional Turing machine

7.8.4   Multihead Turing Machine

A multihead (k-head) Turing machine can be viewed as a Turing machine with a single tape, single finite state control but with multiple independent read-write heads. The heads are numbered 1 through k. A basic move of such a machine depends on the current state and on the symbol scanned by each head. It is to be noted that, the heads may each move independently left, right, or remain stationary.

The formal definition of a k-head Turing machine is 7-tuple $M = (S, \Sigma, \Gamma, \delta, s_0, B, A)$ consisting of

S = Non-empty finite set of internal states

Γ = Non-empty finite set of tape symbols

B = Special symbol in Γ, called blank

Σ = Non-empty finite set of input symbols, which is a subset of Γ but not including B, i.e.,
$\Sigma \subseteq \Gamma - \{B\}$

s_0= Initial or start state, i.e., $s_0 \in S$

A = Set of accepting states, i.e., $A \subseteq S$

δ = Transition function from $S \times \Gamma_T \to S \times \Gamma_T \times \{L, R\}^k$

It is to be noted that $\Gamma_T = (\Gamma \times \Gamma \times \Gamma \times \cdots \times \Gamma)$ (k-times), where k is the number of read-write heads. However a k-head Turing machine can be simulated to a standard Turing machine. A model of three-head Turing machine is given below in Figure 7.5.

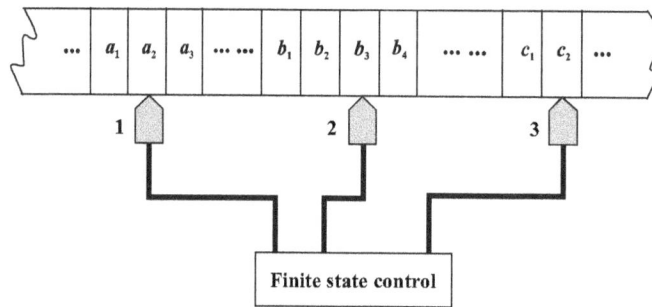

Figure 7.5   Three-head Turing machine

7.8.5   Off-line Turing Machine

A multitape Turing machine whose input tape is read-only is an off-line Turing machine. Usually, the Turing machine is not allowed to move the input tape head off the region between the end markers ϕ and $\$$. A model of an off-line Turing machine is shown below in Figure 7.6.

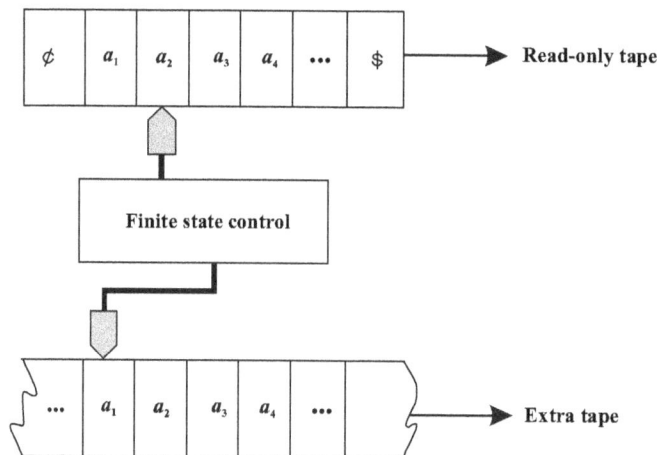

Figure 7.6   Off-line Turing machine

This machine is no more efficient than standard Turing machine as it is a special case of multitape Turing machine. On the other hand, an off-line Turing machine can simulate a standard Turing machine M by introducing one more tape than M. This is because the off-line Turing machine makes a copy of its own input into the extra tape, and then it simulates standard Turing machine M as the extra tape has only M's input.

7.8.6 Multitrack Machine

Multitrack machine is a simple generalization involving the format of the tapes used by Turing machine. In many cases it is observed that, it is convenient to work with machines whose tapes are divided lengthwise into multi-layers or tracks. A two-track machine is shown below in Figure 7.7.

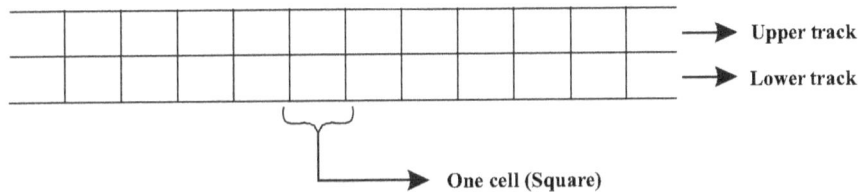

Figure 7.7 Two-track machine

Each square consists of two halves, each of which may be independently inscribed with a single symbol. We assume that the read-write head scans both halves of a given tape square simultaneously. A basic move consists in examining the symbols obtained in the two halves of the scanned square, writing a new symbol in each half, moving the tape one square to the left or right, and entering a new state. The machines that operate on two track tapes is known as two-track machine. However, this idea can be generalized to multitrack machines.

	$\frac{B}{B}$	$\frac{B}{a}$	$\frac{a}{B}$	$\frac{a}{a}$
s_0	$\frac{B}{B}\ Rs_0$	$\frac{B}{a}\ Rs_3$	$\frac{a}{a}\ Rs_1$	$\frac{a}{a}\ Rs_0$
s_0	$\frac{B}{a}\ Ls_2$	$\frac{B}{a}\ Rs_1$	$\frac{a}{B}\ Rs_1$	$\frac{a}{a}\ Rs_1$
s_2	$\frac{B}{B}\ Ls_2$	$\frac{B}{a}\ Ls_2$	$\frac{a}{B}\ Ls_2$	$\frac{a}{a}\ Rs_0$
s_3		$\frac{B}{a}\ Rs_3$	$\frac{a}{B}\ Rs_3$	$\frac{a}{a}\ Rs_3$

It is necessary to define exactly what move the machine will make for every possible combination of the machines current state, the symbol found in the upper half of the scanned square, and the symbol found in the lower half of the scanned square to describe the behaviour of a two-track machine. Let us consider the example given above. The above table shows the state table description of a two-track machine in which each half of a square may either contain a symbol a or else a blank B.

In this context, the notation $\frac{B}{a}$ represents the situation in which the upper half of the scanned square is blank and the lower half contains a symbol a, and so on. In general, if a two-track machine in which the upper half of each square may contain any one of k_1 (non-blank) symbols and the lower half may contain any one of k_2 (non-blank) symbols, the state table description of that machine will require $(k_1 + 1)(k_2 + 1)$ columns.

Any two-track machine can be viewed as a standard Turing machine (one track) whose tape symbols are order pairs. If each order pair is replaced by an individual symbol, then the resulting machine cannot be differentiated from that of a standard Turing machine. For example, replacing the symbol pairs $\frac{B}{B}, \frac{a}{B}, \frac{B}{a}, \frac{a}{a}$ by the respective symbols B, a, b, c converts the above two-track machine into the following standard Turing machine.

	B	b	a	c
s_0	$B\,Rs_0$	$b\,Rs_3$	$c\,Rs_1$	$c\,Rs_0$
s_1	$b\,Ls_2$	$b\,Rs_1$	$a\,Rs_1$	$c\,Rs_1$
s_2	$B\,Ls_2$	$b\,Ls_2$	$a\,Ls_2$	$c\,Rs_0$
s_3		$c\,Rs_3$	$a\,Rs_3$	$c\,Rs_3$

7.9   UNIVERSAL TURING MACHINE

In the preceding subsections we studied different variations of a standard Turing machine and how a suitably chosen machine of one kind could be made to simulate a particular machine of another kind. Now we consider the possibility of designing a single machine of a specified kind that is capable of simulating any other machine of that same kind. A machine having this feature is called a universal Turing machine. So, a Turing machine is said to be an universal Turing machine if it can accept two things: the input data and an algorithm for computation. Since any variation of Turing machine can be simulated to a standard Turing machine, we restrict our attention to machines that operate on single infinite tape.

If a fixed Turing machine is to replicate the behaviour of an arbitrarily machine, it must be supplied with two things: a description of the machine that is to be simulated, and the initial tape pattern of that machine. Once these descriptions have been agreed upon, we can design the universal Turing machine. In general, we now describe our universal Turing machine. For convenience, we refer U as the universal Turing machine and T as the machine whose behaviour is to be simulated.

The tape of the universal Turing machine is divided into three regions as shown in the Figure 7.8 given above. The leftmost cell of the tape is permanently inscribed with the special symbol F. This cell F is followed by a buffer region. This buffer region contains at least $\left(|S| + |\Sigma| + 2\right)$ cells, where

S is the set of states and Σ is the set of symbols used by the machine *T*. The buffer region is followed by a machine description region, which is further followed by a tape description region. The role of buffer region is to store a representation of *T*'s current state and scanned symbol whereas the machine description region is to store a permanent copy of the quintuple description of the machine *T*. The tape description region is used to store the current tape pattern of *T*.

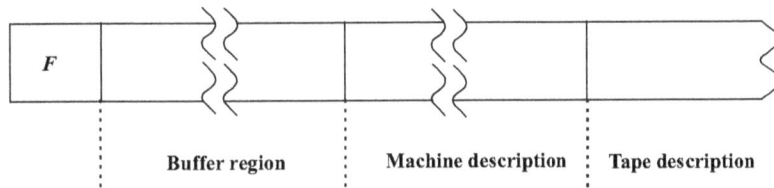

Figure 7.8 Universal Turing machine

The computation of universal Turing machine with its tape inscribed is as follows. The buffer region is initially blank. The encoded description of the Turing machine *T* that is to be simulated is contained in the machine description region. This description is preceded by a special symbol *C* and followed by 3 blank cells. As a result, the symbol *C* immediately precedes a block of 1's corresponding to the starting state of *T*. The encoded description of *T*'s initial tape pattern is contained in tape description region with a special marking symbol *D* immediately preceding the block corresponding to *T*'s initially scanned tape symbol. The read-write head of universal Turing machine *U* initially scans the cell containing *C*. The following series of steps are used to carry out the computations of *U*, each of which is corresponding to one move in the computation of *T*. At each step, the universal machine *U* uses its description of *T* for updating the representation of *T*'s tape pattern and to keep track of *T*'s internal state. If *T* halts eventually, then *U* will halt during the corresponding step of its simulation.

Step 1   The machine *U* copies the block marked by the symbol *C* into the initial portion of the buffer region. Then it immediately writes an auxiliary *F* after the new block, erases the symbol *C*, and moves in the direction of right until it reaches the symbol *D*.

Step 2   The machine *U* next copies the block marked by the symbol *D* into the portion of the buffer region immediately following the auxiliary *F*.

Step 3   The machine *U* erases the auxiliary *F* from the buffer region. It immediately writes a *C* to the left of the machine description. It is to be noted that the buffer region now contains the unary representations of *T*'s current state and scanned symbol in order.

Step 4   The machine *U* next searches through the quintuples of *T*'s description, looking for one whose first two components match the state and symbol representations currently inscribed in the buffer. Generally it is carried out as follows. The machine begins comparing one at a time, the first two cells to the right of the symbol *C* with the two cells in the buffer. If these cells match with each other, then the desired quintuple has been found. If they do not match, *U* moves the *C* marker to the beginning of the next quintuple and repeats the process. If end of the machine description is reached without getting a match, the machine *T* must halt during the move being simulated, and hence *U* also halts.

Step 5 Once getting the quintuple that specifies *T*'s current move, machine *U* erases the contents of the buffer and shifts *C* marker one cell to the right, so that it stands in front of the new symbol component of the quintuple.

Step 6 The machine *U* then updates its description of *T*'s tape pattern by substituting the new symbol cell following the *C* marker for the cell following the *D* marker in the tape description. With this substitution, *U* erases the *C* marker and moves to the right until it reaches the directional component of the quintuple under consideration.

Step 7 The machine *U* counts the number of 1's in the directional component and immediately writes *C* after it. It then shifts the *D* marker one cell in either direction (left or right), as required. If the movement is to the left, *U* must determine whether *T* runs off its tape; and then the machine *U* halts. If the movement is to the right, *U* must determine whether it has reached the end of the encoded part of *T*'s tape pattern; if so, it must add a new cell representing the next blank cell on *T*'s tape.

The step by step simulation process will be continued until *T* may either halt or run off its tape. The design approach that is discussed here does not represent the only way of designing a universal machine. The existence of universal machine is important for at least two reasons. First, the universal machine establishes certain important properties of Turing machine and the functions they compute. Second, the fact that one machine can replicate the behaviour of any other machine. The detailed operation of the universal machine is beyond the scope of this book.

7.10 HALTING PROBLEM OF TURING MACHINE

In this section we will discuss about halting problem. The halting problem is a decision problem which can be stated as follows: given a description of a program and a finite input, decide whether the program finishes running or will run forever, given that input. Alan Turing proved in 1936 that a general algorithm to solve the halting problem for all possible program-input pairs cannot exist. We say that the halting problem is undecidable over Turing machines. It was one of the first problems to be proved undecidable. One such consequence of the halting problem's undecidability is that there cannot be a general algorithm that decides whether a given statement about natural numbers is true or not. The reason for this is that the proposition stating that a certain algorithm will halt given a certain input can be converted into an equivalent statement about natural numbers. If we had an algorithm that could solve every statement about natural numbers, it could certainly solve this one; but that would determine whether the original program halts, which is impossible, since the halting problem is undecidable.

The halting problem of Turing machine can be expressed as follows. For a given Turing machine *M* and a given input '*w*', does the Turing machine *M* perform a computation that eventually halts, while Turing machine starts in the initial configuration s_0, w? It can also be asked whether or not Turing machine *M* can be applied to input '*w*', or simply (*M, w*) halts or does not halt. The halting problem of Turing machine is undecidable problem, but we can run the machine with a given input string. If the machine halts, then a specific case is answered. At the same time if the machine does not halt, then the conclusion is that the machine will never halt by recognizing some pattern within the computation.

7.11 LINEAR-BOUNDED AUTOMATA

The model of computation which we shall examine here is the linear-bounded automaton. The power of the standard Turing machine cannot be extended by complicating the structure of tape. It is possible to limit it by restricting the way in which the tape can be used. The way of limiting the tape uses it to allow the machine to use only that part of the tape occupied by the input. As a result, more space will be available for long input strings than short strings, generating another class of machines called linear-bounded automata (LBA).

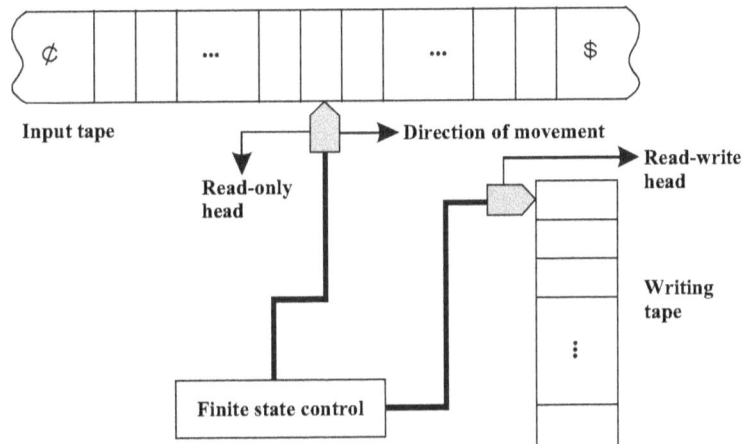

Figure 7.9 Model of linear-bounded automata

The model of linear-bounded automata consists of an input tape and a working tape (Figure 7.9). A read-only head is associated with the input tape and is allowed to move only in right direction. Another read-write is associated with the working tape. It can modify the contents of working tape in any way without any restriction. The left end and right end of the input tape is marked with ¢ and $ respectively. The end markers ¢ and $ on the input tape decide the length of input tape and never allow the machine to go past them. This will ensure that the storage bounds are maintained and help keep our machines from leaving their tapes. If the input tape contains n-cells, then the length of the input string 'w' will be of maximum $(n-2)$. The string 'w' will be accepted by a linear-bounded automaton if there is a Turing machine that accepts 'w' using at most kn cells of input tape, where k is a constant specified in the description of linear-bounded automaton and it is completely independent of the input string. Some authors define a linear-bounded automaton to use only the portion of the tape that is occupied by the input; that is, $k=1$ in all cases. The definitions lead to equivalent classes of machines, because we can compensate for the shorter tape by having a larger tape alphabet. Note that the following definition of linear-bounded automaton is assumed to be essentially nondeterministic.

Formally, a linear-bounded automaton is a nondeterministic Turing machine which has only one tape, and this tape is exactly the same length as the input. We define a linear-bounded automaton as eight-tuple $M = (S, \Sigma, \Gamma, \delta, s_0, \text{¢}, \text{\$}, A)$, consisting of

S = Non-empty finite set of internal states

Γ = Non-empty finite set of tape symbols

Σ = Non-empty finite set of input symbols, including end markers $\textcent$ and $\textdollar$ which is a subset of Γ, i.e., $\Sigma \subseteq \Gamma - \{B\}$.

s_0 = Initial or start state, i.e., $s_0 \in S$.

A = Set of accepting states, i.e., $A \subseteq S$.

δ = Transition function from $S \times \Gamma \rightarrow S \times \Gamma \times \{L, R\}$

The language accepted by the linear-bounded automata M, $L(M)$ is defined as

$$L(M) = \{w \mid w \in (\Sigma - \{\textcent, \textdollar\})^* \text{ and } s_0 \textcent w \textdollar \vdash^* \alpha s \beta; \alpha, \beta \in \Gamma^*, s \in A\}$$

It differs from a Turing machine in that while the tape is initially considered to have unbounded length, only a finite contiguous portion of the tape, whose length is a linear function of the length of the initial input, can be accessed by the read/write head. This limitation makes an LBA a more accurate model of computers that actually exist than a Turing machine in some respects.

Linear-bounded automata are acceptors for the class of context-sensitive languages. The only restriction placed on grammars for such languages is that no production maps a string to a shorter string. Thus no derivation of a string in a context-sensitive language can contain a sentential form longer than the string itself. Since there is a one-to-one correspondence between linear-bounded automata and such grammars, no more tape than that occupied by the original string is necessary for the string to be recognized by the automaton.

7.12 POST MACHINE

Emil L. Post in 1936 developed a mathematical model of computation that is independent of Alan Turing's mathematical model but was essentially equivalent to the Turing machine model. It is a model of equivalent power but increasing complexity. This model is sometimes called as Post's Machine or a Post-Turing machine. A Post machine is a program formulation of an especially simple type of Turing machine comprising a variant of Emil Post's Turing-equivalent model. Post's model and Turing's model, though very similar to one another, were developed independently. A Post-Turing machine uses a binary alphabet, an infinite sequence of binary storage locations, and a primitive programming language with instructions for bi-directional movement among the storage locations and alteration of their contents one at a time.

Post's machine employs a "symbol space" consisting of a two-way infinite sequence of squares or boxes, each box capable of being in either of two possible conditions, namely marked and unmarked. Initially, finitely-many of the boxes are marked, the rest being unmarked. A worker is then to move among the squares, being in and operating in only one square at a time, according to a fixed finite set of instructions, which are numbered in order (1,2,3,..., n). Beginning at a square singled out as the starting point, the worker is to follow the set of instructions one at a time, beginning with the first instruction.

Formally, we define the Post machine (PM) as five-tuple (S, Σ, Q, R, F) consisting of

S = Non-empty finite set of states including initial (ACCEPT) and some halt (REJECT) states.

Σ = Alphabet of input letters including the special symbol #.

Q = Linear storage location called store in the form of a queue.

R = Set of read states that remove the left-most character from the store.

F = Set of ADD states that concentrate a character on to the right end of the string in the store.

It is to be noted that ADD states are different than PUSH states of pushdown automata and it has no such PUSH states. A Post machine is similar to a pushdown automata, but with the following differences. A Post machine is deterministic. It has an auxilliary queue instead of a stack. The input string is assumed to have been previously loaded on to the queue. For an example, if the input string is 101, then the symbol currently at the front of the queue is 1. Items can be inserted only to the rear end of the queue, and is deleted only from the front end.

SOLVED EXAMPLES

Example 1

Design a Turing machine to accept the language $L = \{a^n b^n : n \geq 1\}$.

Solution Initially, the tape of Turing machine M contains $a^n b^n$ followed by infinity of blanks B. Repeatedly, the Turing machine M replaces the leftmost 'a' by another symbol X, moves right to the leftmost 'b' and replace it by new symbol Y, moves left to get the rightmost X and then moves one cell right to the leftmost a, and repeats the cycle.

The Turing machine M halts without accepting the string if M finds a blank B, when searching for a symbol 'b'. If, after changing a 'b' to a Y, Turing machine M finds no more a's, then M checks that no more remain, accepting if there are none.

We define Turing machine M as $M = (S, \Sigma, \Gamma, \delta, s_0, B, A)$ consisting of $\Sigma = \{a, b\}$, $S = \{s_0, s_1, s_2, s_3, s_4\}$, $\Gamma = \{a, b, X, Y, B\}$ and $A = \{s_4\}$. The transitions δ are defined as below.

$\delta(s_0, a) = (s_1, X, R)$; $\delta(s_1, a) = (s_1, a, R)$; $\delta(s_1, Y) = (s_1, Y, R)$; $\delta(s_2, X) = (s_0, X, R)$;

$\delta(s_0, Y) = (s_3, Y, R)$; $\delta(s_1, b) = (s_2, Y, L)$; $\delta(s_2, a) = (s_2, a, L)$; $\delta(s_2, Y) = (s_2, Y, L)$;

$\delta(s_3, Y) = (s_3, Y, R)$; $\delta(s_3, B) = (s_4, B, R)$

Let us consider a string $w = aabb$. The sample computation of Turing machine M is as follows:

$$s_0aabb \vdash Xs_1abb$$
$$\vdash Xas_1bb \vdash Xs_2aYb \vdash s_2XaYb \vdash Xs_0aYb$$
$$\vdash XXs_1Yb \vdash XXYs_1b \vdash XXs_2YY \vdash Xs_2XYY$$
$$\vdash XXs_0YY \vdash XXYs_3Y \vdash XXYYs_3B \vdash XXYYBs_4$$

Therefore, the given string $w = aabb$ is accepted by the above Turing machine M.

Example 2

Design a Turing machine for proper subtraction. Proper subtraction $k \dot- l$ is defined as $(k - l)$ if $k \geq l$ and 0 if $k < l$.

Solution   Proper subtraction $k \dot- l$ is defined as below.

$$k \dot- l = \begin{cases} k - l & \text{if } k \geq l \\ 0 & \text{if } k < l \end{cases}$$

Initially, the tape of Turing machine M contains $0^k 1 0^l$ on its tape and halts with $0^{k \dot- l}$ on its tape. M repeatedly replaces its leading 0 by blank B, then searches right for a 1 followed by a 0 and replaces the 0 by 1. Secondly, M moves left until it encounters a blank and then repeats the cycle. The process terminates if either of the cases occur.

i.   Searching right for a 0, M encounters a blank. Then, the l 0's in $0^k 1 0^l$ have all been replaced to 1's, and $(l + 1)$ of the k 0's have been replaced to B. M replaces the $(l + 1)$ 1's by a 0 and l B's, leaving $(k - 1)$ 0's on its tape.

ii. In the beginning of a cycle, M cannot find a 0 to replace to a blank B, as first k 0's already have been replaced. In this case $k \leq l$ and so $k \dot- l = 0$. Therefore, M replaces all remaining 1's and 0's by B.

We define Turing machine M as $M = (S, \Sigma, \Gamma, \delta, s_0, B, \phi)$ consisting of $\Sigma = \{0,1\}$, $S = \{s_0, s_1, s_2, s_3, s_4, s_5, s_6\}$ and $\Gamma = \{0, 1, B\}$. The transitions δ are defined as below.

1.   Begin the cycle and replace the leading 0 by B, i.e., $\delta(s_0, 0) = (s_1, B, R)$.

2.   Searching right to get the first 1, i.e., $\delta(s_1, 0) = (s_1, 0, R)$; $\delta(s_1, 1) = (s_2, 1, R)$.

3.   Search right past 1's until encountering a 0 and then change it to 1, i.e., $\delta(s_2, 1) = (s_2, 1, R)$; $\delta(s_2, 0) = (s_3, 1, L)$.

4.   Move left to blank and enter state s_0 to repeat the cycle. i.e., $\delta(s_3, 1) = (s_3, 1, L)$; $\delta(s_3, 0) = (s_3, 0, L)$; $\delta(s_3, B) = (s_0, B, R)$.

5.   If in state s_2 a B is encountered before a 0, then it is described above in case (i). In such case, enter state s_4 and move left, changing all 1's to B's until encountering a B. This B is changed back to a 0 and state s_6 is entered. The machine M halts.

   $\delta(s_2, B) = (s_4, B, L)$; $\delta(s_4, B) = (s_6, 0, R)$; $\delta(s_4, 1) = (s_4, B, L)$; $\delta(s_4, 0) = (s_4, 0, L)$.

6. If, 1 is encountered instead of a 0 while in state s_0, the first block of 0's has been exhausted as discussed in case (ii). In such case M enters the state s_6 and the Turing machine M halts. The transitions are defined as below. $\delta(s_0,1) = (s_5,B,R)$; $\delta(s_5,0) = (s_5,B,R)$; $\delta(s_5,1) = (s_5,B,R)$; $\delta(s_5,B) = (s_6,B,R)$.

For example, consider the input string $w = 00010$ to compute $(3 \div 1)$. A sample computation of Turing machine M on the input string is given below.

$$s_0 00010 \vdash B s_1 0010$$
$$\vdash B0s_1 010 \vdash B00s_1 10 \vdash B001s_2 0 \vdash B00s_3 11$$
$$\vdash B0s_3 011 \vdash Bs_3 0011 \vdash s_3 B0011 \vdash Bs_0 0011$$
$$\vdash BBs_1 011 \vdash BB0s_1 11 \vdash BB01s_2 1 \vdash BB011s_2 B$$
$$\vdash BB01s_4 1B \vdash BB0s_4 1BB \vdash BBs_4 0BBB \vdash Bs_4 B0BB$$
$$\vdash B0s_6 0BBB$$

Therefore, $(3 \div 1) = 2$. Similarly, to compute the proper subtraction $(1 \div 2)$, we consider the input string $w = 0100$. A sample computation of M on the input string is given below.

$$s_0 0100 \vdash Bs_1 100$$
$$\vdash B1s_2 00 \vdash Bs_3 110 \vdash s_3 B110 \vdash Bs_0 110$$
$$\vdash BBs_5 10 \vdash BBBs_5 0 \vdash BBBBs_5$$
$$\vdash BBBBBs_6$$

Therefore, $(1 \div 2) = 0$.

Example 3

Design a Turing machine to recognize all strings containing even number of 0's.

Solution We construct the Turing machine M that recognizes all strings containing even number of 0's as follows.

(a) Initially, the Turing machine M is at initial state s_0. M enters state s_1 on scanning 0, replaces 0 by B and move right, i.e., $\delta(s_0,0) = (s_1,B,R)$.

(b) If in state s_1 a 0 is encountered, then it enters state s_0, move right and replace 0 by B, i.e., $\delta(s_1,0) = (s_0,B,R)$.

(c) The only accepting state of M is s_0.

Now, we define Turing machine M as $M = (S,\Sigma,\Gamma,\delta,s_0,B,A)$ consisting of $\Sigma = \{0\}$, $S = \{s_0,s_1\}$, $\Gamma = \{0,B\}$ and $A = \{s_0\}$. The transitions δ are defined as $\delta(s_0,0) = (s_1,B,R)$ and $\delta(s_1,0) = (s_0,B,R)$.

Let us consider a string $w = 0000$. A sample computation of M on input string 0000 is given below.

$$s_0 0000 \vdash Bs_1 000$$
$$\vdash BBs_0 00 \vdash BBBs_1 0 \vdash BBBBs_0 B$$

Similarly, consider the string $w = 000$. A sample computation of M on input string 000 is given below.

$$s_0 000 \vdash Bs_1 00$$
$$\vdash BBs_0 0 \vdash BBBs_1 B$$

From the above sample computation it is clear that the string $w = 0000$ is accepted by the Turing machine whereas the string $w = 000$ is not getting accepted.

Example 4

Design a Turing machine to recognize all strings containing any number of 1's.

Solution   We construct the Turing machine M that recognizes all strings containing any number of 1's as follows.

(a) Initially, the Turing machine M is at initial state s_0. M enters state s_1 on scanning 1, replaces 1 by B and move right, i.e., $\delta(s_0, 1) = (s_1, B, R)$.

(b) If in state s_1 a 1 is encountered, then it enters state s_0, move right and replace 1 by B, i.e., $\delta(s_1, 1) = (s_0, B, R)$.

(c) If in state s_1 a B is encountered instead of a 1, then the machine enters state s_2 and move left by accepting the given string, i.e., $\delta(s_1, B) = (s_2, B, L)$.

(d) If, B is encountered instead of a 1 while in state s_0, then the machine M enters state s_2 and move right by accepting the given string. The machine M halts, i.e., $\delta(s_0, B) = (s_2, B, R)$.

(e) The only accepting state of M is s_2.

Now, we define Turing machine M as $M = (S, \Sigma, \Gamma, \delta, s_0, B, A)$ consisting of $\Sigma = \{1\}$, $S = \{s_0, s_1, s_2\}$, $\Gamma = \{1, B\}$ and $A = \{s_2\}$. The transitions δ are defined as below.

$$\delta(s_0, 1) = (s_1, B, R); \quad \delta(s_1, 1) = (s_0, B, R);$$
$$\delta(s_0, B) = (s_2, B, R); \quad \delta(s_1, B) = (s_2, B, L).$$

Let us consider strings 111 and 1111. A sample computation of M on input strings 111 and 1111 are given below.

$$s_0 111 \vdash Bs_1 11$$
$$\vdash BBs_0 1 \vdash BBBs_1 B \vdash BBs_2 BB$$

$$s_0 1111 \vdash B s_1 111$$
$$\vdash BB s_0 11 \vdash BBB s_1 1 \vdash BBBB s_0 B$$
$$\vdash BBBBB s_2$$

From the above sample computation it is clear that the machine accept strings containing any number of 1's.

Example 5

Design a Turing machine that recognizes all strings containing even number of 0's and even number of 1's.

Solution   We construct the Turing machine M that recognizes all strings having even number of 0's and even number of 1's as follows.

i.   Initially, the Turing machine M is at initial state s_0. If in state s_0 a 0 is encountered, then the machine enters state s_1 and move right without changing the input symbol. On the other hand, if in state s_0 a 1 is encountered instead of 0, then the machine enters state s_2 and move right without affecting the input symbol.

ii. If 0 is encountered while in state s_1, then the machine again enters state s_0 and move right without changing the input symbol. If in state s_1 a 1 is encountered, then the machine M enters state s_3 and move right without changing the input symbol.

iii. If in state s_2 a 1 is encountered, then the machine enters state s_0 without affecting the input symbol. If 0 is encountered while in state s_2 instead of 1, then the machine enters state s_3 without changing the input symbol.

iv. If 0 is encountered while in state s_3, then the machine M changes its state to s_2 without affecting the input symbol and move right. On the other hand if 1 is encountered in state s_3, then the machine M changes its state to s_1 without changing the input symbol and move right.

v.   If in state s_0 a blank B is encountered, then the machine M enters final state s_4, move right and halts without changing the blank B.

vi. If a blank B is encountered while in either of states s_1, s_2 or s_3 then the machine M halts without entering into the final state s_4. In such cases, the Turing machine M halts without accepting the string.

Now, we define Turing machine M as $M = (S, \Sigma, \Gamma, \delta, s_0, B, A)$ consisting of $\Sigma = \{0,1\}$, $S = \{s_0, s_1, s_2, s_3, s_4\}$, $\Gamma = \{0,1,B\}$, $A = \{s_4\}$ and transitions δ are defined as below.

$$\delta(s_0,0) = (s_1,0,R); \quad \delta(s_1,0) = (s_0,0,R); \quad \delta(s_2,1) = (s_0,1,R);$$
$$\delta(s_0,1) = (s_2,1,R); \quad \delta(s_1,1) = (s_3,1,R); \quad \delta(s_3,0) = (s_2,0,R);$$
$$\delta(s_0,B) = (s_4,B,R); \quad \delta(s_2,0) = (s_3,0,R); \quad \delta(s_3,1) = (s_1,1,R).$$

Let us consider a string $w = 01101$. A sample computation is as follows:

$$s_0 01101 \vdash 0s_1 1101$$
$$\vdash 01s_3 101 \vdash 011s_1 01 \vdash 0110s_0 1$$
$$\vdash 01101s_2 B$$

Similarly, consider the string $w = 1001$. A sample computation on M is given below.

$$s_0 1001 \vdash 1s_2 001$$
$$\vdash 10s_3 01 \vdash 100s_2 1 \vdash 1001s_0 B$$
$$\vdash 1001Bs_4$$

From the above analysis it is clear that the string $w = 01101$ is not accepted by the machine M whereas the string $w = 1001$ is accepted by the machine M.

Example 6

Design a Turing machine to accept the following language L, where $L = \{a^n b^n c^n : n \geq 1\}$.

Solution   Initially, the tape of Turing machine M contains $a^n b^n c^n$ followed by infinity of blanks B. Repeatedly, the Turing machine M replaces the leftmost a by another symbol X, moves right to the leftmost b and replace it by another symbol Y, further move right to the leftmost c and replace it by new symbol Z, moves left to get the rightmost X and then moves one cell right to the leftmost a, and repeats the cycle.

We define Turing machine M as $M = (S, \Sigma, \Gamma, \delta, s_0, B, A)$ consisting of $\Sigma = \{a, b, c\}$, $S = \{s_0, s_1, s_2, s_3, s_4, s_5\}$, $\Gamma = \{a, b, c, X, Y, Z, B\}$ and $A = \{s_5\}$. The transitions δ are defined as below.

$\delta(s_0, a) = (s_1, X, R);$ $\delta(s_1, Y) = (s_1, Y, R);$ $\delta(s_3, a) = (s_3, a, L);$ $\delta(s_3, Z) = (s_3, Z, L);$

$\delta(s_0, Y) = (s_4, Y, R);$ $\delta(s_2, b) = (s_2, b, R);$ $\delta(s_3, b) = (s_3, b, L);$ $\delta(s_4, Y) = (s_4, Y, R);$

$\delta(s_1, a) = (s_1, a, R);$ $\delta(s_2, c) = (s_3, Z, L);$ $\delta(s_3, X) = (s_0, X, R);$ $\delta(s_4, Z) = (s_4, Z, R);$

$\delta(s_1, b) = (s_2, Y, R);$ $\delta(s_2, Z) = (s_2, Z, R);$ $\delta(s_3, Y) = (s_3, Y, L);$ $\delta(s_4, B) = (s_5, B, R);$

Let us consider a string $w = a^2 b^2 c^2$. A sample computation of M is given below.

$$s_0 a^2 b^2 c^2 \vdash Xs_1 ab^2 c^2$$
$$\vdash Xas_1 b^2 c^2 \vdash XaYs_2 bc^2 \vdash XaYbs_2 c^2 \vdash XaYs_3 bZc$$
$$\vdash Xas_3 YbZc \vdash Xs_3 aYbZc \vdash s_3 XaYbZc \vdash Xs_0 aYbZc$$
$$\vdash XXs_1 YbZc \vdash XXYs_1 bZc \vdash XXYYs_2 Zc \vdash XXYYZs_2 c$$
$$\vdash XXYYs_3 ZZ \vdash XXYs_3 YZZ \vdash XXs_3 YYZZ \vdash Xs_3 XYYZZ$$
$$\vdash XXs_0 YYZZ \vdash XXYs_4 YZZ \vdash XXYYs_4 ZZ \vdash XXYYZs_4 Z$$
$$\vdash XXYYZZs_4 B \vdash XXYYZZBs_5$$

From the above analysis it is clear that the string $w = a^2b^2c^2$ is accepted by the Turing machine M. On the other hand it can be shown that strings of the set $(\Sigma^* - L)$ are not accepted by the Turing machine M.

Example 7

Construct a Turing machine to recognize the set of all even length palindromes over $\{0, 1\}$.

Solution   Initially, the tape of Turing machine M contains the even length palindrome string followed by infinity of blanks B. The different steps used to recognize the set of all even length palindromes over $\{0, 1\}$ are given below.

(a) Initially, the Turing machine M is at initial state s_0. M enters state s_1 (or s_2) on scanning 0 (or 1), replaces 0 (or 1) by B and move right.

(b) The Turing machine M scans the remaining part of the string without changing the tape symbol until it encounters a blank B and move right.

(c) Once it encounters a blank B, the machine M moves to the left. If the rightmost symbol matches with the leftmost symbol, then the rightmost symbol will be replaced by B and move left until it encounters a blank B.

(d) Steps (a), (b) and (c) are repeated after changing the states suitably.

We define Turing machine M as $M = (S, \Sigma, \Gamma, \delta, s_0, B, A)$ consisting of $\Sigma = \{0, 1\}$, $S = \{s_0, s_1, s_2, s_3, s_4, s_5, s_6, s_7\}$, $\Gamma = \{0, 1, B\}$ and $A = \{s_7\}$. The transitions δ are defined as below.

$$\delta(s_0, 0) = (s_1, B, R); \quad \delta(s_1, 0) = (s_1, 0, R); \quad \delta(s_2, 0) = (s_2, 0, R); \quad \delta(s_3, 0) = (s_5, B, L);$$
$$\delta(s_0, B) = (s_7, B, R); \quad \delta(s_1, 1) = (s_1, 1, R); \quad \delta(s_2, 1) = (s_2, 1, R); \quad \delta(s_4, 1) = (s_6, B, L);$$
$$\delta(s_0, 1) = (s_2, B, R); \quad \delta(s_1, B) = (s_3, B, L); \quad \delta(s_2, B) = (s_4, B, L); \quad \delta(s_5, 0) = (s_5, 0, L);$$
$$\delta(s_5, 1) = (s_5, 1, L); \quad \delta(s_5, B) = (s_0, B, R); \quad \delta(s_6, 0) = (s_6, 0, L); \quad \delta(s_6, 1) = (s_6, 1, L);$$
$$\delta(s_6, B) = (s_0, B, R).$$

For example, consider a input string $w = 0110$. A sample computation of M on the input string is given below.

$$s_0 0110 \vdash Bs_1 110$$
$$\vdash B1s_1 10 \vdash B11s_1 0 \vdash B110s_1 B \vdash B11s_3 0B$$
$$\vdash B1s_5 1BB \vdash Bs_5 11BB \vdash s_5 B11BB \vdash Bs_0 11BB$$
$$\vdash BBs_2 1BB \vdash BB1s_2 BB \vdash BBs_4 1BB \vdash Bs_6 BBBB$$
$$\vdash BBs_0 BBB \vdash BBBs_7 BB$$

Therefore, the given input string $w = 0110$ is accepted by the Turing machine M. Similarly, it can be shown that odd length palindromes are not accepted by the Turing machine.

Example 8

Construct a Turing machine that can accept the strings over {a, b} containing even number of b's.

Solution   We define the Turing machine M that recognize the strings over (a,b) containing even number of b's as follows.

(a) The machine M initially is at initial state s_0. The machine M remains on the same state s_0 while it encounters the symbol a. It moves right without changing the input symbol.

(b) If in state s_0 a symbol b is encountered, then it enters state s_1 in order to count the number of b's present in the string. It moves right without changing the symbol.

(c) If a symbol a is encountered while in state s_1, then the machine M remains at the same state without changing the symbol, but it moves right. If in state s_1 a symbol is encountered, then the machine enters state s_0. It moves right without changing symbol. In this case we get strings having even number of b's.

(d) The machine M halts, when a blank B is encountered while in state s_0. In this case, it moves right and enters the final state s_2.

We define Turing machine M as $M = (S, \Sigma, \Gamma, \delta, s_0, B, A)$ consisting of $\Sigma = \{a, b\}$, $S = \{s_0, s_1, s_2\}$, $\Gamma = \{a, b, B\}$ and $A = \{s_2\}$. The transitions δ are defined as below.

$$\delta(s_0, a) = (s_0, a, R); \qquad \delta(s_1, a) = (s_1, a, R); \qquad \delta(s_0, B) = (s_2, B, R).$$
$$\delta(s_0, b) = (s_1, b, R); \qquad \delta(s_1, b) = (s_0, b, R);$$

Let us consider a input string $w = abab$. A sample computation of M on the input string is given below.

$$s_0abab \vdash as_0bab$$
$$\vdash abs_1ab \vdash abas_1b \vdash ababs_0B$$
$$\vdash ababBs_2B$$

Similarly, consider the input string $w = aabbb$. A sample computation of the Turing machine M on the string is given below.

$$s_0aabbb \vdash as_0abbb$$
$$\vdash aas_0bbb \vdash aabs_1bb \vdash aabbs_0b$$
$$\vdash aabbbs_1B$$

Therefore, it is clear that the input string $w = abab$ is accepted by the Turing machine M whereas the string $w = aabbb$ is not accepted by the Turing machine.

Example 9

Construct a Turing machine to recognize the language given below.

$$L = \{w \in \{0,1\}^* : w \text{ ends with } 010\}$$

Solution   Given that $L = \{w \in \{0,1\}^* : w \text{ ends with } 010\}$. We construct the Turing machine M as follows.

   i.   Initially the machine is at state s_0 and moves to extreme right until it encounters a blank B without changing its state and input symbol. Once it encounters a blank B, the machine M starts moving left and enters state s_1 without changing B.

   ii. If, 0 is encountered while in state s_1, then the machine M enters state s_2 and move left without changing the symbol 0. If, 1 is encountered instead of a 0 while in state s_1, then the machine M halts without accepting the string.

   iii. If, 1 is encountered while in state s_2, then the machine M enters state s_3 and move left without changing the symbol 1. If, 0 is encountered instead of a 1 while in state s_2, then the machine M halts without accepting the string.

   iv. If, 0 is encountered while in state s_3, then the machine M enters final state s_4 and move right without changing the symbol 0. In this case, the machine M accepts the given string. If, 1 is encountered instead of a 0 while in state s_3, then the machine M halts without accepting the string.

We define Turing machine M as $M = (S, \Sigma, \Gamma, \delta, s_0, B, A)$ consisting of $\Sigma = \{0,1\}$, $S = \{s_0, s_1, s_2, s_3, s_4\}$, $\Gamma = \{0,1,B\}$ and $A = \{s_4\}$. The transitions δ are defined as below.

$$\delta(s_0, 0) = (s_0, 0, R); \qquad \delta(s_0, B) = (s_1, B, L); \qquad \delta(s_2, 1) = (s_3, 1, L).$$
$$\delta(s_0, 1) = (s_0, 1, R); \qquad \delta(s_1, 0) = (s_2, 0, L); \qquad \delta(s_3, 0) = (s_4, 0, R)$$

Let us consider a input string $w = 11010$. A sample computation of M on the input string is given below.

$$s_0 11010 \vdash 1 s_0 1010$$
$$\vdash 11 s_0 010 \vdash 110 s_0 10 \vdash 1101 s_0 0 \vdash 11010 s_0 B$$
$$\vdash 1101 s_1 0 B \vdash 110 s_2 10 B \vdash 11 s_3 010 B \vdash 110 s_4 10 B$$

Therefore, it is clear that the input string $w = 11010$ is accepted by the Turing machine M. Similarly, consider the string $w = 110110$. In this case, the machine M halts without accepting the string. A sample computation is given below.

$$s_0 110110 \vdash 1 s_0 10110$$
$$\vdash 11 s_0 0110 \vdash 110 s_0 110 \vdash 1101 s_0 10 \vdash 11011 s_0 0B$$
$$\vdash 110110 s_0 B \vdash 11011 s_1 0B \vdash 1101 s_2 10B \vdash 110 s_3 110B$$

Example 10

Construct a Turing machine that can compute concatenation over $\Sigma = \{a\}$.

Solution   We have to construct a Turing machine that can compute concatenation over $\Sigma = \{a\}$. It implies that if w_1 and w_2 are two input strings, then the output string is $w_1 w_2$.

Assume that, the two strings w_1 and w_2 are written initially on the input tape separated by a blank B. The output $w_1 w_2$ can be obtained by removing the blank B in between w_1 and w_2. This can be done by using the following steps.

    i.   Initially the machine M is at initial state s_0 and moves to right until it encounters a blank B. It replace the blank B by a, enters state s_0 and move right.

    ii.   If, a is encountered while in state s_1, then move right without changing the input symbol and current state. Repeat this step until a B is encountered. If, B is encountered instead of a while in state s_1 then the machine M enters state s_2 and move left without changing blank B.

    iii.   If, a is encountered while in state s_2, then a is replaced by B, move right and the machine M halts.

We define Turing machine M as $M = (S, \Sigma, \Gamma, \delta, s_0, B, A)$ consisting of $\Sigma = \{a\}$, $S = \{s_0, s_1, s_2\}$, $\Gamma = \{a, B\}$ and $A = \{s_2\}$. The transitions δ are defined as below.

$$\delta(s_0, a) = (s_0, a, R); \qquad \delta(s_1, a) = (s_1, a, R); \qquad \delta(s_2, a) = (s_2, B, R).$$
$$\delta(s_0, B) = (s_1, a, R); \qquad \delta(s_1, B) = (s_2, B, L);$$

Let us consider $w_1 = aaa$ and $w_2 = aa$. Therefore, we have

$$s_0 w_1 B w_2 = s_0 aaaBaa \; s \; as_0 aaBaa$$
$$\vdash aas_0 aBaa \vdash aaas_0 Baa \vdash aaaas_1 aa$$
$$\vdash aaaaas_1 a \vdash aaaaaas_1 B \vdash aaaaas_2 aB$$
$$\vdash aaaaaBs_2 B \vdash w_1 w_2 Bs_2 B$$

Example 11

Design a Turing machine that computes $(k + l)$ for positive integers k and l.

Solution   Our aim is to compute $(k + l)$ for positive integers k and l. Initially, the tape of Turing machine M contains $0^k 1 0^l$ on its tape. The Turing machine halts with 0^{k+l} on its tape. The output 0^{k+l} can be obtained by removing the symbol 1 in between 0^k and 0^l. It can be done by using the following steps.

    (a)   The machine M is at initial state s_0 and move to right until it encounters a 1. It replaces 1 by 0, enters state s_1 and move right.

(b) If, 0 is encountered while in state s_1, then move right without changing the input symbol and current state. Repeat this step until a blank B is encountered. If, B is encountered instead of 0 while in state s_1, then the machine M enters state s_2 and move left without changing blank B.

(c) If, 0 is encountered while in state s_2, then 0 is erased by B, move right and halts.

We define Turing machine M as $M = (S, \Sigma, \Gamma, \delta, s_0, B, A)$ consisting of $\Sigma = \{0, 1\}$, $S = \{s_0, s_1, s_2\}$, $\Gamma = \{0, 1, B\}$ and $A = \{s_2\}$. The transitions δ are defined as below.

$$\delta(s_0, 0) = (s_0, 0, R); \qquad \delta(s_1, 0) = (s_1, 0, R); \qquad \delta(s_2, 0) = (s_2, B, L);$$
$$\delta(s_0, 1) = (s_1, 0, R); \qquad \delta(s_1, B) = (s_2, B, L).$$

For example, consider the input string 00010 to compute $(3 + 1)$. A sample computation of M on the input string is given below.

$$s_0 00010 \vdash 0 s_0 0010$$
$$\vdash 00 s_0 010 \vdash 000 s_0 10 \vdash 0000 s_1 0$$
$$\vdash 00000 s_1 B \vdash 0000 s_2 0B \vdash 0000 B s_2 B$$

From the above computation it is clear that $(3 + 1) = 4$.

Example 12

Design a Turing machine that accepts aa^* over $\Sigma = \{a\}$.

Solution We construct the Turing machine M that accepts aa^* over $\Sigma = \{a\}$ as follows. The different steps involved are given below.

i. Initially, the Turing machine is at initial state s_0. It enters state s_1, when it reads a first time and move right.

ii. If, a is encountered while in state s_1, then the machine M remains on the same state and move right without changing the input symbol. If, B is encountered while in state s_1, then the machine M enters final state s_2 without affecting B and halts.

We define Turing machine M as $M = (S, \Sigma, \Gamma, \delta, s_0, B, A)$ consisting of $\Sigma = \{a\}$, $S = \{s_0, s_1, s_2\}$, $\Gamma = \{a, B\}$ and $A = \{s_2\}$. The transitions δ of the machine are defined as $\delta(s_0, a) = (s_1, a, R)$; $\delta(s_1, a) = (s_1, a, R)$; $\delta(s_1, B) = (s_2, B, R)$.

For example, consider the string $w = aaa$. A sample computation of the Turing machine M on the input string is given below.

$$s_0 aaa \vdash a s_1 aa$$
$$\vdash aa s_1 a \vdash aaa s_1 B \vdash aaa B s_2$$

It implies that, the machine M accepts $w = aaa$ and hence it accepts aa^*.

Example 13

Design a Turing machine that copies string w for any $w \in \{a\}^+$.

Solution   We construct the machine that copies string for any $w \in \{a\}^+$ as below.

(a) Initially, the Turing machine is at state s_0 and move right by replacing every input symbol a by X. Once a blank B is encountered at state s_0, then it enters state s_1 and move left without affecting B.

(b) In state s_1, the machine M finds the rightmost X, move right and enters state by replacing the rightmost X with a. If, a is encountered while in state s_1, then it moves left without affecting the state and input symbol.

(c) In state s_2, if a blank B is encountered, then it enters state s_1, move left and replace B by a. If, a is encountered while in state s_2, then it moves right without affecting state and input symbol.

(d) The Turing machine M halts when the machine encounters a blank B while in state s_1. In such case, the machine M enters the final state s_3 and halts.

We define Turing machine M as $M = (S, \Sigma, \Gamma, \delta, s_0, B, A)$ consisting of $\Sigma = \{a\}$, $S = \{s_0, s_1, s_2, s_3\}$, $\Gamma = \{a, X, B\}$ and $A = \{s_3\}$. The transitions δ are defined below.

$$\delta(s_0, a) = (s_0, X, R); \quad \delta(s_1, a) = (s_1, a, L); \quad \delta(s_1, B) = (s_3, B, R); \quad \delta(s_2, B) = (s_1, a, L);$$
$$\delta(s_0, B) = (s_1, B, L); \quad \delta(s_1, X) = (s_2, a, R); \quad \delta(s_2, a) = (s_2, a, R).$$

For example, consider the string $w = aaa$. A sample computation of the Turing machine M on the input string is given below.

$$s_0 w = s_0 aaa \vdash X s_0 aa$$
$$\vdash XX s_0 a \vdash XXX s_0 B \vdash XX s_1 XB$$
$$\vdash XX a s_2 B \vdash XX s_1 aa \vdash X s_1 Xaa$$
$$\vdash X a s_2 aa \vdash X a a s_2 a \vdash X a a a s_2 B$$
$$\vdash X a a s_1 aa \vdash X a s_1 aaa \vdash X s_1 aaaa$$
$$\vdash s_1 Xaaaa \vdash a s_2 aaaa \vdash a a s_2 aaa$$
$$\vdash a a a s_2 aa \vdash a a a a s_2 a \vdash a a a a a s_2 B$$
$$\vdash a a a a s_1 aa \vdash a a a s_1 aaa \vdash a a s_1 aaaa$$
$$\vdash a s_1 aaaaa \vdash s_1 aaaaaa \vdash s_1 Baaaaaa$$
$$\vdash B s_3 aaaaaa = B s_3 ww$$

Example 14

Construct a Turing machine that accept the following language L on $\{0,1\}$, where $L = L(010^*1)$.

Solution Given that $L = L(010^*1)$. Our aim is to construct Turing machine that accept L on $\{0,1\}$. We define the Turing machine M as $M = (S, \Sigma, \Gamma, \delta, s_0, B, A)$ consisting of $S = \{s_0, s_1, s_2, s_3\}$, $\Sigma = \{0,1\}$, $\Gamma = \{0,1,B\}$ and $A = \{s_3\}$. The transitions δ are defined as $\delta(s_0, 0) = (s_1, 0, R)$; $\delta(s_1, 1) = (s_2, 1, R)$; $\delta(s_2, 0) = (s_2, 0, R)$ and $\delta(s_2, 1) = (s_3, 1, R)$.

For example, consider the string $w = 01001$. A sample computation of the Turing machine is given below.

$$s_0 01001 \vdash 0s_1 1001$$
$$\vdash 01s_2 001 \vdash 010s_2 01 \vdash 0100s_2 1$$
$$\vdash 01001s_3 B$$

Therefore, it is clear that the string $w = 01001$ is accepted by the Turing machine M.

Example 15

Construct a Turing machine that accept the following language L on $\{a,b\}$, where $L = \{w : |w| \text{ is even}\}$.

Solution We construct the Turing machine M that recognizes all even length strings w on $\{a,b\}$ as follows. The different steps involved are given below.

(a) The Turing machine M is initially at initial state s_0. M enters state s_1 on scanning either a or b and move right by replacing a with X and b with Y. The machine M halts in final state s_2 when a blank B is encountered while in state s_0.

(b) In state s_1, the machine M enters state s_0 on scanning either a or b and move right by replacing a with X and b with Y.

We define the Turing machine M as $M = (S, \Sigma, \Gamma, \delta, s_0, B, A)$ consisting of $\Sigma = \{a,b\}$, $S = \{s_0, s_1, s_2\}$, $\Gamma = \{a,b,X,Y,B\}$ and $A = \{s_2\}$. The transitions δ are given below.

$$\delta(s_0, a) = (s_1, X, R); \qquad \delta(s_1, a) = (s_0, X, R); \qquad \delta(s_0, B) = (s_2, B, R);$$
$$\delta(s_0, b) = (s_1, Y, R); \qquad \delta(s_1, b) = (s_0, Y, R).$$

For example, consider a string of even length $w = abba$. A sample computation of the Turing machine M is given below.

$$s_0 abba \vdash Xs_1 bba$$
$$\vdash XYs_0 ba \vdash XYYs_1 a \vdash XYYXs_0 B \vdash XYYXBs_2$$

Similarly, consider a string of odd length $w = aab$. A sample computation of the Turing machine M is given below. It is also clear that the constructed machine accepts all strings of even length on $\{a,b\}$.

$$s_0aab \vdash Xs_1ab$$
$$\vdash XXs_0b \vdash XXYs_1B$$

Example 16

Construct a Turing machine that accept the following language L on $\{a,b\}$, where $L = \{w : |w| \text{ is a multiple of } 3\}$.

Solution   We construct the Turing machine M that recognize all strings of length multiple of 3 on $\{a,b\}$ as follows. The different steps involved are given below.

(a) Initially, the Turing machine M is at initial state s_0. It enters state s_1 on reading either a or b and move right by replacing a with X and b with Y. The Turing machine M halts in final state s_3 when a blank B is encountered while in state s_0.

(b) The machine M enters state s_2 from the current state s_1 on scanning either a or b and move right by replacing a with X and b with Y.

(c) In state s_2, the machine M enters state s_0 on scanning either a or b and move right by replacing a with X and b with Y.

We define the Turing machine M as $M = (S, \Sigma, \Gamma, \delta, s_0, B, A)$ consisting of $\Sigma = \{a,b\}$, $S = \{s_0, s_1, s_2, s_3\}$, $\Gamma = \{a,b,X,Y,B\}$ and $A = \{s_3\}$. The transitions δ are given below.

$$\delta(s_0,a) = (s_1,X,R); \qquad \delta(s_1,a) = (s_2,X,R); \qquad \delta(s_2,a) = (s_0,X,R);$$
$$\delta(s_0,b) = (s_1,Y,R); \qquad \delta(s_1,b) = (s_2,Y,R); \qquad \delta(s_2,b) = (s_0,Y,R);$$
$$\delta(s_0,B) = (s_3,B,R).$$

To show the correctness by example, consider a string $w = babbab$. A sample computation of the Turing machine M is given below.

$$s_0babbab \vdash Ys_1abbab$$
$$\vdash YXs_2bbab \vdash YXYs_0bab \vdash YXYYs_1ab$$
$$\vdash YXYYXs_2b \vdash YXYYXYs_0B \vdash YXYYXYBs_3B$$

Similarly, consider a string $w = abaa$. Therefore, we get

$$s_0abaa \vdash Xs_1baa$$
$$\vdash XYs_2aa \vdash XYXs_0a \vdash XYXXs_1B$$

From the sample computation it is clear that the designed Turing machine M given above accepts all strings on $\{a,b\}$ that are having length as multiple of 3.

Example 17

Design a Turing machine that accepts the following language L on $\{a,b\}$, where $L = \{a^n b^m : n \geq 1, n \neq m\}$.

Solution Given that $L = \{a^n b^m : n \geq 1, n \neq m\}$. We construct the Turing machine that recognizes L as follows. The different steps involved are listed below.

(a) If in state s_0, the symbol a is encountered, then the machine enters state s_1 and moves right without changing the input symbol a.

(b) If, a is encountered while in state s_1, then the machine remains on the same state and move right without changing the current symbol. Same transition also holds if b is encountered while in state s_1.

(c) The Turing machine M halts in the following cases. If a blank B is encountered while in state s_1, then the machine enters the final state s_2 and move right without changing the blank B.

We define the Turing machine M as $M = (S, \Sigma, \Gamma, \delta, s_0, B, A)$ consisting of $\Sigma = \{a,b\}$, $S = \{s_0, s_1, s_2\}$, $\Gamma = \{a, b, B\}$ and $A = \{s_2\}$. The transitions δ of the machine M are defined as $\delta(s_0, a) = (s_1, a, R)$, $\delta(s_1, a) = (s_1, a, R)$, $\delta(s_1, b) = (s_1, b, R)$ and $\delta(s_1, B) = (s_2, B, R)$.

For example, consider a string $w = aabbb$. A sample computation of the Turing machine is given below.

$$s_0 aabbb \vdash a s_1 abbb$$
$$\vdash aa s_1 bbb \vdash aab s_1 bb \vdash aabb s_1 b$$
$$\vdash aabbb s_1 B \vdash aabbb B s_2 B$$

Similarly, consider a string $w = aaa$. A sample computation is given below.

$$s_0 aaa \vdash a s_1 aa$$
$$\vdash aa s_1 a \vdash aaa s_1 B \vdash aaa B s_2$$

Example 18

Represent the Turing machine by transition diagram that can accept the strings over $\{0,1\}$ containing even number of 1's.

Solution We define the Turing machine M as $M = (S, \Sigma, \Gamma, \delta, s_0, B, A)$ consisting of $S = \{s_0, s_1, s_2\}$, $\Sigma = \{a,b\}$, $\Gamma = \{a, b, B\}$ and $A = \{s_2\}$. The transitions δ are defined as $\delta(s_0, a) = (s_0, a, R), \delta(s_0, b) = (s_1, b, R), \delta(s_1, a) = (s_1, a, R), \delta(s_1, b) = (s_0, b, R)$ and $\delta(s_0, B) = (s_2, B, R)$.

The transitions graph representation of the Turing machine is as follows:

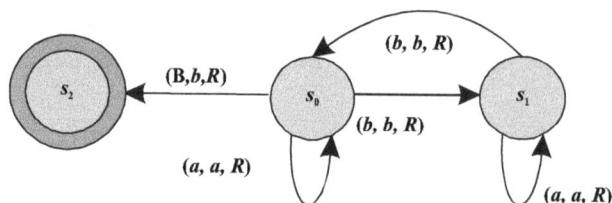

Example 19

Represent the Turing machine by transition graph that recognize all strings containing even number of a's and even number of b's.

Solution   The Turing machine $M = (S, \Sigma, \Gamma, \delta, s_0, B, A)$ that recognize all strings containing even number of a's and even number of b's is defined as $S = \{s_0, s_1, s_2, s_3, s_4\}$, $\Sigma = \{a, b\}$, $\Gamma = \{a, b, B\}$ and $A = \{s_4\}$. The transitions δ are given as $\delta(s_0, a) = (s_1, a, R)$, $\delta(s_0, b) = (s_2, b, R)$, $\delta(s_1, a) = (s_0, a, R)$, $\delta(s_1, b) = (s_3, b, R)$, $\delta(s_2, a) = (s_3, a, R)$, $\delta(s_2, b) = (s_0, b, R)$, $\delta(s_3, a) = (s_2, a, R)$, $\delta(s_3, b) = (s_1, b, R)$ and $\delta(s_0, B) = (s_4, B, R)$. The transition graph representation of the Turing machine is given below.

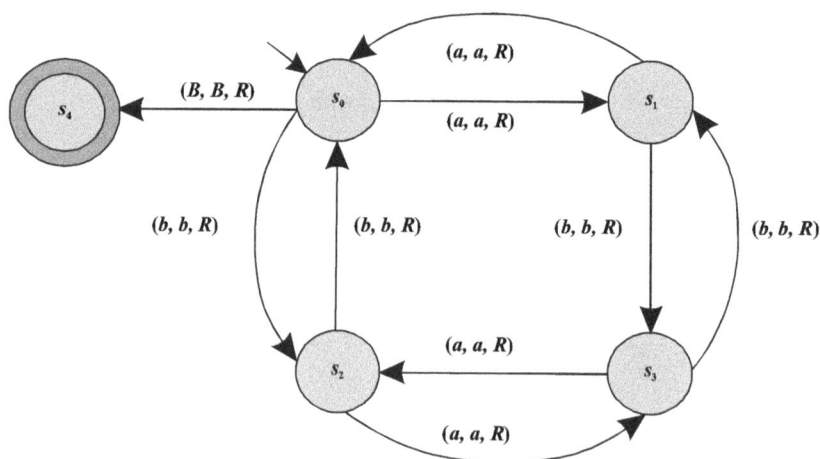

Example 20

Represent the Turing machine by transition table that accept the language $L = \{a^n b^n c^n : n \geq 1\}$.

Solution   The Turing machine that accepts the language $L = \{a^n b^n c^n : n \geq 1\}$ is designed in example 6. The transition table representation of the Turing machine is given below.

State	a	b	c	X	Y	Z	B
s_0	XRs_1				YRs_4		
s_1	aRs_1	YRs_2			YRs_1		
s_2		bRs_2	ZLs_3			ZRs_2	
s_3	aLs_3	bLs_3		XRs_0	YLs_3	ZLs_3	
s_4					YRs_4	ZRs_4	BRs_5
s_5							

Example 21

Represent the Turing machine by transition table that accept the language $L = \{w \in \{0,1\}^* : w \text{ ends with } 010\}$.

Solution   Given that $L = \{w \in \{0,1\}^* : w \text{ ends with } 010\}$. The Turing machine that accepts the language L is designed in example 9. The transition table representation of the Turing machine is given below.

State	0	1	B
s_0	$0Rs_0$	$1Rs_0$	BLs_1
s_1	$0Ls_2$		
s_2		$1Ls_3$	
s_3	$0Rs_4$		
s_4			

Example 22

Consider the Turing machine described by the transition table given below. Obtain the inverse production rules.

State	¢	a	X	B
s_0	$¢Rs_0$	XRs_0		BLs_1
s_1		aLs_1	aRs_2	BRs_3
s_2		aRs_2		aLs_1
s_3				

Solution   In the above transition table, s_0 is the initial state and s_3 is the final state. We obtain the inverse production rules by using the following steps.

(a) Productions corresponding to right moves are:

$$s_0 \cent \to \cent s_0; \ s_0 a \to X s_0; \ s_1 X \to a s_2; \ s_1 B \to B s_3; \ s_2 a \to a s_2.$$

(b) Productions corresponding to left moves are:

$$as_0 B \to s_1 aB; \quad as_1 a \to s_1 aa; \quad as_2 B \to s_1 aa;$$
$$Xs_0 B \to s_1 XB; \quad Xs_1 a \to s_1 Xa; \quad Xs_2 B \to s_1 Xa;$$
$$Bs_0 B \to s_1 BB; \quad Bs_1 a \to s_1 Ba; \quad Bs_2 B \to s_1 Ba.$$

(c) Productions corresponding to right end are:

$$aB] \to a]; \quad XB] \to X]; \quad BB] \to B]; \quad s_0] \to s_0 B]; \quad s_1] \to s_1 B]; \quad s_2] \to s_2 B]; \quad s_3] \to s_3 B]$$

(d) Productions corresponding to left end are:

$$[s_0 B \to [s_1 BB; \quad [s_1 a \to [s_1 Ba; \quad [s_2 B \to [s_1 Ba; \quad [B \to [$$

(e) Productions corresponding to end markers are:

$$a \to [s_0 \cent a; \quad X \to [s_0 \cent X; \quad a \to a\$]; \quad X \to X\$]; \quad [s_3, B] \to Q_0$$

where Q_0 is the newly introduced symbol called the start symbol.

The inverse productions are obtained by reversing the arrows of the above productions. These productions are defined below.

$$\begin{array}{lllll}
\cent s_0 \to s_0 \cent; & s_1 aB \to as_0 B; & s_1 Ba \to Bs_1 a; & X] \to XB]; & s_3 B] \to s_3]; \\
Xs_0 \to s_0 a; & s_1 XB \to Xs_0 B; & s_1 aa \to as_2 B; & B] \to BB]; & [s_1 BB \to [s_0 B; \\
as_2 \to s_1 X; & s_1 BB \to Bs_0 B; & s_1 Xa \to Xs_2 B; & s_0 B] \to s_0]; & [s_1 Ba \to [s_1 a; \\
Bs_3 \to s_1 B; & s_1 aa \to as_1 a; & s_1 Ba \to Bs_2 B; & s_1 B] \to s_1]; & [s_1 Ba \to [s_2 B; \\
as_2 \to s_2 a; & s_1 Xa \to Xs_1 a; & a] \to aB]; & s_2 B] \to s_2]; & [\to [B; \\
[s_0 \cent a \to a; & [s_0 \cent X \to X; & a\$] \to a; & X\$] \to X; & Q_0 \to [s_3, B].
\end{array}$$

This is the type-0 grammars corresponding to the given Turing machine.

Example 23

Consider the Turing machine described by the transition table given below. Obtain the type-0 grammars corresponding to the given Turing machine.

State	Tape symbol			
	¢	0	1	B
s_0	$¢Rs_0$	$0Rs_0$	$1Rs_0$	BLs_1
s_1		$0Ls_2$		
s_2			$1Ls_3$	
s_3		$0Rs_4$		
s_4				

Solution In the above transition table, s_0 is the initial state and s_4 is the final state. We obtain the type-0 grammars by using the following steps.

(a) Productions corresponding to right moves are:

$s_0¢ \rightarrow ¢s_0$; $s_00 \rightarrow 0s_0$; $s_01 \rightarrow 1s_0$; $s_30 \rightarrow 0s_4$

(b) Productions corresponding to left moves are:

$0s_0B \rightarrow s_10B$; $0s_10 \rightarrow s_200$; $0s_21 \rightarrow s_301$;
$1s_0B \rightarrow s_11B$; $1s_10 \rightarrow s_210$; $1s_21 \rightarrow s_311$;
$Bs_0B \rightarrow s_1BB$; $Bs_10 \rightarrow s_2B0$; $Bs_21 \rightarrow s_3B1$.

(c) Productions corresponding to right end are:

$0B] \rightarrow 0]$; $BB] \rightarrow B]$; $s_1] \rightarrow s_1B]$; $s_3] \rightarrow s_3B]$;
$1B] \rightarrow 1]$; $s_0] \rightarrow s_0B]$; $s_2] \rightarrow s_2B]$; $s_4] \rightarrow s_4B]$.

(d) Productions corresponding to left end are:

$[s_0B \rightarrow [s_1BB$; $[s_10 \rightarrow [s_2B0$; $[s_21 \rightarrow [s_3B1$; $[B \rightarrow [$.

(e) Productions corresponding to end markers are:

$0 \rightarrow [s_0¢0$; $1 \rightarrow [s_0¢1$; $0 \rightarrow 0$]; $1 \rightarrow 1$]; $[s_4,B] \rightarrow Q_0$

where Q_0 is the newly introduced symbol called the start symbol.

The inverse productions are obtained by reversing the arrows of the above productions. These productions are defined as follows:

$\cent s_0 \rightarrow s_0 \cent;$ $s_1 BB \rightarrow Bs_0 B;$ $s_3 B1 \rightarrow Bs_2 1;$ $s_2 B] \rightarrow s_2];$ $[\rightarrow [B;$

$0s_0 \rightarrow s_0 0;$ $s_2 00 \rightarrow 0s_1 0;$ $0] \rightarrow 0B];$ $s_3 B] \rightarrow s_3];$ $[s_0 \cent 0 \rightarrow 0;$

$1s_0 \rightarrow s_0 1;$ $s_2 10 \rightarrow 1s_1 0;$ $1] \rightarrow 1B];$ $s_4 B] \rightarrow s_4];$ $[s_0 \cent 1 \rightarrow 1;$

$0s_4 \rightarrow s_3 0;$ $s_2 B0 \rightarrow Bs_1 0;$ $B] \rightarrow BB];$ $[s_1 BB \rightarrow [s_0 B;$ $0\$] \rightarrow 0;$

$s_1 0B \rightarrow 0s_0 B;$ $s_3 01 \rightarrow 0s_2 1;$ $s_0 B] \rightarrow s_0];$ $[s_2 B0 \rightarrow [s_1 0;$ $1\$] \rightarrow 1;$

$s_1 1B \rightarrow 1s_0 B;$ $s_3 11 \rightarrow 1s_2 1;$ $s_1 B] \rightarrow s_1];$ $[s_3 B1 \rightarrow [s_2 1;$ $Q_0 \rightarrow [s_4, B].$

Example 24

Consider the Turing machine described by the transition table given below. Find the processing of the strings (a) *aabb* and (b) *baabb*.

State	Tape symbol		
	a	*b*	*B*
$\rightarrow s_0$	BRs_1	BRs_1	BRs_1
s_1	aRs_1	bRs_1	BLs_2
s_2	BLs_3	BLs_4	BRs_5
s_3	bLs_3		aRs_5
s_4		aLs_3	bRs_5

Solution   Consider the Turing machine described by the transition table given above. The processing of the strings is given below.

(a) $s_0 aabb \vdash Bs_1 abb$

       $\vdash Bas_1 bb \vdash Babs_1 b \vdash Babbs_1 B$

       $\vdash Babs_2 bB \vdash Bas_4 bBB \vdash Bs_3 aaBB$

       $\vdash s_3 BbaBB \vdash as_5 baBB$

(b) $s_0 baabb \vdash Bs_1 aabb$

     $\vdash Bas_1 abb \vdash Baas_1 bb \vdash Baabs_1 b$

     $\vdash Baabbs_1 B \vdash Baabs_2 bB \vdash Baas_4 bBB$

     $\vdash Bas_3 aaBB \vdash Bs_3 abaBB \vdash s_3 BbbaBB$

     $\vdash as_5 bbaBB$

Example 25

Design a Turing machine that accept the language L given below, where $L = \{ww : w \in \{0,1\}^*\}$.

Solution Given that $L = \{ww : w \in \{0,1\}^*\}$. Our aim is to construct a Turing machine that accepts L. Here, we have to process the string in two parts. First, we have to identify the mid position of the string and then to distinguish the symbols in the second part from those in the first part.

The Turing machine M change symbols 0 and 1 to X and Y respectively from both ends. Once the symbols are changed to X and Y, it erases the matching X and Y symbols. It is clear that, if the length of 'ww' is odd, then the input string can be rejected. It also reject a string of even length when the first part fails to match the corresponding symbol in the second part.

We define the Turing machine M as $M = (S, \Sigma, \Gamma, \delta, s_0, B, A)$, consisting of $\Sigma = \{0,1\}$, $S = \{s_0, s_1, s_2, s_3, s_4, s_5, s_6, s_7, s_8, s_9\}$, $\Gamma = \{0, 1, X, Y, B\}$ and $A = \{s_9\}$. The transitions δ are defined as below.

$$\delta(s_0, 0) = (s_1, X, R); \qquad \delta(s_0, 1) = (s_1, Y, R); \qquad \delta(s_0, B) = (s_9, B, L);$$
$$\delta(s_0, X) = (s_4, X, L); \qquad \delta(s_0, Y) = (s_4, Y, L); \qquad \delta(s_1, 0) = (s_1, 0, R);$$
$$\delta(s_1, 1) = (s_1, 1, R); \qquad \delta(s_1, B) = (s_2, B, L); \qquad \delta(s_1, X) = (s_2, X, L);$$
$$\delta(s_1, Y) = (s_2, Y, L); \qquad \delta(s_2, 0) = (s_3, X, L); \qquad \delta(s_2, 1) = (s_3, Y, L);$$
$$\delta(s_3, 0) = (s_3, 0, L); \qquad \delta(s_3, 1) = (s_3, 1, L); \qquad \delta(s_3, X) = (s_0, X, R);$$
$$\delta(s_3, Y) = (s_0, Y, R); \qquad \delta(s_4, B) = (s_5, B, R); \qquad \delta(s_4, X) = (s_4, 0, L);$$

For example, consider the string $w = 010010$. A sample computation of the Turing machine M is given below.

$s_0 010010 \vdash X s_1 10010$

$\vdash X 1 s_1 0010 \vdash X 10 s_1 010 \vdash X 100 s_1 10 \vdash X 1001 s_1 0$

$\vdash X 10010 s_1 B \vdash X 1001 s_2 0B \vdash X 100 s_3 1XB \vdash X 10 s_3 01XB$

$\vdash X 1 s_3 001XB \vdash X s_3 1001XB \vdash s_3 X 1001XB \vdash X s_0 1001XB$

$\vdash X Y s_1 001XB \vdash X Y 0 s_1 01XB \vdash X Y 00 s_1 1XB \vdash X Y 001 s_1 XB$

$\vdash X Y 00 s_2 1XB \vdash X Y 0 s_3 0YXB \vdash X Y s_3 00YXB \vdash X s_3 Y00YXB$

$\vdash X Y s_0 00YXB \vdash X Y X s_1 0YXB \vdash X Y X 0 s_1 YXB \vdash X Y X s_2 0YXB$

$\vdash X Y s_3 XXYXB \vdash X Y X s_0 XYXB \vdash X Y s_4 XXYXB \vdash X s_4 Y0XYXB$

$\vdash s_4 X10XYXB \vdash s_4 B010XYXB \vdash B s_5 010XYXB \vdash X s_7 10XYXB$

$\vdash X 1 s_7 0XYXB \vdash X 10 s_7 XYXB \vdash X 1 s_8 0BYXB \vdash X s_8 10BYXB$

$\vdash s_8 X10BYXB \vdash X s_5 10BYXB \vdash X Y s_6 0BYXB \vdash X Y 0 s_6 BYXB$

$\vdash X Y 0B s_6 YXB \vdash X Y 0 s_8 BBXB \vdash X Y s_8 0BBXB \vdash X s_8 Y0BBXB$

$$\vdash XYs_5 0BBXB \vdash XYXs_7 BBXB \vdash XYXBs_7 BXB \vdash XYXBBs_7 XB$$
$$\vdash XYXBs_8 BBB \vdash XYXs_8 B \vdash XYs_8 XB \vdash XYXs_5 B$$
$$\vdash XYXBs_9$$

From the above analysis it is clear that the string $w = 010010$ is accepted. Similarly, it can be shown that the strings $w = 010$ and $w = 0100$ are not accepted by the above Turing machine.

Example 26

Design a Turing machine that compute $n \bmod 2$, where n is a natural number, i.e., $n \in N$.

Solution   The function $n \bmod 2$; $n \in N$ is defined as

$$n \bmod 2 = \begin{cases} 0 & \text{if } n \text{ is a multiple of } 2 \\ 1 & \text{otherwise} \end{cases}$$

Initially, the tape of the Turing machine M contains a^n on its tape. The Turing machine halts with $a^{n \bmod 2}$ on its tape. We construct the Turing machine as follows:

i.   Initially the machine is at initial state s_0 and move right until it encounters a blank B. If in state s_0, a blank B is encountered, then the machine enters state s_1 and move left without changing B.

ii.   If in state s_1, the symbol a is encountered, then the machine enters state s_2, move left and replace a by B. If a blank B is encountered in state s_1, then it enters final state s_3 and halts without changing B.

iii.   If in state s_2, the symbol a is encountered, then the machine enters state s_1, move left and replace a by B. If a blank B is encountered in state s_2, then the machine enters final state s_3, move right and replace B by the symbol a.

Now, we define Turing machine M as $M = (S, \Sigma, \Gamma, \delta, s_0, B, A)$ consisting of $\Sigma = \{a\}$, $S = \{s_0, s_1, s_2, s_3\}$, $\Gamma = \{a, B\}$ and $A = \{s_3\}$. The transitions δ are defined as below.

$$\delta(s_0, a) = (s_0, a, R); \qquad \delta(s_0, B) = (s_1, B, L); \qquad \delta(s_1, a) = (s_2, B, L);$$
$$\delta(s_1, B) = (s_3, B, R); \qquad \delta(s_2, a) = (s_1, B, L); \qquad \delta(s_2, B) = (s_3, a, R).$$

For example, consider the string $w = a^4$. A sample computation of the string on the Turing machine is given below.

$$s_0 a^4 \vdash a s_0 aaa \vdash aa s_0 aa \vdash aaa s_0 a \vdash aaaa s_0 B$$
$$\vdash aaa s_1 aB \vdash aa s_2 aB \vdash a s_1 aB \vdash s_2 aB$$
$$\vdash s_1 B \vdash s_3 B$$

Therefore, $4 \bmod 2 = 0$. Similarly, consider the string $w = a^5$. A sample computation of the string on the Turing machine is given below.

$$s_0 a^5 \vdash as_0 aaaa \vdash aas_0 aaa \vdash aaas_0 aa \vdash aaaas_0 aB \vdash aaaaas_0 B$$
$$\vdash aaaas_1 a \vdash aaas_2 aB \vdash aas_1 aB \vdash as_2 aB \vdash s_1 aB \vdash s_2 B \vdash as_3 B$$

Therefore, 5 mod 2 = 1 and hence the Turing machine M halts with $a^{5 \bmod 2} = a$ on its tape.

REVIEW QUESTIONS

1. Define a Turing machine model.
2. Write the different representations of a Turing machine.
3. What is instantaneous description? Explain the move relation using instantaneous description.
4. State Minsky's theorem.
5. Explain the construction of type-0 grammar for a Turing machine.
6. Write a note on variations of Turing machines.
7. Write a note on the following:
    i.   Nondeterministic Turing machine
    ii. Multitape Turing machine
    iii. Multihead Turing machine
    iv. Multitrack machine
    v.   Multidimensional Turing machine
8. Explain Universal Turing machine.
9. What do you mean by equivalence of two-stack pushdown automata and Turing machine?
10. Write the basic guidelines for designing a Turing machine.
11. Explain Turing machine can be viewed as language acceptors.
12. State and discuss halting problem of Turing machine.
13. Explain Post's machine in detail.
14. Define linear-bounded automata in the context of computing.
15. Write a note on off-line Turing machine.

PROBLEMS

1. Construct Turing machines that will accept the following languages on $\{0, 1\}$.

   (a) $L = L(001^*0)$

   (b) $L = \{w : |w| \text{ is odd}\}$

   (c) $L = \{w : |w| \text{ is a multiple of 5}\}$

   (d) $L = \{0^m 1^n : m \neq n, n \geq 1\}$

2. Design Turing machines that will accept the following languages on $\{a, b\}$.

   (a) $L = L(a^* b)$

   (b) $L = \{w : w \text{ ends with } aba\ \}$

   (c) $L = \{a^n b^{2n} : n \geq 1\}$

   (d) $L = \{w : n_a(w) = n_b(w)\}$

   It is given that $n_a(w)$ represents the number of a's in the string w.

3. Design Turing machines that will accept the following languages on $\{0, 1\}$.

   (a) $L = \{0^n 1^n 0^n 1^n : n \geq 1\}$

   (b) $L = \{0^n 1^m 0^{n+m} : n \geq 0, m \geq 1\}$

   (c) $L = \{w : n_0(w) = n_1(w) + 1\}$

   It is given that $n_0(w)$ represents the number of 0's in the string w.

4. What language is accepted by the Turing machine $M = (S, \Sigma, \Gamma, \delta, s_0, B, A)$ consisting of $S = \{s_0, s_1, s_2, s_3\}$, $\Sigma = \{0,1\}$, $\Gamma = \{0,1,B\}$, $A = \{s_3\}$ with transitions

   $$\delta(s_0, 0) = (s_1, 0, R); \quad \delta(s_0, 1) = (s_2, 1, R); \quad \delta(s_1, 1) = (s_1, 1, R);$$
   $$\delta(s_1, B) = (s_3, B, R); \quad \delta(s_2, 1) = (s_2, 1, R); \quad \delta(s_2, 0) = (s_3, 0, R).$$

5. Consider the Turing machine M given below. Find the language accepted by the Turing machine. It is given that $M = (S, \Sigma, \Gamma, \delta, s_0, B, A)$ consisting of $S = \{s_0, s_1, s_2, s_3, s_4\}$, $\Sigma = \{a, b\}$, $\Gamma = \{a, b, B\}$, $A = \{s_4\}$ with transitions

   $$\delta(s_0, a) = (s_0, a, R); \quad \delta(s_0, b) = (s_0, b, R); \quad \delta(s_0, B) = (s_1, B, L);$$
   $$\delta(s_1, a) = (s_2, a, L); \quad \delta(s_2, b) = (s_3, b, L); \quad \delta(s_3, a) = (s_4, a, R).$$

6. Find the language obtained by the Turing machine $M = (S, \Sigma, \Gamma, \delta, s_0, B, A)$ consisting of $S = \{s_0, s_1, s_2, s_3, s_4\}$, $\Sigma = \{a, b\}$, $\Gamma = \{a, b, B\}$, $A = \{s_4\}$ with transitions

   $$\delta(s_0, b) = (s_1, b, R); \quad \delta(s_1, a) = (s_2, a, R); \quad \delta(s_2, b) = (s_3, b, R);$$
   $$\delta(s_3, a) = (s_3, a, R); \quad \delta(s_3, b) = (s_3, b, R); \quad \delta(s_3, B) = (s_4, B, R).$$

7. Represent the Turing machines obtained in problem 1 in terms of a transition table.

8. Represent the Turing machines obtained in problem 2 in terms of a transition graph.

9. Construct a Turing machine to compute the language $L = \{ww : w \in \{a,b\}^*\}$.

10. Design Turing machines to compute the following functions.

    (a) $f(n) = n + 1;$ $n \in N$
    (b) $f(n) = n + 3;$ $n \in N$
    (c) $f(m,n) = m + n;$ $m, n \in N$

11. Show that there exists a Turing machine for the given language $L = \{ww^R : w \in \{a,b\}^*\}$.

12. Construct a Turing machine that accept the language $L = \{waw : w \in \{b,c\}^*\}$.

13. Design a Turing machine to compute $n \mod 3$ for $n \in N$.

14. Construct a Turing machine that delete the last symbol from a string $w \in \{a,b\}^*$ using only two states.

15. Express the Turing machine obtained in problem 10 by transition table and transition graph.

16. Design a Turing machine to compute the function $f(n) = n - 2$, when $n \geq 2$.

17. Find a Turing machine M, which recognizes the language $L = \{01^n : n > 0\}$ over $\Sigma = \{0, 1\}$.

18. Construct a Turing machine that recognizes the languages $L = \{a^{2n}b^n : n \geq 0\}$.

19. Consider the Turing machine M described by the transition table given below. Explain the processing of the strings (a) *aabba*, (b) *ababa* and (c) *abba* if the accepting state is s_3.

States	Tape symbols		
	a	*b*	*B*
$\rightarrow s_0$	aRs_1	BRs_1	
s_1	aRs_1	BRs_2	
s_2	aLs_2	BRs_2	BLs_3
s_3			

20. Write the computation sequence for the string (i) $a^2b^2c^2$, (ii) a^2bc^2 and (iii) abc^2 for the Turing machine given in solved example 6.

21. Write the computation sequence for strings (a)001001, (b)11001 and (c) 0101101 for the Turing machine given in solved example 7.

22. Design a Turing machine that recognizes the language $L = \{0^n1^m2^{nm} : n, m \geq 1\}$.

23. Construct a Turing machine that recognizes the set of all bit strings which end with a 1.

24. What can you say about the Turing machine given below by transition table, where s_2 is the final state.

States	Tape symbols		
	a	*b*	*B*
$\to s_0$	aRs_0	bRs_1	BRs_2
s_1	aRs_1	bRs_0	
s_2			

25. Design a Turing machine M that accepts the language given by $\{a,b\}^* bab$.

26. Consider the Turing machine described by problem 24. Obtain the corresponding inverse production rules.

27. Show that there exists a Turing machine M for which the halting problem is unsolvable.

28. Obtain the type-0 grammar for the Turing machine M described in problem 5.

29. Construct the type-0 grammar for the Turing machine M described below by transition table, where s_3 is the final state.

States	Tape symbols			
	$\cent$	*a*	*b*	*B*
$\to s_0$	$\cent Rs_0$	bLs_1	aRs_1	
s_1	$\cent Rs_3$	bRs_2	bLs_0	
s_2		bRs_2	bRs_2	BLs_0
s_3		aLs_3	aRs_3	

30. Find linear-bounded automata for the language $L = \{a^{m^2} : m \geq 1\}$.

31. Construct a Turing machine that can compute concatenation over $\Sigma = \{a,b\}$.

8

LR(k) AND *LL(k)* GRAMMAR

8.0   INTRODUCTION

In computer science, one of the chief goals is to develop efficient compilers. Basically, these compilers are developed by implementation of different parsers. Therefore, it is essential to study the fundamental concepts of grammars that help us to design a parser for a compiler. Parser is a technique to construct a derivation tree or parsed tree for a given sentence Q_0 in a context-free language. In the design of a compiler, the subclasses of context-free grammars play an important role. This subclass of context-free grammars is known as $LR(k)$ grammar. In 1965, Kunth studied the equivalence of deterministic pushdown automata to $LR(1)$ grammar and later he generalized subclasses of context-free grammars to efficient parsing algorithm. In 1970, Grahm showed that a number of other classes of grammar such as SLR (simple LR), LALR (Look ahead LR), and canonical LR defined exactly in context-free languages. The decidability of equivalence is extended to a proper subset of $LR(k)$ grammars known as $LL(k)$ grammar. In this chapter, we study these grammars that are widely used in briefly.

8.1   *LR(k)* GRAMMAR

Generally, we study different parsing techniques in compiler design. Before we study these techniques, it is essential to know what parsing is. Parsing is a syntactical structure of a string, which is the first step in understanding the meaning of the sentence. A grammar for which we can construct a parsing table is said to be a LR grammar. It is to be noted that some context-free grammars are not LR grammars, and these grammars can be avoided for typical programming languages construct. In a LR grammar, when LR parser is implemented using a stack, left to right shift reduce parser should be able to recognize handles when they appear on top of the stack. We study handles and left to right shift reduce parser in due course in this chapter.

In $LR(k)$ grammars L stands for left to right scanning of input string whereas R stands for producing rightmost derivation, and k is number of next symbols used in the input sting. To find a derivation tree for a given sentence 'w' we can start with 'w' and replace a substring w_1 of w, with the help of a variable Q_1 and with the production $Q_1 \rightarrow w_1$. This process is repeated until we get the start symbol Q_0. However, this process is not easy because of many choices available at each stage.

For certain subclasses of context-free grammars, the derivation in the reverse order for a given string in a deterministic way is possible to perform the abovesaid process.

Let us consider a context-free grammar G and some sentential form $\alpha\beta w$ of G, where $\alpha, \beta \in (V_N \cup T)^*$ and $w \in T^*$. Further let us aim at finding the production that is applied in the last step of the derivation for $\alpha\beta w$. Before we find the last step of derivation, we use the symbols $\underset{RM}{\Rightarrow}$ and $\underset{RM}{\overset{*}{\Rightarrow}}$ to denote single rightmost derivation and rightmost derivations, respectively. A sentential form that can be derived by using rightmost derivation is known as right sentential form. If $Q \rightarrow \beta$ is a production, then it is used in last step of rightmost derivation of $\alpha\beta w$ if there are k symbols after β in $\alpha\beta w$. In such a case we call G as $LR(k)$ grammar. The production $Q \rightarrow \beta$ that could be introduced at the last step in a rightmost derivation of $\alpha\beta w$ is known as handle production whereas β is known as a handle. It is to be noted that in this context, the position of β within $\alpha\beta w$ is important as we have discussed above.

Now we define formally the $LR(k)$ grammar. A context-free grammar $G = (V_N, T, P, Q_0)$ in which $Q_0 \overset{n}{\Rightarrow} Q_0$ for $n = 0$, is said to be a $LR(k)$ grammar $(k \geq 0)$ if the following conditions hold.

   i. $Q_0 \underset{RM}{\overset{*}{\Rightarrow}} \alpha Q w \underset{RM}{\overset{*}{\Rightarrow}} \alpha \beta w$, where $\alpha, \beta \in V_N^*$ and $w \in T^*$,

   ii. $Q_0 \underset{RM}{\overset{*}{\Rightarrow}} \alpha_1 Q_1 w_1 \underset{RM}{\Rightarrow} \alpha_1 \beta_1 w_1$, where $\alpha_1, \beta_1 \in V_N^*$ and $w_1 \in T^*$, and

   iii.   The first $|\alpha\beta| + k$ symbols of "$\alpha\beta w$" and "$\alpha_1 \beta_1 w_1$" coincide, then $\alpha = \alpha_1$, $Q = Q_1$ and $\beta = \beta_1$.

Note We note here the following important points regarding the $LR(k)$ grammar.

   (*a*) If the number of symbols in $\alpha\beta w$ or $\alpha_1 \beta_1 w_1$ are less than $|\alpha\beta| + k$, then we add some blank symbols on the right and compare.

   (*b*) If we encounter $\alpha\beta w$ in the sentential form, then we can get a rightmost derivation of βw the following way:

If $Q \rightarrow \beta$ is a production, then we have to decide whether $Q \rightarrow \beta$ is used in last step of rightmost derivation of $\alpha\beta w$. This can be identified by looking k symbols after β in $\alpha\beta w$. We can apply the production $Q_1 \rightarrow \beta_1$ in the last step of rightmost derivation of $\alpha_1 \beta_1 w_1$, if $\alpha_1 \beta_1 w_1$ is the sentential form satisfying final condition (iii) of the definition given above. But by definition, we have $Q = Q_1, \beta = \beta_1$ and $\alpha = \alpha_1$. Therefore, $Q \rightarrow \beta$ is only the production we can apply and are able to identify by looking k symbols after β. The process is repeated until we get the start symbol Q_0.

   (*c*) If a grammar G is $LR(k)$, then we say the grammar G is also $LR(k_1), \forall\, k_1 > k$.

8.1.1   *LR* Items

In this section we discuss some preliminary definitions that are useful in defining $LR(0)$ grammar. An item for a given context-free grammar is a production with a dot anyplace in the right side of the production. The dot can be included at the beginning or at the end also. In case of null production $Q \to \wedge$ or $Q \to \varepsilon$, the only LR item is $Q \to \cdot$. This is because the length of $\wedge$ is zero. A viable prefix of a right sentential form $\alpha\beta w$ is any prefix of $\alpha\beta w$ ending no farther right than the right end of a handle of $\alpha\beta w$. We say an item $Q \to \alpha \cdot \beta$ is valid for a viable prefix γ if there is a rightmost derivation

$$Q_0 \underset{RM}{\overset{*}{\Rightarrow}} \delta Q w \underset{RM}{\Rightarrow} \delta\alpha\beta w$$

and $\delta\alpha = \gamma$. An LR item is said to be complete if the dot is the rightmost symbol in the item. For example, $Q \to \alpha \cdot$ is a complete item.

For example, consider a grammar G with productions $Q_0 \to Qa$, $Q \to QA \big| A$, and $A \to bQc \big| bc$. The LR items corresponding to the production $Q_0 \to Qa$ are $Q_0 \to \cdot Qa$, $Q_0 \to Q \cdot a$, and $Q_0 \to Qa \cdot$. The LR items corresponding to the production $Q \to QA \big| A$ are $Q \to \cdot QA$, $Q \to Q \cdot A$, $Q \to QA \cdot$, $Q \to \cdot A$, and $Q \to A \cdot$. Similarly, the LR items corresponding to the production $A \to bQc \big| bc$ can be obtained. For the above grammar G consider a rightmost derivation

$$Q_0 \Rightarrow Qa \Rightarrow QAa \Rightarrow QbQca.$$

Therefore, it is clear that $QbQca$ is a right sentential form with the handle bQc. The viable prefixes corresponding to the right sentential form $QbQca$ are $\wedge, Q, Qb, QbQ$ and $QbQc$.

8.1.2   **$LR(0)$ Grammar**

In $LR(k)$ grammar, when $k = 0$, it reduces to $LR(0)$ grammar. In $LR(0)$ grammar L stands for left to right scanning of input string whereas R stands for producing rightmost derivation with 0 (zero) number of next symbols in the input sting. It indicates that $LR(0)$ is a restricted type of context-free grammar and first in the family collectively called LR grammars. It ($LR(0)$) defines exactly the deterministic context-free language (DCFL) having the prefix property. The prefix property is not a major restriction. This is because; we can convert any DCFL into a DCFL with prefix property just by introducing an end marker. For example, if L is a DCFL, then $L\$ = \{w\$: w \in L\}$ is a DCFL with prefix property.

Now, we define formally the $LR(0)$ grammar. A grammar G is said to be $LR(0)$ grammar if

(i)   its start symbol does not appear on the right side of any production, and

(ii) for every feasible prefix β of G, whenever $Q \rightarrow \alpha.$ is a complete item valid for β, then neither a complete item nor any other item with a terminal of the right of the dot is valid for β.

It is to be noted that the only item that could be valid simultaneously with $Q \rightarrow \alpha.$ are productions with a nonterminal to the right of the dot. This can occur only if $\alpha = \wedge$, otherwise another violation of the $LR(0)$ condition can be shown to occur.

8.1.3   Computing Sets of Valid Items

The method of accepting the language $L(G)$ for $LR(0)$ grammar G by deterministic pushdown automaton and the definition of $LR(0)$ grammars each depend on knowing the set of valid items for each viable prefix γ. It implies that, the set of viable prefixes is a regular set for every context-free grammar G, and this regular set is accepted by a nondeterministic finite automaton (NFA) whose states are the items for the grammar G. An equivalent deterministic finite automaton (DFA) can be obtained by applying subset construction on this nondeterministic finite automaton (NFA). The states of constructed deterministic finite automata (DFA) are the set of valid items for γ in response to the viable prefix γ. The nondeterministic finite automata (NFA) M recognizing the viable prefixes for context-free grammar $G = (V_N, T, P, Q_0)$ is defined as $M = (S, \Sigma, \delta, s_0, A)$, where S is the set of items for G including the state s_0, which is not an item, $\Sigma = V_N \cup T$, and $A = S$. The transition function δ is defined as follows:

1. $\delta(s_0, \wedge) = \{Q \rightarrow \cdot \alpha \mid Q \rightarrow \alpha$ is a production$\}$
2. $\delta(A \rightarrow \alpha \cdot B\beta, \wedge) = \{B \rightarrow \cdot \gamma \mid B \rightarrow \gamma$ is a production$\}$
3. $\delta(A \rightarrow \alpha \cdot X\beta, X) = \{A \rightarrow \alpha X \cdot \beta\}$

The above rule (2) allows expansion of a variable B appearing immediately to the right of the dot. The rule (3) allows moving the dot over any grammar symbol X if X is the next input symbol.

Theorem   *The nondeterministic finite automata (NFA) M defined above has the property that $\delta(s_0, \gamma)$ contains $A \rightarrow \alpha \cdot \beta$ if and only if $A \rightarrow \alpha \cdot \beta$ is valid for γ.*

Proof   (Only if part)   First of all, we must prove that each item $A \rightarrow \alpha \cdot \beta$ contained in $\delta(s_0, \gamma)$ is valid for γ. We prove this by the method of induction on the length of the shortest path labelled γ from s_0 to $A \rightarrow \alpha \cdot \beta$ in the transition graph for M. The only paths of length one from s_0 are labelled $\wedge$ and to items of the form $Q_0 \rightarrow \cdot \alpha$. Each of these items is valid for $\wedge$ because of the rightmost derivation $Q_0 \underset{RM}{\Rightarrow} \alpha$. Thus, the basis is true.

In order to show the induction step, let us assume that the result is true for paths smaller than k, and further let there be a path of length k labelled γ from s_0 to $A \to \alpha \cdot \beta$. Hence, there arise two cases and this is due to the last edge. Therefore, the last edge must be labelled with either $\wedge$ or not.

Case 1 The last edge is labelled X, for X in $\Sigma = V_N \cup T$. The edge must come from a state $A \to \alpha' \cdot X\beta$, where $\alpha = \alpha' X$. Thus, by inductive hypothesis, $A \to \alpha' \cdot X\beta$ is valid for γ', where $\gamma = \gamma' X$. So, there is a rightmost derivation

$$Q_0 \underset{RM}{\overset{*}{\Rightarrow}} \delta A w \underset{RM}{\Rightarrow} \delta \alpha' X\beta w$$

where $\delta \alpha' = \gamma'$. In the same way it can be shown that $A \to \alpha' X \cdot \beta$, i.e., $A \to \alpha \cdot \beta$ is valid for γ.

Case 2 If the last edge is labelled $\wedge$, then α must be $\wedge$, and $A \to \alpha \cdot \beta$ is really $A \to \cdot \beta$. The item in the previous state is of the form $B \to \alpha_1 \cdot A\beta_1$, and is valid for γ. Therefore, we have a derivation

$$Q_0 \overset{*}{\Rightarrow} \delta B w \Rightarrow \delta \alpha_1 A \beta_1 w$$

where $\gamma = \delta \alpha_1$. Let us assume $\beta_1 \overset{*}{\Rightarrow} x$ for some terminal string x. Thus, the derivation

$$Q_0 \underset{RM}{\overset{*}{\Rightarrow}} \delta B w \underset{RM}{\Rightarrow} \delta \alpha_1 A \beta_1 w \underset{RM}{\overset{*}{\Rightarrow}} \delta \alpha_1 A x w \underset{RM}{\Rightarrow} \delta \alpha_1 \beta x w$$

can be written as $Q_0 \underset{RM}{\overset{*}{\Rightarrow}} \delta a_1 A x w \underset{RM}{\Rightarrow} \delta \alpha_1 \beta x w$. So, $A \to \cdot \beta$ is valid for γ, as $\gamma = \delta \alpha_1$.

(If part) assume that $A \to \alpha \cdot \beta$ is valid for γ. Then we have

$$Q_0 \overset{*}{\Rightarrow} \gamma_1 A w \underset{RM}{\Rightarrow} \gamma_1 \alpha \beta w \tag{1}$$

where $\gamma_1 \alpha = \gamma$. If we show that $\delta(s_0, \gamma_1)$ contains $A \to \cdot \alpha\beta$, then we know that $\delta(s_0, \gamma)$ contains $A \to \alpha \cdot \beta$. This is because of rule (3) defined earlier in this section. Therefore, we show by mathematical induction on the length of derivation (1) that $\delta(s_0, \gamma_1)$ contains $A \to \cdot \alpha\beta$.

The basis follows from rule (1). In order to show the induction step, assume the step $Q_0 \underset{RM}{\overset{*}{\Rightarrow}} \gamma_1 A w$ in which A was introduced. Therefore, we write $Q_0 \underset{RM}{\overset{*}{\Rightarrow}} \gamma_1 A w$ as

$$Q_0 \underset{RM}{\overset{*}{\Rightarrow}} \gamma_2 B x \underset{RM}{\Rightarrow} \gamma_2 \gamma_3 A \gamma_4 x \underset{RM}{\overset{*}{\Rightarrow}} \gamma_2 \gamma_3 A y x$$

where $\gamma_2 \gamma_3 = \gamma_1$ and $yx = w$. Therefore, by the mathematical induction hypothesis applied to the derivation $Q_0 \underset{RM}{\overset{*}{\Rightarrow}} \gamma_2 B x \underset{RM}{\Rightarrow} \gamma_2 \gamma_3 A \gamma_4 x$, it is clear that $B \to \cdot \gamma_3 A \gamma_4$ is in $\delta(s_0, \gamma_2)$. Thus we have by rule (3) $B \to \gamma_3 \cdot A \gamma_4$ is in $\delta(s_0, \gamma_2 \gamma_3)$. Again we have by rule (2) $A \to \cdot \alpha\beta$ is in $\delta(s_0, \gamma_2 \gamma_3)$. Hence, we have proved the induction hypothesis, since $\gamma_2 \gamma_3 = \gamma_1$.

Note The above theorem provides a way for computing the sets of valid items for any viable prefix. Just convert the nondeterministic finite automaton (NFA) whose states are items to a deterministic finite automaton (DFA). In case of DFA, the path from the start state labelled γ leads to the state that is the set of valid items for γ. Just inspect each state to see if a violation of the $LR(0)$ condition occurs in the constructed DFA.

8.1.4   Properties of *LR(k)* Grammar

We have discussed in detail the pushdown automata (PDA) in chapter 6, where we have discussed the equivalence of pushdown automata and context-free languages. It indicates that for any context-free language, we can generate a pushdown automaton. At the same time the language accepted by pushdown automaton is a context-free language. In this section we state some of the important properties of $LR(k)$ grammar that play an important role in parser generation, and provide relation between $LR(k)$ grammars and PDA.

Property 1   If G is an $LR(k)$ grammar, then there exists a deterministic pushdown automaton that accept $L(G)$.

Property 2   If M is a deterministic pushdown automaton, then there is an $LR(1)$ grammar G such that $L(G) = N(M)$, where $N(M)$ is the set accepted by null stack.

Property 3   If G is an $LR(k)$ grammar where $k > 1$, then there exists an equivalent grammar G' which is $LR(1)$.

Property 4   The grammar defined by $LR(0)$ is exactly known as deterministic context-free language (DCFL). The deterministic context-free languages are closed under complementation but not under union and intersection.

Property 5   The class of deterministic context-free languages is a proper subclass of the class of context-free languages.

Property 6   A context-free language is generated by a $LR(0)$ grammar if and only if it is accepted by a deterministic pushdown automaton and has prefix property.

Property 7   There is an algorithm decide whether a given context-free grammar is $LR(k)$ for a given natural number k.

Property 8   Every $LR(k)$ grammar G is unambiguous.

Proof   Let $G = (V_N, T, P, Q_0)$ be an $LR(k)$ grammar. Our aim is to show that the grammar $LR(k)$ is unambiguous. Therefore, for any $w \in T^*$, there exists only one rightmost derivation. We prove it by the method of contradiction. Assume that the grammar G is ambiguous. Therefore, there exist two distinct rightmost derivations for $w \in T^*$ such that

$$Q_0 \underset{RM}{\overset{*}{\Rightarrow}} \alpha Q x \underset{RM}{\Rightarrow} \alpha \beta x = w \tag{1}$$

and
$$Q_0 \underset{RM}{\overset{*}{\Rightarrow}} \alpha_1 Q_1 x_1 \underset{RM}{\Rightarrow} \alpha_1 \beta_1 x_1 = w \tag{2}$$

As $\alpha \beta x = \alpha_1 \beta_1 x_1$, from the definition of $LR(k)$ grammar it follows that $\alpha = \alpha_1, Q = Q_1$ and $\beta = \beta_1$. It implies that $x = x_1$ and so $\alpha Q x = \alpha_1 Q_1 x_1$. Therefore, the last step of derivation in both the equations (1) and (2) are same. The same argument also holds for the other sentential forms derived in the course of (1) and (2). It indicates that both equations (1) and (2) are same.

Hence, it leads to a contradiction and so our assumption is wrong. Therefore, the $LR(k)$ grammar G is unambiguous. This completes the proof.

8.2   *LR*(0) GRAMMAR AND DETERMINISTIC PDA

In this section we show that the language generated by every $LR(0)$ grammar is a deterministic context-free language (DCFL), and every deterministic context-free language with the prefix property has an $LR(0)$ grammar. Also, we have an exact characterization of the deterministic CFL; namely L is a deterministic context-free language if and only if $L\$$ is an $LR(0)$ grammar. This is because of every language with a $LR(0)$ grammar will be shown to have the prefix property. In this section we discuss the conversion of grammar to deterministic pushdown automata and conversely.

8.2.1   Conversion of *LR*(0) Grammar to Deterministic PDA

The way of construction of a deterministic PDA from an $LR(0)$ grammar is different then the way of construction of nondeterministic PDA from an arbitrary context-free language. Here, we shall trace out a rightmost derivation, in reverse, by using a stack to hold a viable prefix of a right sentential form that include all variables of that right sentential form, allowing the reminder of the form to appear on the input.

To simulate rightmost derivations in an $LR(0)$ grammar, we do keep a record of viable prefix on the stack. At the same time, for every symbol we keep a state of the DFA recognizing viable prefixes. If viable prefix $X_1, X_2, X_3, \dots, X_k$ is on the stack, then the complete stack contents will be

$$q_0 X_1 q_1 X_2 q_2 X_3 \cdots X_k q_k$$

where $q_i = \delta(s_0, X_1 X_2 \cdots X_i)$ and δ is the transition function of the DFA. The top state q_k provides the valid items for $X_1 X_2 X_3 \cdots X_k$.

If q_k contains $A \to \alpha \cdot$, then $A \to \alpha \cdot$ is valid for $X_1 X_2 X_3 \cdots X_k$. Therefore, α is a suffix of $X_1 X_2 X_3 \cdots X_k$. Let us say $\alpha = X_{i+1} X_{i+2} \cdots X_k$. It is to be noted that α may be $\wedge$ and in this case we have $i = k$. Moreover, there is some string w such that $X_1 X_2 \cdots X_k w$ is the right sentential form and there is a derivation

$$Q_0 \underset{RM}{\overset{*}{\Rightarrow}} X_1 X_2 X_3 \cdots X_i A w \underset{RM}{\Rightarrow} X_1 X_2 X_3 \cdots X_k w.$$

Now to obtain the right sequential form previous to $X_1 X_2 X_3 \cdots X_k w$ in a rightmost derivation we reduce α to A, by replacing $X_{i+1} X_{i+2} \cdots X_k$ on top of the stack by A. That is, by a sequence of pop operations followed by a push operation that pushes A. The correct covering state onto the stack, the DPDA will enter a sequence of ID's

$$(s, q_0 X_1 q_1 X_2 \cdots q_{k-1} X_k q_k, w) \overset{*}{\vdash} (s, q_0 X_1 q_1 X_2 \cdots q_{i-1} X_i q_i A q, w) \tag{1}$$

where $q = \delta(q_i, A)$. It is to be noted that if the grammar is $LR(0)$, then q_k contains only $A \to \alpha\cdot$, unless $\alpha = \wedge$, in which case q_k may contain some incomplete items. However, by definition of $LR(0)$, none of these items have a terminal to the right of the dot, or are complete. Therefore, for any y such that $X_1 X_2 \cdots X_k y$ is a right sentential form, $X_1 X_2 \cdots X_i A y$ must be the previous right sentential form, therefore reduction of α to A is correct regardless of the current input.

Now let us consider the case when q_k contains only incomplete items. Then the right sentential form previous to $X_1 X_2 \cdots X_k w$ could not be formed by reducing a suffix for $X_1 X_2 \cdots X_k$ to some variable, rather there would be a complete item valid for $X_1 X_2 \cdots X_k$. Since $X_1 X_2 \cdots X_k$ is a viable prefix, there must be a handle ending to the right of X_k in $X_1 X_2 \cdots X_k w$. Therefore, the only appropriate action for the deterministic PDA is to shift the next input symbols onto the stack. That is,

$$(s, q_0 X_1 q_1 X_2 \cdots q_{k-1} X_k q_k, ay) \overset{*}{\vdash} (s, q_0 X_1 q_1 X_2 \cdots q_{k-1} X_k q_k a t, y) \tag{2}$$

where $t = \delta(q_k, a)$. If t is not the empty set of items, $X_1 X_2 \cdots X_k a$ is a viable prefix. If t is empty, then we shall prove there is no possible previous right sentential form for $X_1 X_2 \cdots X_k a y$. Therefore, the original input is not in the grammar's language and the deterministic pushdown automata dies instead of making the move as given in equation (2).

Theorem *If L is L(G) for an LR(0) grammar G, then L is N(M) for a deterministic pushdown automata M.*

Proof   Initially, we construct a deterministic finite automata D for the grammar G, with the transition function δ that recognizes viable prefixes of grammar G. Assume that the pushdown (stack) symbols of PDA M are the symbols of grammar G, and states are as in DFA D. Let the deterministic PDA has an initial state s, along with additional states used to perform reductions by sequence of moves such as (1) defined above.

If a grammar G is $LR(0)$, then the reductions are the only possible way to get the previous right sentential form when the state of the deterministic finite automata on the top of M's pushdown store that contains a complete item. When deterministic PDA M starts with string $w \in L(G)$ on its input and only q_0 on its stack, it will construct a rightmost derivation for string w in reverse order. The remaining part requiring this proof is that when a shift operation is called for, as in equation (2), there could not be a handle among, grammar symbols $X_1 X_2 \cdots X_k$ found on the pushdown store at that time. If there were such handle, then definitely some DFA state on the pushdown store below the top would have a complete item.

Assume that there were such a state containing $A \to \alpha\cdot$. Note that each state, when it is first put on the pushdown store either by equation (1) or (2), is on top of the stack. Thus, it will immediately call for reduction of α to a nonterminal A. If $\alpha \neq \wedge$, then $A \to \alpha\cdot$ is removed from the pushdown store. If $\alpha = \wedge$, then reduction of $\wedge$ to A takes place by equation (1), causing A to be put on the stack above $X_1 X_2 \cdots X_k$. In such a case there will always be a variable above X_k on the pushdown store as long as $X_1 X_2 \cdots X_k$ occupies the bottom positions on the pushdown store. At the same time $A \to \wedge$ is at position k and could not be the handle of any right sentential form $X_1 X_2 \cdots X_k \beta$, where β contains a nonterminal. Finally, we have to show the acceptance by the grammar G.

Here, we show the acceptance by the grammar G. If the top state on the pushdown store is $Q_0 \to \alpha\cdot$, where Q_0 is the start symbol of the grammar G, then G pops its pushdown store by accepting the state. Thus, the reverse of a rightmost derivation of the original input string have completed. Since Q_0 does not appear on the right side of any production, it is not possible that there is an item of form $A \to Q_0 \cdot \alpha$ valid for viable prefix Q_0. Therefore, it is not necessary to shift additional input symbols when the start symbol Q_0 alone appears on the pushdown store. Otherwise, we can say that, $L(G)$ always has the prefix property if the grammar G is $LR(0)$.

Thus we have proved that, if $w \in L(G)$, then the deterministic PDA M finds a rightmost derivation of the string w, reduces w to Q_0 and accepts, where Q_0 is start symbol of grammar G. Conversely, if deterministic PDA M accepts w, then the sequence of right sentential forms represented by the instantaneous descriptions (ID) of M gives a derivation of the string w from the start symbol Q_0. Therefore, we can say that $N(M) = L(G)$. This completes the proof.

8.2.2   Conversion of Deterministic PDA to *LR(0)* Grammar

In this section we discuss how to obtain an $LR(0)$ grammar from a given deterministic pushdown automata. If a language L is $N(M)$ for a deterministic PDA M, then L has an $LR(0)$ grammar. We define the grammar $LR(0)$ as follows:

Let $M = (S, \Sigma, \Gamma, \delta, s_0, z_0, \phi)$ be a deterministic PDA. We define a grammar $G = (V_N, T, P, Q_0)$ such that $N(M) = L(G)$, where V_N contains the start symbol Q_0, the symbol $[sXp]$ for $s, p \in S$ and $X \in \Gamma$, and the symbols A_{saY} for $s \in S$, $a \in \Sigma \cup \{\wedge\}$ and $Y \in \Gamma$. The start symbol Q_0 and $[sXp]$'s play the same role as in case of equivalence of pushdown automata and context-free language. The symbol A_{saY} indicates that the production is obtained from the move of M in $\delta(s, a, Y)$. The productions of grammar G is defined by following rules:

R_1 : $Q_0 \rightarrow [s_0 z_0 p]$ $\forall$ $p \in S$.

R_2 :   If $\delta(s, a, Y) = (p, \wedge)$, then there is a production $[sYp] \rightarrow A_{saY}$.

R_3 :   If $\delta(s, a, Y) = (p_1, X_1 X_2 \cdots X_k)$ for $k \geq 1$, then for each sequence of states $p_1, p_2, \cdots, p_{k+1}$ there is a production

$[sYp_{k+1}] \rightarrow A_{saY}[p_1 X_1 p_2][p_2 X_2 p_3] \cdots [p_k X_k p_{k+1}].$

R_4 :   For all s, a, and Y, there is a production $A_{saY} \rightarrow a$.

8.3   *LL(k)* GRAMMAR

The grammar that is widely used in top-down parsing by applying certain techniques to certain subclasses of context-free languages is represented as $LL(k)$ grammar. The grammar G that look ahead of k symbols, left to right scanning of input sting in $L(G)$, and leftmost derivation is called an $LL(k)$ grammar. On considering $k = 1$, the grammar that looks ahead of 1 symbol, left to right scanning of input string in $L(G)$ and leftmost derivation is called an $LL(1)$ grammar.

Alternatively, a grammar is said to be $LL(k)$ if looking ahead k symbols input is always enough to choose the subsequent move of the pushdown automata. Such a grammar always allows the construction of a deterministic top-down parser and there are systematic methods of obtaining a context-free grammar $LL(k)$. However, there is a significant difference between LL and LR grammars. In $LR(k)$, we must be able to recognize the occurrence of the right side of production, with k input look ahead symbols whereas the requirement of $LL(k)$ grammars is quite different, where we recognize the use of a production looking only the first k symbols of what its right side derives. Thus, LR grammar describe more languages than LL grammar however LL grammar has its own importance. Here, we also define some important terms that is widely used in LL grammar. These are left factoring and left recursion.

Left factoring   It is a grammar transformation that is generally used for producing a grammar in a suitable form for predictive parsing. For example, the production $Q \rightarrow \alpha\beta_1 | \alpha\beta_2$ in a grammar G can be rewritten by using left factoring as below, where Q' is a newly introduced variable. The new set of productions is $Q \rightarrow \alpha Q'$ and $Q' \rightarrow \beta_1 | \beta_2$.

Left recursion A grammar G is said to be left recursive if it has a nonterminal Q such that there is a derivation $Q \overset{+}{\Rightarrow} Q\alpha$ for some string a. However, it is necessary to eliminate left recursion, as top-down parsing methods cannot handle left recursion grammar. For example, a left recursive pair of productions $Q \to Q\alpha | \beta$ can be replaced by a set of non-recursive productions as $Q \to \beta Q'$, and $Q' \to \alpha Q' | \wedge$, without changing the set of strings that is derivable from Q.

8.3.1 *LL*(1) Grammar

The first L in $LL(1)$ indicates the left-to-right scanning of input string whereas the second L stands for leftmost derivation. The integer l in the grammar is used for looking one input symbol at each step of parsing action. It is to be noted that, no ambiguous or left-recursive grammar can be $LL(1)$. So, in order to get the $LL(1)$ grammar, it is essential to remove left recursion from the grammar and it can be removed as discussed earlier. Alternatively, a grammar is said to be $LL(1)$ if its parsing table has no multiple defined entries. A parse table is a blueprint for the creation of a $LL(1)$ parser. The rows in the parse table are labelled with the nonterminals of the grammar, and columns of the parse table are labelled with the terminals of the grammar. Each entry in the parse table is either empty, or contains a production rule. The rule located at row Q, column a of a parse table tells us to apply the rule when we are trying to parse the nonterminal Q, and the next symbol in the input string is an a.

SOLVED EXAMPLES

Example 1

Find the items for the following grammar G having productions $Q_0 \to Qa$, $Q \to aAb$, and $A \to aA | ab$.

Solution Given productions are $Q_0 \to Qa$, $Q \to aAb$, and $A \to aA | ab$. Our aim is to find the items for the grammar. The items corresponding to the production $Q_0 \to Qa$ are $Q_0 \to \cdot Qa$, $Q_0 \to Q \cdot a$, and $Q_0 \to Qa \cdot$. The items corresponding to the production $Q \to aAb$ are $Q \to \cdot aAb$, $Q \to a \cdot Ab$, $Q \to aA \cdot b$, and $Q \to aAb \cdot$. The items corresponding to the production $A \to aA | ab$ are $A \to \cdot aA$, $A \to a \cdot A$, $A \to aA \cdot$, $A \to \cdot ab$, $A \to a \cdot b$, and $A \to ab \cdot$.

Example 2

Consider the grammar G with productions $Q_0 \to Qa$, $Q \to aAb$, and $A \to aA | ab$. Find the handle and viable prefixes of the right sentential form $aaAba$.

Solution Given productions are $Q_0 \to Qa$, $Q \to aAb$, and $A \to aA | ab$. The rightmost derivation of the right sentential form $aaAba$ is given as below.

$$Q_0 \Rightarrow Qa \Rightarrow aAba \Rightarrow aaAba$$

From the derivation it is clear that the handle is aA. The viable prefixes of the right sentential form $aaAba$ are $\wedge$, a ,aa, and aaA.

Example 3

Consider the grammar G with productions $Q_0 \to Qc, Q \to QA|A$, and $A \to aQb|ab$. Construct nondeterministic finite automata for the above grammar G.

Solution

Given productions are $Q_0 \to Qc, Q \to QA|A$, and $A \to aQb|ab$. The items for the grammar are given below. The state graph of the NFA is given below.

$$\begin{array}{llllll}
Q_0 \to \cdot Qc & Q \to \cdot QA & Q \to \cdot A & A \to a \cdot Qb & A \to \cdot ab \\
Q_0 \to Q \cdot c & Q \to Q \cdot A & Q \to A \cdot & A \to aQ \cdot b & A \to a \cdot b \\
Q_0 \to Qc \cdot & Q \to QA \cdot & A \to \cdot aQb & A \to aQb \cdot & A \to ab \cdot
\end{array}$$

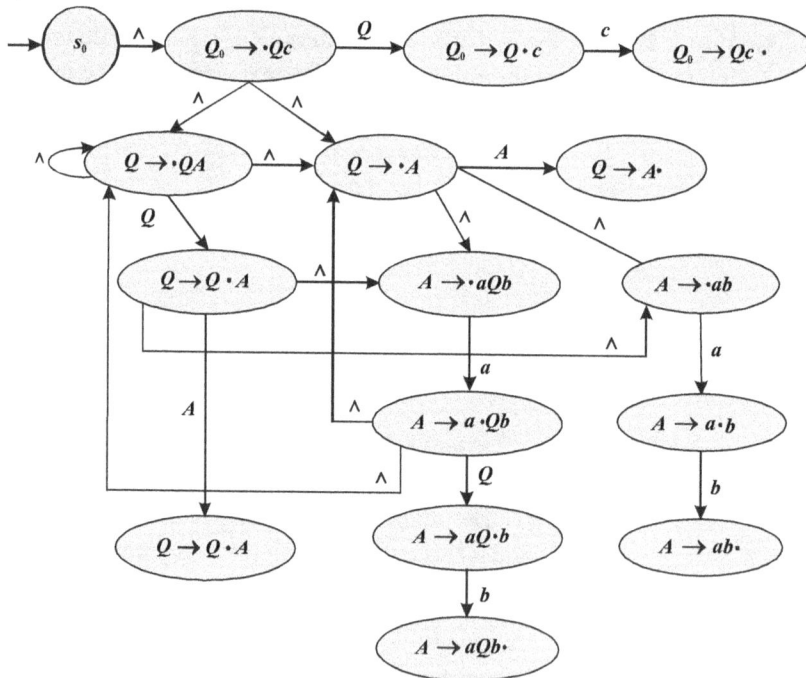

Example 4

Is the grammar G with productions, $Q_0 \to AB|aB$, $A \to aA|a$ and $B \to b$ $LR(k)$ for some k?

Solution Given productions are $Q_0 \to AB|aB$, $A \to aA|a$ and $B \to b$. Let us consider a string $w = ab$. Therefore, the string $w = ab$ can be derived in two distinct ways as

$Q_0 \Rightarrow AB \Rightarrow aB \Rightarrow ab$ and $Q_0 \Rightarrow aB \Rightarrow ab$. It indicates that the given grammar G is ambiguous. Therefore, we conclude that it is not an $LR(k)$ grammar.

Example 5

Write the items for the CFL L, where $L = \{a^n b^n \,|\, n \geq 1\}$.

Solution The context-free grammar for the language $L = \{a^n b^n \,|\, n \geq 1\}$ is given as $G = (V_N, T, P, Q_0)$, where $P = \{Q_0 \to aQ_0b\,|\,ab\}$. The items for above grammar are $Q_0 \to \cdot aQ_0b$, $Q_0 \to a \cdot Q_0b$, $Q_0 \to aQ_0 \cdot b$, $Q_0 \to aQ_0b \cdot$, $Q_0 \to \cdot ab$, $Q_0 \to a \cdot b$, and $Q_0 \to ab \cdot$.

Example 6

Consider the grammar G with productions $Q_0 \to Qc, Q \to QA|A$, and $A \to aQb|ab$. Construct deterministic finite automata (DFA) for the grammar G.

Solution The transition graph of the NFA for the grammar is given in example 3. We construct the DFA whose states are the sets of valid items for the grammar.

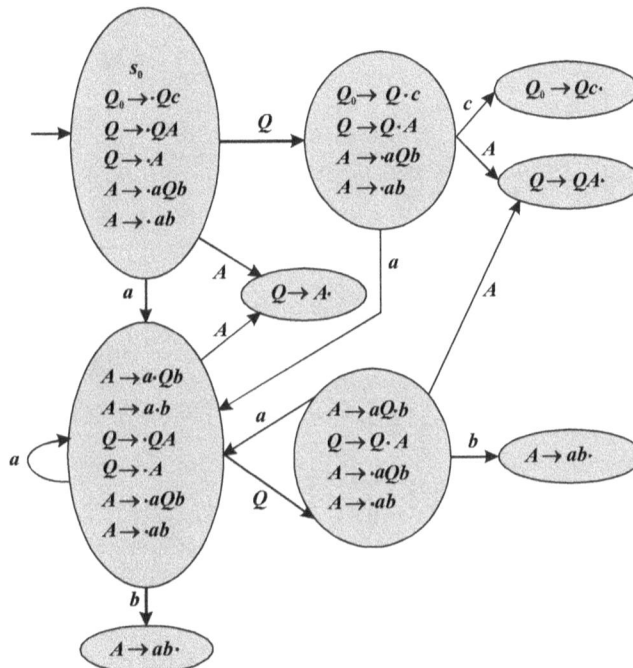

Example 7

Consider the context-free grammar $Q_0 \to A + Q_0$, $Q_0 \to A * Q_0|A$, and $A \to b$. Let $w = b + b * b$, $w \in L(G)$. Show that G is an $LL(2)$ grammar.

Solution   In case of *LL* grammar, we construct a top-down parser. Let us consider the string $w = b + b * b$. It is clear that, looking ahead for one symbol is not enough in constructing the string. By looking at the string, we can think of applying production $A \rightarrow b$ but it is not suitable as b is followed by $+$ or $*$. Therefore, it is necessary to look ahead for 2 symbols.

If we start with Q_0, then we have three productions $Q_0 \rightarrow A + Q_0$, $Q_0 \rightarrow A * Q_0$, and $Q_0 \rightarrow A$. The first two symbols in $w = b + b * b$ are $b +$. This implies us to apply only the production $Q_0 \rightarrow A + Q_0$. Now, we apply the production $A \rightarrow b$ to get $b + Q_0$. The remaining part of the string is $b * b$ and the first two symbols are $b *$. So, it is essential to apply the production $Q_0 \rightarrow A * Q_0$. The third symbol in the string $w = b + b * b$ is b. Thus, we apply the production $A \rightarrow b$ to get $b + b * Q_0$. Finally, Q_0 will be replaced by the production $Q_0 \rightarrow A$ and then A will be replaced by b.

A leftmost derivation of the string $w = b + b * b$ is given below.

$$
\begin{aligned}
Q_0 &\Rightarrow A + Q_0 \\
&\Rightarrow b + Q_0 &&[A \rightarrow b] \\
&\Rightarrow b + A * Q_0 &&[Q_0 \rightarrow A * Q_0] \\
&\Rightarrow b + b * Q_0 &&[A \rightarrow b] \\
&\Rightarrow b + b * A &&[Q_0 \rightarrow A] \\
&\Rightarrow b + b * b &&[A \rightarrow b]
\end{aligned}
$$

As it is essential to look ahead 2 symbols at a time to derive the string, so, the grammar is an *LL*(2) grammar.

Example 8

Prove that the grammar G with productions $Q_0 \rightarrow Ab$, $A \rightarrow Aaa | a$ is an *LR*(0) grammar.

Solution   Given productions of the grammar G are $Q_0 \rightarrow Ab$, and $A \rightarrow Aaa | a$. From the productions it is clear that $L(G) = \{a^{2n+1} b | n \geq 0\}$. The rightmost derivation of the right sentential form $a^{2n+1}b$ is given as below.

$$Q_0 \Rightarrow Ab \Rightarrow Aaab \Rightarrow \cdots \Rightarrow Aa^{2n}b \Rightarrow a^{2n+1}b$$

From the derivation it is clear that the handle production is $A \rightarrow a$. The handle is a without looking at any symbol to the right of a. Similarly, the other sentential forms of G are Ab, and $Aa^{2n}b$. As Ab, Aaa, and a are the possible right-hand sides of productions, only $A \rightarrow a$ can be the last production. We are able to decide this without looking at any symbol to the right of a. Similarly, in the right sentential form $Aa^{2n}b$, the last production is $A \rightarrow Aaa$ whereas $Q_0 \rightarrow Ab$ is the last production for the right sentential form Ab. Therefore, we are able to say that $A \rightarrow Aaa$ is the last production for any sentential form $Aa^{2n}b$ for all $n \geq 1$. Therefore, G is an *LR*(0) grammar.

Example 9

Show that the grammar $Q_0 \to aAb$, $A \to aAb|a$ is not an $LR(0)$ grammar.

Solution   From the productions it is clear that the language generated by the given grammar is $L = \{a^{n+1}b^n \,|\, n \geq 1\}$. The rightmost derivation of the right sentential form $a^{n+1}b^n$ is given as below.

$$Q_0 \Rightarrow aAb \Rightarrow aaAbb \Rightarrow \cdots \Rightarrow a^{n-1}aAbb^{n-1} \Rightarrow a^n ab^n$$

From the derivation it is clear that the handle production is $A \to a$. The handle is a if and only if the symbol to the right of a is scanned and found to be b. Similarly, we can get another handle production $A \to aAb$ if and only if the symbol to the right of aAb is only b. Also, $Q_0 \to aAb$ is a handle production if and only if the symbol to the right of aAb is $\wedge$. Therefore, it is clear that the given grammar is $LR(1)$. Thus, we conclude that the grammar is not an $LR(0)$ grammar.

Example 10

Construct an $LL(1)$ grammar for the grammar $G = (V_N, T, P, E)$, where $V_N = \{E, F, H\}$, $T = \{a, 1, 2, +, *, (,)\}$, and productions P consists of

$$\begin{array}{llll} E \to E+H & H \to H*F & F \to (E) & \\ E \to H & H \to F & F \to a1 & F \to a2 \end{array}$$

Solution   In the production $E \to E + H$, the nonterminal E of the LHS is repeated as the first symbol of the right-hand side and so we have a left recursion. Thus, the production $E \to E + H$ can be replaced by the productions $E \to HE'$, and $E' \to +HE' | \wedge$, where E' is a newly introduced nonterminal.

Similarly, in the production $H \to H * F$, the nonterminal H of the LHS is repeated as the first symbol of the right-hand side and so we have a left recursion. Thus, the production $H \to H * F$ can be replaced by the productions $H \to FH'$, and $H' \to *FH' | \wedge$, where H' is a newly introduced nonterminal.

Now, consider the left factoring productions $F \to a1$ and $F \to a2$. The above productions can be replaced by the productions $F \to aF'$, and $F' \to 1 | 2$, where F' is a newly introduced nonterminal.

Therefore, the resulting equivalent grammar is given as $G' = (V_N', T, P', E)$, where $V_N' = \{E, F, H, E', H', F'\}$, $T = \{a, 1, 2, +, *, (,)\}$, and P' consists of the following productions.

$$\begin{array}{llll} E \to HE' & E' \to +HE' | \wedge & H \to F & F \to (E) & F' \to 1 \\ E \to H & H \to FH' & H' \to *FH' | \wedge & F \to aF' & F' \to 2 \end{array}$$

The above grammar G' obtained is known as $LL(1)$ grammar.

REVIEW QUESTIONS

1. Discuss in detail the $LR(k)$ grammar.

2. State the different properties of $LR(k)$ grammar.

3. Define LR item and explain the concept by considering a suitable example.

4. Write the transition function for the nondeterministic finite automata recognizing the viable prefixes for context-free grammar.

5. Define $LR(0)$ grammar and give an example.

6. Prove that the NFA that recognizes viable prefixes for context-free grammar has the property that $\delta(s_0, \gamma)$ contains $A \to \alpha \cdot \beta$ if and only if $A \to \alpha \cdot \beta$ is valid for γ, where the symbols have their usual meaning.

7. Write the rules that convert deterministic PDA to $LR(0)$ grammar.

8. Explain the conversion of $LR(0)$ grammar to deterministic PDA.

9. Prove that, if L is $L(G)$ for an $LR(0)$ grammar G, then L is $N(M)$ for a deterministic pushdown automata M.

10. Write a note on $LL(k)$ grammar and state the difference between $LR(k)$ and $LL(k)$ grammar.

11. Define the parse table for $LL(1)$ grammar and explain with an example.

PROBLEMS

1. Is the grammar $Q_0 \to 0 \mid 01Q_01 \mid 0A1$, $A \to 1Q_0 \mid 0AA1$ $LR(k)$ for some k?

2. Show that the context-free grammar G with productions $Q_0 \to Q_0aQ_0 \mid b$ is not $LR(k)$ for some k.

3. Write the items for the context-free language L, where $L = \{a^n b^{2n} \mid n \geq 1\}$.

4. Eliminate left recursion from the grammar G having productions $Q_0 \to Q_0 * Q_1 \mid Q_1$, $Q_1 \to Q_1 + Q_2 \mid Q_2$, and $Q_2 \to (Q_0) \mid a$.

5. Obtain the items for the context-free language L, where $L = \{a^m b^n \mid m \leq 2n\}$.

6. Write a grammar for the context-free language L, and obtain the items, where $L = \{a^n b^{3n} \mid n \geq 1\} \cup \{(ab)^n \mid n \geq 1\}$.

7. Write the items for the grammar $Q_0 \rightarrow Q_0 + Q_1 | Q_1$, $Q_1 \rightarrow Q_1 * Q_2 | Q_2$, and $Q_2 \rightarrow (Q_0) | a$.

8. Consider the grammar G with productions $Q_0 \rightarrow AaB$, $A \rightarrow aAb | b$ and $B \rightarrow aB | ab$. Find the handle and viable prefixes of the sentential form $aa\,AbbaB$.

9. Consider the grammar G with productions $Q_0 \rightarrow Ac$, $A \rightarrow AB | B$, and $B \rightarrow bAa | aAb | ba$. Construct nondeterministic finite automata for the above grammar G.

10. Construct deterministic finite automata for the grammar G given in problem 9.

11. Show that the grammar $Q_0 \rightarrow aAb$, $A \rightarrow cAc | c$, is not an $LR(0)$ grammar.

12. Prove that the language $\{1^m 0^m 2^n \mid m,n \geq 1\} \cup \{1^m 0^n 2^n \mid m,n \geq 1\}$ cannot be generated by an $LR(k)$ grammar for any k.

13. Show that the grammar G with productions $Q_0 \rightarrow Q_1 Q_2$, $Q_1 \rightarrow 0Q_1 2 | \wedge$, and $Q_2 \rightarrow Q_2 b | b$ is an $LL(1)$ grammar.

14. Eliminate left factoring from the grammar having productions $Q_0 \rightarrow aAbQ_0 | aAbQ_0 cQ_0 | 1$, and $A \rightarrow 2$.

15. Obtain an $LL(1)$ grammar for $G = (V_N, T, P, E)$, where $V_N = \{E, F, H\}$, $T = \{a, 1, 2, +, *, (,)\}$, and productions P consists of $E \rightarrow E * F | H$, $H \rightarrow H + F$, $H \rightarrow F, F \rightarrow a + 1$ and $F \rightarrow a + 2$.

16. Show that the deterministic context-free languages are not closed under union, concatenation or Kleene closure.

17. Prove that the grammar $Q_0 \rightarrow Q_0 Q_0 | aQ_0 b | bQ_0 a | \wedge$ generating the language L, $L = \{w \mid n_a(w) = n_b(w)\}$ is not $LL(k)$.

18. Obtain an LL grammar for the language L, $L = \{1^m 0^n 2^{m+n} \mid m,n \geq 0\}$.

19. Construct an LL grammar for the language $L(b^*ab) \cup L(baaa^*)$.

20. Give an example of a context-free language that is not a deterministic context-free language.

21. Is all unambiguous grammar $LL(1)$? If not, then give a counter example.

9

COMPUTABILITY AND UNDECIDABILITY

9.0   INTRODUCTION

It is observed that certain problems are unsolvable. Turing machines cannot compute all functions even those with precise definitions. One such problem is halting problem. It can be stated as follows: Given a description of a program and finite input, decide whether the program finishes running or will run forever, given that input. Virtually speaking, the behaviour of programs for a non-trivial problem is undecidable.

In 1930, it was a great attempt to define computability and algorithms. Gödel pointed out in 1934 that, primitive recursive functions can be computed by a finite procedure and any function computable by a finite procedure can be specified by a recursive function. Here, in this chapter we will study some undecidable problems. Apart from this we will concentrate on numerical functions and try to find a way to characterize the ones that can be actually computed.

9.1   UNSOLVABLE PROBLEMS

It is stated that a function f on a certain domain is said to be computable if there exists a Turing machine that computes the value of f for all arguments in its domain. On the other hand if there is no such Turing machine that computes the value of f for all arguments in its domain, we say that the function is uncomputable. Therefore it is clear that, a problem may be decidable on some domain but not on another. Here we concern somewhat simplified setting where the result of computation is a simple yes or no. It leads about a problem being decidable (solvable) or undecidable (unsolvable).

A decision problem is said to be decidable if there is a Turing machine that will always halt in a finite amount of time, producing a yes or no answer. On the other hand a decision problem is undecidable if a Turing machine may run forever without giving an answer. In this section, we discuss some undecidable problems such as Post correspondence problem and Rice theorem.

9.1.1   Post Correspondence Problem (PCP)

Here we discuss a combinatorial problem called Post correspondence problem formulated by Emil L. Post in 1946. Later it was observed to have many applications in the theory of formal languages.

The Post correspondence problem can be stated as follows: Given two sequences of *n* strings on some alphabet Σ, say

$$X = u_1, u_2, u_3, \cdots, u_{n-1}, u_n \quad \text{and}$$
$$Y = v_1, v_2, v_3, \cdots, v_{n-1}, v_n$$

we say that there exists a Post correspondence solution (PC Solution) for the pair (X,Y) if there is a non-empty sequence of integers $i_1, i_2, i_3, \cdots, i_m$, such that

$$u_{i_1} u_{i_2} u_{i_3} \cdots u_{i_m} = v_{i_1} v_{i_2} v_{i_3} \cdots v_{i_m}$$

The Post correspondence problem is to devise an algorithm by which we can tell, for any (X,Y), whether or not there exists a PC solution. It is observed that, in some instances we may be able to show by explicit construction that a pair (X,Y) permits a solution or we may be able to argue that no such solution can exist. In general there is no such algorithm for deciding this question under all circumstances. Therefore, the Post correspondence problem is said to be undecidable.

The above concept may be visualized clearly with the help of following Figure 9.1, where the top half of the *i*th group represents a string u_i of the sequence X and the bottom half of the *i*th group represents a string v_i of the sequence Y. Finding a solution for this instance means lining up one or more groups in a horizontal row, each one positioned vertically, so that the string formed by their top halves matches the string formed by their bottom halves as shown in Figure 9.1.

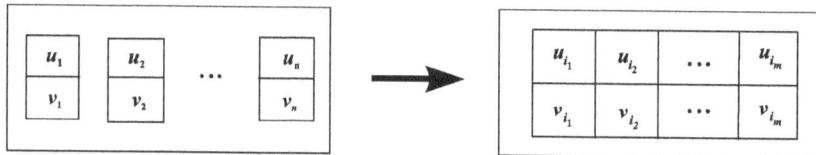

Figure 9.1   Post correspondence problem

For example, consider $X = (ab, bba, b, baa)$ and $Y = (abb, ba, aab, b)$ defined over $\Sigma = \{a,b\}$. For this case, there exists a Post correspondence solution. This can be visualized by the following Figure 9.2, where the sequence is $\{1, 2, 4, 3\}$.

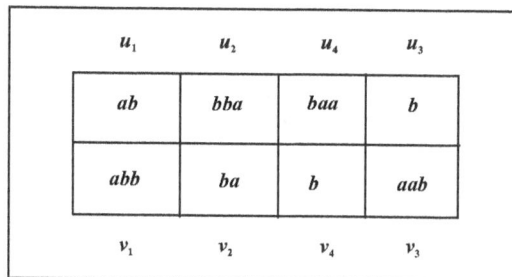

u_1	u_2	u_4	u_3
ab	*bba*	*baa*	*b*
abb	*ba*	*b*	*aab*
v_1	v_2	v_4	v_3

Figure 9.2   An Instance of PCP

Note If the first substring used in Post correspondence problem is always u_1 and v_1, then it is known as modified Post correspondence problem. We say that the pair (X,Y) has a modified Post correspondence solution (MPC-Solution) if there exists a sequence $i_1, i_2, i_3, \cdots, i_m$ such that

$$u_1 u_{i_1} u_{i_2} u_{i_3} \cdots u_{i_m} = v_1 v_{i_1} v_{i_2} v_{i_3} \cdots v_{i_m}$$

Therefore, it is clear that an MPC solution must start with u_1 on the left side and with v_1 on the right side. It is to be noted that, if the pair (X,Y) has MPC solution, then the pair has a PC solution, but the converse is not true.

9.1.2 Rice's Theorem

In the theory of computable functions, Rice's theorem is an important result in computer science. Before we state Rice's theorem, a property of partial functions is called trivial if it holds for all partial computable functions or for none. Rice theorem states that, for any non-trivial property of partial functions, the question of whether a given algorithm computes a partial function with this property is undecidable. Another way of stating this problem that is more useful in computability theory is discussed below. The above theorem is named after Henry Gordon Rice and is otherwise known as Rice–Myhill–Shapiro theorem.

Statement Any non-trivial property about the language recognized by a Turing machine is undecidable. A property about Turing machines can be represented as the language of all Turing machines, encoded as strings, which satisfy that property. The property P is about the language recognized by Turing machines if whenever $L(M) = L(N)$ then P contains M if and only if it contains N. The property is non-trivial if there is at least one Turing machine that has the property, and at least one Turing machine that hasn't.

Proof Here we assume that a Turing machine that recognizes the empty language does not have the property P. If it does, we just take the complement of P and the undecidability of that complement would immediately imply the undecidability of P. We prove this by the method of contradiction.

In order to get a contradiction, assume that P is decidable. It implies that there is a halting Turing machine M_1 that recognizes the descriptions of Turing machines that satisfy P. Using Turing machine M_1 we can construct a Turing machine M_2 that accepts the language

{$(M, w)|M$ is the description of a Turing machine that accepts the string 'w'}.

Since the latter problem is undecidable, this will show that M_1 cannot exist and thus P must be undecidable as well.

Let M_p be a Turing machine that satisfies P (as P is nontrivial). Now the Turing machine M_2 operates as follows:

1. On input (M, w), create a Turing machine $C(M,w)$ as follows:
   (a) On input x, let the Turing machine M run on the string w until it accepts. Equivalently if M does not accept, then the Turing machine $C(M,w)$ will run forever.

(*b*) Next Turing machine M_p runs on x. (Accept if and only if M_p does). It is to be noted that Turing machine $C(M,w)$ accepts the same language as M_p if M accepts w and $C(M,w)$ accepts the empty language if M does not accept w.

Therefore, if M accepts w the Turing machine $C(M,w)$ has the property P, and otherwise it does not.

2. The description of $C(M,w)$ is given to Turing machine M_1. If M_1 accepts, then the input (M,w) is accepted otherwise rejected.

Note   It is to be noted that Rice's theorem does not tell anything about properties of machines, only of functions and languages.

9.2   HILBERT'S PROBLEM

Hilbert's problems are a set of twenty-three problems in mathematics proposed by David Hilbert in 1900 at the International Congress of Mathematicians in Paris. The problems were all unsolved at the time, and several of them were designed to serve as examples for the kinds of problems whose solutions would lead to the furthering of disciplines in mathematics. Here, we discuss the tenth problem. This problem does not ask whether there exists a universal algorithm for deciding the solvability of Diophantine equations, but rather asks for the construction of such an algorithm. Before describing that problem, we briefly discuss polynomials.

A polynomial is a sum of terms, where each term is a product of certain variables and a constant called a coefficient. For example, $2 \cdot x \cdot x \cdot y \cdot z = 2x^2yz$ is a term with coefficient 2, and $2x^2yz + 3xz^2 - 2xy^3 - 18$ is a polynomial containing four terms over the variables x, y, and z.

A root of a polynomial is an assignment of values to its variables such that the value of the polynomial is zero. If all the variables are assigned integer values, then the root is said to be an integral root. It is also known that some polynomials have an integral root and some polynomials do not. For example, an integral root of above polynomial is at $x = 2$, $y = 3$ and $z = 3$.

Hilbert's tenth problem was to devise an universal algorithm that tests whether a polynomial has an integral root. He did not use the term algorithm but rather " a process according to which it can be determined by a finite number of operations." He assumed that such an algorithm exists but someone need only to find it.

As we know now, there is no algorithm to determine whether a given polynomial Diophantine equation with integer coefficients has an integer solution. For mathematicians of that period it was impossible to devise a process according to which it can be determined in a finite number of operations whether the equation is solvable in rational integers. This was due to lack of clear definition of an algorithm. In 1936, Church–Turing defined a clear definition of an algorithm. Church used a notational system called the λ-calculus to define algorithms and Turing proved it with his machine. These two definitions were shown to be equivalent. The connection between the informal notion of algorithm and the precise definition leads to the Church–Turing thesis. It provides the definition of algorithm necessary to solve Hilbert's tenth problem. In 1970, Yuri Matijiaseiè, showed that there is no algorithm to test whether a polynomial has integral roots.

Let us define Hilbert's tenth problem as below.

Let $D = \{p \mid p$ is a polynomial with an integral root$\}$.

Hilbert's tenth problem asks whether the set D is decidable. We can show that D is Turing recognizable. Now, consider a small problem that have only a single variable, such as $2x^3 - 7x^2 - 24x + 45$.

Let $D_1 = \{p \mid p$ is a polynomial over x with an integral root$\}$.

The Turing machine M_1 that recognizes D_1 is given below:

M_1 = "The input is a polynomial p over the variable x. Evaluate p with successive values of x such as 0, 1, –1, 2, –2, 3, –3, … If at any value the polynomial evaluates to 0, then accept."

If the polynomial p has an integral root, eventually the Turing machine M_1 will find it and accept. On the other hand M_1 will run forever, if the polynomial p does not have an integral root. We can develop a similar Turing machine M for the multivariable polynomial p that recognizes D. In this case, M goes through all possible combination of its variables to integral values.

We can say both M_1 and M are recognizers but not deciders. This is because of the following reason. We can convert M_1 to be a decider for D_1 as we can compute bounds within which the roots of a single variable polynomial lie. Once the bounds are calculated, we can restrict the search to these bounds. If a root is not found within these bounds, the machine rejects. It is proved that the roots of a single variable polynomial must lie between the values

$$\pm\ k\frac{c_{max}}{c_1},$$

where c_{max} is the coefficient with largest absolute value, c_1 is the coefficient of the highest order term, and k is the number of terms in the polynomial. However, Matijaseviè proved that computation of such bounds is not possible for multivariable polynomials. Therefore, M cannot be a decider for D.

9.3 CHURCH THESIS

The Church–Turing thesis concerns the notion of an effective method in logic and mathematics. A method, M for achieving some desired result is called effective just in case:

(*a*) M is set out in terms of a finite number of exact instructions;

(*b*) M will produce the desired result in a finite number of steps without error;

(*c*) M in principle can be carried out by a human being unaided by any machinery save paper and pencil;

(*d*) M demands no insight on the part of the human being carrying it out.

The notion of an effective method is an informal one, and attempts to characterize effectiveness, such as the above, lack rigour, for the key requirement that the method demand no insight is left unexplicated. The Church–Turing thesis is the assertion that this set contains every function whose values can be obtained by a method satisfying the above conditions for effectiveness. We discuss Church–Turing thesis in brief as below.

Church–Turing thesis or Church's thesis in computability theory is a combined hypothesis about the nature of effectively computable functions by recursion, by mechanical device equivalent to a Turing machine or by use of Church's λ-calculus. It states that "Every effectively computable function is general recursive". Informally it states that if an algorithm exists, then there is an equivalent Turing machine or applicable λ-function for that algorithm. It encompasses more kinds of computations than those originally envisioned, such as those involving cellular automata, register machines and substitution systems. It also applies to other kinds of computations found in theoretical computer science such as quantum computing and probabilistic computing. The importance of the Church–Turing thesis is that it allows us to consider computability results without loss of generality.

9.4   COMPUTABILITY

Computability theory is the branch of the theory of computation that studies problems that are computationally solvable using different models of computation. The goal of computability theory is to define a formal sense in which we can understand how hard a particular problem is to solve on a computer. It differs from computational complexity theory that deals with the question of how efficiently a problem can be solved, rather than whether it is solvable at all.

In this section, we proceed with the definitions of various functions that include partial function, total function, recursive function, Ackermann's function, primitive recursive function, etc.

9.4.1   Partial Function

A relation $f : X \rightarrow Y$ is said to be a partial function if and only if the domain of f is a subset of X and each element of the domain of f corresponds to an unique element of Y, the range of f. Mathematically, a relation $f : X \rightarrow Y$ is said to be a partial function if and only if

   i.   Dom $f \subseteq X$ and

   ii. If $(x, y) \in f$ and $(x, z) \in f$, then $y = z$

9.4.2   Total Function

A relation $f : X \rightarrow Y$ is said to be a total function or function if and only if it satisfies the following characteristics.

   i.   Dom $f = X$ and

   ii. If $(x, y) \in f$ and $(x, z) \in f$, then $y = z$

Therefore, it is clear that every total function is a partial function but the converse is not true. A partial or total function $f : X^k \to X$ is called a function of k variables and is denoted by $f(x_1, x_2, \cdots, x_k)$, where $x_1, x_2, \cdots x_k$, are arguments.

For example, $f(x_1, x_2, x_3) = x_1 + 2x_2 + x_3$ is a function of three variables with arguments x_1, x_2 and x_3.

9.4.3 Ackermann's Function

Ackermann's function is a function from $(I \times I) \to I$, defined by the rules

$$A(0, y) = y + 1$$
$$A(x, 0) = A(x - 1, 1)$$
$$A(x, y) = A(x - 1, A(x, y - 1))$$

It is clear that, Ackermann's function $A(x, y)$ can be computed for every (x, y); $x, y \in I$. Therefore, $A(x, y)$ is a total computable function. It is also easy to write a recursive computer program for computation of Ackermann's function, but it is not primitive recursive. We will prove this at the later stage of this chapter.

9.4.4 Primitive Recursive Functions

In computability theory, primitive recursive functions are a class of functions on the naturals which form an important building block on the way to a full formalization of computability. In this context, we start with basic functions known as initial functions.

9.4.5 Initial Functions

To keep our discussion simple, we will consider functions of one or two variables, whose domain is either a set of natural numbers N or $N \times N$, and range is in N. In this context, we define the initial functions over N as follows:

The zero function   The zero function returns zero regardless of its argument. Therefore, it is hard to imagine a more intuitively computable function. Analytically, it is defined as

$$z(x) = 0 \text{ for all } x \in N.$$

The successor function   The successor function returns the successor of its argument. It implies that, the successor function $s(x)$ returns the next integer value $(x + 1)$ in sequence to x. Analytically, it is defined as

$$s(x) = x + 1; x \in N$$

The projector function   The projection function p_i^n returns the ith value from given n-arguments. This function is designed to take any number of arguments. The superscript of this function indicates the number of arguments it takes whereas the subscript argument will be returned by the function. This projector function is otherwise known as identity function. This function is more effective when it takes more than one argument and returns one of them. Analytically, it is defined as

$$p_i^n(x_1, x_2, x_3, \cdots, x_i, \cdots, x_n) = x_i$$

For example, $p_3^5(1, 15, 2, 3, 8) = 2$. Here, the superscript 5 represents the total number of arguments present in the function whereas the 3rd value, i.e., 2 is returned by the function.

9.4.6 Building Operation

There are two ways by which we can construct more complicated functions. These are composition and primitive recursion.

Composition This is the way by which one can build more complicated functions from initial functions such as zero function, successor function and projector function. This composition is otherwise known as substitution. If h, g_1 and g_2 are known computable functions, then analytically it is defined as

$$f(x, y) = h(g_1(x, y), g_2(x, y))$$

Primitive recursion This is the way by which a function can be defined recursively through

$$f(x, 0) = g_1(x)$$
$$f(x, y+1) = h(g_2(x, y), f(x, y))$$

First remember that g_1 and h are known computable functions. Primitive recursion is a method of defining a new function $f(x, y)$ through old functions h and g. Apart from this we have two equations. If the second argument is zero, then the first equation applies whereas the second equation is applicable when the second argument is not equal to zero.

For example, the function **add**(x,y) adds two integers x and y. It is defined as

$$\mathbf{add}(x, 0) = x$$
$$\mathbf{add}(x, y+1) = 1 + \mathbf{add}(x, y)$$

In order to add two integers 2 and 4, we apply these rules successively as follows:

$$\mathbf{add}(2, 4) = 1 + \mathbf{add}(2, 3)$$
$$= 1 + (1 + \mathbf{add}(2, 2))$$
$$= 1 + (1 + (1 + \mathbf{add}(2, 1)))$$
$$= 1 + (1 + (1 + (1 + \mathbf{add}(2, 0))))$$
$$= 1 + (1 + (1 + (1 + 2)))$$
$$= 6$$

9.4.7   Primitive Recursive

A function is called primitive recursive if and only if it can be constructed from the initial functions z, s and p_i^n by successive composition and primitive recursion.

Theorem   *Ackermann's function is not primitive recursive.*

Proof   Let f be any primitive recursive function. Then there exists some integer n such that

$$f(i) < A(n,i) \quad \forall\, i = n, n+1, \cdots$$

Let us consider $g(i) = A(i,i)$

Assume that Ackermann's function A is primitive recursive. Therefore, g is also primitive recursive. Thus, according to the above result, there exists an n such that

$$g(i) < A(n,i) \qquad \forall\ i$$

On taking $i = n$, we get

$$g(n) < A(n,n)$$
$$\text{i.e.,}\quad A(n,n) < A(n,n) \qquad [\because g(n) = A(n,n)]$$

Therefore, we get a contradiction. So our assumption is wrong. Hence, we conclude that Ackermann's function is not primitive recursive.

9.4.8 μ-Recursive Functions

Here we extend our idea of recursive functions to include Ackermann's function and other computable functions. One way it is possible by introducing minimalization operator μ, defined by

$$\mu_y(f(x,y)) = \text{Smallest } y \text{ such that } f(x,y) = 0$$

Here, we assume that $f(x,y)$ is a total function. For example, consider the total function $f(x,y)$ as

$$f(x,y) = 2x + y - 6$$

If $x \le 3$, then $y = 6 - 2x$ is the result of minimalization. If $x > 3$, then there is no $y \in N$ such that $2x + y - 6 = 0$. Therefore,

$$\mu_y(f(x,y)) = \begin{cases} 6 - 2x; & \text{if } x \le 3 \\ \text{undefined}; & \text{if } x > 3 \end{cases}$$

It implies that $\mu_y(f(x,y))$ is a partial function even if $f(x,y)$ is a total function. Now, we formally define the set S of μ-recursive partial functions.

   i.   All initial functions are μ-recursive functions and are elements of S.

   ii. Any function obtained from elements of set S by building operation is also an element of S.

iii.   Any function that is obtained by applications of μ-operator is also an element of *S*.

iv.   No other functions are in *S*.

It is to be noted that all μ-recursive functions are computable and all computable functions are μ-recursive. Therefore, μ-recursive functions provide us another model for algorithmic computation.

9.5   TURING COMPUTABLE FUNCTIONS

Before we discuss the Turing computable functions, let us think of the Turing machine that concatenates two strings w_1 and w_2. The development of the Turing machine is discussed in Chapter 7, Example 10. In this case initially, w_1 and w_2 appear on the input tape separated by a blank *B*. Finally, we get the concatenated string $w_1 w_2$ on the input tape. A slight modification of this method can be used to compute the function $f(x_1, x_2, x_3, \cdots, x_n)$ over *N* for given arguments $a_1, a_2, a_3, \cdots, a_n$. Initially, the arguments $a_1, a_2, a_3, \cdots, a_n$ appear on the input tape. The input argument values are separated by markers $x_1, x_2, x_3, \cdots, x_n$. Finally, the computed value of the function, $f(a_1, a_2, a_3, \cdots, a_n) = c$, appears on the input tape. Also we introduce another marker *y* to locate the computed value *c*. The computed value *c* appears to the right of x_n and to the left of *y*.

Here, we use tally notation to make our construction simpler. In tally notation, a string of *B*'s is represented as 0 whereas a string of *n* 1's is represented as a positive integer $n \in N$. Therefore, the initial tape position is given as $1^{a_1} x_1 1^{a_2} x_2 1^{a_3} x_3 \cdots 1^{a_n} x_n B y$. The initial instantaneous description (ID) is given as $s_0 1^{a_1} x_1 1^{a_2} x_2 1^{a_3} x_3 \cdots 1^{a_n} x_n B y$, where s_0 is the initial state of the Turing machine. As a result of computation, the initial ID is changed to the terminal ID $1^{a_1} x_1 1^{a_2} x_2 1^{a_3} x_3 \cdots 1^{a_n} x_n 1^c s' y$ for some $s' \in S$, where *S* is the non-empty finite set of internal states of the Turing machine. The computed value is found in between x_n and *y*.

We say that a function $f(x_1, x_2, x_3, \cdots, x_n)$ is Turing computable for given arguments $a_1, a_2, a_3, \cdots, a_n$ if there is a Turing machine for which

$$s_0 1^{a_1} x_1 1^{a_2} x_2 1^{a_3} x_3 \cdots 1^{a_n} x_n B y \overset{*}{\vdash} 1^{a_1} x_1 1^{a_2} x_2 1^{a_3} x_3 \cdots 1^{a_n} x_n 1^c s' y .$$

9.6   UNLIMITED REGISTER MACHINE

In this section, we give possible formalizations of the concept of computable function which is fairly close to the most common programming languages like BASIC. There is lot of other provably equivalent formalizations having their own merits. This coincidence of formalizing the notion of computable function supports Church–Turing thesis saying that the notion of computable function coincides with any of these mathematical formalizations. A promised formalization of computable function is the Unlimited Register Machine (URM). An Unlimited Register Machine is an abstract model of a digital computer, similar in aim to the Turing machine. It consists of countable infinite sequence of registers labelled $R_1, R_2, R_3, \cdots$. At any time the register R_n could store any natural number say r_n.

In other words, the number of registers is potentially infinite, and each register's "size" is infinite. A block diagram of URM is given below (Figure 9.3).

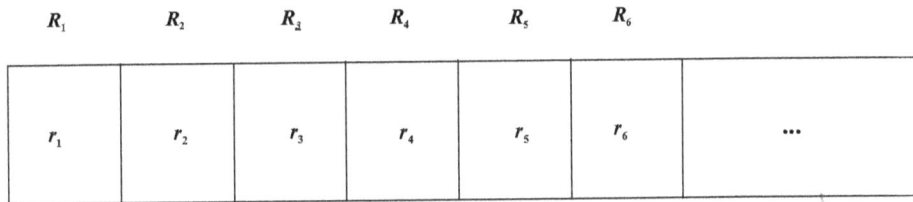

Figure 9.3   Block diagram of a URM

9.6.1   URM Instructions

The unlimited register machine offers the following instructions. Descriptions of these instructions are given below.

Zero instruction   The zero instruction $Z(n)$ replaces the content of nth register by zero (0). In otherwords this instruction clears the register n.

Successor instruction   The successor instruction $S(n)$ replaces the content of nth register by its successor. In otherwords this instruction increment the content of register n by one.

Transfer instruction   The transfer instruction $T(m, n)$ replaces the content of the register by the content of nth register but does nothing to the mth register. In otherwords this instruction copy content of register m to register n without changing the content of register m.

Jump instruction   The jump instruction is denoted as $J(m, n, p)$. If the content of mth register is equal to the content of nth register, then machine goes to p^{th} instruction otherwise goes to next instruction.

9.6.2   URM Computable

A URM program is a finite sequence of instructions, each of which is one of the four basic URM instructions. Each URM computation using a given program starts with instruction number 1 on the list and carries out the rest in numerical order unless told to jump. A computation will halt if it runs out of instructions to obey. In order to input a k-tuple $(x_1, x_2, x_3, \cdots, x_k)$, we start with $x_1, x_2, x_3, \cdots, x_k$ in registers $R_1, R_2, R_3, \cdots, R_k$ respectively, and with 0 (zero) in all the other registers. If a computation halts, then the output is the number stored in register R_1, otherwise there is no output. For knowing that a function $f : N^k \to N$ is computable one does not need a definition of what is computable in principle simply because one recognizes an algorithm whenever one sees it. However, for showing that f is not computable one definitely has to outline a priori the collection of functions that are computable in principle.

A URM program P computes the function $f:N^k \to N$ if and only if for all $(x_1,x_2,x_3,\cdots,x_k) \in N^k$, the computation with input $(x_1,x_2,x_3,\cdots,x_k)$ using program P halts with output value $f(x_1,x_2,x_3,\cdots,x_k)$. A function f is URM computable if and only if there is a URM program which computes f.

For example, consider the function that add two natural numbers, $f(x_1,x_2)=x_1+x_2$. The URM program P that computes the said function is given below:

1. $J(2,3,5)$
2. $S(1)$
3. $S(3)$
4. $J(1,1,1)$

Notice that in the trace table for input $(5,2)$ given below, each instruction in the left column acts on the contents of the registers shown to its left. The instruction $J(1,1,1)$ is known as an unconditional jump instruction. It is clear that P carries out addition.

Trace table for input value $(5,2)$:

Instruction	R_1	R_2	R_3
1	5	2	0
2	5	2	0
3	6	2	0
4	6	2	1
1	6	2	1
2	6	2	1
3	7	2	1
4	7	2	2
1	7	2	2
Halt			

In the trace table it is easy to see that the output is the content of the register R_1, i.e., 7. The flow chart corresponding to this program P is given as follows:

Start

$r_2 = r_3$

Yes

Halt

No

$r_1 = r_1 + 1$

$r_3 = r_3 + 1$

Yes

$r_1 = r_1$

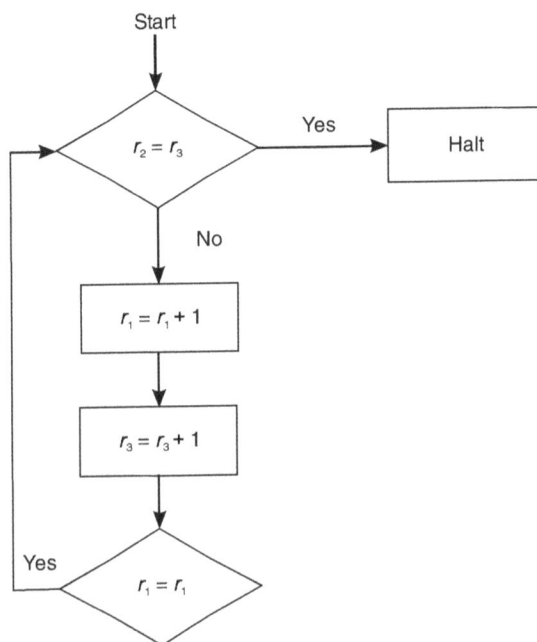

9.7 MAPPING REDUCIBILITY

We can use the notion of Turing machine and computable functions to refine our understanding of reducibility in order to apply it to a broader class of problems. In this section we discuss the notion of simple type of reducibility called mapping reducibility or many-one reducibility. It allows us to prove certain languages are not Turing recognizable. Also it helps for applications in complexity theory.

In this case, we reduce problem P_1 to problem P_2 by using mapping reducibility. It means that there exists a computable function that converts instances of problem P_1 to instances of problem P_2. If we have such a conversion function, we can solve P_1 with a solver for P_2. The conversion function we use is called a reduction. We formally define the mapping reducibility as below.

Before we define a formal definition of mapping reducibility, it is necessary to discuss a computable function. A Turing machine computes a function by starting with the input to the function on the tape and halting with the output of the function on the tape. A function $f : \Sigma^* \to \Sigma^*$ is said to be a computable function if some Turing machine M, halts with just $f(w)$ on its tape, for every input w.

Language L_1 is mapping reducible to language L_2, written $L_1 \leq_m L_2$, if there is a computable function $f : \Sigma^* \to \Sigma^*$, where for every w, $w \in L_1 \Leftrightarrow f(w) \in L_2$. The conversion function f is called the reduction of L_1 to L_2.

A mapping reduction of L_1 to L_2 is an approach to convert membership testing in L_1 to membership testing in L_2. In this approach we use the conversion function f to map w to $f(w)$ and test whether $f(w) \in L_2$. If one problem is reducible to a previously solved second problem, then we can obtain a solution to the original problem. This is exactly what mapping reduction tells. A pictorial representation of mapping reducibility is given below.

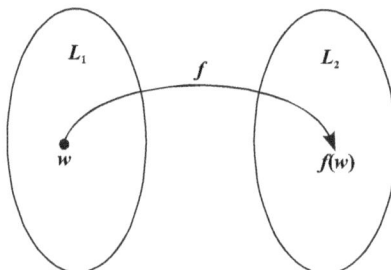

9.7.1 Properties of Mapping Reducibility

From the definition L_1 to L_2 of mapping reducibility, we can obtain several properties. The last two properties show that mapping reducibility can also be used to show the recognizability or unrecognizability of languages. We prove some of them over here.

1. For every problem P, $P \leq_m P$. Every problem can be mapping-reduced to itself, by the identity function represented by a Turing machine that immediately halts and leaves its input unchanged.

2. If $P_1 \leq_m P_2$, and $P_2 \leq_m P_3$, then $P_1 \leq_m P_3$. This property implies that mapping-reducibility is a transitive relation.

3. If $P_1 \leq_m P_2$, then the complement of P_1 is mapping reducible to complement of P_2, written $\overline{P_1} \leq_m \overline{P_2}$. From the definition, if $w \in P_1 \Leftrightarrow f(w) \in P_2$, then also $w \notin P_1 \Leftrightarrow f(w) \notin P_2$ or $w \in \overline{P_1} \Leftrightarrow f(w) \in \overline{P_2}$.

4. If $P_1 \leq_m P_2$, and P_2 is decidable, then P_1 is decidable. Since mapping reducibility is a special case of reducibility, this rule still applies.

   Proof   Assume that P_2 is decidable. Let M be the decider for P_2. Let f be the reduction from P_1 to P_2. We describe a decider N for P_1 as follows.

   N = "On input w:

   1. Compute $f(w)$.

   2. Run M on input $f(w)$ and output whatever M outputs."

   Clearly, if $w \in P_1$, then $f(w) \in P_2$ because f is a reduction from P_1 to P_2. Thus, M accepts $f(w)$ whenever $w \in P_1$ or N accepts P_1. Therefore, P_1 is decidable.

5. If $P_1 \leq_m P_2$, and P_1 is undecidable, then P_2 is undecidable. This is the corollary of the above property. It has been our main tool for proving undecidability.

   Proof Assume that P_2 is decidable. Let M be the decider for P_2. Let f be the reduction from P_1 to P_2. We describe a decider N for P_1 as follows.

   $N =$ "On input w:

   1. Compute $f(w)$.
   2. Run M on input $f(w)$ and output whatever M outputs."

   Clearly, if $w \in P_1$, then $f(w) \in P_2$ because f is a reduction from P_1 to P_2. But, since P_1 is undecidable by assumption this machine cannot exist and therefore our assumption that P_2 is decidable must be wrong. Therefore, P_2 is undecidable.

6. If $P_1 \leq_m P_2$ and P_2 is Turing recognizable, then P_1 is Turing recognizable. Since $w \in P_1 \Leftrightarrow f(w) \in P_2$, if there is a recognizer for P_2, then P_1 is recognized by running the recognizer for P_2 on input $f(w)$.

7. If $P_1 \leq_m P_2$ and P_1 is not Turing recognizable, then P_2 is not Turing recognizable. This is the corollary of the above property.

9.8 GÖDEL NUMBER

In mathematical logic, a Gödel numbering is a function that assigns to each symbol and strings of some formal language a unique natural number called its Gödel number. This is necessary when considering models of computation such as Turing machines that manipulate strings rather than numbers. This type of encoding process is called as Gödel numbering and it is named after the logician Kurt Gödel. It is important to note that, there are many ways in which these numbers could be defined and thus a Gödel numbering is not unique. In the theory of computability, the term Gödel numbering is used in settings more general than the one described above. In general, it refers to the following:

1. An assignment of elements from a countable mathematical object, to natural numbers to allow algorithmic manipulation of the mathematical object.

2. Any assignment of the elements of a formal language to natural numbers in such a way that the numbers can be manipulated by an algorithm. It helps to simulate manipulation of elements of the formal language.

Now, we discuss the correspondence formally in our particular application. Let Σ be any alphabet. Let, $\beta = |\Sigma| + 1$. Fix some ordering of the symbols in Σ such as $\Sigma = \{b_1, b_2, b_3, \cdots, b_{\beta-1}\}$. Therefore, each string in Σ^* can be viewed as an integer in base β notation. We define a function $g_n : \Sigma^* \to N$ as follows: If $w = b_{i_1} b_{i_2} b_{i_3} \cdots b_{i_n}$, where $n \geq 0$ and $1 \leq i_j \leq \beta - 1$ for $j = 1, 2, 3, \cdots, n$, then $g_n(w)$ is defined as

$$g_n(w) = \beta^{n-1} \cdot i_1 + \beta^{n-2} \cdot i_2 + \beta^{n-3} \cdot i_3 + \cdots + \beta^1 \cdot i_{n-1} + \beta^0 \cdot i_n$$

In particular, $g_n(\wedge) = g_n(e) = 0$, where e is considered as an empty string. We say that $g_n(w)$ is the Gödel number of the string w. From the above relation it is clear that every string corresponds to exactly one number and no number corresponds to more than one string. Therefore, we say that g_n is a true encoding of strings as numbers.

However, the function g_n is not a bijection. This is because certain numbers do not correspond to any string. For example, say β. In order to make g_n as bijective, we introduce a new symbol b_0 to correspond to the digit 0. Therefore, g_n can be expressed as a bijection from $\{e\} \cup \Sigma(\Sigma \cup \{b_0\})^*$ to N. Let $\mathfrak{D} = \{e\} \cup \Sigma(\Sigma \cup \{b_0\})^*$. Thus, $g_n : \mathfrak{D} \to N$ is a bijection and hence g_n^{-1} is a well-defined function from N to $\mathfrak{D}$.

9.8.1   Gödel Numbering of Sequence of Natural Numbers

By using these numbers Gödel was able to describe logical relations in the formal language of arithmetic he was dealing with. Gödel used the following system to encode an entire formula, which is a sequence of symbols. Given a sequence $x_1 x_2 x_3 \cdots x_n$ of positive integers, the Gödel encoding of the sequence is the product of the first n primes raised to their corresponding values in the sequence. Analytically, this can be expressed as $g_n(x_1, x_2, x_3, \cdots, x_n)$, such that

$$g_n(x_1, x_2, x_3, \cdots, x_n) = 2^{x_1} \cdot 3^{x_2} \cdot 5^{x_3} \cdot \cdots \cdot (P_n)^{x_n},$$

where P_n is the nth prime number. It is also clear from the above relation that, the first position from the left can be a 2, the second can be a 3, the third can be a 5, the fourth a 7, etc. It is to be noted that these are the primes from lowest upwards. Therefore, the Gödel number of any sequence is greater than or equal to 1, and every integer greater or equal to 1 is the Gödel number of a sequence. Also we have,

$$g_n(x_1, x_2, x_3, \cdots, x_n) = g_n(y_1, y_2, y_3, \cdots, y_n, y_{n+1}, y_{n+2}, \cdots, y_{n+k})$$

i.e., $\prod_{i=1}^{n}(P_i)^{x_i} = \prod_{i=1}^{n+k}(P_i)^{y_i}$

It implies that, the function g_n is not one to one. Consider the following example:

$$g_n(1, 3, 4) = g_n(1, 3, 4, 0, 0, 0)$$

i.e., $2^1 \cdot 3^3 \cdot 5^4 = 2^1 \cdot 3^3 \cdot 5^4 \cdot 7^0 \cdot 11^0 \cdot 13^0$

Therefore it is clear that, if two sequences have the same Gödel number, then they are identical with the difference that they may end with a different number of zeros. According to the fundamental theorem of arithmetic, any number obtained this way can be uniquely factored into prime factors, so it is possible to effectively recover the original sequence from its Gödel number. For example, the Gödel number 1078 has the sequence (1, 0, 0, 2, 1). This is because,

$$1078 = 2 \cdot 7^2 \cdot 11 = 2^1 \cdot 3^0 \cdot 5^0 \cdot 7^2 \cdot 11^1 = g_n(1, 0, 0, 2, 1)$$

9.9 DECIDABILITY OF LOGICAL THEORIES

The branch of mathematics that investigates mathematics itself is mathematical logic. In mathematical logic, the term decidable refers to the existence of an effective procedure for determining membership in a set of formulas. Propositional logic or logical systems are decidable if membership in their set of logically valid formulas can be effectively determined. A theory in a fixed logical system is decidable if there is an effective procedure for determining whether arbitrary formulas are included in the theory. The definition of decidablility of logical theories can be given either in terms of effective methods or in terms of computable functions. In general, these are considered equivalent per Church's thesis. In fact, the proof that a logical theory is undecidable will use the formal definition of computability to show that an appropriate set is not a decidable set, and then invoke Church's thesis to show that the logical theory is not decidable by any effective procedure.

Here, we just provided an idea of a decidable and undecidable theory, however more detailed study on these topics is beyond the scope of this book.

9.9.1 A Decidable Theory

A theory is a set of formulas that is assumed to be closed under logical consequence. The decidability for a theory is whether there is an effective procedure that decides whether the formula is a member of the theory or not. This problem arises in general when a theory is defined as the set of logical consequences of a fixed set of axioms. There are several basic results about decidability of theories. Every inconsistent theory is decidable. This is because, every formula in the signature of the theory will be a logical consequence of, and thus member of, the theory. Every complete recursively enumerable first-order theory is decidable whereas extension of a decidable theory may not be decidable.

9.9.2 An Undecidable Theory

A theory is said to be undecidable if there is no effective procedure that decides whether the formula is a member of the theory or not. In such cases, no algorithm exists for deciding the truth or falsity of mathematical statements. The interpretability method is often used to establish undecidability of theories. If an undecidable theory T_1 is interpretable in a consistent theory T_2, then T_2 is also undecidable. This is closely related to the concept of a many-one reduction in computability theory. A consistent theory which has the property that every consistent extension is undecidable is said to be essentially undecidable. In fact, every consistent extension will be essentially undecidable.

9.10 TURING REDUCIBILITY

In the previous section, we discussed the concept of reducibility that is used as a way of using a solution to one problem to solve other problems. Also, we have introduced mapping reducibility, a specific form of reducibility. However, mapping reducibility does not capture our intuitive concept of reducibility in general. Therefore, it is essential to present a very general form of reducibility, known as Turing reducibility. Before, we define Turing reducibility it is essential to define oracle for a language. An oracle for a language L is an external device that is capable of determining whether any string $w \in L$. An oracle Turing machine is a modified Turing machine that has the additional capability of querying an oracle. We denote M_L as the oracle Turing machine that has an oracle for language L.

Language L_1 is Turing reducible to language L_2, if L_1 is decidable relative to L_2 and we write $L_1 \leq_T L_2$. We say L_1 is Turing equivalent to L_2 and write $L_1 \equiv_T L_2$ if both $L_1 \leq_T L_2$ and $L_2 \leq_T L_1$. The equivalence classes of Turing equivalent sets are called Turing degrees. The Turing degree of a set A is written deg (A). Also, if $L_1 \leq_T L_2$ and L_2 is decidable, then L_1 is decidable. This is because of the following reason. If L_2 is decidable, then the oracle for L_2 may be replaced by an actual algorithm that decides L_2. As a result, we may replace the oracle Turing machine that decides L_1 by a standard Turing machine that decides L_1. Therefore, L_1 is decidable.

9.10.1 Properties of Turing Reducibility

From the definition of Turing reducibility, we can obtain several properties. Here, we list some of them.

1. There are pairs of sets (L_1, L_2) such that L_1 is not Turing reducible to L_2 and L_2 is not Turing reducible to L_1. Thus $\leq_T$ is not a linear order.

2. Every computable set is Turing reducible to every other computable set. Because these sets can be computed with no oracle, they can be computed by an oracle machine that ignores the oracle it is given.

3. The relation $\leq_T$ is transitive: if $L_1 \leq_T L_2$ and $L_2 \leq_T L_3$, then $L_1 \leq_T L_3$. Moreover $L \leq_T L$ holds for every set L, and thus the relation $\leq_T$ is a pre-order. It is not a partial order because $L_1 \leq_T L_2$ and $L_2 \leq_T L_1$ does not necessarily imply $L_1 = L_2$.

4. There are infinite decreasing sequences of sets under $\leq_T$. Thus this relation is not well-founded.

5. Every set is Turing reducible to its own Turing jump, but the Turing jump of a set is never Turing reducible to the original set.

6. Every set is Turing equivalent to its complement.

SOLVED EXAMPLES

Example 1

By giving suitable URM programs show that the following functions are URM computable.

(a) $f(x,y) = \begin{cases} 0 & \text{if } x = y \\ 1 & \text{if } x \neq y \end{cases}$

(b) $f(x,y) = \begin{cases} 0 & \text{if } x \leq y \\ 1 & \text{if } x > y \end{cases}$

Solution

(a) The URM program P that computes the said function is given below:

1. $J(1, 2, 3)$
2. $S(3)$
3. $T(3,1)$
4. $J(1, 1, 1)$

(b) The URM program P that computes the said function is given below:

1. $J(1, 2, 8)$
2. $J(1, 3, 8)$
3. $S(3)$
4. $J(2, 4, 7)$
5. $S(4)$
6. $J(1, 1, 2)$
7. $S(6)$
8. $T(6,1)$

Example 2

Compute the Gödel number of the sequence $(1, 3, 4, 2)$.

Solution The Gödel number of the sequence $(1, 3, 4, 2)$ is given as:

$$g_n(1, 3, 4, 2) = 2^1 \cdot 3^3 \cdot 5^4 \cdot 7^2$$
$$= 1653750$$

Example 3

Consider the URM program given below. Describe the computation of P for input $(3, 1)$.

1. $T(1, 3)$
2. $T(2, 5)$
3. $S(5)$
4. $S(6)$
5. $J(1, 5, 8)$

6. $S(3)$

7. $J(1, 1, 3)$

8. $T(6,1)$

Solution   The trace table for computation of URM program P for input $(3, 1)$ is given below.

Instruction	R_1	R_2	R_3	R_4	R_5	R_6
1	3	1	0	0	0	0
2	3	1	3	0	0	0
3	3	1	3	0	1	0
4	3	1	3	0	2	0
5	3	1	3	0	2	1
6	3	1	3	0	2	1
7	3	1	4	0	2	1
3	3	1	4	0	2	1
4	3	1	4	0	3	1
5	3	1	4	0	3	2
8	2	1	4	0	3	2
Halt						

In the trace table it is easy to see that the output is the content of the register R_1, i.e., 2.

Example 4

Show that the function $f(x, y) = \min(x, y)$, i.e., minimum of x and y, is primitive recursive.

Solution   The function $f(x, y) = \min(x, y)$ can be expressed as below.

$$f(x, y) = \min(x, y) = x \dot{-} (x \dot{-} y)$$

Also, we have shown that $f(x, y) = (x \dot{-} y)$ is primitive recursive and hence the function $f(x, y) = \min(x, y)$ is primitive recursive.

Example 5

Consider the URM program given below. Give the trace table of the computation for this program for the input value 3.

1. $J(1, 4, 9)$

2. $S(3)$

3. $J(1, 3, 7)$

4. $S(2)$

5. $S(3)$

6. $J(1, 1, 3)$

7. $T(2,1)$

Solution Trace table for input value 3 for the URM program is given below.

Instruction	R_1	R_2	R_3	R_4
1	3	0	0	0
2	3	0	0	0
3	3	0	1	0
4	3	0	1	0
5	3	1	1	0
6	3	1	2	0
3	3	1	2	0
4	3	1	2	0
5	3	2	2	0
6	3	2	3	0
3	3	2	3	0
7	2	2	3	0
Halt				

In the trace table it is easy to see that the output is the content of the register R_1, i.e., 2.

Example 6

By giving suitable URM programs show that the following function is URM computable.

$$f(x) = \begin{cases} \frac{x}{3} & \text{if } x \text{ is a multiple of } 3 \\ 0 & \text{otherwise} \end{cases}$$

Solution The URM program for the function is given below.

1. $J(1, 2, 11)$

2. $S(3)$

3. $J(1, 2, 9)$

4. $S(1)$

5. $J(1, 2, 9)$

6. $S(1)$

7. $S(3)$

8. $J(1, 1, 1)$

9. $Z(1)$

10. $J(1,1,12)$

11. $T(3,1)$

Example 7

Show that the function $f(x, y) = x^y$ is primitive recursive and hence compute $f(2,3)$.

Solution   The function $f(x,y) = x^y$ can be expressed as below.

$$f(x,0) = 1$$
$$f(x, y+1) = x \cdot f(x, y)$$

Therefore, $f(x, y) = x^y$ is primitive recursive. Now,

$$f(2,3) = 2 \cdot f(2,2) = 2 \cdot 2 \cdot f(2,1)$$
$$= 2 \cdot 2 \cdot 2 \cdot f(2,0) = 8$$

Example 8

Show that Post correspondence system with following two instances $X = \{01, 10, 0, 00, 11\}$ and $Y = \{0100, 100, 10, 001, 111\}$ has no solution?

Solution   Given that $X = \{01, 10, 0, 00, 11\}$ and $Y = \{0100, 100, 10, 001, 111\}$. Therefore, we have $u_1 = 01, u_2 = 10, u_3 = 0, u_4 = 00, u_5 = 11, \ v_1 = 0100, v_2 = 100, v_3 = 10, v_4 = 001$ and $v_5 = 111$.

It is clear that, for this instance of PCP, we have $|u_i| \leq |v_i|$ for all $i = 1, 2, 3, 4, 5$. Thus, the length of the string generated by the sequence of substrings of instance X is less than the string generated by the sequence of substrings of instance Y. Hence, the correspondence system has no solution.

Example 9

Draw the flow chart corresponding to the following program.

1. $J(1, 2, 6)$

2. $S(3)$

3. $S(2)$

4. $S(2)$

5. $J(1, 1, 1)$

6. $T(3,1)$

Solution   The flow chart corresponding to this program P is given below.

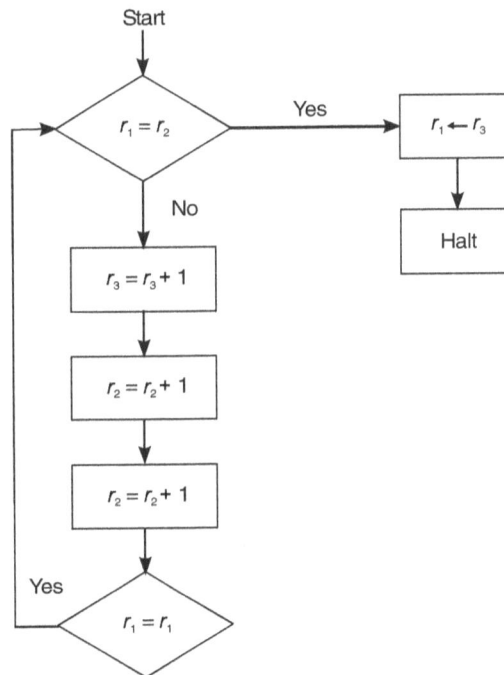

Example 10

Give a suitable URM program for the following function and draw a flow chart of the program.

$$f(x) = \begin{cases} 0; & \text{if } x = 0 \\ 1; & \text{if } x \neq 0 \end{cases}$$

Solution   The URM program for the function is as follows:

1. $J(1, 2, 3)$

2. $S(3)$

3. $T(3,1)$

The flow chart corresponding to this program P is given below.

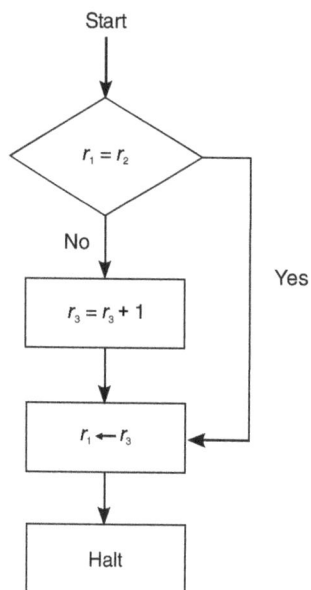

Example 11

Prove that $x \neq 3$ is a decidable predicate.

Solution   The characteristic function is given as

$$f(x) = \begin{cases} 1 & \text{if } x \neq 3 \\ 0 & \text{otherwise} \end{cases}$$

The URM program for the above characteristic function is given below.

1. $S(2)$

2. $S(2)$

3. $S(2)$

4. $S(3)$

5. $J(1, 2, 8)$

6. $T(3,1)$

7. $J(1, 1, 10)$

8. $Z(1)$

Example 12

Prove that $x \neq 2k, k \in N$ is a decidable predicate.

Solution   The characteristic function is given as

$$f(x) = \begin{cases} 1 & \text{if } x \text{ is even} \\ 0 & \text{if } x \text{ is odd} \end{cases}$$

The URM program for the above characteristic function is given below.

1. $J(1,2,8)$

2. $S(2)$

3. $J(1,2,6)$

4. $S(2)$

5. $J(1, 1, 1)$

6. $T(3,1)$

7. $J(1, 1, 10)$

8. $S(4)$

9. $T(4,1)$

Example 13

Prove that $x < y; x, y \in N$ is a decidable predicate. Justify your answer by using URM program.

Solution   The characteristic function is given as

$$f(x, y) = \begin{cases} 1 & \text{if } x < y \\ 0 & \text{otherwise} \end{cases}$$

The URM program for the above characteristic function is given below.

1. $J(1,2,9)$

2. $J(1,3,11)$

3. $J(2,3,9)$

4. $S(3)$

5. $S(4)$

6. $J(1,3,11)$

7. $J(2,4,9)$

8. $J(1,1,4)$

9. $Z(1)$

10. $J(1,1,13)$

11. $Z(1)$

12. $S(1)$

Example 14

Compute the Gödel number of the sequence $(1,0,3,0,1)$.

Solution   The Gödel number of the sequence $(1,0,3,0,1)$ is given as:

$$g_n(1,0,3,0,1) = 2^1 \cdot 3^0 \cdot 5^3 \cdot 7^0 \cdot 11^1$$
$$= 2750$$

Example 15

Write a URM program and draw a flow chart that computes the function $f(x) = \left\lceil \frac{x}{3} \right\rceil$.

Solution   The URM program of the function $f(x) = \left\lceil \frac{x}{3} \right\rceil$ is given below.

1. $J(1,3,9)$

2. $S(2)$

3. $S(3)$

4. $J(1,3,9)$

5. $S(3)$

6. $J(1,3,9)$

7. $S(3)$

8. $J(1,1,1)$

9. $T(2,1)$

The flow chart of the above URM program is as follows:

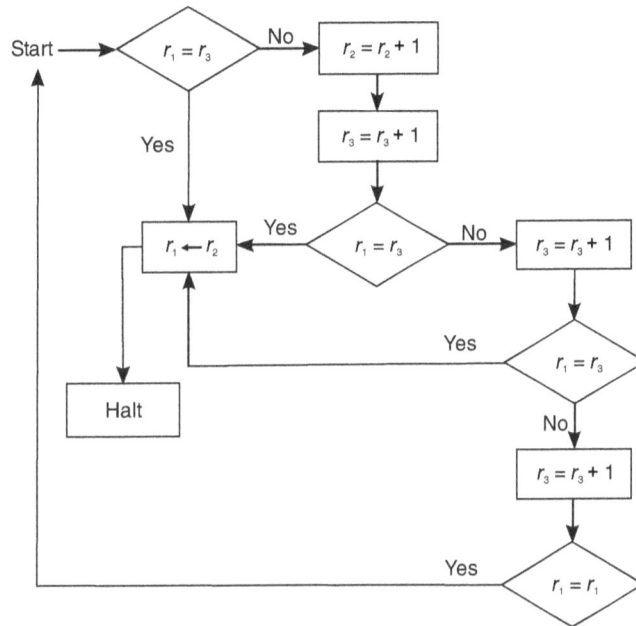

Example 16

Write a URM program that computes the greatest integer function $f(x) = \left[\frac{x}{3}\right]$. Give the trace table of the computation with this program for the input 4.

Solution The URM program of the function $f(x) = \left[\frac{x}{3}\right]$ is given below.

1. $J(1,3,9)$
2. $S(3)$
3. $J(1,3,9)$
4. $S(3)$
5. $J(1,3,9)$
6. $S(3)$
7. $S(2)$
8. $J(1,1,1)$
9. $T(2,1)$

The trace table for computation of URM program P for input 4 is as follows:

Instruction	R_1	R_2	R_3
1	4	0	0
2	4	0	0
3	4	0	1
4	4	0	1
5	4	0	2
6	4	0	2
7	4	0	3
8	4	1	3
1	4	1	3
2	4	1	3
3	4	1	4
9	1	1	4
Halt			

In the trace table it is easy to see that the output is the content of the register R_1, i.e., 1.

Example 17

Write a URM program and draw a flow chart that computes the function $f(x) = \left\lfloor \frac{x}{4} \right\rfloor$.

Solution   The URM program of the function $f(x) = \left\lfloor \frac{x}{4} \right\rfloor$ is given below.

1. $J(1,3,11)$
2. $S(3)$
3. $J(1,3,11)$
4. $S(3)$
5. $J(1,3,11)$
6. $S(3)$
7. $J(1,3,11)$
8. $S(3)$
9. $S(2)$
10. $J(1,1,1)$
11. $T(2,1)$

The flow chart of the above URM program is given below.

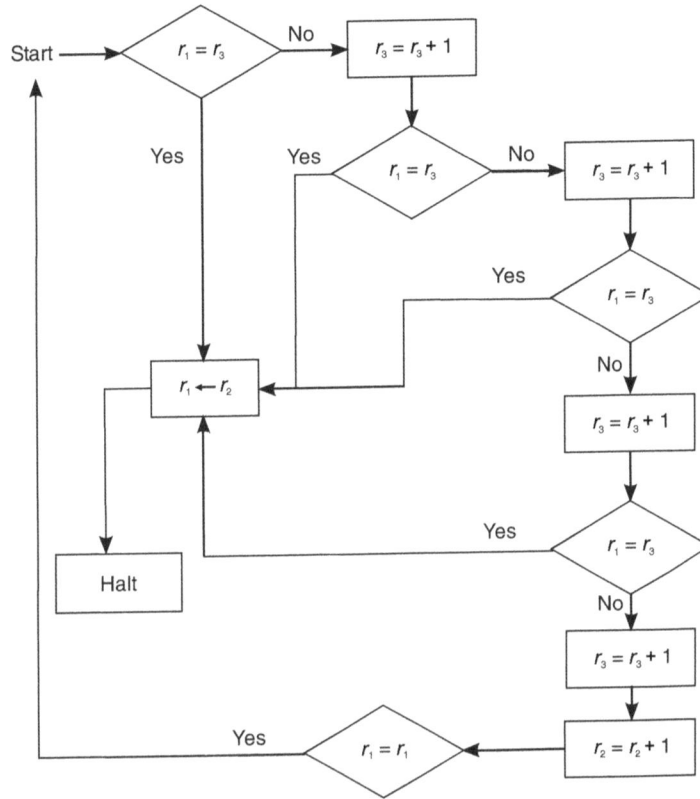

Example 18

Write a URM program that computes $f(x) = 3x$ and draw a flow chart of it.

Solution   The URM program that computes $f(x) = 3x$ is given below.

1. $J(1,3,7)$
2. $S(2)$
3. $S(2)$
4. $S(2)$
5. $S(3)$
6. $J(1,1,1)$
7. $T(2,1)$

The flow chart corresponding to the above URM program is given below.

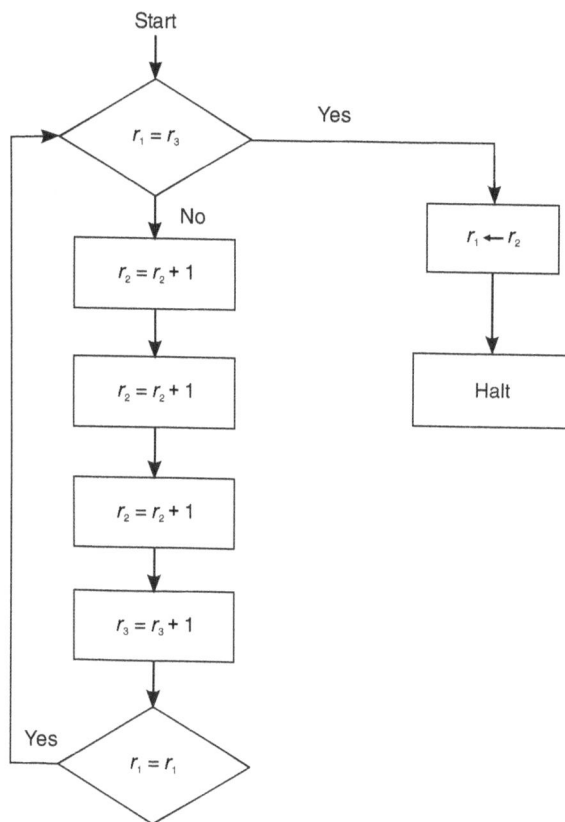

Example 19

Show that the function $f(x, y) = xy$ is computable. Support your answer by giving a URM program.

Solution   The URM program that computes $f(x, y) = xy$ is given below.

1. $J(1, 3, 17)$

2. $J(2, 3, 14)$

3. $S(3)$

4. $J(1, 3, 16)$

5. $J(2, 3, 17)$

6. $T(1, 5)$

7. $J(2, 3, 17)$

8. $Z(4)$

9. $S(3)$

10. $S(4)$

11. $S(1)$

12. $J(4,5,7)$

13. $J(1,1,10)$

14. $Z(1)$

15. $J(1,1,17)$

16. $T(2,1)$

Example 20

Write a URM program for the function $f(x) = \left[\frac{2x}{3}\right]$.

Solution   The URM program that computes $f(x) = \left[\frac{2x}{3}\right]$ is given below.

1. $J(1,2,15)$
2. $T(1,3)$
3. $S(1)$
4. $S(2)$
5. $J(2,3,7)$
6. $J(1,1,3)$
7. $J(1,4,15)$
8. $S(4)$
9. $J(1,4,15)$
10. $S(4)$
11. $J(1,4,15)$
12. $S(4)$
13. $S(5)$
14. $J(1,1,7)$
15. $T(5,1)$

Example 21

Find the Gödel sequence of the number 24200.

Solution   The natural number 24200 can be expressed as

$$24200 = 2^3 \cdot 5^2 \cdot 11^2$$
$$= 2^3 \cdot 3^0 \cdot 5^2 \cdot 7^0 \cdot 11^2$$

Therefore, the Gödel sequence of the number 24200 is given as $(3, 0, 2, 0, 2)$.

Example 22

Find the Gödel sequence of the number 14300.

Solution   The natural number 14300 can be expressed as

$$14300 = 2^2 \cdot 5^2 \cdot 11 \cdot 13$$
$$= 2^2 \cdot 3^0 \cdot 5^2 \cdot 7^0 \cdot 11^1 \cdot 13^1$$

Example 23

Consider the correspondence system described by the following instances $X = \{ab, ba, b, abb, a\}$ and $Y = \{aba, abb, ab, b, bab\}$. Does this system has a Post correspondence solution?

Solution   Given that $X = \{ab, ba, b, abb, a\}$ and $Y = \{aba, abb, ab, b, bab\}$. Therefore, we have $u_1 = ab, u_2 = ba, u_3 = b, u_4 = abb, u_5 = a$, $v_1 = aba, v_2 = abb, v_3 = ab, v_4 = b$, and $v_5 = bab$.

It is clear that, for this instance of PCP, the sequence must start with 1 because both substrings u_1 and v_1 begin with same symbol a. Here we see that, for the sequence $\{1, 5, 2, 3, 4, 4, 3, 4\}$.

$$u_1 u_5 u_2 u_3 u_4 u_4 u_3 u_4 = abababbabbbbabb = v_1 v_5 v_2 v_3 v_4 v_4 v_3 v_4.$$

Therefore, the given correspondence system has a PC solution.

Example 24

Does Post correspondence system with following two instances $X = \{01, 010, 001, 100, 011\}$ and $Y = \{0100, 100, 10, 00, 111\}$ has a solution?

Solution   Given that $X = \{01, 010, 001, 100, 011\}$ and $Y = \{0100, 100, 10, 00, 111\}$. Therefore, we have $u_1 = 01, u_2 = 010, u_3 = 001$, $u_4 = 100, u_5 = 011$, $v_1 = 0100, v_2 = 100$, $v_3 = 10, v_4 = 00$ and $v_5 = 111$.

It is clear that, for this instance of PCP, the sequence must start with 1 because both substrings u_1 and v_1 begin with same symbol 0 followed by 1. Here we see that, for the sequence $\{1, 3, 2, 5, 4\}$

$$u_1 u_3 u_2 u_5 u_4 = 01001010011100 = v_1 v_3 v_2 v_5 v_4.$$

Therefore, the given correspondence system has a PC solution.

Example 25

Compute Ackermann's function $A(1, 1)$ and $A(1, 6)$.

Solution   On using the definition of Ackermann's function we have:

$$A(1, 1) = A(0, A(1, 0))$$
$$= 1 + A(1, 0)$$
$$= 1 + A(0, 1)$$
$$= 1 + (1 + 1)$$
$$= 3$$

$$A(1, 6) = A(0, A(1, 5))$$
$$= 1 + A(1, 5)$$
$$= 1 + A(0, A(1, 4))$$
$$= 1 + 1 + A(1, 4)$$
$$= 1 + 1 + A(0, A(1, 3))$$
$$= 1 + 1 + 1 + A(1, 3)$$
$$= 1 + 1 + 1 + A(0, A(1, 2))$$
$$= 1 + 1 + 1 + 1 + A(1, 2)$$
$$= 1 + 1 + 1 + 1 + A(0, A(1,1))$$
$$= 1 + 1 + 1 + 1 + 1 + A(1,1)$$
$$= 1 + 1 + 1 + 1 + 1 + 3$$
$$= 8$$

Example 26

Compute Gödel number for the following strings defined over the alphabet $\{a, b, c\}$.

    (a) *aabbc*   (b) *bcb*   (c) *abc*   (d) *abbc*

Solution   Given that $\Sigma = \{a, b, c\}$. Then $\beta = |\Sigma| + 1 = 4$. Let us define an ordering of symbols $b_1 = a, b_2 = b$ and $b_3 = c$. The Gödel number of different strings is computed as below.

    (a) $g_n(aabbc) = g_n(b_1 b_1 b_2 b_2 b_3)$
$$= 4^4 \cdot 1 + 4^3 \cdot 1 + 4^2 \cdot 2 + 4^1 \cdot 2 + 3 = 363$$

(b) $g_n(bcb) = g_n(b_2 b_3 b_2)$

$$= 4^2 \cdot 2 + 4^1 \cdot 3 + 2 = 46$$

(c) $g_n(abc) = g_n(b_1 b_2 b_3)$

$$= 4^2 \cdot 1 + 4^1 \cdot 2 + 3 = 27$$

(d) $g_n(abbc) = g_n(b_1 b_2 b_2 b_3)$

$$= 4^3 \cdot 1 + 4^2 \cdot 2 + 4^1 \cdot 2 + 3 = 107$$

Example 27

Compute Gödel number for the following sequence of natural numbers.

(a) $(2,1,3,1)$ (b) $(2,2,2)$ (c) $(1,2,3,4)$ (d) $(2,5)$

Solution   The Gödel number for the sequence of natural numbers is given below.

(a) $g_n(2,1,3,1) = 2^2 \cdot 3^1 \cdot 5^3 \cdot 7^1 = 10500$

(b) $g_n(2,2,2) = 2^2 \cdot 3^2 \cdot 5^2 = 900$

(c) $g_n(1,2,3,4) = 2^1 \cdot 3^2 \cdot 5^3 \cdot 7^4 = 5402250$

(d) $g_n(2,5) = 2^2 \cdot 3^5 = 972$

Example 28

Let $\Sigma = \{a,b,c\}$. Compute the inverse of Gödel function, $g_n^{-1}(m)$ for $m \in N, 0 \le m \le 8$.

Solution   Given that $\Sigma = \{a,b,c\}$. Then $\beta = |\Sigma| + 1 = 4$. Let us define an ordering of symbols $b_1 = a, b_2 = b$ and $b_3 = c$. Let us compute few Gödel numbers as follows.

$$g_n(\wedge) = 0$$
$$g_n(a) = g_n(b_1) = 4^0 \cdot 1 = 1$$
$$g_n(b) = g_n(b_2) = 4^0 \cdot 2 = 2$$
$$g_n(c) = g_n(b_3) = 4^0 \cdot 3 = 3$$
$$g_n(ab_0) = g_n(b_1 b_0) = 4^1 \cdot 1 + 4^0 \cdot 0 = 4$$
$$g_n(aa) = g_n(b_1 b_1) = 4^1 \cdot 1 + 4^0 \cdot 1 = 5$$
$$g_n(ab) = g_n(b_1 b_2) = 4^1 \cdot 1 + 4^0 \cdot 2 = 6$$
$$g_n(ac) = g_n(b_1 b_3) = 4^1 \cdot 1 + 4^0 \cdot 3 = 7$$
$$g_n(bb_0) = g_n(b_2 b_0) = 4^1 \cdot 2 + 4^0 \cdot 0 = 8$$

The values of $g_n^{-1}(m)$ for $m \in N, 0 \le m \le 8$ are given as follows:

$$g_n^{-1}(0) = \wedge$$
$$g_n^{-1}(1) = b_1 = a$$
$$g_n^{-1}(2) = b_2 = b$$
$$g_n^{-1}(3) = b_3 = c$$
$$g_n^{-1}(4) = b_1 b_0 = ab_0$$
$$g_n^{-1}(5) = b_1 b_1 = aa$$
$$g_n^{-1}(6) = b_1 b_2 = ab$$
$$g_n^{-1}(7) = b_1 b_3 = ac$$
$$g_n^{-1}(8) = b_2 b_0 = bb_0$$

Example 29

Show that the function $f(x, y) = x \cdot y$ is primitive recursive and hence compute $f(2, 4)$.

Solution The function $f(x, y) = x \cdot y$ can be expressed as below.

$$f(x, 0) = 0$$
$$f(x, y + 1) = x + f(x, y)$$

Therefore, $f(x, y) = x \cdot y$ is primitive recursive. Now,

$$f(2, 4) = 2 + f(2, 3) = 2 + 2 + f(2, 2)$$
$$= 2 + 2 + 2 + f(2, 1) = 2 + 2 + 2 + 2 + f(2, 0)$$
$$= 8 + 0 = 8$$

Example 30

Show that the sign function, $f(x) = Sg(x)$, is primitive recursive, where

$$Sg(x) = \begin{cases} 0 & \text{if } x = 0 \\ 1 & \text{if } x > 0 \end{cases}$$

Solution The sign function $f(x) = Sg(x)$ can be expressed as below:

$$f(0) = 0$$
$$f(x + 1) = 1$$

Therefore, $f(x) = Sg(x)$ is primitive recursive.

Example 31

Show that the predecessor function, $f(x) = v(x)$, is primitive recursive, where

$$v(x) = \begin{cases} 0 & \text{if } x = 0 \\ x - 1 & \text{if } x > 0 \end{cases}$$

Solution   The predecessor function $f(x) = v(x)$ can be expressed as below:

$$f(0) = 0$$
$$f(x+1) = x$$

Therefore, $f(x) = v(x)$ is primitive recursive.

Example 32

Show that the function $f(x_1, x_2, x_3) = x_1 + x_2 + x_3$, is primitive recursive and hence compute $f(3,2,2)$.

Solution   The function $f(x_1, x_2, x_3) = x_1 + x_2 + x_3$ can be expressed as below:

$$f(x_1, x_2, x_3) = g(x_1, g(x_2, x_3))$$

The function $g(x, y)$ can be expressed as below.

$$g(x, 0) = x$$
$$g(x, y+1) = 1 + g(x, y)$$

Therefore, the function $g(x, y)$ is primitive recursive. Here, the function $f(x_1, x_2, x_3)$ is obtained from a primitive recursive function $g(x, y)$ and hence the function f is primitive recursive. Now,

$$f(3,2,2) = g(3, g(2,2)) = g(3, 1 + g(2,1))$$
$$= g(3, 1+1+g(2,0)) = g(3, 1+1+2)$$
$$= g(3, 4) = 7$$

Example 33

Construct a Turing machine that can compute the zero function, i.e., $z(x) = 0$ for all $x \geq 0$.

Solution   The zero function is defined as $z(a_1) = 0$ for all $a_1 \geq 0$.

Therefore, the initial instantaneous description is $s_0 1^{a_1} x_1 By$. Our aim is to get the computed value $z(a_1) = 0$. So, we require 0 (zero) to appear to the left of y. Therefore, it is essential to halt the machine without changing the input. This is because, B is represented as 0 in tally notation.

Therefore, we define Turing machine M as $(S, \Sigma, \Gamma, \delta, s_0, B, A)$, where $S = \{s_0, s_1\}$, $\Gamma = \{1, B, x_1, y\}$. The transition δ is defined as follows:

$$\delta(s_0, B) = (s_0, B, R); \qquad \delta(s_0, 1) = (s_0, 1, R); \qquad \delta(s_0, x_1) = (s_1, x_1, R).$$

From the transitions it is clear that the read-write head move to the right until x_1 is encountered. Once x_1 is encountered, the Turing machine enter state s_1 and halts. Therefore, in terms of instantaneous description we have

$$s_0 1^{a_1} x_1 By \overset{*}{\vdash} 1^{a_1} s_0 x_1 By \overset{*}{\vdash} 1^{a_1} x_1 s_1 By$$

Example 34

Show that the proper subtraction function, $f(x, y) = x \doteq y$, is primitive recursive and hence compute $f(3, 2)$, where

$$x \doteq y = \begin{cases} 0 & \text{if } x \leq y \\ x - y & \text{if } x > y \end{cases}$$

Solution The predecessor function $f(x) = v(x)$ can be expressed as below:

$$f(0) = v(0) = 0$$
$$f(x+1) = v(x+1) = x$$

Therefore, $f(x) = v(x)$ is primitive recursive.

The proper subtraction function $f(x, y) = x \doteq y$ can be expressed as below.

$$f(x, 0) = x$$
$$f(x, y+1) = v(x \doteq y) = v(f(x, y))$$

Therefore, $f(x, y) = x \doteq y$ is primitive recursive. Now,

$$f(3, 2) = v(f(3, 1)) = v(v(f(3, 0)))$$
$$= v(v(3)) = v(2) = 1$$

Example 35

Show that the function $f(x, y) = |x - y|$ is primitive recursive.

Solution The function $f(x, y) = |x - y|$ can be expressed as below.

$$f(x, y) = |x - y| = (x \doteq y) + (y \doteq x)$$

Also, we have shown that $f(x, y) = (x \doteq y)$ is primitive recursive and hence the function $f(x, y) = |x - y|$ is primitive recursive.

Example 36

Construct a Turing machine that can compute the successor function, i.e., $s(x) = x + 1; x \in N$.

Solution The successor function is defined as $s(a_1) = a_1 + 1$ for all $a_1 \in N$.

Therefore, the initial instantaneous description is $s_0 1^{a_1} x_1 B y$. Our aim is to get the computed value $s(a_1) = a_1 + 1$. So, we require 1^{a_1+1} to appear to the left of y. Therefore, we define Turing machine M as $(S, \Sigma, \Gamma, \delta, s_0, B, A)$, where $S = \{s_0, s_1, s_2, s_3, s_4, s_5\}$, $\Gamma = \{1, B, x_1, y\}$. The transition δ is defined as below:

$$\delta(s_0, B) = (s_0, B, R); \qquad \delta(s_0, 1) = (s_1, B, R); \qquad \delta(s_0, x_1) = (s_5, x_1, R);$$

$$\delta(s_1, 1) = (s_1, 1, R); \qquad \delta(s_1, x_1) = (s_1, x_1, R); \qquad \delta(s_1, B) = (s_1, B, R);$$

$$\delta(s_1, y) = (s_2, 1, R); \qquad \delta(s_2, B) = (s_3, y, L); \qquad \delta(s_3, 1) = (s_3, 1, L);$$

$$\delta(s_3, B) = (s_4, B, L); \qquad \delta(s_4, 1) = (s_4, 1, L); \qquad \delta(s_4, x_1) = (s_4, x_1, L);$$

$$\delta(s_4, B) = (s_0, 1, R); \qquad \delta(s_5, B) = (s_5, 1, R); \qquad \delta(s_5, 1) = (s_5, 1, R);$$

For example, let us compute $s(2)$. Therefore, the initial instantaneous description is $s_0 11 x_1 B y$. The computation has the following moves:

$$s_0 11 x_1 B y \vdash B s_1 1 x_1 B y \vdash B 1 s_1 x_1 B y \vdash B 1 x_1 s_1 B y$$
$$\vdash B 1 x_1 B s_1 y \vdash B 1 x_1 B 1 s_2 B \vdash B 1 x_1 B s_3 1 y$$
$$\vdash B 1 x_1 s_3 B 1 y \vdash B 1 s_4 x_1 B 1 y \vdash B s_4 1 x_1 B 1 y$$
$$\vdash s_4 B 1 x_1 B 1 y \vdash 1 s_0 1 x_1 B 1 y \vdash 1 B s_1 x_1 B 1 y$$
$$\vdash 1 B x_1 s_1 B 1 y \vdash 1 B x_1 B s_1 1 y \vdash 1 B x_1 B 1 s_1 y$$
$$\vdash 1 B x_1 B 1 1 s_2 B \vdash 1 B x_1 B 1 s_3 1 y \vdash 1 B x_1 B s_3 1 1 y$$
$$\vdash 1 B x_1 s_3 B 1 1 y \vdash 1 B s_4 x_1 B 1 1 y \vdash 1 s_4 B x_1 B 1 1 y$$
$$\vdash 1 1 s_0 x_1 B 1 1 y \vdash 1 1 x_1 s_5 B 1 1 y \vdash 1 1 x_1 1 s_5 1 1 y$$
$$\vdash 1 1 x_1 1 1 s_5 1 y \vdash 1 1 x_1 1 1 1 s_5 y$$

Thus, the Turing machine M halts and $s(2) = 3$. This is because of 111 to the left of y.

REVIEW QUESTIONS

1. Explain Post's correspondence problem with a suitable example.
2. State and prove Rice theorem.
3. State Church–Turing thesis.
4. Discuss in detail Hilbert's 10th problem and show that the roots of a single variable polynomial must lie between the values $\pm\ k\dfrac{c_{max}}{c_1}$, where c_{max} is the coefficient with largest absolute value, c_1 is the coefficient of the highest order term, and k is the number of terms in the polynomial.
5. Define initial functions and show that Ackermann's function is not primitive recursive.
6. What do you mean by Turing computability? Explain with an example.
7. Explain the working principle of unlimited register machine.
8. Define mapping reducibility and state different properties of mapping reducibility.
9. Define Gödel number for sequence of natural numbers. Show with an example that Gödel function that computes Gödel number is not one-one.
10. Write a note on decidable and undecidable theory.
11. Discuss Turing reducibility in the context of computability theory.
12. If $L_1 \leq_T L_2$ and L_1 is undecidable, then L_2 is undecidable.

PROBLEMS

1. Let P be the URM program given below. Describe the computation of P for input $(4, 4)$ and $(7, 2)$.

   i. $J(1, 2, 3)$

   ii. $S(3)$

   iii. $T(3,1)$

   iv. $J(1, 1, 1)$

2. Consider the URM program given below.

   i. $J(1, 2, 8)$

   ii. $J(1, 3, 8)$

   iii. $S(3)$

    iv. $J(2, 4, 7)$

    v. $S(4)$

    vi. $J(1, 1, 2)$

    vii. $S(6)$

    viii. $T(6,1)$

    (a) Describe the computation of P for input $(5, 7)$.

    (b) Draw the flow chart corresponding to this program.

    (c) Give the trace table of the computation with this program for the input $(7, 3)$.

    (d) Say which function is computed by this program?

3   By giving suitable URM programs show that the following functions are URM computable.

    (a) $f(x) = x \div 1 = \begin{cases} x-1 & \text{if } x > 0 \\ 0 & \text{if } x = 0 \end{cases}$ (b) $f(x,y) = \begin{cases} x-y & \text{if } x \geq y \\ 0 & \text{otherwise} \end{cases}$

4. For the above functions given in problem 3, draw the flow chart corresponding to the URM program.

5. Give a suitable program for the following function to show URM computable.

$$f(x) = \begin{cases} \frac{x}{2}; & \text{if } x \text{ is even} \\ \text{Undefined}; & \text{if } x \text{ is odd} \end{cases}$$

    (a) Describe the computation of P for input 5 and 6.

    (b) Draw the flow chart corresponding to this program.

6. Show that the following function is URM computable.

$$f(x) = \begin{cases} \frac{x}{3} & \text{if } x \text{ is a multiple of 3} \\ 0 & \text{otherwise} \end{cases}$$

7. For the following URM program describe the trace table for the input 5. Also, say which function is computed by this program?

    1. $J(1, 2, 3)$

    2. $S(3)$

    3. $T(3,1)$

    (a) Describe the computation of P for input value 9 and 5.

    (b) Draw the flow chart corresponding to this program.

8. Write a URM program P for the following functions. Describe the trace table of P for the input value 3. Draw flow chart for each of the program.

(a) $f(x) = \begin{cases} 1 & \text{if } x \neq 3 \\ 0 & \text{otherwise} \end{cases}$ (b) $f(x) = \begin{cases} 1 & \text{if } x \text{ is even} \\ 0 & \text{if } x \text{ is odd} \end{cases}$

9. Use URM program to show that $x > y, x, y \in N$ is a decidable predicate.

10. Write a URM program and draw a flow chart for each of the following functions.

(a) $f(x) = \left\lfloor \frac{x}{3} \right\rfloor$

(b) $f(x) = \left\lceil \frac{x}{4} \right\rceil$

(c) $f(x) = \left\lfloor \frac{x}{4} \right\rfloor$

Describe the trace table for $x = 9$, $x = 5$ and $x = 12$.

11. Write a URM program for each of the following greatest integer functions. Give the trace table of the computation with this program for the input 4.

(a) $f(x) = \left\lfloor \frac{x}{4} \right\rfloor$

(b) $f(x) = \left\lceil \frac{x}{2} \right\rceil$

(c) $f(x) = \left\lceil \frac{x}{5} \right\rceil$

12. Write a URM program that computes $f(x) = 5x$ and give a trace table for the input value 3. Draw a flow chart of it.

13. For the URM program given in solved example 14, give a trace table for $f(2,4)$. Draw a flow chart of it.

14. Compute the Gödel number of the following sequence of numbers.
   (a) (2, 0, 2, 1, 1)   (b)   (1, 2, 3, 0, 1)
   (c) (2, 2, 2, 3)   (d)   (1, 2, 2, 1)
   (e) (3, 0, 0, 1, 1)   (f)   (1, 2, 0, 1)

15. Write a URM program for the function $f(x) = \left\lceil \frac{5x}{3} \right\rceil$. Give a trace table for $f(3)$ and draw a flow chart.

16. Obtain the Gödel sequence of the following natural numbers.
   (a) 4200   (b) 175175   (c) 2156
   (d) 4851   (e) 52   (f) 7735

17. Show that Post correspondence system with following two instances $X = \{01, 100, 0, 110\}$ and $Y = \{100, 01, 11, 0100\}$ has no solution?

18. Show that Post correspondence system with following two instances $X = \{100, 010, 11, 101\}$ and $Y = \{10, 01, 1, 0\}$ has no solution?

19. Does the Post correspondence system with two instances $X = \{ab, aa, b, bab\}$ and $Y = \{abb, a, ba, ab\}$ has a solution?

20. Consider the following Post correspondence system with two instances $X = \{ab, aba, ba, bba, b\}$ and $Y = \{abb, b, ab, baa, ab\}$. Does the given PCP has a solution?

21. Compute Ackermann's function $A(2,3)$, $A(3,2)$, $A(4,3)$ and $A(3,3)$.

22. Construct a Turing machine that can compute the projector function, i.e., p_i^n .

10

NP-COMPLETENESS

10.0 INTRODUCTION

In computational complexity theory it is observed that all algorithms are not polynomial-time algorithms. This is because Turing's famous halting problem cannot be solved by any computer, no matter how much time is provided. In this chapter, we discuss an interesting class of problems called *NP*-complete problems whose status is unknown. Here, *NP* is abbreviated as nondeterministic polynomial time. The complexity class *NP*-complete is a class of problems having two properties.

1. Any given solution to the problem can be verified quickly in polynomial time. The set of problems with this property is called *NP*.

2. If the problem can be solved quickly, then so can every problem in *NP*.

Most computer scientists believe that the *NP*-complete problems are intractable. This is because if any single *NP*-complete problem can be solved in polynomial time, then every *NP*-complete problem has a polynomial-time algorithm. Till date no polynomial-time algorithm has yet been discovered and so $P \neq NP$. As a consequence, determining whether or not it is possible to solve these problems is one of the principal unsolved problems in computer science today. Before we discuss *NP*-complete problems our objective is to present the basics of time complexity theory. Later we try to classify problems according to the amount of time required.

10.1 MEASURING COMPLEXITY

Analysing an algorithm has come to mean predicting the resources that the algorithm requires. Though, resources such as memory, logic gates or communication bandwidth are of primary reason, but most often we measure the computational time. The computational time in general depends on the number of steps used in the algorithm. The number of steps that an algorithm uses on a particular input may depend on several parameters. For example, in case of a graph, the number of steps may depend on number of edges, the number of nodes, and maximum degree of the graph, or some combination of these and other factors. We analyse the algorithm for Turing machine M deciding a language L to determine how much time it uses. For simplicity we compute the complexity of an algorithm purely as a function of the length of the string representing the input and do not consider any other parameters. The worst-case running time is the longest running time for any input of size n whereas we consider the average of all the running times of inputs of a particular length.

10.1.1 Asymptotic Notation

Although we can sometimes determine the exact running time of an algorithm, but for large enough inputs it is very complex to determine the exact running time. This is because the multiplicative constants and lower order terms of an exact running time are dominated by the effects of the input size itself. Therefore, we usually estimate the running time. One such way of estimation is known as asymptotic analysis. In this section we define several types of asymptotic notation for analysing algorithms.

Definition (θ-Notation)   Let f and g be two functions from $N \to R^+$. We say that $f(n) = \theta(g(n))$ if there exists positive constants c_1, c_2 and n_0 such that

$$c_1 g(n) \leq f(n) \leq c_2 g(n) \text{ for all } n \geq n_0.$$

In other words, for all $n \geq n_0$, the function $f(n) = g(n)$ to within a constant factor. We say that $g(n)$ is an asymptotically tight bound for $f(n)$. The definition of $\theta(g(n))$ requires that every member of $\theta(g(n))$ be asymptotically non-negative, i.e., $f(n)$ is non-negative whenever n is sufficiently large. Consequently, the function $g(n)$ itself must be asymptotically non-negative or else the set $\theta(g(n))$ is empty.

Definition (O-Notation)   Let f and g be two functions from $N \to R^+$. We say that $f(n) = O(g(n))$ if there exists positive constants c, and n_0 such that

$$f(n) \leq c g(n) \text{ for all } n \geq n_0.$$

We use big O-notation to give an upper bound on a function. More precisely $g(n)$ is an asymptotic tight upper bound for $f(n)$, to emphasize that we are suppressing constant factors.

Definition (Ω-Notation)   Let f and g be two functions from $N \to R^+$. We say that $f(n) = \Omega(g(n))$ if there exists positive constants c, and n_0 such that

$$f(n) \geq c g(n) \text{ for all } n \geq n_0.$$

We use Ω-notation to give a tight lower bound on a function. More precisely $g(n)$ is an asymptotic tight lower bound for $f(n)$, to emphasize that we are suppressing constant factors. Also it is clear that, for any two functions $f(n)$ and $g(n)$, $f(n) = \theta(g(n))$ if and only if $f(n) = O(g(n))$ and $f(n) = \Omega(g(n))$.

Definition (o-Notation)   The asymptotic upper bound provided by O-notation may or may not be asymptotically tight. We use little o-notation to denote an upper bound that is not asymptotically tight. Let f and g be two functions from $N \to R^+$. We say that $f(n) = o(g(n))$ if there exists positive constants c, and n_0 such that

$$f(n) < cg(n) \text{ for all } n \geq n_0.$$

We use little-o-notation to give an upper bound rather than a tight upper bound on a function. Intuitively, in the o-notation, the function $f(n)$ becomes insignificant relative to $g(n)$ as n approaches infinity. Mathematically, it can be defined as $f(n) = o(g(n))$ if

$$\lim_{n \to \infty} \frac{f(n)}{g(n)} = 0 \, .$$

Definition (ω-Notation)   The asymptotic lower bound provided by Ω-notation may or may not be asymptotically tight. We use little ω-notation to denote a lower bound that is not asymptotically tight. Let f and g be two functions from $N \to R^+$. We say that $f(n) = \omega(g(n))$ if there exists positive constants c, and n_0 such that

$$f(n) > cg(n) \text{ for all } n \geq n_0 \, .$$

We use little ω-notation to give a lower bound rather than a tight lower bound on a function. Intuitively, in the ω-notation, the function $g(n)$ becomes insignificant relative to $f(n)$ as n approaches infinity. Mathematically, it can be defined as $f(n) = \omega(g(n))$ if

$$\lim_{n \to \infty} \frac{g(n)}{f(n)} = 0 \, .$$

10.1.2 Analysis of Algorithms

In this section we analyse the language L given by $L = \{a^k b^k \mid k \geq 0\}$. It is clear that L is a decidable language. Our aim is to analyse the following algorithm for Turing machine M deciding the language L to determine how much time it uses. Before analysing we construct a Turing machine M for an input string w as follows. The Turing machine has the following low level steps. We also include the head motion on the tape, so as to count the number of steps it takes.

1. The machine scans across the tape and rejects the string if a symbol a is found to the right of a symbol b.

2. The machine repeats the following if both a's and b's remain on the tape.

3. The machine scans across the tape, crossing off a single symbol a and a single symbol b.

4. If a's still remain after all the b's have been crossed off, or if b's still remain after all the a's have been crossed off, then the machine rejects the string. Otherwise, if neither a's nor b's remain on the tape, then the machine accepts the string.

In order to analyse the algorithm, we consider separately each of the three stages. In stage 1, the Turing machine scans across the tape to verify that the input string is of the form 0^*1^*. As the length of the input string is n, so stage 1 takes n steps. Once the verification is over, the head is to be repositioned at the left-hand end of the tape. It takes another n steps. Therefore, total number of steps used in this stage is equal to $2n$. In stage 2 and 3, the Turing machine scans the tape repeatedly and crosses off an '*a*' and '*b*' on each scan. So, each scan crosses off two symbols in at most $\frac{n}{2}$ number of steps whereas each scan uses n steps. Thus the total number of steps required in stage 2 and 3 is equal to $n(\frac{n}{2})$. The Turing machine makes a single scan to decide whether to accept or reject the string in stage 4. This requires at most n steps.

Therefore, the total running time $= O(n) + O(n^2) + O(n) = O(n^2)$.

10.2   THE CLASS *P*

A problem is assigned to the class *P*, if there is at least one algorithm to solve that problem such that the number of steps of the algorithm is bounded by a polynomial in *n*, where *n* is the length of the input string. More specifically, these problems can be solved in time $O(n^k)$ for some non-negative integer *k*. In other words the class *P* is defined as the set of all languages which can be decided by a deterministic polynomial-time Turing machine. A deterministic polynomial-time Turing machine is a deterministic Turing machine *M* which satisfies the following two conditions:

1. *M* halts on all input *w*; and

2. there exists $k \in N$ such that $T(n) \in O(n^k)$, where $T(n) = \max\{t(w) : w \in \Sigma^*, |w| = n\}$, and $t(w)$ is the number of steps that the machine *M* takes to halt on input *w*.

Therefore, we formally define the class *P* as:

$P = \{L : L = L(M)$ for some deterministic polynomial-time Turing machine *M*}, where $L(M) = \{w \in \Sigma^* : M$ accepts $w\}$.

10.2.1   Polynomial Time

In computer science, polynomial time refers to the running time of an algorithm. It refers to the number of computation steps an abstract machine requires to evaluate the algorithm. Problems for which a polynomial time algorithm exists belong to the complexity class *P*, which is central in the field of computational complexity theory. More specifically, let $T(n)$ be the running time of the algorithm on inputs of size at most *n*. Then the algorithm is polynomial time if there exists a polynomial $p(n)$ such that, for all input sizes *n*, the running time $T(n)$ is smaller than $p(n)$.

Polynomial time are of two types. These are strongly polynomial time and weakly polynomial time. If the algorithm's running time depends only on the size of the input, regardless of the numerical values said input is known as strongly polynomial time. On the other hand the running time that is polynomial not in the size of the input, but in the numerical value of the input, which may be exponentially larger than its representation is known as weakly polynomial time.

For example, an algorithm which could sort *n* integers each less than *k* in time $O(n^2)$ would be strongly polynomial, whereas an algorithm sorting them in time $O(nk)$ would be weakly polynomial. This is because an integer less than *k* can be represented in size logarithmic in *k*.

10.3 THE CLASS *NP*

The class *NP* is defined as the set of all decision problems solvable by a nondeterministic algorithm in polynomial time. Deterministic algorithms are treated as a special case of nondeterministic algorithms and so $P \subseteq NP$. The class *NP* consists of those problems that are verifiable in polynomial time. If a problem is in complexity class *NP* and is hard as any problem in class *NP* is known as *NP*-complete. It was introduced by Stephen Cook in 1971. Now we formally define the class *NP*.

The class *NP* is defined as the set of languages over some finite alphabet, Σ, that have a verifier that runs in polynomial time. The notion of verifier is defined as follows: Let *L* be a language defined over some finite alphabet, Σ. The language $L \in NP$ if, and only if, there exists a binary relation $R \subset \Sigma^* \times \Sigma^*$ and a positive integer *k* such that the following two conditions are satisfied:

1. For all $x \in \Sigma^*, x \in L$ there exists $y \in \Sigma^*$ such that $(x, y) \in R$ and $|y| \in O(|x|^k)$; and

2. the language $L_R = \{x \# y : (x, y) \in R\}$ over $\Sigma \cup \{\#\}$ is decidable by a Turing machine in polynomial time.

A Turing machine that decides L_R is called a verifier for *L* and a *y* such that $(x, y) \in R$ is called a certificate of membership of *x* in *L*. In general, a verifier does not have to be polynomial-time. However, for a language *L* to be in *NP*, there must be a verifier that runs in polynomial time.

10.3.1 *NP*-Completeness

NP-complete are a subset of the class *NP*. The set of all decision problems whose solutions can be solved and verified in polynomial time on a nondeterministic Turing machine is known as *NP*-complete. A problem *P* in *NP* is also in complexity class *NP*-complete if and only if every other problem in *NP* can be transformed into *P* in polynomial time. However, we define *NP*-completeness as follows. A language *L* is said to be *NP*-complete if it satisfies the following two properties.

1. *L* is in *NP*, i.e., $L \in NP$ and

2. For every language $L' \in NP$, $L' \leq_P L$, i.e., L' is reducible to *L* in polynomial time.

The decision problem *C* can be shown to be in *NP* by demonstrating that a candidate solution to *C* can be verified in polynomial time. The problems in the class *NP*-complete are known as *NP*-complete problems.

10.3.2 *NP*-Hard

The concept of *NP*-hardness plays a vital role in the discussion about the relationship between the complexity classes *P* and *NP*. It is also often defined as complexity class *NP*-complete is the intersection of *NP* and *NP*-hard. As a result the complexity class *NP*-hard can be understood as the class of problems that are *NP*-complete or harder. The most computer scientists view the relationships among *P*, *NP*, *NP*-complete (*NPC*), and *NP*-hard is shown in the Figure 10.1 . From the figure given below it is clear that both *P* and *NPC* are wholly contained within *NP* with $P \cap NPC = \phi$.

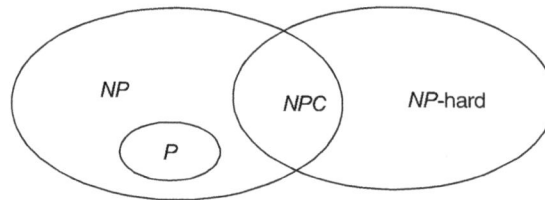

Figure 10.1   The class *NP*

Now we formally define *NP*-hard as the language *L* that satisfies the second property of *NP*-completeness but not necessarily the property 1. In other words a language is *NP*-hard if it can be translated into one for solving any *NP*-problem.

10.4   *NP*-COMPLETE PROBLEMS

There are several *NP*-complete problems in various fields such as graph theory, propositional calculus, operations research, etc. In this section we discuss some of such *NP*-complete problems.

10.4.1   Boolean Satisfiability Problem

In complexity theory, the Boolean satisfiability problem is a decision problem, whose instance is a Boolean expression written using only logical connectives AND, OR, NOT, variables, and parentheses. Does there exists a choice of truth values for the variables for which the given expression assumes the truth value 1. The propositional satisfiability problem, which decides whether a given propositional formula is satisfiable, is of central importance in various areas of computer science, including theoretical computer science, algorithmics, and artificial intelligence. The Boolean satisfiability problem is *NP*-complete, and in fact, this was the first decision problem proved to be *NP*-complete. It was proved by Cook–Levin in 1971.

10.4.2   Independent Set Problem

Given a graph *G*, an independent set is a subset of its vertices that are pairwise not adjacent. In other words, the subgraph induced by these vertices has no edges, only isolated vertices. Does a given graph *G* have an independent set, of cardinality at least *k* for a positive integer *k*? The maximum independent set problem is an optimization problem that attempts to find the largest independent set in a graph. It is easy to see that the problem is in *NP*, since if we have a subset of vertices, we can check to make sure there are no edges between any two of them in polynomial time. This independent set problem is *NP*-hard.

10.4.3   Clique Problem

Independent set problems and clique problems may be easily translated into each other: An independent set in a graph *G* is a clique in the complement graph of *G*, and vice versa. A clique in a graph is a set of pairwise adjacent vertices, or in other words, an induced subgraph which is a complete graph.

Consider the following example: In the graph given below, vertices v_1, v_2, and v_6 form a clique, because each has an edge to all the others.

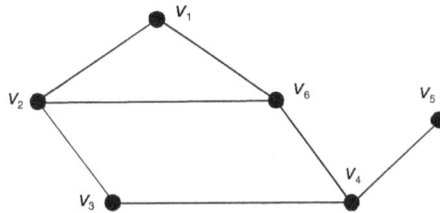

The clique problem is to determine a clique of at least a given size k for a given graph G. Once we have located k or more vertices which form a clique, it is easy to verify that they do. This is the main reason for clique problem to be *NP*. Finding the largest clique in a graph G is the corresponding optimization problem, known as maximum clique problem. The *NP*-completeness of the clique problem follows directly from the *NP*-completeness of the independent set problem, because there is a clique of size at least k if and only if there is an independent set of size at least k in the complement graph for a positive integer k. It is easy to see, since if a subgraph is complete, its complement subgraph has no edges at all.

10.4.4 Vertex Cover Problem

A vertex cover for an undirected graph $G = (V, E)$ is a subset S of its vertices such that each edge has at least one endpoint in S. In other words, for each edge $(v_i, v_j) \in E$, one of v_i or v_j must be an element of S. The size of a vertex cover is the number of vertices in it. Consider the following example: In the graph given below, $S = \{v_1, v_2, v_3, v_5\}$ is an example of a vertex cover of size 4. However, it is not a smallest vertex cover as there exist vertex covers of size 3, such as $\{v_1, v_2, v_5\}$ and $\{v_2, v_3, v_4\}$. Finding the smallest vertex cover in a graph $G = (V, E)$ is the corresponding optimization problem, known as minimum vertex cover problem.

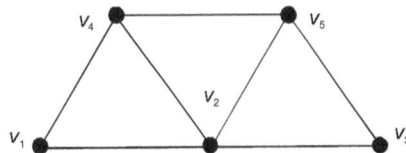

The vertex cover problem is *NP*-complete, which means that, it is unlikely to get an efficient algorithm to solve it exactly. Vertex cover remains *NP*-complete even in cubic graphs and even in planar graphs of degree at most 3.

10.4.5 Graph Colouring Problem

In graph theory, graph colouring is a way of colouring the vertices of a graph such that no two adjacent vertices share the same colour. This is otherwise known as vertex colouring. A k-coloring of an undirected graph $G = (V, E)$ is a function $c : V \rightarrow \{1, 2, \cdots, k\}$ such that $c(v_i) \neq c(v_j)$ for every edge $(v_i, v_j) \in E$. In otherwords, the numbers $1, 2, \cdots, k$ represent the k colours, and adjacent vertices must have different colours. The corresponding optimization problem is to determine the minimum number of colours needed to colour a given graph. The minimum colour required to colour the graph is known as the chromatic number.

The convention of using colours originates from colouring the countries of a map, where each face is literally coloured. This was generalized to colouring the faces of a graph embedded in the plane. The graph colouring problem is *NP*-complete and still it is a very active field of reearch.

10.4.6 Hamiltonian Cycle Problem

Given an undirected graph $G = (V, E)$, a Hamiltonian cycle is a simple cycle that contains each vertex in V. A graph G is said to be Hamiltonian if it contains a Hamiltonian cycle. We can define the Hamiltonian cycle problem as: "Does a graph G have a Hamiltonian cycle?" The Hamiltonian cycle problem is *NP*-complete.

10.4.7 Travelling Salesman Problem

The travelling salesman problem is closely related to Hamiltonian cycle problem. The job of a travelling salesman is to visit all the cities linked with roads in a particular territory and finishing at the city he starts from. He has to visit all the cities exactly once and to finish at the starting city in such a manner that the total distance travelled by him is minimum, where the total distance is the sum of the individual distances along the edges of the tour. In graph theory we denote nodes as cities joined by a weighted edge if and only if road connects them which does not pass through any of the other cities. In travelling salesman problem, we find a minimum Hamiltonian cycle. The travelling salesman problem is *NP*-complete as we are finding a Hamiltonian cycle on it.

10.5 COOK'S THEOREM

The statement that the satisfiability problem is *NP*-complete is known as Cook's theorem. The proof of this theorem is beyond the scope of this book. However, we prove some intermediate results in this section.

Lemma 1   Let $L_1, L_2 \in \{0,1\}^*$ be two languages such that L_1 is polynomial time reducible to L_2, i.e., $L_1 \leq_P L_2$. If L_2 is solvable in polynomial time ($L_2 \in P$), then L_1 is also solvable in polynomial time ($L_1 \in P$).

Proof   Suppose A_2 is a polynomial-time algorithm that decides L_2. Let F is a polynomial-time reduction algorithm that computes the reduction function f. Let us define a polynomial-time algorithm A_1 as follows:

Let $w \in \{0,1\}^*$ be an input string. The algorithm A_1 uses polynomial-time reduction algorithm F and transforms w into $f(w)$ and then it uses polynomial-time algorithm A_2 to test whether $f(w) \in L_2$.

The value produced by A_2 is the required output from the polynomial-time algorithm A_1. The algorithm A_1 solves in polynomial time because both F and A_2 solves in polynomial time. The above construction is shown in the Figure 10.2.

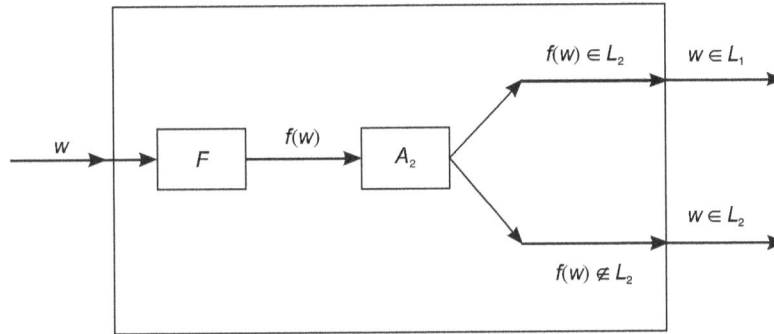

Figure 10.2 Polynomial-time algorithm A_1

Lemma 2 Let L_1, L_2, L_3 be languages. If $L_1 \leq_P L_2$ and $L_2 \leq_P L_3$, then $L_1 \leq_P L_3$.

Proof Suppose that $L_1 \leq_P L_2$ and $L_2 \leq_P L_3$. Let the corresponding polynomial time computable functions be f and g. Therefore, it is clear that $w \in L_1$ implies $f(w) \in L_2$. Again $f(w) \in L_2$ implies $g(f(w)) \in L_3$. Let $h(w) = g(f(w))$ and so h is a polynomial time computable function. Therefore, $L_1 \leq_P L_3$.

Theorem *If any NP-complete problem is polynomial time solvable, then P = NP.*

Proof Let L be in P and also L be in NP-complete. For any $L' \in NP$, by the definition of NP-completeness we have $L' \leq_P L$. Therefore, by Lemma 1 $L' \in P$. It proves that if any NP-complete problem is polynomial time solvable, then $P = NP$.

10.6 THE BLUM AXIOMS

In computational complexity theory the Blum axioms are axioms which specify desirable properties of complexity measures on the set of computable functions. A Blum complexity measure is defined using computable functions without any reference to a specific model of computation. It was first defined by Manuel Blum in 1967. In order to make the definition more accessible we define the Blum axioms in terms of Turing machines.

A Blum complexity measure is a function Φ from pairs (M, x), to the natural numbers union infinity, where M is the Turing machine with input string x to M. Furthermore, Φ should satisfy the following two axioms:

1. $\Phi(M, x)$ is finite if and only if $M(x)$ halts.

2. There is an algorithm which, on input (M, x, n) decides if $\Phi(M, x) = n$.

For example, suppose $\Phi(M, x)$ gives the number of time steps that the machine M runs for on input x before halting. The first axiom is clear whereas the second follows because a universal Turing machine can simulate M on x while counting its steps. If the Turing machine M exceeds the number of steps n, then the Turing machine M halts and rejects the input string x. Therefore, there is no need to determine if M halts on the input x.

10.7 THE GAP THEOREM

Let $g(x, y)$ be any recursive function such that $g(x, y) > y$ and let C be any complexity measure. Then, there is a recursive function $t(x)$ such that if $x > i$ and $C_i(x) < g(x, t(x))$, then $C_i(x) \leq t(x)$. We omit the proof of the theorem as it is beyond the scope of this book.

10.8 THE SPEED-UP THEOREM

Let $g(x, y)$ be any given recursive function and let C be any complexity measure. Then, there is a recursive function $f(x)$ such that $f(x) < x$ and whenever $\Phi_i = f$, there is a j such that

1. $\Phi_j(x) = f(x)$ almost everywhere and

2. $g(x, C_j(x)) \leq C_i(x)$ almost everywhere.

We omit the proof of the theorem as it is beyond the scope of this book.

SOLVED EXAMPLES

Example 1

Show that $7n^2 - 5n + 6 = \theta(n^2)$

Solution   Let $f(n) = 7n^2 - 5n + 6$. Let us take $g(n) = n^2$. Thus, we have

$$\lim_{n \to \infty} \frac{f(n)}{g(n)} = \lim_{n \to \infty} \frac{7n^2 - 5n + 6}{n^2} = 7 > 0$$

Therefore, we have $f(n) = \theta(g(n))$. So, we get $7n^2 - 5n + 6 = \theta(n^2)$.

Example 2

Show that $5n - 3 = o(n^2)$.

Solution   Let $f(n) = 5n - 3$. Let us take $g(n) = n^2$. Thus, we have

$$\lim_{n \to \infty} \frac{f(n)}{g(n)} = \lim_{n \to \infty} \frac{5n - 3}{n^2} = 0$$

Therefore, we have $f(n) = o(g(n))$. So, we get $5n - 3 = o(n^2)$.

Example 3

Show that $7n^3 - 5n^2 - 3n + 2 = \Omega(n^3)$

Solution Let $f(n) = 7n^3 - 5n^2 - 3n + 2$. Let us take $g(n) = n^3$. Thus, we have

$$\lim_{n \to \infty} \frac{f(n)}{g(n)} = \lim_{n \to \infty} \frac{7n^3 - 5n^2 - 3n + 2}{n^3} = 7 > 0$$

Therefore, we have $f(n) = \Omega(g(n))$. So, we get $7n^3 - 5n^2 - 3n + 2 = \Omega(n^3)$.

Example 4

Show that $3n^3 + 2n^2 + 3n + 1 = O(n^3)$.

Solution Let $f(n) = 3n^3 + 2n^2 + 3n + 1$. Then, selecting the highest order term $3n^3$ and disregarding its coefficient 3 gives $f(n) = O(n^3)$.

In order to verify the result with the formal definition let us take $c = 4$ and $n_0 = 3$. Then we have $3n^3 + 2n^2 + 3n + 1 \le 4n^3$ for every $n \ge 3$. In addition to this, we can also say that $f(n) = O(n^4)$ because n^4 is larger than n^3 and so is still an asymptotic upper bound on $f(n)$.

Example 5

Show that $1 + 2 + 3 + \cdots + n = o(n^3)$.

Solution Let $f(n) = 1 + 2 + 3 + \cdots + n = \frac{n(n+1)}{2}$. Let us take $g(n) = n^3$. Now,

$$\lim_{n \to \infty} \frac{f(n)}{g(n)} = \lim_{n \to \infty} \frac{1 + 2 + 3 + \cdots + n}{n^3} = \lim_{n \to \infty} \frac{n(n+1)}{2n^3} = 0$$

Therefore, we have $f(n) = o(g(n))$. So, we get $1 + 2 + \cdots + n = o(n^3)$.

Example 6

Design a low level algorithm for the language $L = \{a^m b^m \mid m \ge 0\}$ having running time $O(n \log n)$.

Solution Given that $L = \{a^m b^m \mid m \ge 0\}$ is M. Before analysing we construct a Turing machine M for an input string w as follows. The Turing machine has the following low level steps. We also include the head motion on the tape, so as to count the number of steps it takes.

1. The machine scans across the tape and rejects the string if a symbol a is found to the right of a symbol b.

2. The machine repeats the following if some a's and some b's remain on the tape.

3. The machine scans across the tape and counts the total number of a's and b's. If the total number is odd, then the machine rejects the string.

4. It scans again across the tape, crossing off every other a starting with the first a, and then crossing off every other b starting with the first b.

5. The machine accepts the string if no a's and no b's remain on the tape, otherwise reject.

On analysing the above algorithm, it can be shown that the running time is $O(n \log n)$.

Example 7

Show that the Boolean formula $f = (\overline{x} \wedge y) \vee (x \wedge \overline{z})$ is satisfiable.

Solution   Given Boolean formula is $f = (\overline{x} \wedge y) \vee (x \wedge \overline{z})$. On taking $x = 1$, $y = 0$, and $z = 0$ we get

$$f = (\overline{x} \wedge y) \vee (x \wedge \overline{z}) = (0 \wedge 0) \vee (1 \wedge 1)$$
$$= 0 \vee 1 = 1$$

Therefore, it is clear that some assignment of 0's and 1's to the variables makes the formula evaluate to 1. Hence, the given Boolean formula is satisfiable.

Example 8

Show that the computing time for the language $L = \{w : w \in \{a,b\}^*\}$ is $O(n^2)$.

Solution   Consider the language $L = \{w : w \in \{a,b\}^*\}$. Let M be the Turing machine that accepts the above language. The machine does so by reading the input string from right to left and acting as a finite automaton, while at the same time erasing the input. Therefore, the minimum number of writing steps is $|w| + 1$ whereas the total number of moving steps is $|w| + 3$. If $|w| = n$, then the total number of steps is $2|w| + 4 = 2n + 4$. Again we know that any linear polynomial is $O(n^2)$. Therefore, the computing time for the above language is $O(n^2)$.

REVIEW QUESTIONS

1. State Boolean satisfiability problem. Explain the problem with an example.
2. Write a note on P vs NP.
3. Define all asymptotic notations.
4. Explain how travelling salesman problem reduces to Hamiltonian cycle problem.
5. Write the difference between NP-complete and NP-hard.
6. Explain with a suitable example why analysis of algorithms is important.
7. State graph colouring problem. Explain it with an example.
8. Explain vertex cover problem with a suitable example.
9. State the Blum's axioms.
10. State and prove Cook's theorem.
11. Is there any relation between vertex cover problem and clique problem? If yes, explain with a suitable example.
12. Give statements for Gap theorem and Speed-up theorem.

PROBLEMS

1. Show that the following equalities are correct.

   (a) $5n^2 + 2n - 1 = \theta(n^2)$

   (b) $3n^3 + 2n^2 - 22n + 10 = \Omega(n^3)$

   (c) $2n^3 + 5n + 3 = o(n^3)$

   (d) $7n^2 + 2n + 2 = \omega(n)$

   (e) $6n^3/(\log n + 1) = O(n^3)$

   (f) $n^{2^n} + 5 \cdot 2^n = \theta(n^{2^n})$

2. If $f(n) = c_m n^m + c_{m-1} n^{m-1} + \cdots + c_1 n + c_0$ and $c_m > 0$, then show that $f(n) = \theta(n^m)$.

3. If $f(n) = c_m n^m + c_{m-1} n^{m-1} + \cdots + c_1 n + c_0$ and $c_m > 0$, then show that $f(n) = \Omega(n^m)$.

4. Show that the following inequalities are incorrect.

   (a) $3n^2 + 3n - 5 = \theta(n)$

   (b) $5n^3 + 3n^2 - 25n + 15 = \Omega(n^2)$

(c) $n^2 \log n = \theta(n^2)$

(d) $2n^2 + 5n + 1 = \omega(n^2)$

(e) $n^2/\log n = \theta(n^2)$

(f) $n^3 2^n + 5n^2 3^n = O(n^3 2^n)$

5. Let $f(n)$ and $g(n)$ be asymptotically non-negative functions. Using basic definition, show that $\max(f(n), g(n)) = \theta(f(n) + g(n))$.

6. Prove that for any real constants a and b, $(n + a)^b = \theta(n^b)$, where $b > 0$.

7. Consider the statement: The running time of any algorithm is at least $O(n^2)$. Is this statement true? If yes, explain why this statement is context-free.

8. Prove that Boolean satisfiability problem is *NP*-complete.

9. Show that the Boolean formula $f = (\overline{x} \vee \overline{y}) \wedge (x \vee z)$ is satisfiable.

10. Show that the Boolean formula $f = (x \wedge y) \wedge (\sim(x \vee y))$ is not satisfiable.

11. Show that the clique problem is *NP*-complete.

12. Prove that vertex cover problem is *NP*-complete.

13. Show that travelling salesman problem is *NP*-complete.

14. Prove that graph colouring is *NP*-complete.

15. Give an efficient algorithm to determine a 2-colouring of a graph if one exists.

16. Find the computing time for the language $L = \{w \in \{a,b\}^* : w \text{ ends with } aba\}$ on a single tape Turing machine.

17. Compute the running time for the language L that contains even number of 0's and 1's on a single tape Turing machine.

18. Determine the running time for the following languages on a single tape Turing machine. It is given that $n_a(w)$ represents the number of a's in the string w.

    (a) $L = \{a^n b^{2n} : n \geq 1\}$

    (b) $L = \{w : n_a(w) = n_b(w)\}$

19. Find the computing time of the language $L = \{ww : w \in \{a,b\}^*\}$ on a single tape Turing machine.

REFERENCES

Bridges, D.S. (1994). *Computability*: *A Mathematical Sketchbook*. Springer Verlag, Inc., New York.

Church, A. (1941). *The Calculi of Lambda-Conversion*. Princeton University Press, Princeton.

Church, A . (1936). "Note on the Entscheidungsproblem." *Journal of Symbolic Logic*. Vol. 1. pp. 40–41.

Cook, S. (1971). "The Complexity of Theorem-proving procedures," In: Proceedings of the 3rd ACM Symposium on the Theory of Computing, ACM, NY. pp. 151–158.

Daniel, I. and Cohen, A. (2008). *Introduction to Computer Theory*, 2nd edn. Wiley India, New Delhi.

David Galles. (2005). *Modern Compiler Design*. Addison-Wesley, Inc., USA.

Donald, E. Knuth. (1973). *The Art of Computer Programming*, 2nd edn. Vol. 1, Addison-Wesley, Inc., USA.

Fred Hennei. (1977). *Introduction to Computability*. Addison-Wesley Publishing Inc., USA.

Grzegorz Rozenberg and Arto Salomaa. (1997). *Handbook of Formal Languages*. Vol. I-III. Springer, Inc., New York.

Hamilton, A.G. (1978). *Logic for Mathematicians*. Cambridge University Press.

Helena Rasiowa and Roman Sikorski. (1970). *The Mathematics of Metamathematics*, 3rd edn. Polish Scientific Publishers (PWN), Warsaw, Poland.

Hopcroft, J.E., Ullman, J.D. and Sethi, R. (1999). *Compilers Principles*, *Techniques* and *Tools*. Pearson Education Singapore Pvt. Ltd.

Horowitz, E., Sahni, S. and Rajasekaran S. (2000). *Fundamentals of Computer Algorithms*. Galgotia Publications Pvt. Ltd., New Delhi.

John, C. Martin. (2003). *Introduction to Languages and the Theory of Computation*, 3rd edn. McGraw-Hill, New York.

John, E. Hopcroft and Jeffrey, D. Ullman. (1979). *Introduction to Automata Theory, Languages and Computation*. Addison-Wesley Publishing Inc., USA.

Kleene, S.C. (1935). "A theory of positive integers in formal Logic." *American Journal of Mathematics*. Vol. 57, pp. 153–173.

Kleene, S.C. (1936). "Lambda-definability and recursiveness." *Duke Mathematical Journal.* Vol. 2, pp. 340–353.

Kohavi, Z. (1986). *Switching and Finite Automata Theory*. McGraw-Hill, New York.

Kolman, B., Busby, R.C. and Ross, S.C. (2004). *Discrete Mathematical Structures*, 5th edn. Prentice-Hall, Inc.

Lewis, H. R. and Papadimitriou, C.H. (2003). *Elements of the Theory of Computation*, 2nd edn. Prentice Hall, Inc.

Martin, D. Davis, Ron Sigal and Elaine, J. Weyuker. (1994). *Computability, Complexity, and Languages–Fundamentals of Theoretical Computer Science*, 2nd edn. Academic Press, Inc., New York, USA.

Michael, A. Harrison. (1978). *Introduction to Formal Language Theory*. Addison-Wesley.

Michael, O. Albertson and Joan, P. Hutchinson. (2001). *Discrete Mathematics with Algorithms*. John Wiley and Sons, NJ, USA.

Michael Sipser. (1996). *Introduction to the Theory of Computation*, PWS Publishing, London.

Peter Linz. (2007). *An Introduction to Formal Languages and Automata*, 4th edn. Narosa Publishing House, New Delhi.

Rabin, M. O. and Scott, D. (1959). "Finite Automata and their Decision Problems." *IBM Journal of Research and Development*. Vol.3, No.2. pp. 114–125.

Richard Johnsonbaugh. (2001). *Discrete Mathematics*, 5th edn. Pearson Education, New Delhi.

Rosen, K.H. (1999). *Discrete Mathematics and Applications*. McGraw-Hill, New York.

Rozemberg, G. and Salomaa, A. (1979). (eds.). *Introduction to Automata Theory, Languages, and Computation*. Addison-Wesley Publishing Inc., USA.

Seymour Ginsburg. (1975). *Algebraic and Automata Theoretic Properties of Formal Languages*. Elsevier Science, Inc., New York, USA.

Thomas, H. Cormen, Charles, E. Leiserson and Ronald, L. Rivest. (2000). *Introduction to Algorithms*. MIT Press, Cambridge.

Tremblay, J. P. and Manohar, R. (2003). *Discrete Mathematical Structures with Applications to Computer Science*. McGraw-Hill Publishing Company Limited, New York, USA.

INDEX